Nanomaterials to Enhance Food Quality, Safety, and Health Impact

Nanomaterials to Enhance Food Quality, Safety, and Health Impact

Special Issue Editors

Jose María Lagaron
Sergio Torres-Giner
Cristina Prieto

MDPI • Basel • Beijing • Wuhan • Barcelona • Belgrade

Special Issue Editors

Jose María Lagaron
Novel Materials and Nanotechnology Group, Institute of Agrochemistry and Food Technology (IATA), Spanish Council for Scientific Research (CSIC)
Spain

Sergio Torres-Giner
Novel Materials and Nanotechnology Group, Institute of Agrochemistry and Food Technology (IATA), Spanish Council for Scientific Research (CSIC)
Spain

Cristina Prieto
Novel Materials and Nanotechnology Group, Institute of Agrochemistry and Food Technology (IATA), Spanish Council for Scientific Research (CSIC)
Spain

Editorial Office
MDPI
St. Alban-Anlage 66
4052 Basel, Switzerland

This is a reprint of articles from the Special Issue published online in the open access journal *Nanomaterials* (ISSN 2079-4991) from 2018 to 2020 (available at: https://www.mdpi.com/journal/nanomaterials/special_issues/Nanomaterials_Food_Safety_Preservation).

For citation purposes, cite each article independently as indicated on the article page online and as indicated below:

LastName, A.A.; LastName, B.B.; LastName, C.C. Article Title. *Journal Name* **Year**, *Article Number*, Page Range.

ISBN 978-3-03936-334-6 (Pbk)
ISBN 978-3-03936-335-3 (PDF)

Cover image courtesy of Jose Maria Lagaron.

Contents

About the Special Issue Editors

Jose María Lagaron received a M.S. in Chemical Sciences and a Ph.D. in Polymer Physics from the University of Valladolid, Spain. He worked for several years at DSM Research and BP Chemicals and then joined the Spanish Council of Scientific Research, where he founded the Novel Materials and Nanotechnology Group. He has published over 300 papers and is inventor of more than 50 patents, some of which relate to high-throughput electrospinning technologies.

Sergio Torres-Giner received in 2003 a Dipl-Ing in Chemical Engineering from the Polytechnic University of Valencia (UPV), Spain. In 2004, he achieved a MSc in Process Systems Technology at Cranfield University, England, followed by an MBA in Industrial Management in 2005 at Catholic University of Valencia 'San Vincente Mártir', Spain. He was able to complete his PhD in 2010 in Food Science at the University of Valencia, Spain. He is a scientist in the field of macromolecular science of application interest in food packaging technology. He has more than 15 years of experience in both public research agencies and industrial R&D organizations. He has published over 70 peer-reviewed scientific papers indexed in JCR, 10 book and encyclopedia chapters, and 4 patents. His research activity has strongly contributed to advancing the knowledge of biopolymers and to transferring it into applications and products for food-related applications.

Cristina Prieto graduated in Chemical Engineering (2010) at the University of Santiago de Compostela (Spain), received a M.S. in Industrial Processes Engineering and a P.hD. in Chemical Engineering at the University Complutense of Madrid (Spain). Currently, she is a postdoctoral researcher at the Novel Materials and Nanotechnology Group (Spanish Council of Scientific Research). She has received the Extraordinary Doctoral Thesis Award, the Flucomp Best Thesis Award, the UCM University Entrepreneurship Award, as well as several awards for the quality of her scientific communications. She has published more than 17 scientific papers in peer-reviewed journals, 1 licensed patent, 8 chapters, and 32 abstracts in international conferences. Her expertise involves emerging and conventional encapsulation technologies with application in the food and pharmaceutical industries.

Preface to "Nanomaterials to Enhance Food Quality, Safety, and Health Impact"

Nanomaterials to Enhance Food Quality, Safety, and Health Impact provides overviews of the most recent fundamental and oriented efforts by multidisciplinary researchers and technologists in the application of nanoscience and nanotechnology to generate new added value solutions for the food industry.

Nanotechnology has significant potential to secure or even enhance food quality, safety, and health impact. Nanomaterials produced by nanofabrication or nanoencapsulation techniques provide alternatives to food fortification and to support their quality and safety by being added directly into a food matrix or into food contact materials, such as food packaging. In this sense, nanomaterials can be combined with technologies such as melt compounding, lamination, or electrohydrodynamic processing (EHDP) in the design of food packaging to promote passive, active, and even bioactive properties such as barrier, antimicrobial, antioxidant, and oxygen scavenging roles and the controlled release of functional ingredients. These attributes can be produced either by the intended or non-intended migration of the nanomaterials or by the active substances they may carry. Lastly, nanomaterials can be advantageously applied to provide unique opportunities in Circular Bioeconomy strategies in relation to the valorization of, for instance, agro-industrial wastes, and food processing by-products.

This book is divided into 15 chapters that aim to contribute to advancing knowledge about how nanomaterials can improve food quality, safety, and health. The first three chapters deal with nanocomposites prepared with cellulose nanomaterials, which are an alternative to conventional technologies for improving biopolymer passive properties. It includes two research articles devoted to the isolation of cellulose nanomaterials and their application in food packaging as well as one review focused on their inhalation risks. The next seven chapters discuss the use of nanomaterials to develop active packaging technologies based on both scavenging and releasing systems. All these research articles describe the intentional incorporation of active agents, such as antioxidants and antimicrobials, into packaging material that is thereafter released into and/or absorb substances from the packaged food or the environment surrounding the food. The book continues with three more chapters that present the use of nanotechnology in the field of intelligent or smart packaging. This part of the book includes two research articles and one review that describe the development of various nanomaterials that can allow food packaging to contain, evaluate, and transmit relevant information in a near future. The last two book chapters demonstrate the potential of nanotechnology to encapsulate and release bioactives or functional ingredients to provide health benefits and reduce the risk of diseases. This section is composed of an overview of the potential of nanomaterial applications in dietary supplements and foods for special medical purposes and a research article dealing with the nanoencapsulation of bioactives.

This book, which could be of interest to food scientists, food technologists, and food engineers, provides a source of up-to-date information and broadens the reader's horizons on the use of emergent nanotechnologies to enhance food quality, safety, and health.

Jose María Lagaron, Sergio Torres-Giner, Cristina Prieto
Special Issue Editors

nanomaterials

MDPI

Editorial

Nanomaterials to Enhance Food Quality, Safety, and Health Impact

Sergio Torres-Giner *, Cristina Prieto and Jose M. Lagaron *

Novel Materials and Nanotechnology Group, Institute of Agrochemistry and Food Technology (IATA), Spanish Council for Scientific Research (CSIC), Calle Catedrático Agustín Escardino Benlloch 7, 46980 Paterna, Spain; cprieto@iata.csic.es

* Correspondence: storresginer@iata.csic.es (S.T.-G.); lagaron@iata.csic.es (J.M.L.); Tel.: +34-963-900-022 (S.T.-G. & J.M.L.)

Received: 28 April 2020; Accepted: 11 May 2020; Published: 14 May 2020

Abstract: Food quality and safety are key aspects to guarantee that foods reach consumers in optimal conditions from the point of view of freshness and microbiology. Nanotechnology offers significant potential to secure or even enhance these aspects. Novel technologies, such as nanofabrication and nanoencapsulation, can provide new added value solutions for the fortification of foods with bioactives and targeted controlled release in the gut. Nanomaterials can also support food preservation aspects by being added directly into a food matrix or into food contact materials such as packaging. Thus, nanomaterials can be leveraged in the form of nanocomposites in food packaging design by melt compounding, solvent casting, lamination or electrohydrodynamic processing (EHDP) to promote passive, active, and even bioactive properties such as barrier, antimicrobial, antioxidant, and oxygen scavenging roles and the controlled release of functional ingredients. These attributes can be exerted either by the intended or non-intended migration of the nanomaterials or by the active substances they may carry. Lastly, nanomaterials can be advantageously applied to provide unique opportunities in Circular Bioeconomy strategies in relation to the valorization of, for instance, agro-industrial wastes and food processing by-products.

Keywords: antioxidants; oxygen scavengers; antimicrobials; bioactives; barrier; active packaging; controlled release; migration; nanoencapsulation; nanocomposites; Circular Bioeconomy

---

The use of nanotechnology for food applications is a rapidly evolving field, and given the specific properties of nanomaterials and their tremendous potential in food science and packaging technology, an increased number of innovations that contribute to improving food quality, safety, and health impacts are foreseen. This Special Issue compiles twelve articles and three reviews written by researchers and technologists that, together, constitute an interesting and multi-disciplinary approach to addressing all these topics.

In food packaging applications, nanocomposites represent an alternative to conventional technologies for improving biopolymer passive properties by adding nanoparticles for which at least one dimension is in the nanometer range. Cellulose is an appropriate candidate to be used as a reinforcing nanomaterial since it is a fibrous, tough, and water-insoluble material, also being the most abundant renewable biopolymer produced in nature. However, cellulose can be obtained not only from vegetables (e.g., plants and some alga species) but also from microbes (e.g., bacteria). The so-called bacterial cellulose (BC) is constituted of fermented fibers that, in contrast to cellulose plant fibers, consist of pure cellulose, so they display higher crystallinity and show improved properties such as high purity (with the absence of lignin and hemicellulose), ultrafine fibrous structure, low density, high water-retention capacity, and biocompatibility. Wardhono et al. [1] developed a fast, highly-efficient, and eco-friendly preparation method for the extraction of cellulose nanocrystals

(CNCs) from BC. This method consisted of a two-step process, namely the partial depolymerization of BC under ultrasonic irradiation and the extraction of crystalline regions using microwaves, assisted by manganese (II) chloride ($MnCl_2$)-catalyzed hydrolysis, successfully yielding bacterial cellulose nanocrystals (BC-NCs) with similar features to commercially available nanocrystalline cellulose. In another study, Guzman-Puyol et al. [2] prepared all-cellulose nanocomposites with potential applications in compostable food packaging by using a simple method consisting of dissolving cellulose in a trifluoroacetic acid/trifluoroacetic anhydride 2:1 (vol/vol) mixture and subsequently adding different cellulose nanofibers (CNFs) dispersed in chloroform. The best performance was achieved for concentrations of nanofibers ranging from 5 to 9 wt%, maintaining excellent transparency, improving the mechanical properties, and reducing the water permeability. Nevertheless, more insights into the life cycle risk assessment and environmental health and safety roadmap of the potential risks from the inhalation of cellulose nanomaterials are needed to advance the safe commercialization of these materials. To this end, Ede et al. [3] summarized, in a review, the currently available published literature regarding the cellulose nanomaterial inhalation hazard and evaluated the quality of the studies for risk assessment purposes. It was concluded that short-term exposure to cellulose nanomaterials could result in transient inflammation, similarly to other poorly soluble and low toxicity dusts. However, several data gaps still remain, and there is a lack of understanding of the effects of long-term and low-dose exposures that represent realistic workplace conditions, essential for a quantitative assessment of the potential health risks.

Many efforts made to prevent food deterioration and improve the effectiveness of active compounds have triggered recent innovations in the field of food packaging. In this context, the passive role of traditional packaging in the protection and marketing of a food product has evolved into a novel "extra" function as a carrier of active compounds. Active packaging technology is based on the intentional incorporation of active agents into a packaging material that would then be released into and/or absorb substances from the packaged food or the environment surrounding the food. Concerning active packaging materials, these can be classified as either active-scavenging types (e.g., oxygen scavengers) or active-releasing types (e.g., antioxidants and antimicrobials). The development of novel antioxidant packaging using nanotechnology provides an opportunity to extend the freshness of food products by absorbing the compounds that deteriorate the food such as oxygen or free radicals. In a research article, Vera et al. [4] demonstrated that active packaging bags based on selenium nanoparticles (SeNPs) between 50 and 70 nm can prevent the oxidation and extend the shelf life of hazelnuts, walnuts, and potato chips. The metalloid nanoparticles were incorporated in solution into the adhesive layer of a polyethylene terephthalate (PET)/adhesive/low-density polyethylene (LDPE) multilayer, performing as a strong free radical scavenger. The newly developed active packaging material was also tested in food industrial lines where cooked ham and chicken as well as a ready-to-eat vegetable mixture seasoned with butter were industrially packaged with the antioxidant multilayer, and improvements higher than 25% were obtained. This study thus opened the door to new developments of active materials containing nanoparticles in which the nanomaterials are not in direct contact with the food but act to protect the food via the packaging, reducing the formation of lipid off-flavors that cause food product rejection by consumers. In the field of active-scavenging packaging, also, Cherpinski et al. [5] developed oxygen scavenging films consisting of poly(3-hydroxybutyrate) (PHB) containing palladium nanoparticles (PdNPs) by electrospinning and annealing treatment at 160 °C. The PdNPs were pretreated with surfactants permitted for food contact applications to optimize their dispersion and distribution in PHB. The PHB/PdNP nanocomposite films had high oxygen scavenging performance, although the activity was reduced after annealing. In a second study [6], water barrier and oxygen scavenging multilayer films were developed by the coating of paper with electrospun PdNP-containing PHB and poly(ε-caprolactone) (PCL) fiber mats followed by the annealing of the whole bilayer. The PdNPs-containing PCL/paper multilayer exhibited higher oxygen scavenging capacity than the homologous based on PHB due to the higher oxygen permeability of PCL and the higher dispersion of

PdNPs. The active multilayered biopapers developed are of significant relevance to the development of the next generation of fully biodegradable barrier papers of interest in food packaging.

Active-releasing type packaging can additionally provide novel functions, such as aromatic, antioxidant, and long-term antimicrobial properties. In this regard, the electrohydrodynamic processing (EHDP), including both electrospinning and electrospraying techniques, supposes an innovative nanofabrication approach for the development of active food packaging coating and interlayer materials. This technology employs a high-voltage electric field imposed on a polymer solution to create polymer fibers or beads with diameters ranging from below 100 nm to several micrometers. As an example of this, Arrieta et al. [7] developed bio-based and biodegradable bilayer systems with antioxidant properties. The outer layer was based on a compression-molded poly(3-hydroxybutyrate-*co*-3-hydroxyvalerate) (PHBV) film, while the inner active layer was formed by electrospun nanofibers based on polylactide (PLA) and PHB blends (3:1, wt/wt) loaded with 1 and 3 wt% of catechin, a natural flavonoid with antioxidant activity. Moreover, in order to increase the stretchability and to facilitate the electrospinning process, 15 wt% of oligomer of lactic acid (OLA) was added as a plasticizer. The obtained bilayers showed effective capacities to release antioxidants into a fatty food simulant, while the bilayer films showed appropriate disintegration under compost conditions in less than three months (90% of disintegration according to ISO 20200), thus showing their potential as compostable active packaging for fatty food products. In other work, Salević et al. [8] prepared novel active films made of PCL containing a solid dispersion of sage (*Salvia officinalis* L.) extract at loading contents of 5, 10, and 20 wt% by means of the electrospinning technique and subsequent annealing. The thermal post-treatment yielded thin and hydrophobic films with good contact transparency, showing high free radical scavenging ability and also strong activity against foodborne pathogens. The evaluated characteristics of the PCL-based films containing sage extract suggest great potential in active food packaging applications with the aim of preventing oxidation processes and microbiological contamination or growth.

As shown above, the design of active-releasing type packaging materials represents a very dynamic field, and multiple functions can be added by electrospinning. For instance, active packaging systems can not only extend the shelf life of food products and reduce food waste by maintaining the quality of food products for longer but also increase product safety by securing the foods against pathogens. In this regard, antimicrobial packaging applications can improve food microbial safety by preventing the growth of spoilage and/or pathogenic microorganisms. In this area, Melendez-Rodriguez et al. [9] developed electrospun PHBV films with long-term antimicrobial capacities by incorporating mesoporous silica nanoparticles in the 2.5–20 wt% range. The mineral nanoparticles containing 50 wt% of eugenol were loaded into PHBV by electrospinning, and the resultant electrospun composite fibers were annealed at 155 °C to produce continuous films. The electrospun PHBV films with loadings above 10 wt% of mesoporous silica nanoparticles containing eugenol successfully inhibited bacterial growth, and their antimicrobial activity increased after 15 days when stored in hermetically closed systems due to the volatile portion accumulated in the system's headspace and the sustained release capacity of the active films. The resultant biopolymer films were reported candidates to be applied in active food packaging applications to provide shelf life extension and food safety. Similarly, Figueroa-Lopez et al. [10] encapsulated oregano essential oil (OEO), rosemary extract (RE), and green tea extract (GTE) in ultrathin fibers of PHBV derived from fruit waste using solution electrospinning, and the resultant electrospun mats were annealed to produce continuous films. The incorporation of the natural extracts and essential oils resulted in PHBV films with a relatively high contact transparency and hydrophobicity, of which the electrospun OEO-containing PHBV films presented the highest antimicrobial activity against two strains of food-borne bacteria. Furthermore, these active films also showed the most significant antioxidant performance, which was ascribed to the films' high contents of carvacrol and thymol. One can expect that these new materials can be applied as new active coating or interlayer systems in the design of active food packaging structures in the frame of Circular Bioeconomy to prolong the shelf life of foods and delay the proliferation of microorganisms and enzymatic oxidation of foodstuffs.

Nanotechnology also opens up new opportunities in the field of intelligent or smart packaging. This concept relates to future technologies that allow packaging to contain, evaluate, and transmit relevant information. For instance, the use of electrically conductive polymer-based materials can serve to create "smart" labels or tags that can enable the tracking and monitoring of the conditions and quality of the packaged products, such as food freshness, from the production line to the end user. In our recent study [11], graphene nanoplatelets (GNPs) were embedded in poly(ethylene-*co*-vinyl alcohol) (EVOH) nanofibers by electrospinning to create electrically conductive thin layers with nanofiller contents of only 0.5 wt%. The electrospun mats were also thermally post-treated at 158 °C to produce continuous and contact transparent EVOH films to be applied as smart tags in film interlayers. In recent years, the fast development of various nanomaterials has also opened up new possibilities for biosensing signal amplification. Among them, Wang et al. [12] demonstrated the potential of protein-inorganic hybrid nanoflowers to maintain or even increase the activity of the proteins and effectively amplify the detection signals. A new enzyme-free assay was explored to detect *S. typhimurium*, being able to detect bacterial cells at levels as low as 28 CFU/mL within two hours. One of the most interesting and promising nanomaterials for the detection of various species (e.g., bacteria, cells, nucleic acids, molecules, ions, etc.) are colloidal semiconducting quantum dots (QDs). Lesiak et al. [13] reviewed the use of QDs to enable high-throughput and mobile diagnostic platforms for screening for pathogens and toxins immediately in the field. Nevertheless, it was concluded that a universal sensor for different types of food samples is currently a challenge because of the inherent complexity of biological samples.

As nanotechnology provides "a new dimension" accompanied by novel or modified properties conferred to many current materials, it can also be successfully used for the production of a new generation of nanoformulated human dietary supplements, functional foods, or nutraceuticals. A bioactive or functional ingredient is any food or food component providing health benefits beyond basic nutrition and reducing the risk of disease. Nutraceuticals are bioactive substances or a mixture of bioactive compounds based on food, herbal, or other natural products that are used in the form of pharmaceutical formulations such as tablets, capsules, drops, or liquids. In contrast to drugs, in all these cases, the bioactives are present in low concentrations. The intake of nutraceuticals and fortification of edible products with bioactive components have both become increasingly popular in modern society, since they can help to balance the total nutrient profile of a diet, supplement nutrients lost in processing, and, thus, correct or prevent insufficient nutrient intake and associated deficiencies. In this context, the review performed by Jampilek et al. [14] compiles the current state-of-the-art in the development of bio-based nanoscaled delivery systems such as nanocapsules, nanofibers, or nanoparticles, among others, to stabilize and enhance nutraceuticals' functionality in various food products and drinks during food processing or digestion. In the field of the nanoencapsulation of bioactives, the electrospraying process has shown to be highly suitable for entrapping thermolabile compounds due to its high efficiency and the fact that it can be performed under room conditions. In this regard, Ramos-Hernández et al. [15] studied the ability of high degree of polymerization Agave fructans (HDPAF) to form capsules by electrospraying and assessed the viability of this polysaccharide as a nanoencapsulating material for β-carotene by direct electrospraying and by electrospraying coating (EC). The findings showed that the EC method, based on a three-step process, yielded ultrathin particles with higher bioactive/polysaccharide ratios. Moreover, the HDPAF nanoparticles obtained by the EC method showed good photoprotection and improved the stability of β-carotene, being promising for human consumption.

In summary, the Special Issue entitled *Nanomaterials to Enhance Food Quality, Safety, and Health Impact* in *Nanomaterials* reflects the high diversity and creativity of new nanomaterials and nanofabrication processes that are rapidly developing in the research fields of food science and packaging technology. This focus will contribute to the research interest in the field of food quality and safety, providing our readership with a multi-faceted scenario that outlines the importance of food-related nanotechnologies and their wide-ranging applications. It is also expected that the present Special Issue will encourage new multidisciplinary research on nanomaterials to promote health benefits, broadening the range of potential practical uses. However, all nanomaterials applied in the food industry should be used advisedly and only after the in-depth investigation of cytotoxicity due to possible increased toxicity effects induced by their high surface reactivity. In food packaging, future research efforts will, particularly, have to be coupled to new studies devoted to the risk factors associated with the potential migration of nanoparticles from the packages and their impact on both human health and the environment in the long term.

Author Contributions: All the guest editors wrote and reviewed this Editorial Letter. All authors have read and agreed to the published version of the manuscript.

Funding: This research work was funded by the Spanish Ministry of Science and Innovation (MICI) project number RTI2018-097249-B-C21.

Acknowledgments: S.T.-G. acknowledges MICI for his Juan de la Cierva–Incorporación contract (IJCI-2016-29675). The guest editors thank all the authors for submitting their work to the Special Issue and for its successful completion. We also acknowledge all the reviewers participating in the peer-review process of the submitted manuscripts and for enhancing their quality and impact. We are also grateful to Erika Zhao and the editorial assistants who made the entire Special Issue creation a smooth and efficient process.

Conflicts of Interest: The authors declare no conflict of interest.

References

1. Wardhono, E.Y.; Wahyudi, H.; Agustina, S.; Oudet, F.; Pinem, M.P.; Clausse, D.; Saleh, K.; Guénin, E. Ultrasonic Irradiation Coupled with Microwave Treatment for Eco-friendly Process of Isolating Bacterial Cellulose Nanocrystals. *Nanomaterials* **2018**, *8*, 859. [CrossRef] [PubMed]
2. Guzman-Puyol, S.; Ceseracciu, L.; Tedeschi, G.; Marras, S.; Scarpellini, A.; Benítez, J.J.; Athanassiou, A.; Heredia-Guerrero, J.A. Transparent and Robust All-Cellulose Nanocomposite Packaging Materials Prepared in a Mixture of Trifluoroacetic Acid and Trifluoroacetic Anhydride. *Nanomaterials* **2019**, *9*, 368. [CrossRef] [PubMed]
3. Ede, J.D.; Ong, K.J.; Goergen, M.; Rudie, A.; Pomeroy-Carter, C.A.; Shatkin, J.A. Risk Analysis of Cellulose Nanomaterials by Inhalation: Current State of Science. *Nanomaterials* **2019**, *9*, 337. [CrossRef] [PubMed]
4. Vera, P.; Canellas, E.; Nerín, C. New Antioxidant Multilayer Packaging with Nanoselenium to Enhance the Shelf-Life of Market Food Products. *Nanomaterials* **2018**, *8*, 837. [CrossRef]
5. Cherpinski, A.; Gozutok, M.; Sasmazel, H.T.; Torres-Giner, S.; Lagaron, J.M. Electrospun Oxygen Scavenging Films of Poly(3-hydroxybutyrate) Containing Palladium Nanoparticles for Active Packaging Applications. *Nanomaterials* **2018**, *8*, 469. [CrossRef] [PubMed]
6. Cherpinski, A.; Szewczyk, P.K.; Gruszczyński, A.; Stachewicz, U.; Lagaron, J.M. Oxygen-Scavenging Multilayered Biopapers Containing Palladium Nanoparticles Obtained by the Electrospinning Coating Technique. *Nanomaterials* **2019**, *9*, 262. [CrossRef]
7. Arrieta, M.P.; Díez García, A.; López, D.; Fiori, S.; Peponi, L. Antioxidant Bilayers Based on PHBV and Plasticized Electrospun PLA-PHB Fibers Encapsulating Catechin. *Nanomaterials* **2019**, *9*, 346. [CrossRef]
8. Salević, A.; Prieto, C.; Cabedo, L.; Nedović, V.; Lagaron, J.M. Physicochemical, Antioxidant and Antimicrobial Properties of Electrospun Poly(ε-caprolactone) Films Containing a Solid Dispersion of Sage (*Salvia officinalis* L.) Extract. *Nanomaterials* **2019**, *9*, 270. [CrossRef] [PubMed]
9. Melendez-Rodriguez, B.; Figueroa-Lopez, K.J.; Bernardos, A.; Martínez-Máñez, R.; Cabedo, L.; Torres-Giner, S.; Lagaron, J.M. Electrospun Antimicrobial Films of Poly(3-hydroxybutyrate-*co*-3-hydroxyvalerate) Containing Eugenol Essential Oil Encapsulated in Mesoporous Silica Nanoparticles. *Nanomaterials* **2019**, *9*, 227. [CrossRef] [PubMed]
10. Figueroa-Lopez, K.J.; Vicente, A.A.; Reis, M.A.M.; Torres-Giner, S.; Lagaron, J.M. Antimicrobial and Antioxidant Performance of Various Essential Oils and Natural Extracts and Their Incorporation into Biowaste Derived Poly(3-hydroxybutyrate-*co*-3-hydroxyvalerate) Layers Made from Electrospun Ultrathin Fibers. *Nanomaterials* **2019**, *9*, 144. [CrossRef] [PubMed]
11. Torres-Giner, S.; Echegoyen, Y.; Teruel-Juanes, R.; Badia, J.D.; Ribes-Greus, A.; Lagaron, J.M. Electrospun Poly(ethylene-*co*-vinyl alcohol)/Graphene Nanoplatelets Composites of Interest in Intelligent Food Packaging Applications. *Nanomaterials* **2018**, *8*, 745. [CrossRef] [PubMed]

12. Wang, L.; Huo, X.; Guo, R.; Zhang, Q.; Lin, J. Exploring Protein-Inorganic Hybrid Nanoflowers and Immune Magnetic Nanobeads to Detect Salmonella Typhimurium. *Nanomaterials* **2018**, *8*, 1006. [CrossRef] [PubMed]
13. Lesiak, A.; Drzozga, K.; Cabaj, J.; Bański, M.; Malecha, K.; Podhorodecki, A. Optical Sensors Based on II-VI Quantum Dots. *Nanomaterials* **2019**, *9*, 192. [CrossRef] [PubMed]
14. Jampilek, J.; Kos, J.; Kralova, K. Potential of Nanomaterial Applications in Dietary Supplements and Foods for Special Medical Purposes. *Nanomaterials* **2019**, *9*, 296. [CrossRef]
15. Ramos-Hernández, J.A.; Ragazzo-Sánchez, J.A.; Calderón-Santoyo, M.; Ortiz-Basurto, R.I.; Prieto, C.; Lagaron, J.M. Use of Electrosprayed Agave Fructans as Nanoencapsulating Hydrocolloids for Bioactives. *Nanomaterials* **2018**, *8*, 868. [CrossRef]

nanomaterials

MDPI

Article

Ultrasonic Irradiation Coupled with Microwave Treatment for Eco-friendly Process of Isolating Bacterial Cellulose Nanocrystals

Endarto Yudo Wardhono [1,*], Hadi Wahyudi [1], Sri Agustina [1], François Oudet [2], Mekro Permana Pinem [3], Danièle Clausse [3], Khashayar Saleh [3] and Erwann Guénin [3,*]

1 University of Sultan Ageng Tirtayasa, Cilegon 42435, Banten, Indonesia; hadi.wahyudi@untirta.ac.id (H.W); sri.agustina@hotmail.com (S.A)

2 Physico-Chemical Analysis Services, University of Technology of Compiègne, Sorbonne Universities, 60200 Compiègne, France; francois.oudet@utc.fr

3 Integrated Transformations of Renewable Matter Laboratory (EA TIMR 4297 UTC-ESCOM), University of Technology of Compiègne, Sorbonne Universities, 60200 Compiègne, France; mekro-permana.pinem@utc.fr (M.P.P.); daniele.clausse@utc.fr (D.C.); khashayar.saleh@utc.fr (K.S.)

* Correspondence: endarto.wardhono@untirta.ac.id (E.Y.W.); erwann.guenin@utc.fr (E.G.); Tel.: +62-254-395-502 (E.Y.W.); +33-344-234-584 (E.G.)

Received: 17 September 2018; Accepted: 17 October 2018; Published: 20 October 2018

Abstract: The isolation of crystalline regions from fibers cellulose via the hydrolysis route generally requires corrosive chemicals, high-energy demands, and long reaction times, resulting in high economic costs and environmental impact. From this basis, this work seeks to develop environment-friendly processes for the production of Bacterial Cellulose Nanocrystals (BC-NC). To overcome the aforementioned issues, this study proposes a fast, highly-efficient and eco-friendly method for the isolation of cellulose nanocrystals from Bacterial Cellulose, BC. A two-step processes is considered: (1) partial depolymerization of Bacterial Cellulose (DP-BC) under ultrasonic conditions; (2) extraction of crystalline regions (BC-NC) by treatment with diluted HCl catalyzed by metal chlorides ($MnCl_2$ and $FeCl_3.6H_2O$) under microwave irradiation. The effect of ultrasonic time and reactant and catalyst concentrations on the index crystallinity (CrI), chemical structure, thermal properties, and surface morphology of DP-BC and BC-NC were evaluated. The results indicated that the ultrasonic treatment induced depolymerization of BC characterized by an increase of the CrI. The microwave assisted by $MnCl_2$-catalyzed mild acid hydrolysis enhanced the removal of the amorphous regions, yielding BC-NC. A chemical structure analysis demonstrated that the chemical structures of DP-BC and BC-NC remained unchanged after the ultrasonic treatment and $MnCl_2$-catalyzed acid hydrolysis process.

Keywords: bacterial cellulose nanocrystals; ultrasonic irradiation; microwave treatment; catalyzed hydrolysis; crystallinity index

1. Introduction

The most advantageous characteristics of the bio-based edible film are their edibility and inherent biodegradability [1]. Various biopolymers have been explored to reduce the use of non-degradable petroleum-based materials such as cellulose, chitosan, starch, collagen, pectin, etc. [2]. However, problems of strong hydrophilic character, high degradation, and inadequate mechanical properties in moist environments still limit the applications of biopolymers [3,4]. To become more applicable in practice, biopolymers have to be modified in terms of properties and functionalities [5]. In food packaging applications, for instance, the incorporation of reinforcement fillers [6,7] into the biopolymers matrix has shown to be an efficient strategy to overcome some critical issues [8] such as low mechanical resistance [9], hydrophilicity [10], and poor barrier to water vapor [11,12] compared to those of pure polymer or conventional (microscale) composites. More importantly, the process is less expensive compared to the development of new synthetic polymeric materials [13].

Nanocomposites represent an alternative to conventional technologies for improving biopolymer properties, by adding nanoparticles for which at least one dimension is in the nanometer range [14]. Most composite materials consist of one or more discontinuous phases distributed in one continuous phase. Discontinuous phase materials are usually harder and possess superior mechanical properties compared to continuous phase materials. The continuous phase is called the matrix, and the discontinuous one is called reinforcement [15]. The entity of the interactions is strongly affected by the nature of the discontinue phase; this can be maximized by passing from iso-dimensional particles to nanotubes [4]. Preparation of hybrid polymeric materials filled with natural particles also allows the fabrication of films with smart functions, such as antibacterial [16–18] and antioxidant capacities [19–21]. Numerous studies have been done on potential applications of biopolymers. Cellulose is an appropriate candidate used as a reinforcing material. Cellulose is a fibrous, tough, water-insoluble biomaterial that can play a substantial role in blending with different biopolymers to produce various bio-based nanocomposites [22]. Cellulose is the most abundant renewable biopolymer produced in the biosphere, and is obtained mainly from vegetables (plants and some algae species) and microbes (bacteria) [23].

Bacterial Cellulose, BC is constituted of fermented fibers, and is commonly synthesized by bacteria that are members of the Gluconacetobacter genus. Compared to cellulose plant fibers, BC displays higher crystallinity, and possesses improved properties such as high purity (with the absence of lignin and hemicellulose), ultrafine fibrous structure, low density, high water-retention capacity, and biocompatibility [24]. All these features make BC a promising biomaterial for industrial applications [25,26]. BC and plant fibers are both biopolymers that have similar molecular units but present a different structural organization. Depending on the source, plant fibers are mainly composed of three major components: cellulose, hemicellulose, and lignin. In contrast, the fibers made by bacteria are of pure cellulose; therefore, they present different physical properties [27]. Cellulose is a linear polysaccharide which consists of D-anhydro glucopyranose units linked by β-1,4-glycosidic bonds. The cellulose microfibrils have two types of structural regions: (i) the ordered region (crystalline) and (ii) the disordered region (amorphous). The crystalline regions give important mechanical properties to the cellulose fibers. Cellulose crystallinity, the degree of organization of the cellulose lattice, is a parameter describing the relative amount of crystalline content in the cellulose [28]. Crystallinity is a major factor affecting the activity of most celluloses; its values vary depending on the source and the mode of chemical treatment of the fibers [29].

Nanocrystals Cellulose (NCC) can be obtained by removing the amorphous regions while keeping the crystalline regions through partial depolymerization and purification from fiber sources. A comparison of the preparation of NCC from different natural materials and synthesis routes are presented in [30]. The most commonly-employed method to produce NCC is via acid hydrolysis conducted by strong mineral acids such as sulfuric acid, H_2SO_4, or hydrochloric acid, HCl [31]. The reaction involves the preferential hydrolysis of amorphous regions, promoting cleavage of glycosidic bonds. This procedure leads to the removal of the individual crystallites, which are regularly distributed along the microfibers, and drives to the formation of rod-like nanocrystals. The type of acid used determines the characteristics of the obtained NCC. H_2SO_4 will promote sulfonation of the crystallites surface [32] that produces a stable colloidal suspension due to electrostatic repulsion [33]. However, the presence of sulfate groups induces some crystallites to degrade, and reduces the thermostability of NCC [34–36]. It is generally known that low thermal stability may limit the use of nanocellulose and the manufacturing of its nanocomposites at high temperatures [37]. Although residual sulfate can be removed by dialysis, it is a time-consuming process, and particle aggregation is very difficult to avoid [38–40]. On the other hand, HCl produces hydroxyl groups on the surface of crystallites [41]. It generates a low-density surface charge with limited NCC dispersibility, which tends to promote flocculation in aqueous suspensions [42]. HCl is less corrosive than H_2SO_4, and though the yield is lower [41], it permits a significant increase in thermal stability of NCC [43]. To reach high yield value, a highly-concentrated aqueous solution of HCl is needed under hydrothermal conditions at 110 °C for a long period of the reaction [35].

As described in the previous paragraph, high-yield production of NCC is obtained using an excessive amount of mineral acids. Pollution to the environment, corrosion to the equipment, and the difficulty of controlling the reaction are the major limitations to synthesis using acid hydrolysis [44]. To overcome these issues, the aim of this study is to develop a fast, highly-efficient, and eco-friendly preparation method for the extraction of cellulose nanocrystals from Bacterial Cellulose, BC. A two-step process is considered, yielding Bacterial Cellulose NanoCrystals, BC-NC, namely: (1) partial depolymerization of BC under ultrasonic irradiation, (2) extraction of crystalline regions using microwave assisted by $MnCl_2$-catalyzed hydrolysis. The effect of irradiation time on the partial depolymerization process and impact of $MnCl_2$ concentration during the hydrolysis treatment is evaluated on the chemical structure, crystallinity index, thermal properties, and surface morphology of irradiated Depolymerized Bacterial Cellulose DP-BC and extracted BC-NC

2. Materials and Methods

2.1. Materials

Nata de coco (BC pellicles) was collected from a local market, in Cilegon (Banten, Indonesia) region. Sodium hydroxide, NaOH, hydrochloric acid, HCl, and ethanol, C_2H_5OH were obtained from Thermo Fischer Scientific. Metal chlorides $MnCl_2$, $FeCl_3.6H_2O$ were purchased from Merck Indonesia (Jakarta, Indonesia). Commercial Microcrystal Cellulose, MCC was purchased from Sigma-Aldrich (Saint-Quentin, France) and Commercial Nanocrystal Cellulose, NCC from CelluForce, QC, Canada. All the reagents and chemicals are used as a laboratory grade without further purification. Demineralized water (conductivity of 0.06 $mScm^{-1}$) produced by a purification chain was used for all experiments.

2.2. Methods

2.2.1. BC Preparation

BC pellicles were maintained in 0.5% NaOH (w/v) at room temperature for 24 h, followed by rinsing in the drained water until a neutral pH was attained; any chemicals used in the nata de coco production removed. The cellulose was then sun dried for two days, powdered, and sieved through a 149 μm sieve (100 Mesh).

2.2.2. Partial Depolymerization of BC

The ultrasonic irradiation was carried out using an ultrasonic processor (Vibra Cell, Type 72434, 100 Watts, horn diameter: 1.0 mm, Fisher Scientific, Illkirch, France). One half of a gram (1% w/v) of BC powder was introduced into a 100-mL flat-bottom flask with a mixture of water/ethanol (50% w/w). The ultrasonic horn was placed at the center of the suspension, while the temperature was maintained at room temperature with a circulating water condenser. The suspension was constantly stirred at 300 rpm using magnetic stirring bar. All sonication runs were carried out at 20 kHz by varying irradiation times of 10, 20, 30, 60, and 120 min. After that, each sample of DP-BC was immediately washed with water and filtered with Whatman filter paper no. 1 until the filtrate was neutral. The DP-BC was then oven dried at 70 °C for 24 h.

2.2.3. Extraction of Crystalline Regions

After ultrasonic irradiation treatment, the DP-BC (oven dry 0.3 g) was put into a microwave reaction vial (30 mL, G30), and 15 mL water of 0.1 mol/L HCl with a concentration of $MnCl_2$ in the range 0; 1; 2.5; 5% w/v was added. The suspension was then placed into a microwave reactor (Anton Paar, Monowave 300) and heated as follows: (1) heating to 125 °C in 3 min; (2) maintaining the temperature at 125 °C for 30 min; and (3) cooling to 30 °C within 7 min. During the protocol, the suspension was stirred with a magnetic stirring bar at 1200 rpm. Upon completion of the hydrolysis, the vial was removed from the microwave oven and cooled at ambient temperature. The BC-NC suspension was transferred to a 50 mL plastic centrifuge tube and centrifuged at 12,000 rpm for 5 min (Jouan, MR 1812 Refrigerated Centrifuge, MN, USA) to remove residual acid and chemicals. The precipitate was purified by five washing cycles with deionized water followed by centrifugation at 12,000 rpm for 5 min. The BC-NC was then oven dried at 70 °C for 24 h before characterizations.

2.3. *Characterization*

2.3.1. Fourier Transform Infrared (FT-IR)

FT-IR study was conducted to determine the functional groups present in the cellulose. The measurements were performed using a Nicolet iS5 spectrometer (Thermo Scientific, Whatman, MA, USA). Spectra were obtained between 4000 and 400 cm^{-1} at a resolution of 4 cm^{-1} and scanning speed of 20 mm/sat.

2.3.2. X-Ray Diffraction (XRD)

XRD patterns of the cellulose were performed in a D8 Advance (Bruker, Bremen, Germany). Samples were examined with a scanning angle of 2θ from 10° to 40° at a rate of 1°/min with the CuKα filtered radiation. The crystallinity index, CrI was calculated from the diffraction intensity data using deconvolution method [45]. In which the diffraction profile was fitted by Gaussian function to find the contribution of each individual peak relative to the crystallographic planes and the amorphous background. The CrI was calculated Hermans equation as follows:

$$\text{CrI} = \left(\frac{A_{cr}}{A_{cr} + A_{am}}\right) \times 100\% \quad (1)$$

where A_{am} is the amorphous area, and A_{cr} is the sum of the area of the 101, $10\bar{1}$, 002, 040 peaks.

2.3.3. Differential Scanning Calorimetry (DSC)

DSC was carried out to analyze the thermal behavior of the cellulose. The samples were characterized on a DSC Q100 (TA Instruments, DE, USA) under constant nitrogen flow (50 mL/min), from 25 to 400 °C, at a heating rate of 10 °C/min.

2.3.4. Transmission Electron Microscopy (TEM) and Scanning Transmission Electron Microscopy STEM)

The morphology of BC-NC suspension was measured by using the high-resolution JEOL-2100F TEM (Jeol, Akishima, Tokyo, Japan) in TEM and STEM mode. Samples were conventionally deposited on carbon coated copper grids and a negative staining was achieved using uranyLess solution (Delta Microscopies, Toulouse, France). The size and diameter distribution particle were measured by ImageJ (version 1.41 h) and origin pro-8 software.

3. Results

3.1. *Chemical Structure*

FT-IR spectroscopy was used to investigate changes in the chemical structure of cellulose sample before and after the treatments. The spectra displayed the intensity of absorption of the functional groups between 4000 and 400 cm^{-1}, which can identify the chemical bond in the cellulose molecule. The absorption bands for characteristic chemical groups of the raw material (native BC) and the treated celluloses (DP-BC and BC-NC) can be observed in Figure 1, and the typical vibration bands are listed in Table 1. The FT-IR spectra were divided into two parts: (1) H-bonding region from 4000 to 2600 cm^{-1}, and (2) fingerprint region from 1800 to 400 cm^{-1} [46]. The broad peak in the 3650–3000 cm^{-1} bands was assigned to O–H stretching vibrations, which are characteristic of the hydroxyl groups generally present in cellulose, water, and lignin. In this region, intramolecular hydrogen bonds appeared at 3342 cm^{-1} and 3432 cm^{-1}, and were attributed respectively to the two crystalline cellulose allomorphs, cellulose Iα and cellulose Iβ [47].

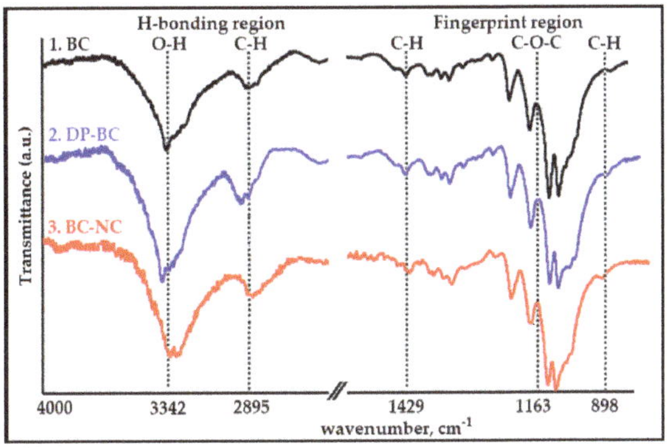

Figure 1. Fourier Transform Infrared (FT-IR) spectra of: 1. native Bacterial Cellulos (BC); 2. depolymerized cellulose, depolymerization of Bacterial Cellulose (DP-BC), (optimum conditions of ultrasonic irradiation step); 3. extracted crystalline regions, bacterial cellulose nanocrystals (BC-NC) (the best results of the catalyzed hydrolysis treatment).

Table 1. Typical vibration bands for the Fourier Transform Infrared (FTIR) spectra of the cellulose samples Bacterial Cellulos (BC), depolymerization of Bacterial Cellulose (DP-BC) and bacterial cellulose nanocrystals (BC-NC).

Sample	Wavenumber, cm^{-1}				
	H-bonding region		**Fingerprint Print Region**		
	Stretching of O–H bonds	**Symmetric C–H stretching vibration**	**Asymmetric Angular Deformation of C–H (Crystalline Regions)**	**Asymmetrical C–O–C Glycoside Bonds**	**Asymmetric Angular Deformation of C–H (Amorphous Region)**
BC DP-BC BC-NC	3342	2895	1429	1163	898

According to Börjesson and Westman [48], these hydroxyl groups were responsible for the stiffness in the polymer chain, and for allowing the linear polymers to form sheet structures. The strong vibration band around 2895 cm^{-1} corresponded to C–H stretching vibrations [49]. This band may be associated with a hydrocarbonate linear chain. Higher values in this specific band are correlated to a decrease in the calculated total crystallinity value [50]. An intense band at 1429 cm^{-1} band can be assigned to the bending of asymmetric angular deformation of C–H bonds. The band found between 1420 to 1430 cm^{-1} was associated with the amount of the cellulose ordered form, while the band appearing at 898 cm^{-1} was assigned to the disordered region [51]. The 1163 cm^{-1} band was assigned to asymmetrical stretching of C–O–C glycoside bonds.

3.2. Crystallinity Index

The crystallinity index of the cellulose was analyzed by X-ray Diffraction analysis. The CrI was calculated by curve-fitting process where individual crystalline peaks were extracted from the diffraction intensity profiles [52,53]. X-ray diffractogram of BC sample was fitted by Gauss function; the results are shown in Figure 2a, and the optimum results of the partial depolymerization step and the extraction of crystalline domains step are presented in Figure 2b. The integrated peak area obtained of each fitting curve is shown in Table 2.

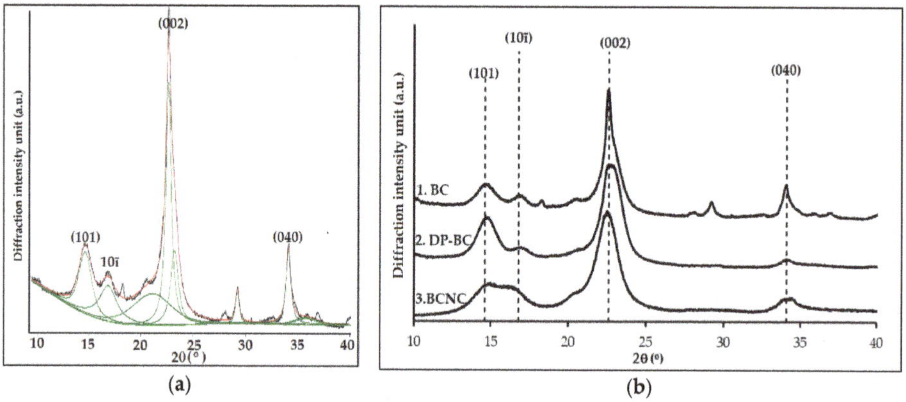

Figure 2. X-ray diffractogram of: (**a**) BC that was fitted by Gauss function; (**b**) XRD pattern of: 1. native BC; 2. depolymerized cellulose, DP-BC, (the optimum result of ultrasonic irradiation step); 3. extracted crystalline regions, BC-NC (the best results of the catalyzed hydrolysis treatment).

Table 2. The integrated peak area obtained of each fitting curve.

Sample	Area (2θ)							
	14.6° (101)	**16.8° (10ī)**	**22.6° (002)**	**23.8°**	**27.2°**	**29.2°**	**34.1° (040)**	**34.8°**
BC	41.8	46.8	324.0	88.3	27.9	125.9	37.6	48.9
DP-BC	95.8	36.5	583.8	26.1	23.0	10.1	33.5	276.6
BC-NC	46.8	38.1	420.6	110.8	17.8	51.9	122	73.6

There were at least eight peaks that had been separated from the diffraction intensity profiles but only four distinct characteristic peaks at 2θ = 14.6°, 16.8°, 22.6°, and 34.1°, which were considered to correspond to 101, 10ī, 002, and 040 crystallographic planes [54,55]. The broad peaks were attributed to the amorphous contribution. The assumption for this analysis was that the amorphous contribution increase was the main contributor to peak broadening. According to Park et al. [45], the other intrinsic factors that influence peak broadening were crystallite size and non-uniform strain within the crystal. These assumptions were then utilized to carry out an investigation of the crystallinity for all the samples. After subtracting the amorphous regions from the whole samples, the CrI was calculated by dividing the remaining diffractogram area due to crystalline cellulose by the total area of the original diffractogram. The samples presented correspond to the optimum results of the partial depolymerization step (DP-BC) which have been further hydrolyzed into the microwave reactor (BC-NC).

3.3. Thermal Properties

Crystallinity is an important parameter, which can greatly affect the physical properties of biodegradable polymers. The identification of neat polymers, copolymers, polymer blends, and composites, as well as the determination of their purity and stability, are generally described by DSC. Amorphous polymers exhibit a glass transition temperature while crystalline or semi-crystalline polymers may possess glass transition temperature, a freezing and melting temperature with various freezing, and melting enthalpies [56]. In this work, the glass transition temperature, T_g, melting point, T_m, and decomposition temperature, T_d were investigated to interpret thermal behaviors of the extracted celluloses. The DSC thermograms for the BC, DP-BC, BC-NC and the reference materials, commercial Microcrystal Cellulose, MCC and Nanocrystal Cellulose, NCC were registered at a heating rate of 10 °C/min and depicted in Figure 3. The thermograms revealed that BC exhibited different pattern from DP-BC and BC-NC. The heat-flow curve of BC displayed a small inflection of the baseline around 105–110 °C, which is the glass transition temperature, T_g, and is followed by an endothermic peak with the onset, T_m = 113.8 °C. In the treated celluloses, DP-BC and BC-NC, with the increased in crystalline content (CrI), the endothermic peak shifted toward a higher temperature. This slightly marked peak could be attributed to the presence of an amorphous region [57]. The curve of DP-BC showed an endothermic peak at 310–370 °C, which appeared to be a melting temperature at T_m = 348.7 °C. The peak was followed by a degradation temperature at T_d = 381.3 °C of the cellulosic material. A similar result was found for BC-NC: an endothermic peak was detected around 260–290 °C with the onset temperature at T_m = 282.8 °C then followed by degradation at T_d = 318.6 °C. The T_g of both samples, DP-BC and BC-NC disappeared or became difficult to detect because of the partial removal of the amorphous regions.

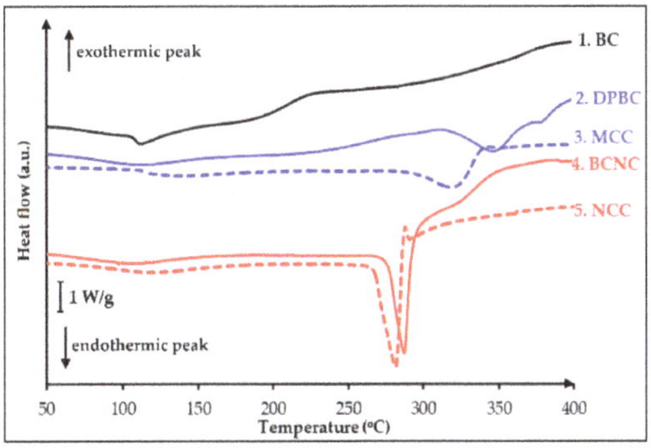

Figure 3. Differential Scanning Calorimetry (DSC)-Thermograms of: 1. BC; 2. depolymerized cellulose, DP-BC; 3. extracted nanocrystalline cellulose, BC-NC.

3.4. Morphology

The morphology of nanocrystals after the hydrolysis treatment was characterized by TEM and STEM observations. Figure 4 presents a comparison of TEM and STEM micrograph of sample BC-NC obtained from a 30 min microwave reaction with $MnCl_2$ catalyst (5% *w*/*w*) at 125 °C to commercial NCC. Both samples were prepared in the same conditions, and as shown, are similarly constituted of a mixture of fibrillated structure with variable length and smaller needles or nano-rods. It seems that the fibrillated structures are in fact constituted of the densely-packed needles. The BC-NC nano-rods are of (164.51 ± 7.56) nm in length with an average diameter of (25.05 ± 2.80) nm. They look like the commercial forms which are (90.94 ± 10.05) nm in length with an average diameter of (12.58 ± 0.87) nm. This result confirmed the crystalline structure already described with a very close similarity to commercially-available nanocrystalline cellulose.

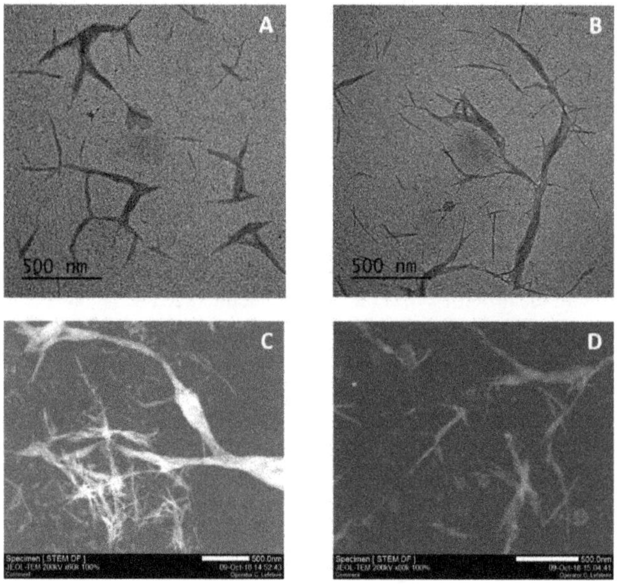

Figure 4. Transmission Electron Microscopy (TEM) and Scanning Transmission Electron Microscopy (STEM) micrograph of BC-NC (**A**,**C**) produced with 0.1M HCl and 5% *w*/*w* of $MnCl_2$ compare to commercial NCC (**B**,**D**).

4. Discussion

4.1. Effect of Ultrasonic Irradiation on The Depolymerization Cellulose

The term "ultrasonic" describes sound waves with a frequency greater than 20 kHz. Many studies have reported the exposure to this wave is responsible for a number of physical and chemical changes. The ultrasonication was the adopted method here to carry out partial depolymerization of native BC into microfibers. The utilization of ultrasonic waves offers a simple and versatile tool for synthesizing micro or nanostructured materials that are often unavailable by conventional methods. In this work, native BC was irradiated into a mixture of water/ethanol (50% *w*/*w*) at constant power of 100 W and frequency of 20 kHz by varying irradiation times. The influence of the length of ultrasonic irradiation period on the CrI is presented in Table 3.

Table 3. Influence of the length of ultrasonic irradiation period on Crystallinity Index, CrI.

Sample	Integrated Area						
	BC	Length of Ultrasonic Irradiation Period (min)					
		10	20	30	60	120	MCC [1]
A_{cr}	450.2	665.3	700.3	749.4	692.6	586.3	637.5
A_{tot}	741.2	1016.0	1032.9	1085.2	979.4	821.1	850.5
CrI	60.7%	65.5%	67.8%	69.1%	70.7%	71.4%	75.0%

[1] Microcrystal Cellulose, MCC (commercial).

From the table, it could be concluded that the increase of CrI is dependent on the irradiation time. Ultrasonic irradiation in water/ethanol induced partial depolymerization of BC with a CrI increase of 8.4% during the first 30 min. The maximum CrI was observed to be 71.4% at 120 min. The results indicated that the irradiation leads to the rupture of amorphous cellulose chains. The disintegration of amorphous regions may be explained by acoustic cavitation. As native BC in a liquid medium was exposed to ultrasonic irradiation, the acoustic waves induce alternating high and low pressure; this creates bubbles (i.e., cavities) and makes them oscillate. A bubble can grow while absorbing the ultrasonic energy at each cycle, until it becomes unstable and finally collapses violently, releasing the energy stored within it, subsequently producing shock waves in the medium [58]. A shear deformation during the collapse of the bubbles is considered to be responsible for the chemical effects which induce disintegration of the amorphous regions of cellulose.

Cavitation occurs over a very wide range of frequencies, from 10 Hz to 10 MHz. Above that frequency regime, the intrinsic viscosity of liquids prevents cavitation from occurring. According to Suslick and coworkers [59], most high intensity ultrasonic horns operate within the range of 20 to 40 kHz. Several factors can affect acoustic cavitation, such as reaction temperature, hydrostatic pressure, frequency, acoustic power, and the type of the solvent medium used. In our study, the amorphous regions degradation increases slightly after 30 min irradiation, and the CrI of cellulose for all sample experiments is still lower compared with that of commercial MCC (75.0%). In our experiments, optimum crystallinity is observed with 30 min irradiation; then, when reaction time is increased, degradation might occurred in both the amorphous and crystalline regions, consecutively reducing the product crystallinity. Concerning the solvent medium, when more volatile solvent is used such as ethanol in water, the mixture is expected to produce more cavitation bubbles, which can significantly promote the reduction rate of amorphous regions. We therefore compare our results (using water/ethanol 50% *w*/*w*) to reactions done in pure water. From XRD spectra, it was calculated that the increase in CrI was less pronounced when the reaction was run in sole water (increase from 60.7 to 67.0%) than in water/ethanol mixture (increase from 60.7 to 69.1%). This result is in accordance with DSC observations, as the endothermic peak of sample that was irradiated with ethanol (the red line in Figure 5B) was detected broader than the one with pure water (blue line in Figure 5B). Both samples have the same melting temperature, i.e., around T_m = 344 °C. The endothermic peak for sample as detected in the DSC curves becomes larger when the crystallinity of the sample increases. According to Ciolacu et al., 2011 [57], the broadening of endothermic peaks detected in the DSC curves of celluloses is in a linear relationship with the percentage value of the amorphous material from their crystalline structure.

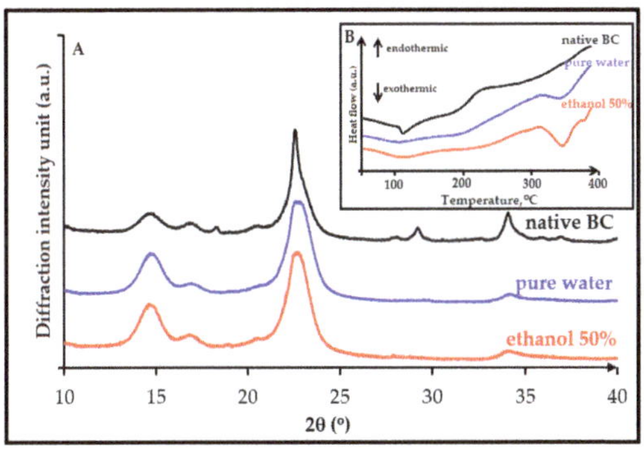

Figure 5. The evolutions of crystalline regions formations for treatment of bacterial cellulose after 30 min irradiation time in pure water (blue curve) or a mixture of ethanol / water 50% (*w*/*w*) (red line) observed by XRD (**A**) and by DSC (**B** in insert) (untreated BC, dark line).

4.2. Effect of $MnCl_2$ Concentration on The Extraction of Crystalline Regions

The microwave treatment assisted by HCl-$MnCl_2$ catalyzed hydrolysis was evaluated to hydrolyze cellulose. In this part, the use of a microwave reactor was performed to get a higher conversion and a shorter reaction time for catalyzed hydrolysis of depolymerized cellulose. In comparison, a conventional heating microwave is a high-frequency radiation that possesses both electrical and magnetic properties [60]. Regarding the addition of catalysts, $MnCl_2$, was utilized to improve the extraction rate of crystalline regions during the hydrolysis. It was already shown that metal chlorides, due to their Lewis acid property, exhibit higher catalytic activity than inorganic acids [61]. The concentration of HCl used in this work was significantly reduced to 0.1 M, instead of 6 M, as used in many reported works.

Table 4 shows the effect of the metal chlorides catalyzed hydrolysis reaction on the crystallinity index. In the absence of a catalyst (0%) in the 0.1M HCl medium, thermal hydrolysis could not occur effectively, and the CrI obtained was only 0.5% higher than for the starting material (DP-BC) from 69.1% to 69.6% for 30 min of reaction. Conversely, with the addition of $MnCl_2$, the CrI was increased to 71.3%, 72.7%, and 79.4% for 1%, 2.5%, and 5% (*w*/*w*), respectively. A similar result was obtained for the use of $FeCl_3 \cdot 6H_2O$, 5% (*w*/*w*) as catalyst with a CrI increased from 69.1% to 77.8%. It was also found that the CrI of all experiments showed a lower value compared with that of the commercial NCC (85.0%). Nevertheless, it appears that the presence of a catalyst plays an important role in the extraction of crystalline regions. In our hypothesis, during partial depolymerization, the ultrasonic treatment leads to the distortion of the amorphous parts and eases the accessibility of chemical reagent to loosen them. Thus, the protons could more easily penetrate into the disordered regions during catalyzed hydrolysis, and as a result, greatly promote the hydrolytic cleavage of glycosidic bonds even in the diluted HCl medium. For this step, the hydrolysis reaction at 0.1M of HCl and 5% *w*/*w* of both metal chlorides ($MnCl_2$ and $FeCl_3 \cdot 6H_2O$) for 30 min reaction can enhance the removal of the amorphous regions, even though the CrI obtained is still less than 80%.

Table 4. Influence of concentration of catalyst $MnCl_2$ and $FeCl_3 \cdot 6H_2O$ on Crystallinity Index, CrI.

	Integrated Area						
	Catalyst Concentration (% *w*/*w*)						
	DP-BC	**$MnCl_2$**				**$FeCl_3 \cdot 6H_2O$**	**NCC [2]**
		0	**1**	**2.5**	**5**		
A_{cr}	749.4	570.4	550.8	508.4	700.2	581.3	1068.1
A_{am}	335.2	249.3	221.7	191.0	181.4	747.2	186.6
CrI	69.1%	69.6%	71.3%	72.7%	79.4%	77.8%	85.1%

[2] NanoCrystal Cellulose, NCC (commercial).

The rapid degradation of amorphous regions during catalyzed hydrolysis can be explained by the Lewis acid character. According to Stein et al., 2010 [62], some metal chlorides such as $FeCl_3$, $AlCl_3$, $CuCl_2$, and $MnCl_2$ could form hydrated complexes in aqueous solution and coordinate the glycosidic oxygen of cellulose. This helps to scissor the glycosidic linkages and to facilitate the hydrolysis process, while the chloride anions attack the hydroxyl atoms [61,63,64]. Introducing metal chloride salts into acid solution can further improve catalytic performance at which the intra- and inter-molecular hydrogen bonds can be broken, and the degradation of the amorphous regions can be accelerated by permeating the internal structure of irradiated cellulose (DP-BC) to acid. Moreover, under microwave and hydrothermal conditions, the easy diffusion of metal cations and chloride anions into the hydrogen bond network, as well as the strong ability of chloride anions to disrupt the hydrogen bond, can be achieved; thus, the hydrolysis rate is greatly enhanced.

Considering the thermal behavior analysis, the summary of DSC results is presented in Table 5 as follows:

Table 5. Characteristic thermal behavior of sample celluloses.

Temperature	Sample (°C)				
	BC	DP-BC	MCC	BC-NC	NCC
T_g	105.0	-	-	-	-
T_m	113.8	348.7	314.1	282.8	278.7
T_d	160.7	381.3	340.1	318.6	286.4

It was found that compared with the DP-BC, the degradation temperature of the BC-NC decreased by approximately 62.7 °C. Similar results were obtained for commercial MCC and NCC sample references. The nano-sized NCC exhibited lower degradation temperature than the micro-sized MCC, i.e., by 53.7 °C. The reason is that the thermal stability of nanocrystals is related to several factors including their dimension, crystallinity, and composition, which in turn depend on extraction conditions [65,66]. So, the NCC with the highest crystallinity would exhibit the highest thermal stability, but smaller dimensions should also cause a decrease of the degradation temperature. The FT-IR analysis demonstrated that the chemical structures of BC-NC remained unchanged after $MnCl_2$-catalyzed hydrolysis process.

5. Conclusions

In this study, BC-NC was conveniently synthesized by sequential ultrasonic irradiation and microwave treatment. A simple and an eco-friendly approach was developed to control the degradation of bacterial cellulose at very low hydrochloric acid concentration. The results demonstrated that ultrasonic irradiation of BC in water/ethanol mixture led to cellulose depolymerization. The CrI increased from 60.7% to 69.1% during the first 30 min of irradiation. Yet, amorphous region distortion remained constant or increased slightly after 2 h, i.e., by 71.4%. It was found that microwave treatment using $MnCl_2$ as Lewis acid exhibited excellent catalytic activity and promoted the hydrolysis in diluted HCl. The reaction rate and the selectivity of BC-NC formation are shown to be optimal at 0.1 M of HCl with 5% *w*/*v* of $MnCl_2$ for 30 min. These conditions can enhance the removal of amorphous regions, yielding BC-NC that possesses an initiating decomposition temperature of 318.6 °C, and led to improve the CrI of up to 79.4%. The BC-NC is typically a mixture of small needles of (164.51 ± 7.56) nm in length and (25.05 ± 2.80) nm diameter that can form fibrillated structures. This procedure yields nanocrystalline bacterial cellulose having similar features to commercially-available nanocrystalline cellulose.

Author Contributions: E.Y.W. took care about Sections 1–5, proposed the subject of the review to all the other authors and took care about the general planning of the work; H.W. and S.A. prepared the figures, tables and organized references; M.P.P. organized Sections 2 and 2.3.1; D.C. prepared Sections 2.3.3, 3.3, 4.1 and 4.2; F.O. took care about Sections 2.3.1 and 3.2; E.G. prepared Sections 2.3.4, 3.1 and 4 wrote the abstract, Section 1, Section 4 and the conclusions, K.S. supervised all the work and Finally, all the authors contributed equally to the general organization of the manuscript and its revision, with helpful suggestions about the content and the style of the text.

Funding: This research was founded by Ministry of Research, Technology and Higher Education of the Republic of Indonesia (Kemenristekdikti).

Acknowledgments: Technology and Higher Education of the Republic of Indonesia (Kemenristekdikti), Universitas Sultan Agung Tirtayasa (UNTIRTA). Within the framework of the chair: "Green Chemistry and Processes" this research was also sponsored by the European Union through FEDER funding, by the Région Haut de France, the Ecole Supérieur de Chimie Organique et Minérale (ESCOM) and Université de Technologie de Compiègne (UTC).

Conflicts of Interest: The authors declare no conflict of interest. Moreover, the funding sponsors had no role in the design of the study; in the collection, analyses, or interpretation of data; in the writing of the manuscript, and in the decision to publish the results. The design of the study; in the collection, analyses, or interpretation of data; in the writing of the manuscript, and in the decision to publish the results.

References

1. Guilbert, S.; Gontard, N.; Gorris, L.G. Prolongation of the shelf-life of perishable food products using biodegradable films and coatings. *LWT-Food Sci. Technol.* **1996**, *29*, 10–17. [CrossRef]
2. Makaremi, M.; Pasbakhsh, P.; Cavallaro, G.; Lazzara, G.; Aw, Y.K.; Lee, S.M.; Milioto, S. Effect of morphology and size of halloysite nanotubes on functional pectin bionanocomposites for food packaging applications. *ACS Appl. Mater. Interfaces* **2017**, *9*, 17476–17488. [CrossRef] [PubMed]
3. Khan, A.; Khan, R.A.; Salmieri, S.; Tien, C.L.; Riedl, B.; Bouchard, J.; Chauve, G.; Tan, V.; Kamal, M.R.; Lacroix, M. Mechanical and barrier properties of nanocrystalline cellulose reinforced chitosan based nanocomposite films. *Carbohydr. Polym.* **2012**, *90*, 1601–1608. [CrossRef] [PubMed]
4. Bertolino, V.; Cavallaro, G.; Lazzara, G.; Merli, M.; Milioto, S.; Parisi, F.; Sciascia, L. Effect of the biopolymer charge and the nanoclay morphology on nanocomposite materials. *Ind. Eng. Chem. Res.* **2016**, *55*, 7373–7380. [CrossRef]
5. Khan, A.; Huq, T.; Saha, M.; Khan, R.A.; Khan, M.A.; Gafur, M.A. Effect of silane treatment on the mechanical and interfacial properties of calcium alginate fiber reinforced polypropylene composite. *J. Compos. Mater.* **2010**, *44*, 2875–2886. [CrossRef]
6. De Silva, R.T.; Pasbakhsh, P.; Goh, K.L.; Chai, S.P.; Chen, J. Synthesis and characterisation of poly (lactic acid)/halloysite bionanocomposite films. *J. Compos. Mater.* **2014**, *48*, 3705–3717. [CrossRef]
7. Gorrasi, G.; Pantani, R.; Murariu, M.; Dubois, P. PLA/H alloysite Nanocomposite Films: Water Vapor Barrier Properties and Specific Key Characteristics. *Macromol. Mater. Eng.* **2014**, *299*, 104–115. [CrossRef]
8. Cataldo, V.A.; Cavallaro, G.; Lazzara, G.; Milioto, S.; Parisi, F. Coffee grounds as filler for pectin: Green composites with competitive performances dependent on the UV irradiation. *Carbohydr. Polym.* **2017**, *170*, 198–205. [CrossRef] [PubMed]
9. Gorrasi, G.; Bugatti, V.; Vittoria, V. Pectins filled with LDH-antimicrobial molecules: Preparation, characterization and physical properties. *Carbohydr. Polym.* **2012**, *89*, 132–137. [CrossRef] [PubMed]
10. Cavallaro, G.; Donato, D.I.; Lazzara, G.; Milioto, S. Films of halloysite nanotubes sandwiched between two layers of biopolymer: From the morphology to the dielectric, thermal, transparency, and wettability properties. *J. Phys. Chem. C.* **2011**, *115*, 20491–20498. [CrossRef]
11. Lagaron, J.M.; Lopez-Rubio, A. Nanotechnology for bioplastics: Opportunities, challenges and strategies. *Trends Food Sci. Technol.* **2011**, *22*, 611–617. [CrossRef]
12. Pantani, R.; Gorrasi, G.; Vigliotta, G.; Murariu, M.; Dubois, P. PLA-ZnO nanocomposite films: Water vapor barrier properties and specific end-use characteristics. *Eur. Polym. J.* **2013**, *49*, 3471–3482. [CrossRef]
13. Vroman, I.; Tighzert, L. Biodegradable polymers. *Materials.* **2009**, *2*, 307–344. [CrossRef]
14. Ruiz-Hitzky, E.; Aranda, P.; Darder, M.; Rytwo, G. Hybrid materials based on clays for environmental and biomedical applications. *J. Mater. Chem.* **2010**, *20*, 9306–9321. [CrossRef]
15. Berthelot, J.M. *Composite materials: Mechanical behavior and structural analysis*; Springer: Berlin, Germany, 2012; ISBN 978-1-4612-6803-1,978-1-4612-0527-2.
16. Lvov, Y.; Wang, W.; Zhang, L.; Fakhrullin, W. Halloysite clay nanotubes for loading and sustained release of functional compounds. *Adv. Mater.* **2016**, *28*, 1227–1250. [CrossRef] [PubMed]
17. Abdullayev, E.; Lvov, Y. Halloysite clay nanotubes as a ceramic 'skeleton' for functional biopolymer composites with sustained drug release. *J. Mater. Chem. B* **2013**, *1*, 2894–2903. [CrossRef]
18. Abdullayev, W.; Sakakibara, K.; Okamoto, K.; Wei, W.; Ariga, K.; Lvov, Y. Natural tubule clay template synthesis of silver nanorods for antibacterial composite coating. *ACS Appl. Mater. Interfaces* **2011**, *3*, 4040–4046. [CrossRef] [PubMed]
19. Biddeci, G.; Cavallaro, G.; Di Blasi, F.; Lazzara, G.; Massaro, M.; Milioto, S.; Parisi, F.; Riela, S.; Spinellia, G. Halloysite nanotubes loaded with peppermint essential oil as filler for functional biopolymer film. *Carbohydr. Polym.* **2016**, *152*, 548–557. [CrossRef] [PubMed]
20. Gorrasi, G. Dispersion of halloysite loaded with natural antimicrobials into pectins: Characterization and controlled release analysis. *Carbohydr. Polym.* **2015**, *127*, 47–53. [CrossRef] [PubMed]
21. Massaro, M.; Riela, S.; Guernelli, S.; Parisi, F.; Lazzara, G.; Baschieri, A.; Valgimigli, L.; Amorati, R. A synergic nanoantioxidant based on covalently modified halloysite–trolox nanotubes with intra-lumen loaded quercetin. *J. Mater. Chem. B* **2016**, *4*, 2229–2241. [CrossRef]

22. Abdulkhani, A.; Marvast, E.H.; Ashori, A.; Hamzeh, Y.; Karimi, A.N. Preparation of cellulose/polyvinyl alcohol biocomposite films using 1-n-butyl-3-methylimidazolium chloride. *Int. J. Biol. Macromol.* **2013**, *62*, 379–386. [CrossRef] [PubMed]
23. Qiu, X.; Hu, S. Smart' materials based on cellulose: A review of the preparations, properties, and applications. *Materials* **2013**, *6*, 738–781. [CrossRef] [PubMed]
24. Paximada, P.; Tsouko, E.; Kopsahelis, N.; Koutinas, A. A.; and Mandala, I. Bacterial cellulose as stabilizer of o/w emulsions. *Food Hydrocoll.* **2016**, *53*, 225–232. [CrossRef]
25. Hu, Y.; Catchmark, J.M.; Zhu, Y.; Abidi, N.; Zhou, X.; Wang, J.; Liang, H. Engineering of porous bacterial cellulose toward human fibroblasts ingrowth for tissue engineering. *J. Mater. Res.* **2014**, *29*, 2682–2693. [CrossRef]
26. Shah, N.; Ul-Islam, M.; Khattak, W. A.; and Park, J. K. Overview of bacterial cellulose composites: A multipurpose advanced material. *Carbohydr. Polym.* **2013**, *98*, 1585–1598. [CrossRef] [PubMed]
27. Brown, A.J. XLIII-On an acetic ferment which forms cellulose. *J. Chem. Soc. Trans.* **1886**, *49*, 432–439. [CrossRef]
28. Li, J.; Zhang, X.; Zhang, M.; Xiu, H.; and He, H. Optimization of selective acid hydrolysis of cellulose for microcrystalline cellulose using $FeCl_3$. *BioResources.* **2014**, *9*, 1334–1345. [CrossRef]
29. George, J.; Ramana, K.V.; Sabapathy, S.N.; Jagannath, J.H.; Bawa, A.S. Characterization of chemically treated bacterial (*Acetobacter xylinum*) biopolymer: Some thermo-mechanical properties. *Int. J. Biol. Macromol.* **2005**, *37*, 189–194. [CrossRef] [PubMed]
30. Thambiraj, S.; Shankaran, D.R. Preparation and physicochemical characterization of cellulose nanocrystals from industrial waste cotton. *Appl. Surf. Sci.* **2017**, *412*, 405–416. [CrossRef]
31. Dufresne, A. *Nanocellulose: From Nature to High Performance Tailored Materials*; Springer: Berlin, Germany, 2012; ISBN 3110254565.
32. Ureña-Benavides, E.E.; Davis, G.; Ao, V.A.; Kitchens, C.L. Rheology and phase behavior of lyotropic cellulose nanocrystal suspensions. *Macromolecules.* **2011**, *44*, 8990–8998.
33. Beck-Candanedo, S.; Roman, M.; Gray, D.G. Effect of reaction conditions on the properties and behavior of wood cellulose nanocrystal suspensions. *Biomacromolecules* **2005**, *6*, 1048–1054. [CrossRef] [PubMed]
34. Reid, M.S.; Villalobos, M.; Cranston, E.D. Benchmarking cellulose nanocrystals: From the laboratory to industrial production. *Langmuir* **2016**, *33*, 1583–1598. [CrossRef] [PubMed]
35. Yu, H.; Qin, Z.; Liang, B.; Liu, N.; Zhou, Z.; Chen, L. Facile extraction of thermally stable cellulose nanocrystals with a high yield of 93% through hydrochloric acid hydrolysis under hydrothermal conditions. *J. Mater. Chem. A* **2013**, *1*, 3938–3944. [CrossRef]
36. Roman, M.; Winter, W.T. Effect of sulfate groups from sulfuric acid hydrolysis on the thermal degradation behavior of bacterial cellulose. *Biomacromolecules* **2004**, *5*, 1671–1677. [CrossRef] [PubMed]
37. Sheltami, R.M.; Kargarzadeh, H.; Abdullah, I.; Ahmad, I. Thermal Properties of Cellulose Nanocomposites. In *Handbook of Nanocellulose and Cellulose Nanocomposites*; Wiley-VCH: New York, NY, USA, 2017; Volume 2, pp. 523–552.
38. Filson, P.B.; Dawson-Andoh, B.E. Sono-chemical preparation of cellulose nanocrystals from lignocellulose derived materials. *Bioresour. Technol.* **2009**, *100*, 2259–2264. [CrossRef] [PubMed]
39. Jiang, F.; Esker, A.R.; Roman, M. Acid-catalyzed and solvolytic desulfation of H_2SO_4-hydrolyzed cellulose nanocrystals. *Langmuir* **2010**, *26*, 17919–17925. [CrossRef] [PubMed]
40. Rosa, M.F.; Medeiros, E.S.; Malmonge, J.A.; Gregorski, K.S.; Wood, D.F.; Mattoso, L.H.C.; Glenn, G.; Orts, W.J.; Imam, S.H. Cellulose nanowhiskers from coconut husk fibers: Effect of preparation conditions on their thermal and morphological behavior. *Carbohydr. Polym.* **2010**, *81*, 83–92. [CrossRef]
41. Araki, J.; Wada, M.; Kuga, S.; Okano, T. Flow properties of microcrystalline cellulose suspension prepared by acid treatment of native cellulose. *Colloids Surf. Physicochem. Eng. Asp.* **1998**, *142*, 75–82. [CrossRef]
42. Martínez-Sanz, M.; Lopez-Rubio, A.; Lagaron, J.M. Optimization of the nanofabrication by acid hydrolysis of bacterial cellulose nanowhiskers. *Carbohydr. Polym.* **2011**, *85*, 228–236. [CrossRef]
43. Yu, H.Y.; Qin, Z.Y.; Liu, L.; Yang, X.G.; Zhou, Y.; Yao, J.M. Comparison of the reinforcing effects for cellulose nanocrystals obtained by sulfuric and hydrochloric acid hydrolysis on the mechanical and thermal properties of bacterial polyester. *Compos. Sci. Technol.* **2013**, *87*, 22–28. [CrossRef]
44. Brinchi, L.; Cotana, F.; Fortunati, E.; Kenny, J.M. Production of nanocrystalline cellulose from lignocellulosic biomass: Technology and applications. *Carbohydr. Polym.* **2013**, *94*, 154–169. [CrossRef] [PubMed]

45. Park, S.; Baker, J.O.; Himmel, M.E.; Parilla, P.A.; Johnson, D.K. Cellulose crystallinity index: Measurement techniques and their impact on interpreting cellulase performance. *Biotechnol. Biofuels* **2010**, *3*, 10. [CrossRef] [PubMed]
46. Kruer-Zerhusen, N.; Cantero-Tubilla, B.; Wilson, D. B. Characterization of cellulose crystallinity after enzymatic treatment using Fourier transform infrared spectroscopy (FTIR). *Cellulose* **2018**, *25*, 37–48. [CrossRef]
47. Kondo, T. The assignment of IR absorption bands due to free hydroxyl groups in cellulose. *Cellulose* **1997**, *4*, 281. [CrossRef]
48. Börjesson, M.; Westman, G. Crystalline nanocellulose: Preparation, modification, and properties. In *Cellulose-Fundamental Aspects and Current Trends*; InTechOpen: London, UK, 2015.
49. Chang, W.S.; and Chen, H.H. Physical properties of bacterial cellulose composites for wound dressings. *Food Hydrocoll.* **2016**, *53*, 75–83. [CrossRef]
50. Ornaghi, H.L.; Poletto, M.; Zattera, A.J.; Amico, S.J. Correlation of the thermal stability and the decomposition kinetics of six different vegetal fibers. *Cellulose* **2014**, *21*, 177–188. [CrossRef]
51. AAkerholm, M.; Hinterstoisser, B.; Salmén, L. Characterization of the crystalline structure of cellulose using static and dynamic FT-IR spectroscopy. *Carbohydr. Res.* **2004**, *339*, 569–578. [CrossRef] [PubMed]
52. Hult, E.L.; Iversen, T.; Sugiyama, J. Characterization of the supermolecular structure of cellulose in wood pulp fibres. *Cellulose* **2003**, *10*, 103–110. [CrossRef]
53. Garvey, C.J.; Parker, I.H.; and Simon, G.P. On the interpretation of X-ray diffraction powder patterns in terms of the nanostructure of cellulose I fibres. *Macromol. Chem. Phys.* **2005**, *206*, 1568–1575. [CrossRef]
54. He, J.; Cui, S.; Wang, S. Preparation and crystalline analysis of high-grade bamboo dissolving pulp for cellulose acetate. *J. Appl. Polym. Sci.* **2008**, *107*, 1029–1038. [CrossRef]
55. Nishiyama, Y.; Sugiyama, J.; Chanzy, H.; Langan, P. Crystal structure and hydrogen bonding system in cellulose Iα from synchrotron X-ray and neutron fiber diffraction. *J. Am. Chem. Soc.* **2003**, *125*, 14300–14306. [CrossRef] [PubMed]
56. Gregorova, A. Application of differential scanning calorimetry to the characterization of biopolymers. In *Applications of Calorimetry in a Wide Context-Differential Scanning Calorimetry, Isothermal Titration Calorimetry and Microcalorimetry, Odile Carisse*; InTechOpen: London, UK, 2013.
57. Ciolacu, D.; Ciolacu, F.; Popa, V. I. Amorphous cellulose—Structure and characterization. *Cellul. Chem. Technol.* **2011**, *45*, 13.
58. Suslick, K.S. Effects of ultrasound on surfaces and solids. *Adv. Sonochem.* **1990**, *1*, 197–230.
59. Bang, J. H.; and Suslick, K. S. Applications of ultrasound to the synthesis of nanostructured materials. *Adv. Mater.* **2010**, *22*, 1039–1059. [CrossRef] [PubMed]
60. Wong, T.W.; Chan, L.W.; Kho, S.B. Heng, P.W.S. Design of controlled-release solid dosage forms of alginate and chitosan using microwave. *J. Control. Release.* **2002**, *84*, 99–114. [CrossRef]
61. Kamireddy, S.R.; Li, J.; Tucker, M.; Degenstein, J.; Ji, Y. Effects and mechanism of metal chloride salts on pretreatment and enzymatic digestibility of corn stover. *Ind. Eng. Chem. Res.* **2013**, *52*, 1775–1782. [CrossRef]
62. vom Stein, T.; Grande, P.; Sibilla, F.; Commandeur, U.; Fischer, R.; Leitner, W.; Domínguez de María, P. Salt-assisted organic-acid-catalyzed depolymerization of cellulose. *Green Chem.* **2010**, *12*, 1844–1849. [CrossRef]
63. Lu, Q.; Tang, L.; Lin, F.; Wang, S.; Chen, Y.; Chen, X.; Huang, B. Preparation and characterization of cellulose nanocrystals via ultrasonication-assisted $FeCl_3$-catalyzed hydrolysis. *Cellulose* **2014**, *21*, 3497–3506. [CrossRef]
64. Ma, Y.; Ji, W.; Zhu, X.; Tian, L.; Wan, X. Effect of extremely low $AlCl_3$ on hydrolysis of cellulose in high temperature liquid water. *Biomass Bioenergy* **2012**, *39*, 106–111. [CrossRef]
65. Henrique, M.A.; Neto, W.P.F.; Silvério, H.A.; Martins, D.F.; Gurgel, L.V.A.; Barud, H.S.; Morais, L.C.; Pasquini, D. Kinetic study of the thermal decomposition of cellulose nanocrystals with different polymorphs, cellulose I and II, extracted from different sources and using different types of acids. *Ind. Crops Prod.* **2015**, *76*, 128–140. [CrossRef]
66. Jonoobi, M.; Oladi, R.; Davoudpour, Y.; Oksman, K.; Dufresne, A.; Hamzeh, Y.; Davoodi, R. Different preparation methods and properties of nanostructured cellulose from various natural resources and residues: A review. *Cellulose* **2015**, *22*, 935–969. [CrossRef]

nanomaterials

MDPI

Article

Transparent and Robust *All*-Cellulose Nanocomposite Packaging Materials Prepared in a Mixture of Trifluoroacetic Acid and Trifluoroacetic Anhydride

Susana Guzman-Puyol [1,*], Luca Ceseracciu [2], Giacomo Tedeschi [1], Sergio Marras [2], Alice Scarpellini [3], José J. Benítez [4], Athanassia Athanassiou [1,*] and José A. Heredia-Guerrero [1,*]

1 Smart Materials, Nanophysics, Istituto Italiano di Tecnologia, Via Morego, 30, 16163 Genova, Italy; giacomo.tedeschi@iit.it

2 Materials Characterization Facility, Istituto Italiano di Tecnologia, Via Morego, 30, 16163 Genova, Italy; luca.ceseracciu@iit.it (L.C.); sergio.marras@iit.it (S.M.)

3 Electron Microscopy Facility, Istituto Italiano di Tecnologia, Via Morego, 30, 16163 Genova, Italy; alice.scarpellini@iit.it

4 Instituto de Ciencia de Materiales de Sevilla, Centro mixto CSIC-Universidad de Sevilla, Calle Americo Vespucio 49, Isla de la Cartuja, 41092 Sevilla, Spain; benitez@icmse.csic.es

* Correspondence: susana.guzman@iit.it (S.G.-P.); athanassia.athanassiou@iit.it (A.A.); jose.heredia-guerrero@iit.it (J.A.H.-G.)

Received: 30 January 2019; Accepted: 24 February 2019; Published: 5 March 2019

Abstract: *All*-cellulose composites with a potential application as food packaging films were prepared by dissolving microcrystalline cellulose in a mixture of trifluoroacetic acid and trifluoroacetic anhydride, adding cellulose nanofibers, and evaporating the solvents. First, the effect of the solvents on the morphology, structure, and thermal properties of the nanofibers was evaluated by atomic force microscopy (AFM), X-ray diffraction (XRD), and thermogravimetric analysis (TGA), respectively. An important reduction in the crystallinity was observed. Then, the optical, morphological, mechanical, and water barrier properties of the nanocomposites were determined. In general, the final properties of the composites depended on the nanocellulose content. Thus, although the transparency decreased with the amount of cellulose nanofibers due to increased light scattering, normalized transmittance values were higher than 80% in all the cases. On the other hand, the best mechanical properties were achieved for concentrations of nanofibers between 5 and 9 wt.%. At higher concentrations, the cellulose nanofibers aggregated and/or folded, decreasing the mechanical parameters as confirmed analytically by modeling of the composite Young's modulus. Finally, regarding the water barrier properties, water uptake was not affected by the presence of cellulose nanofibers while water permeability was reduced because of the higher tortuosity induced by the nanocelluloses. In view of such properties, these materials are suggested as food packaging films.

Keywords: *All*-cellulose nanocomposites; trifluoroacetic acid; trifluoroacetic anhydride; transparency; robustness; packaging material

1. Introduction

The massive use of petroleum-based plastics in disposable food packaging materials has triggered a global social concern, mainly because of the pollution derived from their synthesis and the related littering problems [1–4]. For this reason, a new model of economic activity, namely "circular economy," has emerged [5–7]. Among the different proposals of a circular economy applied to these plastics, the production of bio-based plastics from alternative feedstocks such as agro-food by-products and naturally occurring biopolymers is a strategy that is attracting the attention of researchers [8–15]. Cellulose is the most abundant polymer on Earth with an annual biomass production of about 1.5 ×

10^{12} tons per year, being, hence, one of the most promising bio-renewable resources for reducing and replacing the huge amount of petroleum-based plastics. Cellulose shows full biodegradability in soil and seawater over short times, is lightweight and has excellent mechanical strength [16–18]. Among the vast group of cellulose materials, *all*-cellulose composites (ACCs) are a category of particular interest. These composites are materials in which both the reinforcement and the matrix are cellulose [19,20]. The use of the same material acting as matrix and reinforcement increases the compatibility between the phases and, therefore, the mechanical properties of the composite [21]. Two different procedures have been reported for the preparation of *all*-cellulose composites [20,22]. The first is a two-step method in which cellulose is partially dissolved and then regenerated in the presence of undissolved cellulose. Thus, regenerated and undissolved cellulose fractions may come from different natural origins [23]. The second procedure consists of a one-step method where the surface of cellulose is partially dissolved and regenerated in situ to create a matrix around the non-dissolved portion. The most common solvents employed in these processes are LiCl/DMAc, *N*-Methylmorpholine *N*-oxide (NMMO), NaOH, and the ionic liquid 1-butyl−3-methylimidazolium chloride [20]. Nevertheless, partial and slow cellulose dissolution and non-recyclability have limited the use of these solvents on an industrial scale. In any case, the dissolution process described above for both methods is followed by subsequent solvent removal, cellulose regeneration, and drying [20]. The result of such a process is an *all*-cellulose composite with exceptional mechanical properties [24], optical transparency [25], and improved barrier properties [22] with respect to regenerated cellulose, as well as full biodegradability [26]. *All*-cellulose composites are used in a wide range of applications such as the reinforcement of other polymers, substitution of bone and cartilage materials, the fabrication of electro-active paper, sensors and electrical displays, and the production of biodegradable food packaging materials and mulching films for agriculture [26–37].

Trifluoroacetic acid (TFA) is one of the non-aqueous derivatizing solvents for cellulose [38–44]. The dissolution of cellulose by TFA might not occur in the absence of a chemical reaction [38]: cellulose is trifluoroacetylated selectively in the C6-hydroxyl groups [39]. This derivative is readily hydrolyzed in water, water vapor, or the moisture in the air, forming amorphous and transparent cellulose films [11,38,40]. Trifluoroacetic acid is a naturally occurring organic acid and biodegradable by microbial action [38,45,46]. Moreover, it is recyclable by distillation due to its high volatility and is miscible with many organic solvents and water [11]. TFA has been recently used to fabricate bioplastics from microcrystalline cellulose and plant wastes, as well as blends of cellulose with seaweeds, silk, nylon, poly(methyl methacrylate), and poly(vinyl alcohol) [11,47–50]. When TFA is combined with trifluoroacetic acid anhydride (TFAA), a reactive mixture that allows the acylation of cellulose and cellulose derivatives with carboxylic acids is generated [51,52].

In this work, we prepared *all*-cellulose nanocomposites with potential application in food packaging by using a simple method consisting of dissolving cellulose in a trifluoroacetic acid:trifluoroacetic anhydride (2:1, v:v) mixture and subsequent addition of different cellulose nanofibers dispersed in chloroform. The effect of the solvent in the nanocelluloses was investigated. In addition, the influence of different percentages of these nanofillers on the morphology, optical, mechanical, thermal, and hydrodynamic properties of the nanocomposites was assessed. Furthermore, a model was developed to analyze the mechanical properties.

2. Materials and Methods

2.1. Materials

Trifluoroacetic acid (TFA), trifluoroacetic anhydride (TFAA), microcrystalline cellulose (MCC) from cotton linters (crystallinity ~79%), and chloroform were purchased from Sigma-Aldrich (St. Gallen, Switzerland) and used without additional purification. Cellulose nanofibers (two types: one nominally shorter than the other and labeled as sNF and lNF, respectively) were purchased from Nanografi

(Ankara, Turkey) and used as received. These nanofibers were prepared from wood pulp by using mechanical methods and commercialized as dry powders.

2.2. Fabrication of All-Cellulose Nanocomposites

The preparation of *all*-cellulose nanocomposites was carried out as follows: first, MCC (450 mg) was dissolved in 30 mL of TFA:TFAA (2:1, v:v) in a 50 mL closed flask and stirred at 50 °C until the solution was completely clear (~1 h). Later, cellulose nanofibers (4.5, 22.5, 45, 90 and 135 mg) were mixed with 30 mL chloroform and dispersed by three consecutive 30 s ultrasound cycles using a 3.2 mm diameter tapered microtip at 10% amplitude attached to a VCX 750 ultrasonic processor (Sonics & Materials, Inc., Newtown, CT, USA). Then, both solutions were blended together and cast in glass Petri dishes. The mixture of solvents was completely evaporated after 1 day under an aspirated hood, originating freestanding films. Pure cellulose films were also prepared as a control using the same protocol in order to study the role of the nanofibers as reinforcement of the cellulose matrix. Similarly, to analyze the effect of the solvents on the nanocelluloses, the nanofibers were subjected to the above treatment but without adding MCC. All samples were stored at 44% relative humidity (RH) for 7 days before analysis to ensure the reproducibility of the measurements. Table 1 summarizes the label and the final composition of the different samples.

Table 1. Label and final formulation of the samples.

Label	MCC (wt.%)	Nanocellulose	
		Short Nanofibers (wt.%)	Long Nanofibers (wt.%)
Cellulose	100	-	-
sNF1	99	1	-
sNF5	95	5	-
sNF10	91	9	-
sNF20	83	17	-
sNF30	77	23	-
lNF1	99	-	1
lNF5	95	-	5
lNF10	91	-	9
lNF20	83	-	17
lNF30	77	-	23

2.3. Morphological Characterization

Atomic force microscopy (AFM) images were acquired using a Nanotec microscope (Nanotec, Madrid, Spain) in low amplitude dynamic mode. Levers used were Nanosensors PPP-NCH (NanoWorld AG, Neuchâtel, Switzerland) with a tip radius curvature less than 10 nm and a resonance frequency of 295 kHz (29 N/m force constant). Samples were prepared from sonicated diluted dispersions (0.1 mg/30 mL water) by placing a 10 µL drop on a freshly cleaved mica muscovite piece (~1 cm^2) and allowing to dry overnight inside a Petri dish. The width and length of the nanofibers were measured with WSxM software [53]. Approximately 100 measurements were taken to obtain each width and length distribution.

High-resolution scanning electron microscopy (SEM) imaging was carried out using a JEOL JSM 7500FA (Jeol, Tokyo, Japan) equipped with a cold field-emission gun (FEG), operating at 15 kV acceleration voltage. The samples were coated with a 10 nm thick film of carbon using an Emitech K950X high-vacuum turbo system (Quorum Technologies Ltd., East Sussex, Lewes, UK). Imaging was performed with the secondary electrons to analyze the morphology of the samples.

2.4. Optical Characterization

Transparency was determined as the normalized transmittance according to the standard ASTM D1746 by using a ultraviolet (UV) spectrophotometer Varian Cary 6000i (USA) [54]. For this, samples were cut into rectangular pieces and placed directly in the spectrophotometer test cell. An empty test cell was used as a reference. Five measurements were taken from different samples and the results were averaged to obtain a mean value. Normalized transmittance, in percentage, was calculated as indicated below:

$$Normalized\ transmittance\ (\%) = \frac{\log \%T}{b} \times 100$$

where $\%T$ is the transmittance at 600 nm and b is the thickness of the sample (mm).

2.5. Structural Characterization

X-ray diffraction (XRD) patterns were recorded on a Rigaku SmartLab X-ray powder diffractometer equipped with a 9 kW CuKα rotating anode (Rigaku, Tokyo, Japan), operating at 40 kV and 150 mA. A Göbel mirror was used to convert the divergent X-ray beam into a parallel beam and to suppress the Cu Kβ radiation. The specimens were analyzed at room temperature using a zero diffraction quartz sample holder. XRD data analysis was carried out using PDXL 2.1 software from Rigaku. The crystallinity index (*CrI*) was determined by using the empirical method proposed by Segal et al. [55]:

$$CrI(\%) = \frac{I - I'}{I} \times 100$$

where I is the intensity of the peak assigned to (002) crystal plane of cellulose located at $21-23°$ and I' is the intensity of the diffractogram of the amorphous cellulose at $18-19°$. In addition, the crystallite size of cellulose (D) was estimated by Scherrer's Equation:

$$D = \frac{K\lambda}{\beta \cos\theta}$$

where K is a constant of value 0.94, λ is the X-ray wavelength (0.15418 nm), θ is the diffraction angle for the (200) plane, and β is the peak width at half the maximum intensity (calculated from peak deconvolution when necessary).

2.6. Mechanical Characterization

The mechanical properties of the films were measured by uniaxial tensile tests on a dual column Instron 3365 universal testing machine. Dog-bone-shaped samples were stretched at a rate of 5 mm/min. All the stress–strain curves were recorded at 25 °C and 44% RH. Ten measurements were conducted for each sample and the results were averaged to obtain a mean value. From the stress–strain curves, Young's modulus, yield stress, elongation at the break, and fracture energy (area below the curve) were calculated.

2.7. Thermal Characterization

The thermal degradation behavior of the nanocelluloses was investigated by a standard thermogravimetric analysis (TGA) method using a TGA Q500 from TA Instruments (New Castle, DE, USA). Measurements were performed using 3–5 mg of sample in an aluminum pan under inert N_2 atmosphere with a flow rate of 50 mL/min in a temperature range from 30 to 600 °C with a heating rate of 5 °C/min. The weight loss and its first derivative were recorded simultaneously as a function of time/temperature.

2.8. Water Uptake and Permeability

For water uptake measurements, samples were first dried by conditioning in a desiccator until no change in sample weight was measured. Dry samples were weighed (~30 mg) on a sensitive electronic balance and, then, placed in a 100% relative humidity (RH) chamber at 25 °C. Once the equilibrium was reached, each sample was again weighed and the amount of adsorbed water was calculated as the difference with the initial dry weight. Three measurements were taken and the results were averaged to obtain a mean value. Water uptake, in percentage, was calculated as indicated below:

$$Water\ uptake\ (\%) = \frac{m_f - m_0}{m_0} \times 100$$

where m_f is the sample weight at 100% RH and m_0 is the sample weight at 0% RH.

Water vapor permeability (WVP) of *all*-cellulose nanocomposites was determined at 25 °C and under 100% relative humidity gradient (ΔRH %) according to the ASTM E96 standard method [56,57]. Then, 400 μL of deionized water (which generates 100% RH inside the permeation cell) was placed in each test permeation cell (7 mm inside diameter, 10 mm inner depth). *All*-cellulose composites were cut into circles and mounted on the top of the permeation cells. The permeation cells were placed in a 0% RH desiccator with anhydrous silica gel used as a desiccant agent. The water transferred through the film was determined from the weight change of the permeation cell every hour over 7 h using an electronic balance (0.0001 g accuracy). The weight loss of the permeation cells was plotted as a function of time. The slope of each line was calculated by linear regression and the water vapor transmission rate (WVTR) was determined as below:

$$WVTR\left(g/\left(m^2 \cdot day\right)\right) = \frac{Slope}{Area\ of\ the\ film}$$

WVP measurements were replicated three times for each sample. The WVP value was calculated as follows:

$$WVP\ (g/(m \cdot day \cdot Pa)) = \frac{WVTR \times l \times 100}{p_s \times \Delta RH}$$

where l (m) is the film thickness measured with a micrometer with 0.001 mm accuracy, ΔRH (%) is the percentage relative humidity gradient, and p_s (Pa) is the saturation water vapor pressure at 25 °C (3168 Pa).

3. Results and Discussion

3.1. Effect of TFA/TFAA Mixture on the Cellulose Nanofibers

Short and long nanofibers (sNF and lNF, respectively) before and after TFA:TFAA treatment were morphologically characterized by AFM (Figure 1). Figure 1A shows the AFM topographies of the pristine and treated cellulose nanofibers. Both types of pristine nanocelluloses exhibited a fiber morphology. The distribution of widths and lengths for each kind of nanofiber is displayed in Figure 1B. While the widths of both nanocelluloses were very similar with a maximum at ~53 nm (although sNF showed a narrower distribution), the values of the length were different: the maximums were ~100 and ~175 nm for sNF and lNF, respectively. Over again, the distribution of sNF was narrower than the lNF. Interestingly, the mixture of solvents produced important changes in the morphology of the cellulose nanofibers (Figure 1A). Broadly, agglomerations of the nanoparticles and flat islands of height ~2 nm were observed. Such islands could be produced by a partial solution of the cellulose by the TFA/TFAA mixture and the formation of flat, amorphous cellulose when solvents were evaporated. Similar flat and featureless AFM topography images with roughness <2 nm were obtained for cellulose bioplastics prepared in TFA [11].

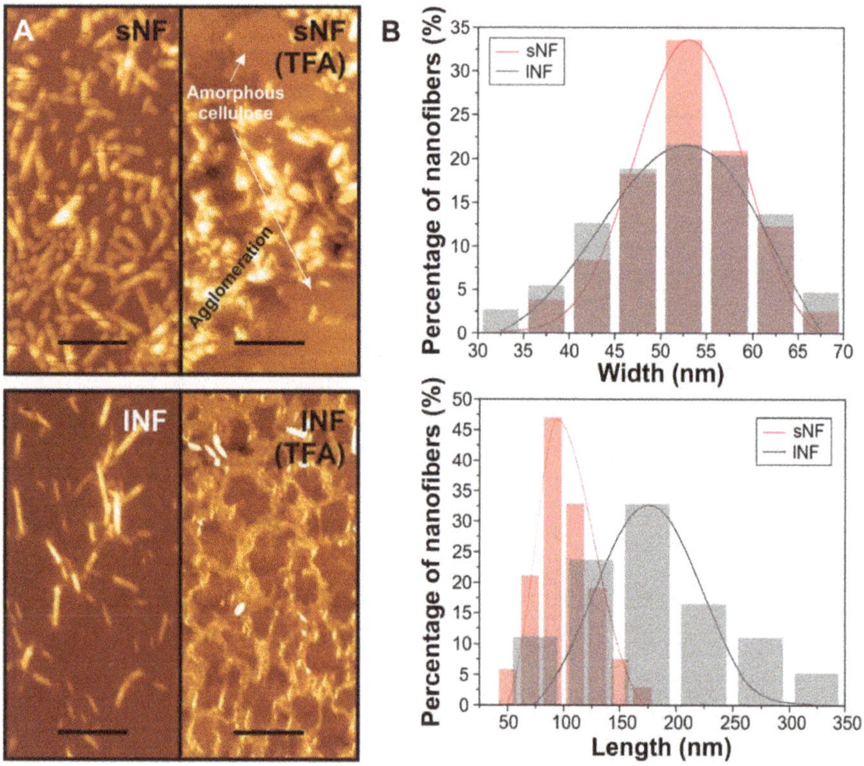

Figure 1. (**A**) Atomic force microscopy (AFM) topographies of short nanofibers (sNF) and long nanofibers (lNF) before and after the solvent treatment. The amorphous domains and agglomerations in the topography of the treated lNF are indicated. Scale bar = 400 nm. (**B**) Histograms showing the width (**top**) and length (**bottom**) distributions of the cellulose nanofibers: sNF (red), lNF (black).

The crystallinity of the cellulose nanofibers was evaluated by XRD (Figure 2A). The pattern of the pristine nanofibers was typical of cellulose I structure [58]. Main peaks were assigned to the following crystalline planes: $(1\bar{1}0)$ at ~15°, (110) at ~17°, (200) at ~23°, and (400) at ~35°, while a minor amorphous contribution was observed at ~21° [59]. After the solvent treatment, crystalline peaks were partially masked by the amorphous one. In fact, the CrI decreased from ~58 and ~45% for pristine sNF and lNF, respectively, to ~13 and ~24% for sNF and lNF after the TFA/TFAA treatment. Moreover, the crystallite size of cellulose was reduced from ~4.0 and ~4.3 nm for pristine sNF and lNF, respectively, to ~2.9 and ~3.1 nm for sNF and lNF after the TFA/TFAA treatment. Hence, the mixture of TFA and TFAA can partially dissolve the cellulose nanofibers, decreasing the crystallinity and the crystallite size of cellulose and originating amorphous cellulose, as observed in AFM images.

The effect of the solvent treatment in the thermal properties of the nanofibers was analyzed by TGA (Figure 2B,C). Pristine sNF and lNF showed a similar behavior with a single weight loss of ~56% at ~275 °C. On the other hand, after the solution in TFA/TFAA, both types of nanocelluloses showed two thermal events: a weight loss of ~30% at ~250 °C and another of ~17% at ~275 °C. The thermal degradation at a lower temperature can be related to the partial hydrolysis of amorphous and lower molecular weight cellulose domains that appear after the solvent treatment [60,61], while the second one can be ascribed to the part of the nanocelluloses unaffected by the acid and the anhydride.

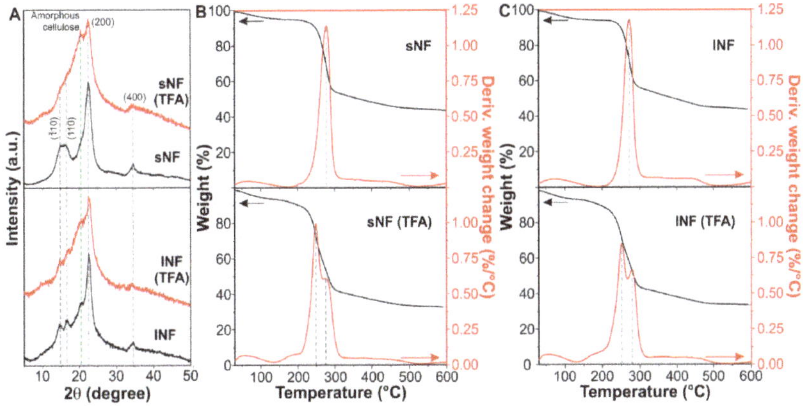

Figure 2. (**A**) X-ray diffraction (XRD) patterns of sNF and lNF before and after the trifluoroacetic acid/trifluoroacetic acid anhydride (TFA/TFAA) treatment. Main assignments are included. (**B**,**C**) Thermogravimetric analysis (TGA) curves and their corresponding derivatives of sNF and lNF, respectively, before and after the TFA/TFAA treatment.

3.2. *Optical and Morphological Characterization of the Nanocomposites*

Transparency is an important feature of food packaging materials since it allows the consumers a visual and direct inspection of the food and it is usually characterized by UV–Vis spectroscopy [10,62]. Figure 3A shows the transparency (i.e., the normalized transmittance calculated from these spectra as the ratio of the corresponding transmittance at 600 nm and the film thickness) for all the samples as a function of the nanocellulose content. In general, the transparency values were higher than 80%, which is considered as the lower limit for good transparency [54]. As observed, there was a relationship between the normalized transmittance and the nanocellulose content independent of the type of cellulose nanofiber used. Values ranged from ~91% for cellulose to ~84 and ~83% for sNF30 and lNF30, respectively. Most likely, this decrease can be related to a higher light scattering induced by the cellulose nanoparticles. To corroborate this, the distribution of the nanocellulose fillers in the cellulose matrix was characterized by HR-SEM (Figure 3B). The cross-sections of lNF30 and sNF30 are shown in Figure 3B. While cellulose displayed a smooth, homogeneous topography (inset Figure 1A), lNF30 and sNF30 exhibited rougher cross-sections with motifs of few tens of nanometers that can be attributed to folded or aggregated nanocelluloses.

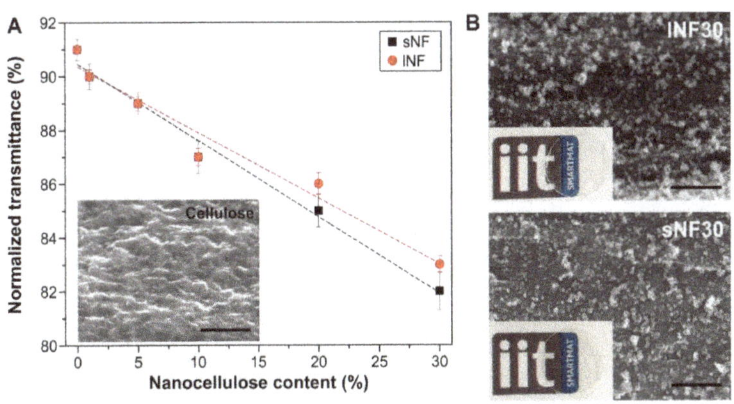

Figure 3. (**A**) Normalized transmittance as a function of nanocellulose content. Inset: HR-SEM cross-section image of a cellulose sample. (**B**) HR-SEM cross-section images of lNF30 and sNF30 samples. Scale bar: 500 nm.

3.3. Mechanical Characterization of the Nanocomposites

Stress–strain curves of *all*-cellulose nanocomposites are shown in Figure 4A,B. In general, the curves were typical of rigid materials with high stresses at the break and low values of elongation at the break. A strong reinforcement effect due to the addition of nanocelluloses was clearly observed. The shape of the curves depended on the amount of nanocellulose and was unrelated to the type of nanofiller used. Figure 4C shows Young's modulus values of the *all*-cellulose films. Initially, Young's modulus increment followed a linear trend from cellulose (~1750 MPa) to a 10 wt.% nanocellulose concentration (~4783 MPa for sNF10 and ~2510 MPa for lNF10) but decreased progressively from that content with either cellulose nanofibers. The composites produced with sNF nanofibers were much stiffer than the longer ones. This is counterintuitive as longer particles are expected to better transfer load from the matrix and to form a more interconnected network. The lower rigidity can be attributed to a higher tendency to aggregation or to lower initial modulus of the longer nanofibers compared to the shorter ones. Both aspects were evaluated by modeling the composite modulus of the lNFs materials through the classic Mallick's model for laminae with randomly dispersed fibers [63]:

$$E_c = \left[\frac{3}{8}\frac{1+2(l/d)\eta_L V_f}{1-\eta_L V_f} + \frac{5}{8}\frac{1+2\eta_T V_f}{1-\eta_T V_f}\right] E_m \tag{1a}$$

$$\eta_L = \frac{E_f/E_m - 1}{E_f/E_m + 2(l/d)} \tag{1b}$$

$$\eta_T = \frac{E_f/E_m - 1}{E_f/E_m + 2} \tag{1c}$$

where E_m, E_f, and E_c are the moduli of the matrix, the filler, and the composite, respectively, V_f is the fibers' relative volume concentration, and l and d are the length and diameter of the fillers. Thus, the composite modulus depends on the filler modulus and on the l/d ratio. We assume here: (i) the geometry of both cellulose nanofibers is the one calculated by AFM and (ii) at low nanofiber concentration, the dispersion is homogeneous. From these assumptions, the value of nanofibers modulus as the only variable in Equation (1a) can be calculated by fitting the first four points measured (cellulose and the nanocomposites containing 1%, 5%, and 9% nanocellulose concentration). For lNF nanocomposites, all points yielded the same values of modulus $E_f \approx 80$ GPa, which is in agreement with reports of bacterial cellulose and indicates that the assumptions above are reliable [64]. From there, the differences between the model and experimental values, which is seen for higher concentrations, can be explained by fiber agglomerations. It should be pointed out that the modulus reduction, which is not fitted with Equation (1a), even for $l/d = 1$ (spherical-like agglomerates), suggests that for such loading, the homogeneous matrix/filler structure was not maintained and the non-continuous fibers could not bear load properly. Therefore, the model applied here can be considered valid only for low nanocellulose concentrations in which the phenomenon of nanofiber aggregation is not predominant. Similar results of the modeling were obtained for the sNF composites, with a slightly higher value of the fitting modulus ($E_f \approx 90$ GPa) and the same discrepancy with experimental data for concentrations above 9%.

Considering other mechanical parameters (Figure 4D–F), yield stress followed a similar trend as Young's modulus in both families of composites. An initial strong increment (from ~20 MPa for cellulose to ~84 MPa for sNF5 and ~73 MPa for lNF10) was followed by a progressive decline. The trend finished at ~56 MPa for both sNF30 and lNF30 nanocomposites, as agglomeration took place. On the other hand, the elongation at the break showed a twofold increment (from ~3.0% for cellulose to ~7.1 and ~8.5% for sNF and lNF films) that was maintained even at high filler concentrations. This was attributed to the bridging effect of fibers that hinder crack propagation with a toughening effect [65]. Direct measurement of the fracture energy confirmed the improvement from ~37 J/cm^3 for cellulose to ~449 and ~510 J/cm^3 for sNF5 and lNF10, respectively, i.e., an increase of ~13 times. These values decreased to ~260 J/cm^3 for the samples with a 30 wt.% of nanocelluloses.

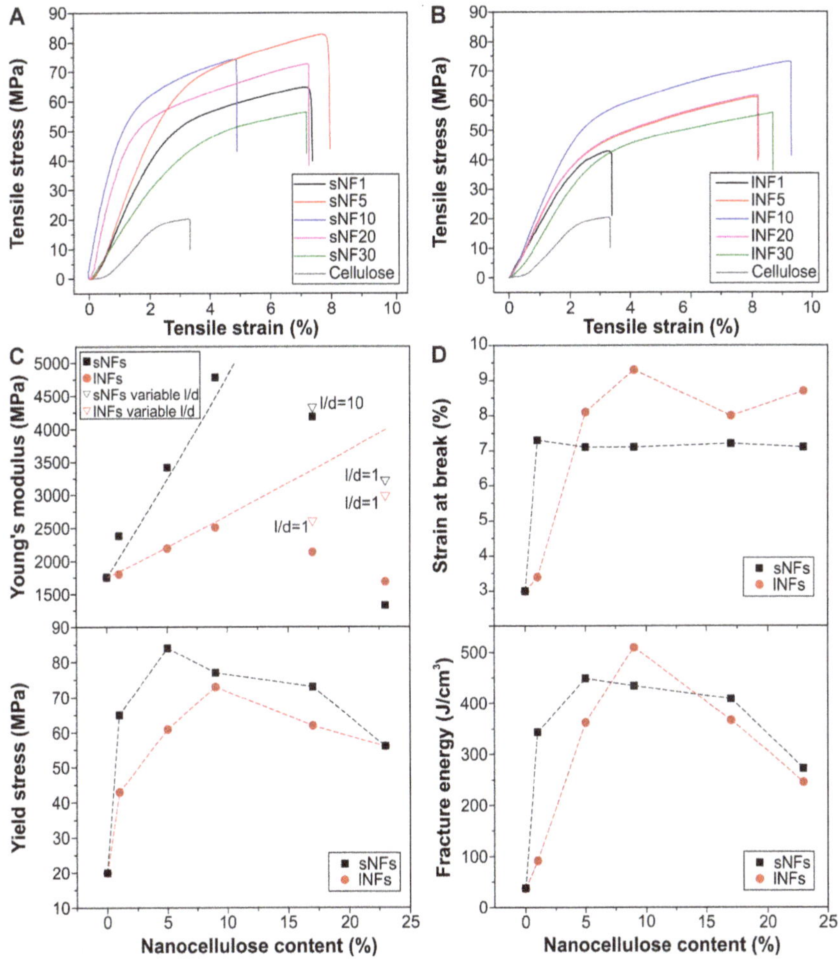

Figure 4. (**A**,**B**) Stress–strain curves of *all*-cellulose nanocomposites prepared with short and long nanofibers, respectively. (**C**) Experimental Young's modulus as a function of nanocellulose content; the dashed line indicates the analytical model for lNF composites. As the nanocellulose content is increased, the model does not fit the experimental data, even if aggregation is accounted for as a variation of the aspect ratio (hollow points). (**D**–**F**) Yield stress, strain at the break, and fracture energy values as a function of nanocellulose content.

3.4. Water Permeability and Uptake of the Nanocomposites

The water permeability was measured for the *all*-cellulose nanocomposites. Figure 5A presents the water permeability values versus the nanocellulose content. Pure cellulose films present a water permeability value of $1.1 \cdot 10^{-3}$ g m^{-1} day^{-1} Pa^{-1} (data not shown). When 1 wt.% cellulose nanofibers were added, the values were ~$2.5 \cdot 10^{-4}$ and $2.9 \cdot 10^{-4}$ g m^{-1} day^{-1} Pa^{-1} for sNF1 and lNF1, respectively. Increasing the nanocellulose content, the values decreased linearly until the final values of $1.5 \cdot 10^{-4}$ and $1.7 \cdot 10^{-4}$ g m^{-1} day^{-1} Pa^{-1} for sNF30 and lNF30, respectively, i.e., a reduction of ~40% for both of them. This decrease can be explained by the increasing tortuosity through the nanocomposite cross-section during water migration. Thus, for the samples containing 1 wt.% nanocellulose, water can easily find a way through the cellulose matrix, which is mainly amorphous [11]. On the other hand, for samples with a 23 wt.% nanocellulose, there are many obstacles—i.e., relatively crystalline, aggregated cellulose nanofibers—that increase the path that water molecules travel to leave the composite. Small differences were found between the two sources of nanocellulose used in this study, being slightly higher values for the films prepared from shorter cellulose nanofibers. This can be explained by a different aggregation and/or folding of these nanocelluloses during the fabrication process, as discussed during the mechanical characterization.

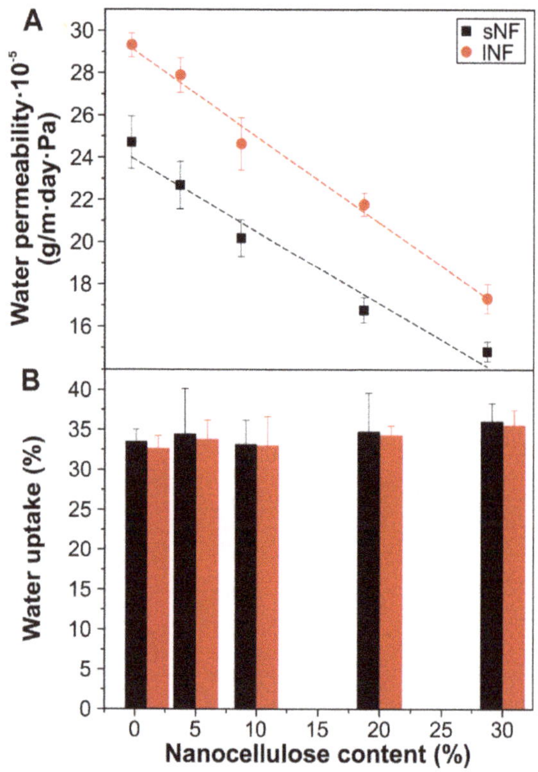

Figure 5. (**A**) Water permeability versus nanocellulose content. (**B**) Water uptake values versus nanocellulose content.

Water uptake was also evaluated for all the samples (Figure 5B). Almost no differences were found with changing the percentage of nanocellulose. The mean water uptake for the nanocomposites was ~34%. This behavior can be explained for the fact that both amorphous cellulose acting as a matrix and nanocelluloses as reinforcements have the typical hydrophilic character of cellulose. Therefore, from a water protection point of view, this material does not provide moisture protection. Nevertheless, further investigations are required to clarify whether water can act as a plasticizer of *all*-cellulose composites in a similar way as described in the literature for other biopolymers [66].

4. Conclusions

In this work, we showed that a mixture of TFA and TFAA can be used as a solvent to produce *all*-cellulose nanocomposites from microcrystalline cellulose and cellulose nanofillers (i.e., short and long nanofibers). Cellulose nanofibers were partially dissolved during the production process, increasing the content of the amorphous phase and reducing the crystallite size of cellulose. This allowed good compatibility with the cellulose matrix. The nanocellulose content affected the final properties of the composites: keeping excellent transparency, improving mechanical properties, and relatively reducing the water permeability. These characteristics can be exploited in their potential application as food packaging films.

Author Contributions: Conceptualization, J.A.H.-G.; Data curation, S.G.-P.; Formal analysis, L.C., S.M., and J.A.H.-G.; Investigation, S.G.-P., G.T., A.S., and J.J.B.; Supervision, A.A. and J.A.H.-G.; Writing–original draft, S.G.-P., L.C., and J.A.H.-G.; Writing–review and editing, S.G.-P., L.C., G.T., S.M., J.J.B., A.A., and J.A.H.-G.

Acknowledgments: The authors would like to thank Lara Marini for her assistance with the thermal measurements.

Conflicts of Interest: The authors declare no conflict of interest.

References

1. Blettler, M.C.M.; Abrial, E.; Khan, F.R.; Sivri, N.; Espinola, L.A. Freshwater plastic pollution: Recognizing research biases and identifying knowledge gaps. *Water Res.* **2018**, *143*, 416–424. [CrossRef] [PubMed]
2. Hamid, F.S.; Bhatti, M.S.; Anuar, N.; Anuar, N.; Mohan, P.; Periathamby, A. Worldwide distribution and abundance of microplastic: How dire is the situation? *Waste Manag. Res.* **2018**, *36*, 873–897. [CrossRef] [PubMed]
3. Law, K.L.; Annual, R. Plastics in the Marine Environment. In *Annual Review of Marine Science*; Annual Reviews; Palo Alto: Santa Clara, CA, USA, 2017; Volume 9, pp. 205–229.
4. Heredia-Guerrero, J.A.; Athanassiou, A. Editorial: Non-polysaccharide Plant Polymeric Materials. *Front. Mater.* **2016**, *3*, 15. [CrossRef]
5. Penca, J. European Plastics Strategy: What promise for global marine litter? *Mar. Pol.* **2018**, *97*, 197–201. [CrossRef]
6. Guillard, V.; Gaucel, S.; Fornaciari, C.; Angellier-Coussy, H.; Buche, P.; Gontard, N. The Next Generation of Sustainable Food Packaging to Preserve Our Environment in a Circular Economy Context. *Front. Nutr.* **2018**, *5*, 13. [CrossRef] [PubMed]
7. Geueke, B.; Groh, K.; Muncke, J. Food packaging in the circular economy: Overview of chemical safety aspects for commonly used materials. *J. Clean Prod.* **2018**, *193*, 491–505. [CrossRef]
8. Thakur, S.; Chaudhary, J.; Sharma, B.; Verma, A.; Tamulevicius, S.; Thakur, V.K. Sustainability of bioplastics: Opportunities and challenges. *Curr. Opin. Green Sustain. Chem.* **2018**, *13*, 68–75. [CrossRef]
9. Tedeschi, G.; Benitez, J.J.; Ceseracciu, L.; Dastmalchi, K.; Itin, B.; Stark, R.E.; Heredia, A.; Athanassiou, A.; Heredia-Guerrero, J.A. Sustainable Fabrication of Plant Cuticle-Like Packaging Films from Tomato Pomace Agro-Waste, Beeswax, and Alginate. *ACS Sustain. Chem. Eng.* **2018**, *6*, 14955–14966. [CrossRef]
10. Heredia-Guerrero, J.A.; Benitez, J.J.; Cataldi, P.; Paul, U.C.; Contardi, M.; Cingolani, R.; Bayer, I.S.; Heredia, A.; Athanassiou, A. All-Natural Sustainable Packaging Materials Inspired by Plant Cuticles. *Adv. Sustain. Syst.* **2017**, *1*, 1600024. [CrossRef]
11. Bayer, I.S.; Guzman-Puyol, S.; Heredia-Guerrero, J.A.; Ceseracciu, L.; Pignatelli, F.; Ruffilli, R.; Cingolani, R.; Athanassiou, A. Direct Transformation of Edible Vegetable Waste into Bioplastics. *Macromolecules* **2014**, *47*, 5135–5143. [CrossRef]
12. Delgado, M.; Felix, M.; Bengoechea, C. Development of bioplastic materials: From rapeseed oil industry by products to added-value biodegradable biocomposite materials. *Ind. Crop. Prod.* **2018**, *125*, 401–407. [CrossRef]
13. Torres-Giner, S.; Hilliou, L.; Melendez-Rodriguez, B.; Figueroa-Lopez, K.J.; Madalena, D.; Cabedo, L.; Covas, J.A.; Vicente, A.A.; Lagaron, J.M. Melt processability, characterization, and antibacterial activity of compression-molded green composite sheets made of poly(3-hydroxybutyrate-co-3-hydroxyvalerate) reinforced with coconut fibers impregnated with oregano essential oil. *Food Packag. Shelf Life* **2018**, *17*, 39–49. [CrossRef]
14. Benítez, J.J.; Castillo, P.M.; del Río, J.C.; León-Camacho, M.; Domínguez, E.; Heredia, A.; Guzmán-Puyol, S.; Athanassiou, A.; Heredia-Guerrero, J.A. Valorization of Tomato Processing by-Products: Fatty Acid Extraction and Production of Bio-Based Materials. *Materials* **2018**, *11*, 2211. [CrossRef] [PubMed]
15. Perotto, G.; Ceseracciu, L.; Simonutti, R.; Paul, U.C.; Guzman-Puyol, S.; Tran, T.-N.; Bayer, I.S.; Athanassiou, A. Bioplastics from vegetable waste via an eco-friendly water-based process. *Green Chem.* **2018**, *20*, 894–902. [CrossRef]
16. Klemm, D.; Heublein, B.; Fink, H.P.; Bohn, A. Cellulose: Fascinating biopolymer and sustainable raw material. *Angew. Chem.-Int. Edit.* **2005**, *44*, 3358–3393. [CrossRef] [PubMed]

17. Siro, I.; Plackett, D. Microfibrillated cellulose and new nanocomposite materials: A review. *Cellulose* **2010**, *17*, 459–494. [CrossRef]
18. Guzman-Puyol, S.; Ceseracciu, L.; Heredia-Guerrero, J.A.; Anyfantis, G.C.; Cingolani, R.; Athanassiou, A.; Bayer, I.S. Effect of trifluoroacetic acid on the properties of polyvinyl alcohol and polyvinyl alcohol-cellulose composites. *Chem. Eng. J.* **2015**, *277*, 242–251. [CrossRef]
19. Nishino, T.; Peijs, T. All-cellulose Composites. In *Handbook of Green Materials, Vol 2: Bionanocomposites: Processing, Characterization and Properties*; Oksman, K., Mathew, A.P., Bismarck, A., Rojas, O., Sain, M., Qvintus, P., Eds.; World Scientific Publ Co Pte Ltd.: Singapore, 2014; Volume 5, pp. 201–216.
20. Huber, T.; Mussig, J.; Curnow, O.; Pang, S.S.; Bickerton, S.; Staiger, M.P. A critical review of all-cellulose composites. *J. Mater. Sci.* **2012**, *47*, 1171–1186. [CrossRef]
21. Wei, X.; Wei, W.; Cui, Y.H.; Lu, T.J.; Jiang, M.; Zhou, Z.W.; Wang, Y. All-cellulose composites with ultra-high mechanical properties prepared through using straw cellulose fiber. *RSC Adv.* **2016**, *6*, 93428–93435. [CrossRef]
22. Li, J.Y.; Nawaz, H.; Wu, J.; Zhang, J.M.; Wan, J.Q.; Mi, Q.Y.; Yu, J.; Zhang, J. All-cellulose composites based on the self-reinforced effect. *Compos. Commun.* **2018**, *9*, 42–53. [CrossRef]
23. Yousefi, H.; Nishino, T.; Faezipour, M.; Ebrahimi, G.; Shakeri, A. Direct fabrication of all-cellulose nanocomposite from cellulose microfibers using ionic liquid-based nanowelding. *Biomacromolecules* **2011**, *12*, 4080–4085. [CrossRef] [PubMed]
24. Tervahartiala, T.; Hildebrandt, N.C.; Piltonen, P.; Schabel, S.; Valkama, J.P. Potential of all-cellulose composites in corrugated board applications: Comparison of chemical pulp raw materials. *Packag. Technol. Sci.* **2018**, *31*, 173–183. [CrossRef]
25. Hu, W.; Chen, G.; Liu, Y.; Liu, Y.Y.; Li, B.; Fang, Z.G. Transparent and Hazy All-Cellulose Composite Films with Superior Mechanical Properties. *ACS Sustain. Chem. Eng.* **2018**, *6*, 6974–6980. [CrossRef]
26. Ghaderi, M.; Mousavi, M.; Yousefi, H.; Labbafi, M. All-cellulose nanocomposite film made from bagasse cellulose nanofibers for food packaging application. *Carbohydr. Polym.* **2014**, *104*, 59–65. [CrossRef] [PubMed]
27. Soykeabkaew, N.; Arimoto, N.; Nishino, T.; Peijs, T. All-cellulose composites by surface selective dissolution of aligned ligno-cellulosic fibres. *Compos. Sci. Technol.* **2008**, *68*, 2201–2207. [CrossRef]
28. Mathew, A.P.; Oksman, K.; Pierron, D.; Harmand, M.F. Fibrous cellulose nanocomposite scaffolds prepared by partial dissolution for potential use as ligament or tendon substitutes. *Carbohydr. Polym.* **2012**, *87*, 2291–2298. [CrossRef]
29. He, X.; Xiao, Q.; Lu, C.H.; Wang, Y.R.; Zhang, X.F.; Zhao, J.Q.; Zhang, W.; Zhang, X.M.; Deng, Y.L. Uniaxially Aligned Electrospun All-Cellulose Nanocomposite Nanofibers Reinforced with Cellulose Nanocrystals: Scaffold for Tissue Engineering. *Biomacromolecules* **2014**, *15*, 618–627. [CrossRef] [PubMed]
30. Kim, J.; Yun, S.; Ounaies, Z. Discovery of cellulose as a smart material. *Macromolecules* **2006**, *39*, 4202–4206. [CrossRef]
31. Nakagaito, A.N.; Nogi, M.; Yano, H. Displays from Transparent Film of Natural Nanofibers. *Mrs Bull.* **2010**, *35*, 214–218. [CrossRef]
32. Khan, A.; Abas, Z.; Kim, H.S.; Kim, J. Recent Progress on Cellulose-Based Electro-Active Paper, Its Hybrid Nanocomposites and Applications. *Sensors* **2016**, *16*, 1172. [CrossRef] [PubMed]
33. Bondeson, D.; Syre, P.; Niska, K.O. All cellulose nanocomposites produced by extrusion. *J. Biobased Mater. Bioenergy* **2007**, *1*, 367–371. [CrossRef]
34. De Azeredo, H.M.C. Nanocomposites for food packaging applications. *Food Res. Int.* **2009**, *42*, 1240–1253. [CrossRef]
35. Hubbe, M.A.; Ferrer, A.; Tyagi, P.; Yin, Y.Y.; Salas, C.; Pal, L.; Rojas, O.J. Nanocellulose in Thin Films, Coatings, and Plies for Packaging Applications: A Review. *Bioresources* **2017**, *12*, 2143–2233. [CrossRef]
36. Li, F.; Mascheroni, E.; Piergiovanni, L. The Potential of NanoCellulose in the Packaging Field: A Review. *Packag. Technol. Sci.* **2015**, *28*, 475–508. [CrossRef]
37. Kumar, T.S.M.; Rajini, N.; Reddy, K.O.; Rajulu, A.V.; Siengchin, S.; Ayrilmis, N. All-cellulose composite films with cellulose matrix and Napier grass cellulose fibril fillers. *Int. J. Biol. Macromol.* **2018**, *112*, 1310–1315. [CrossRef] [PubMed]
38. Geddes, A.L. Interaction of trifluoroacetic acid with cellulose and related compounds. *J. Polym. Sci.* **1956**, *22*, 31–39. [CrossRef]
39. Hasegawa, M.; Isogai, A.; Onabe, F.; Usuda, M. Dissolving states of cellulose and chitosan in trifluoroacetic-acid. *J. Appl. Polym. Sci.* **1992**, *45*, 1857–1863. [CrossRef]

40. Liebert, T.; Schnabelrauch, M.; Klemm, D.; Erler, U. Readily hydrolyzable cellulose esters as intermediates for the regioselective derivatization of cellulose. 2. soluble, highly substituted cellulose trifluoroacetates. *Cellulose* **1994**, *1*, 249–258. [CrossRef]
41. Zhao, H.B.; Holladay, J.E.; Kwak, J.H.; Zhang, Z.C. Inverse temperature-dependent pathway of cellulose decrystallization in trifluoroacetic acid. *J. Phys. Chem. B* **2007**, *111*, 5295–5300. [CrossRef] [PubMed]
42. Zhao, H.B.; Holladay, J.E.; Kwak, J.H.; Zhang, Z.C. A new route to improved glucose yields in cellulose hydrolysis. *J. Biobased Mater. Bioenergy* **2007**, *1*, 210–214. [CrossRef]
43. Heinze, T.; Liebert, T. Unconventional methods in cellulose functionalization. *Prog. Polym. Sci.* **2001**, *26*, 1689–1762. [CrossRef]
44. Liebert, T. Cellulose Solvents—Remarkable History, Bright Future. In *Cellulose Solvents: For Analysis, Shaping and Chemical Modification*; Liebert, T.F., Heinze, T.J., Edgar, K.J., Eds.; American Chemical Society: Washington, DC, USA, 2010; pp. 3–54.
45. Scott, B.F.; Macdonald, R.W.; Kannan, K.; Fisk, A.; Witter, A.; Yamashita, N.; Durham, L.; Spencer, C.; Muir, D.C.G. Trifluoroacetate profiles in the Arctic, Atlantic, and Pacific Oceans. *Environ. Sci. Technol.* **2005**, *39*, 6555–6560. [CrossRef] [PubMed]
46. Kim, B.R.; Suidan, M.T.; Wallington, T.J.; Du, X. Biodegradability of trifluoroacetic acid. *Environ. Eng. Sci.* **2000**, *17*, 337–342. [CrossRef]
47. Guzman-Puyol, S.; Russo, D.; Penna, I.; Ceseracciu, L.; Palazon, F.; Scarpellini, A.; Cingolani, R.; Bertorelli, R.; Bayer, I.S.; Heredia-Guerrero, J.A.; et al. Facile production of seaweed-based biomaterials with antioxidant and anti-inflammatory activities. *Algal Res.* **2017**, *27*, 1–11. [CrossRef]
48. Guzman-Puyol, S.; Heredia-Guerrero, J.A.; Ceseracciu, L.; Hajiali, H.; Canale, C.; Scarpellini, A.; Cingolani, R.; Bayer, I.S.; Athanassiou, A.; Mele, E. Low-Cost and Effective Fabrication of Biocompatible Nanofibers from Silk and Cellulose-Rich Materials. *ACS Biomater. Sci. Eng.* **2016**, *2*, 526–534. [CrossRef]
49. Papadopoulou, E.L.; Heredia-Guerrero, J.A.; Vázquez, M.I.; Benavente, J.; Athanassiou, A.; Bayer, I.S. Self-organized microporous cellulose-nylon membranes. *Polymer* **2017**, *120*, 255–263. [CrossRef]
50. Tran, T.N.; Paul, U.; Heredia-Guerrero, J.A.; Liakos, I.; Marras, S.; Scarpellini, A.; Ayadi, F.; Athanassiou, A.; Bayer, I.S. Transparent and flexible amorphous cellulose-acrylic hybrids. *Chem. Eng. J.* **2016**, *287*, 196–204. [CrossRef]
51. Tedeschi, G.; Guzman-Puyol, S.; Paul, U.C.; Barthel, M.J.; Goldoni, L.; Caputo, G.; Ceseracciu, L.; Athanassiou, A.; Heredia-Guerrero, J.A. Thermoplastic cellulose acetate oleate films with high barrier properties and ductile behaviour. *Chem. Eng. J.* **2018**, *348*, 840–849. [CrossRef]
52. Heredia-Guerrero, J.A.; Goldoni, L.; Benitez, J.J.; Davis, A.; Ceseracciu, L.; Cingolani, R.; Bayer, I.S.; Heinze, T.; Koschella, A.; Heredia, A.; et al. Cellulose-polyhydroxylated fatty acid ester-based bioplastics with tuning properties: Acylation via a mixed anhydride system. *Carbohydr. Polym.* **2017**, *173*, 312–320. [CrossRef] [PubMed]
53. Horcas, I.; Fernández, R.; Gómez-Rodríguez, J.M.; Colchero, J.; Gómez-Herrero, J.; Baro, A.M. WSXM: A software for scanning probe microscopy and a tool for nanotechnology. *Rev. Sci. Instrum.* **2007**, *78*, 013705. [CrossRef] [PubMed]
54. ASTM. *Standard Test Method for Transparency of Plastic Sheeting*; ASTM: West Conshohocken, PA, USA, 2015.
55. Segal, L.; Creely, J.J.; Martin, A.E.; Conrad, C.M. An Empirical Method for Estimating the Degree of Crystallinity of Native Cellulose Using the X-Ray Diffractometer. *Text. Res. J.* **1959**, *29*, 786–794. [CrossRef]
56. Savadekar, N.R.; Mhaske, S.T. Synthesis of nano cellulose fibers and effect on thermoplastics starch based films. *Carbohydr. Polym.* **2012**, *89*, 146–151. [CrossRef] [PubMed]
57. Bedane, A.H.; Eic, M.; Farmahini-Farahani, M.; Xiao, H.N. Water vapor transport properties of regenerated cellulose and nanofibrillated cellulose films. *J. Membr. Sci.* **2015**, *493*, 46–57. [CrossRef]
58. Nishino, T.; Matsuda, I.; Hirao, K. All-Cellulose Composite. *Macromolecules* **2004**, *37*, 7683–7687. [CrossRef]
59. Duchemin, B.J.C.; Newman, R.H.; Staiger, M.P. Structure–property relationship of all-cellulose composites. *Compos. Sci. Technol.* **2009**, *69*, 1225–1230. [CrossRef]
60. Hideno, A. Comparison of the Thermal Degradation Properties of Crystalline and Amorphous Cellulose, as well as Treated Lignocellulosic Biomass. *Bioresources* **2016**, *11*, 6309–6319. [CrossRef]
61. Wang, Z.H.; McDonald, A.G.; Westerhof, R.J.M.; Kersten, S.R.A.; Cuba-Torres, C.M.; Ha, S.; Pecha, B.; Garcia-Perez, M. Effect of cellulose crystallinity on the formation of a liquid intermediate and on product distribution during pyrolysis. *J. Anal. Appl. Pyrolysis* **2013**, *100*, 56–66. [CrossRef]

62. Cazon, P.; Vazquez, M.; Velazquez, G. Cellulose-glycerol-polyvinyl alcohol composite films for food packaging: Evaluation of water adsorption, mechanical properties, light-barrier properties and transparency. *Carbohydr. Polym.* **2018**, *195*, 432–443. [CrossRef] [PubMed]
63. Mallick, P.K. *Fiber-Reinforced Composites: Materials, Manufacturing, and Design*, 2nd ed.; Taylor & Francis: Didcot, UK, 1993.
64. Guhados, G.; Wan, W.K.; Hutter, J.L. Measurement of the elastic modulus of single bacterial cellulose fibers using atomic force microscopy. *Langmuir* **2005**, *21*, 6642–6646. [CrossRef] [PubMed]
65. Norman, D.A.; Robertson, R.E. The effect of fiber orientation on the toughening of short fiber-reinforced polymers. *J. Appl. Polym. Sci.* **2003**, *90*, 2740–2751. [CrossRef]
66. Vieira, M.G.A.; da Silva, M.A.; dos Santos, L.O.; Beppu, M.M. Natural-based plasticizers and biopolymer films: A review. *Eur. Polym. J.* **2011**, *47*, 254–263. [CrossRef]

nanomaterials

MDPI

Review

Risk Analysis of Cellulose Nanomaterials by Inhalation: Current State of Science

James D. Ede [1], Kimberly J. Ong [1], Michael Goergen [2], Alan Rudie [3], Cassidy A. Pomeroy-Carter [1] and Jo Anne Shatkin [1,*]

1 Vireo Advisors, LLC, Boston, MA 02130-4323, USA; jede@vireoadvisors.com (J.D.E.); kong@vireoadvisors.com (K.J.O.); cpomeroy@vireoadvisors.com (C.A.P.-C.)

2 P3Nano, U.S. Endowment for Forestry and Communities, Greenville, SC 29601, USA; michael@usendowment.org

3 Forest Products Laboratory, USDA Forest Service, Madison, WI 53726-2398, USA; arudie@fs.fed.us

* Correspondence: jashatkin@vireoadvisors.com; Tel.: 1-617-612-5906

Received: 27 December 2018; Accepted: 23 February 2019; Published: 2 March 2019

Abstract: Cellulose nanomaterials (CNs) are emerging advanced materials with many unique properties and growing commercial significance. A life-cycle risk assessment and environmental health and safety roadmap identified potential risks from inhalation of powdered CNs in the workplace as a key gap in our understanding of safety and recommended addressing this data gap to advance the safe and successful commercialization of these materials. Here, we (i) summarize the currently available published literature for its contribution to our current understanding of CN inhalation hazard and (ii) evaluate the quality of the studies for risk assessment purposes using published study evaluation tools for nanomaterials to assess the weight of evidence provided. Our analysis found that the quality of the available studies is generally inadequate for risk assessment purposes but is improving over time. There have been some advances in knowledge about the effects of short-term inhalation exposures of CN. The most recent in vivo studies suggest that short-term exposure to CNs results in transient inflammation, similarly to other poorly soluble, low toxicity dusts such as conventional cellulose, but is markedly different from fibers with known toxicity such as certain types of multiwalled carbon nanotubes or asbestos. However, several data gaps remain, and there is still a lack of understanding of the effects from long-term, low-dose exposures that represent realistic workplace conditions, essential for a quantitative assessment of potential health risk. Therefore, taking precautions when handling dry forms of CNs to avoid dust inhalation exposure is warranted.

Keywords: cellulose; nanomaterial; inhalation; risk assessment; safety; nanocellulose; review

1. Introduction

Cellulose nanomaterials (CNs) are emerging materials with numerous applications. They have the potential to be used in high volumes for cement, automotive composites, food packaging, paper and coatings, consumer product packaging, hygiene and absorbent products, and as textiles for clothing, among many other applications [1]. To promote the safe commercialization of these materials, a life-cycle risk assessment and environmental health and safety roadmap identified key knowledge gaps in our understanding of CN safety and prioritized them for development [2]. The assessment found that improving understanding of the risks of inhaling dry CN powders in the workplace is a high priority.

As the commercialization of CNs continues to grow, inhalation of particles into the lung is one of the main routes of exposure, especially, in occupational settings where workers may be exposed to concentrated doses of airborne, dry particulate materials. It is well-understood that inhalation of poorly soluble, low toxicity (PSLT) dusts, such as silica, titanium dioxide, and coal mine dusts, has the potential to irritate the lungs [3] and trigger the immune system; effects may occur when individuals are exposed to either short-term to high levels of PSLT dust, or long-term at low levels that exceed the lowest adverse effect threshold. CNs are bio-based, inert materials that may be similar in risk *potential* to PSLT dusts. However, due to their smaller size and fibrillar form, CNs should be assessed for their potential to be respirable and for their inflammatory effects that may lead to serious health outcomes, such as carcinogenicity. The main forms of wood-based CNs are cellulose nanocrystals (CNCs) and cellulose nanofibrils (CNFs). Both forms are extracted from plants via a purification and homogenization pre-treatment step, followed either by an acid hydrolysis refinement to produce CNCs or mechanical shear to produce CNF. CNCs are smaller and tend to be stiff, with lengths between 50–350 nm and widths of 5–20 nm, while CNFs are flexible, with lengths typically >1 µm and widths of 20–100 nm [4]. In this study, we evaluated the available literature to assess whether, due to their properties, CNs behave differently to conventional cellulose dust, a known respiratory irritant, and, therefore, require different occupational handling approaches.

Performing well-designed inhalation studies for risk assessment purposes is difficult, and testing nanomaterials comes with additional challenges. Considerations for delivery method, dose selection, control groups, and study duration are crucial to identifying outcomes relevant for risk assessment. Repeat, low-level exposures may not be well represented by short-term high-dose studies, but long-term studies require time and extensive resources. A number of groups have highlighted the need for quality assessment of nanotoxicology studies [5–7]. Toxicity studies of nanomaterials require material characterization and special experimental design considerations for the results to meaningfully contribute to an understanding of safety. Recent reviews highlight that a significant number of studies being published on nanomaterials do not meet these criteria, limiting their usefulness for risk assessment [8,9]. This review evaluates the current state of knowledge in relation to the quality of CN inhalation studies for risk assessment purposes using two published approaches: Krug and Wick (2011) [10] and Card and Magnuson (2010) [11]. The criteria do not indicate whether a study is "good" or "bad" but rather if the study was designed and conducted to allow findings to be used for risk assessment purposes to help predict negative biological outcomes as a result of CN inhalation exposure. There is a bias in published toxicity studies, which tend to report on short-term exposures at concentrations that result in negative biological effects and focus on the mechanistic aspects of toxicology, such as determining the mode and mechanism of action. A key challenge for risk assessment is extrapolating the information to assess how these same materials may behave under realistic scenarios, where concentrations are often significantly lower and exposures can be repetitive or prolonged. This analysis focuses on: (1) Reviewing the quality of the studies for risk assessment purposes using published study evaluation tools for nanomaterials to assess their impact on the weight of evidence, and (2) reviewing the results of these papers for their contribution to the current understanding of occupational inhalation hazards of dried CN in the workplace.

2. Literature Review

To date, twelve studies from 2011–2018 have been published on the short-term inhalation toxicity of CNs [12–23]. These studies use both cellular (in vitro) [16–21] and animal (in vivo) [12–16,22,23] models to investigate the effect of CNC and CNF exposure on the lung. The main findings from these studies are summarized here before analyzing the studies using two published approaches for their usefulness for risk assessment purposes.

2.1. Cellular (in vitro) Studies

Acute inhalation toxicity of CN has been investigated in vitro using both 3D triple cell co-culture cell models of human airways and simpler monocultures. Initial investigations using the 3D barrier model examined incubation with aqueous suspensions of CNC (5, 15 and 30 mg/L for 24 h) and found apical cytotoxicity at concentrations of 15 and 30 mg/L, but no basolateral cytotoxicity at any dose examined, and only a small elevation of pro-inflammatory chemokine at the highest dose examined [17]. A more recent study by Endes et al. (2014) exposed the 3D human airway barrier model to aerosolized CNC suspensions [18]. The authors tested nebulized concentrations of CNC from 0.14 to 1.57 $\mu g/cm^2$ and found no significant cytotoxicity, no induction of oxidative stress, and no pro-inflammatory response at any of the concentrations examined, 24 h post-exposure.

In a study using monocultures, Yanamala et al. (2016) exposed aqueous suspensions of different forms of CNC and CNF (5 μg/mL–300 μg/mL) to a human lung epithelial cell line (A549) or a human monocytic (immune) cell line (THP-1) for 24 or 72 h [19]. Bulk microcrystalline cellulose (MCC) was used as a reference material. The authors found responses were cell-type- and material-specific and concluded there was no correlation between cytotoxicity and surface properties. The authors also found elevated pro-inflammatory responses following CN exposure (50 μg/mL; 24 and 72 h) in THP-1 cells, with different responses observed between materials. The authors concluded that at the doses and time points examined, all forms of CNs were nontoxic or less toxic compared to the two positive controls used in the study: asbestos or lipopolysaccharide.

A similar study published by Menas et al. (2017) examined cytotoxicity, oxidative stress, and cytokine secretion in A549 cells following exposure (1.5, 15 or 45 $\mu g/cm^2$) to various forms of CNC or CNF for 24 or 72 h [20]. Chitin and carbon nanofibers were used as reference materials. Generally, cell viability was significantly decreased at all doses examined following CNF, but not CNC, exposure for 72 h. CNC and CNF exposure generally resulted in significant oxidative stress responses at both time points; although, some differences were noted between materials. The authors also found that exposure to 45 $\mu g/cm^2$ CNC for 72 h significantly increased the secretion of several pro-inflammatory cytokines in A549 cells.

Two more recent studies have further investigated cell responses to CNF exposure. Lopes et al. (2017) examined cytotoxicity, oxidative stress, and cytokine secretion following exposure to different surface-functionalized CNFs in human dermal, lung, and immune cells (human dermal fibroblasts, lung MRC-5 fibroblast and THP-1 monocytes), using MCC as a reference material [21]. The authors found that CNF exposure did not induce cytotoxicity in any of the cell lines examined up to 500 μg/mL for 24 h of exposure. The authors also found CNF exposure up to 500 μg/mL: (1) did not induce oxidative stress; and (2) did not result in internalization or morphological changes in THP-1 monocytes. Two pro-inflammatory cytokines were elevated in THP-1 cells exposed to unmodified CNF for 24 h starting at concentrations of 250 μg/mL; these effects were not observed for modified-CNF or the MCC reference material.

A similarly designed study by Ilves et al. (2018) examined the cytotoxicity and pro-inflammatory cytokine production in THP-1 cells following exposure to four surface-modified CNFs and compared these responses to conventional cellulose [16]. The authors found that one of the unmodified CNFs reduced cell viability and triggered pro-inflammatory cytokine production; the remaining three CNF materials had no significant changes and were similar to the response observed for bulk-sized cellulose. Cytotoxic effects of the unmodified CNF were observed after 3, 6, and 24 h of exposure, starting at concentrations of 10 μg/mL. Similarly, increased expression and protein production of pro-inflammatory cytokines were observed after 3, 6, and 24 h of exposure, starting at concentrations of 10 μg/mL.

2.2. Animal (In Vivo) Studies

Acute inhalation studies of CN have been conducted with both mice and rats. O'Connor et al. (2014) exposed rats to aerosolized CNC for four hours and monitored the animals for 14 days [22]. Researchers were only able to achieve a maximum test concentration of 0.26 mg/L and found no mortality, gross toxicity, adverse effects, or behavioral changes at the highest concentration tested. Yanamala et al. (2014) examined the effect of pharyngeal aspiration of two forms of CNC in C57BL/6 mice [12]. Mice were exposed to 100 or 200 μg CNC for four hours and monitored for 24 h. Analysis of bronchial alveolar lavage (BAL) fluid following CNC exposure found pulmonary damage, elevated cytokine and chemokine levels, and recruitment of inflammatory cells.

Two additional studies examined potential sex differences and reproductive effects in mice following pharyngeal aspiration to suspensions of CNC (40 µg/mouse/day; two times per week, for three weeks; cumulative dose of 240 µg/mouse) and observed responses three months post-exposure. Shvedova et al. (2016) reported pulmonary damage and an elevated oxidative stress response in BAL from exposed mice [14]. The authors also reported impaired pulmonary function and global changes in gene expression following CNC exposure. For many of these endpoints, the authors conclude that effects were markedly more pronounced in female compared to male mice, suggesting sex differences in response to CNC exposure, though these results may also be due to weight differences between sexes [24]. In a second study, Farcas et al., (2016) used the same mice to examine potential male reproductive effects following CNC exposure [13]. The authors found significant changes to sperm three months post-exposure. Evaluation of the testes found elevated oxidative stress, inflammatory cytokines, and myeloperoxidase (MPO) activity, and histopathological analysis found damage to testicular structure. Significant changes in hormone levels were also reported.

Two recent studies examined the acute inhalation toxicity of CNF in mice. Catalán et al. (2017) exposed mice to a one-time exposure of CNF via pharyngeal aspiration (10, 40, 80, or 200 µg/mouse) and found an acute inflammatory response and DNA damage 24 h post-exposure [15]. Histological analysis of lung tissue confirmed deposition and accumulation of CNF in the bronchi and alveoli, as well as internalization in macrophages. The authors also found DNA damage in isolated lung cells, though no dose–response relationship was observed. No DNA damage was found in cells isolated from BAL, and no chromosome damage was found in bone marrow erythrocytes.

Park et al. (2018) compared the pulmonary effects of four materials—CNF, CNC, single-walled carbon nanotubes (SWCNTs), and crocidolite asbestos. BALB/c mice were exposed to a single dose of 40 µg/mouse of SWCNT or asbestos, and 40 µg or 80 µg of CNF or CNC by pharyngeal aspiration and evaluated for pulmonary inflammation and immune response 14 days post-exposure. By day 14, all mice showed some cellular alterations indicative of an inflammatory response, such as increased total cell count, mononuclear phagocytes, and polymorphonuclear leukocytes, and lymphocytes in the BAL, increased expression of cytokines and chemokines, and increased lactic acid dehydrogenase (LDH) activity. CNF and CNC responses were generally dose-dependent, with higher doses activating a greater response. However, the immune response induced by asbestos was indicative of chronic inflammation, whereas the SWCNT, CNF, and CNC induced much weaker responses, dissimilar to asbestos, the positive control.

Ilves et al. (2018) recently published a study of one-time exposures using pharyngeal aspiration (10 or 40 µg/mouse) to one of four surface-modified CNFs in mice [16]. The authors included two observation timepoints (24 h and 28 days post-exposure), used positive and negative controls, and compared results to relevant reference materials—multiwalled carbon nanotubes (MWCNTs) and bulk-size cellulose. Twenty-four hours post-CNF exposure, the authors reported recruitment of inflammatory cells in BAL, with similar responses observed for both MWCNT and bulk cellulose exposures; however, differences between CNF materials were noted. CNF exposure enhanced mRNA expression of several pro-inflammatory cytokines 24 h post-exposure, though differences between CNF materials were observed; generally, similar cytokine responses were observed for both bulk-cellulose and MWCNT reference materials. Importantly, only modest immune reactions were observed 28 days post-exposure, with effects reduced compared to 24 h post-exposure for CNF and similar to those triggered by bulk cellulose. In comparison to MWCNTs, the effects of CNF 28 days post-exposure were more minor. The authors also reported that the CNs persisted in the lung 28 days post-exposure.

3. Study Evaluations—Krug and Wick and Card and Magnuson

Two sets of criteria were used to evaluate the studies: One set developed by authors Krug and Wick (2011) [10] and another by Card and Magnuson (2010) [11]. Both sets of criteria were originally developed to assess the quality of nanotoxicity studies, recognizing that various studies have different objectives. Researchers may aim to determine the mechanisms of toxicity (i.e., how are these materials causing their effects at high doses), find the lowest observed adverse effect level (i.e., at what concentration do effects start to occur) or conduct a risk assessment, which considers whether a hazard might cause harm to exposed persons under realistic exposure scenarios. Here, we have adapted the Krug and Wick and Card and Magnuson criteria sets to specifically assess the quality of studies examining short-term CN inhalation for risk assessment purposes and used them to evaluate seven animal (in vivo) studies [12–16,22,23] and six cellular (in vitro) studies [16–21] (Table 1). Studies were evaluated that specifically examined exposure to CN using in vivo or in vitro models of inhalation; studies examining inhalation exposure to bulk cellulose were not included, except when included in the study design as reference materials.

Table 1. Published studies examining effects of cellulose nanomaterial (CN) inhalation.

In Vivo Studies			In Vitro Studies		
First Author	**Year**	**Material**	**First Author**	**Year**	**Material**
Ilves	2018 [16]	cellulose nanofibrils (CNF)	Ilves	2018 [16]	CNF
Park	2018 [23]	CNF, cellulose nanocrystals (CNC)	Menas	2017 [20]	CNC, CNF
Catalan	2017 [15]	CNF	Lopes	2017 [21]	CNF
Shvedova	2016 [14]	CNC	Yanamala	2016 [19]	CNC, CNF
Farcas	2016 [13]	CNC	Endes	2014 [18]	CNC
Yanamala	2014 [12]	CNC	Clift	2011 [17]	CNC
O'Connor	2014 [22]	CNC			

3.1. *The Krug and Wick Approach*

For the past decade, Krug and Wick (2011) [10] have been working toward enhancing the quality and reliability of nanotoxicity studies. They stipulate that sufficient characterization and relevant information on the validity and suitability of the selected test methods should be required for nanotoxicology publications to ensure comparable studies, leading to reliable discussion and the ability to make a conclusive evaluation of the risks associated with exposure to certain nanomaterials.

Krug and Wick developed a set of criteria (Table 2) to evaluate nanomaterial toxicity studies based on (i) the extent of physical and chemical characterization and (ii) the overall study design (e.g., doses administered, exposure route) [10]. For our analysis, we quantified these criteria by assigning a category of 2 if fully met, 1 if partially met, and 0 if not met. Scores were then calculated for a final 'nanomaterial characterization score', out of 14, and a 'study design score', out of 20 (or 18, as one criterion only applies to in vivo studies). The criteria were adapted for evaluating inhalation studies with CN; therefore, the criteria for the octanol-water partition coefficient, solubility, and criteria for ecotoxicological studies were deemed not applicable and excluded.

Table 2. Summary of Krug and Wick study evaluation.

	Criteria	In Vivo Studies									In Vitro Studies							
		O'Connor 2014	**Yanamala et al. 2014**	**Farcas et al. 2016**	**Shvedova et al. 2016**	**Catalan et al. 2017**	**Park et al. 2018**	**Ilves et al. 2018**	**TOTAL**	**%**	**Clift et al. 2011**	**Endes et al. 2014**	**Yanamala et al. 2016**	**Menas et al. 2017**	**Lopes et al. 2017**	**Ilves et al. 2018**	**TOTAL**	**%**
Nanomaterial Characterization Score	Chemical composition, purity, impurities	0	1	1	2	0	1	2	7	50%	0	1	1	1	1	2	6	50%
	Particle size and size distribution	2	2	2	2	2	2	2	14	100%	2	2	2	2	2	2	12	100%
	Specific surface	2	0	0	0	0	0	0	2	14%	0	0	0	0	0	0	0	0%
	Morphology (crystalline/amorphous, shape)	2	2	0	2	2	2	2	12	86%	2	2	0	2	2	2	10	83%
	Surface chemistry, coating, functionalization	1	0	0	1	2	0	1	5	36%	0	1	0	0	2	1	4	33%
	Degree of agglomeration/aggregation and particle size distribution under experimental conditions (for example, media with/without proteins)	0	0	0	0	0	0	0	0	0%	0	1	0	0	2	0	3	25%
	Surface reactivity and/or surface load (zeta potential)	0	0	0	2	2	0	2	6	43%	0	0	0	0	2	2	4	33%
	Characterization Score (out of 14)	7	5	3	9	8	5	9		47%	4	7	3	5	11	9		46%
	Criteria	In vivo studies									In vitro studies							
		O'Connor 2014	**Yanamala et al. 2014**	**Farcas et al. 2016**	**Shvedova et al. 2016**	**Catalan et al. 2017**	**Park et al. 2018**	**Ilves et al. 2018**	**TOTAL**	**%**	**Clift et al. 2011**	**Endes et al. 2014**	**Yanamala et al. 2016**	**Menas et al. 2017**	**Lopes et al. 2017**	**Ilves et al. 2018**	**TOTAL**	**%**
Study Design Score	Applied concentration/dose, to be given in more than one unit.	0	0	0	0	0	0	2	2	17%	0	2	2	2	0	2	8	67%
	Doses should be clearly marked as "overload" or "non-overload".	0	0	0	0	0	1	0	1	8%	NA	NA	NA	NA	NA	NA		
	At least two different tests for each biological end point.	1	1	1	1	2	1	1	8	67%	0	0	1	1	1	1	4	33%
	Study should contain data on the dose–effect relationship of the acute toxic effects.	0	0	0	0	1	0	0	1	8%	1	1	0	1	1	1	5	42%
	Interference of the nanomaterials with the test system.	0	0	0	0	0	0	1	1	8%	2	0	0	0	0	1	3	25%
	Evaluation that contaminants or solvents not responsible for observed toxicity	0	0	0	0	1	0	1	2	17%	0	2	1	1	1	2	7	58%
	Are doses relevant to human exposures?	0	0	0	0	0	1	0	1	8%	0	0	0	1	0	0	1	8%
	Positive Control	0	2	0	0	2	2	2	8	67%	2	2	2	2	2	2	12	100%
	Negative Control	0	0	0	0	2	0	2	4	33%	2	0	0	1	2	2	7	58%
	Vehicle Control	0	0	0	0	2	2	2	6	50%	2	2	0	0	0	2	6	50%
	Study Design Score (out of 20/18)	1	3	1	1	10	7	11		28%	9	9	6	9	7	13		49%

3.2. The Card and Magnuson Approach

Card and Magnuson (2010) proposed a quantitative two-step method to assess nanotoxicity studies for quality [11]. First, a 'study design score' is calculated that assesses the adequacy and documentation of study design, methods, materials, and results using the Toxicology Data Reliability Assessment Tool (ToxRTool). For our analysis, the study design was assessed from a risk assessment perspective. The criterion, "Is the study design chosen appropriate for obtaining the substance-specific data aimed at?" was evaluated according to risk assessment principles, including appropriate exposure delivery, realistic dose and duration of exposure, evaluation of a dose–response relationship, and the inclusion of control groups [24]. Based on the criteria groups in the ToxRTool, the study design is scored as a: 1 (reliable study without restrictions), 2 (reliable study with restrictions), or 3 (unreliable study). In step 2, a 'nanomaterial characterization score' is calculated, based on the completeness of the physical and chemical characterization of the nanomaterial (Table 3). A score of 0 indicates limited or no characterization was completed; a score of 10 indicates thorough characterization. The results are graph-based on the score in both of these categories to depict the overall quality of the study for risk assessment. The authors envisioned that this approach could be used as a standardized method to assess manuscript quality or for regulatory review of nanotoxicity studies. Our calculation of the study design score specifically evaluated study design and reporting as it relates to risk assessment of dry CN powder via inhalation. As with the Krug and Wick approach, the user bears responsibility for critically evaluating and determining the relevance of the information it provides.

Table 3. Summary of Card and Magnuson nanomaterial score evaluation.

	Criteria	In Vivo Studies							In Vitro Studies						Total	%
		O'Connor 2014	**Yanamala 2014**	**Farcas 2016**	**Shvedova 2016**	**Catalán 2017**	**Park 2018**	**Ilves 2018**	**Clift 2011**	**Endes 2014**	**Yanamala 2016**	**Menas 2017**	**Lopes 2017**	**Ilves 2018**		
Nanomaterial Characterization Score	1. Agglomeration and/or aggregation	0	0	0	0	0	0	0	0	1	0	0	1	0	2	15%
	2. Chemical composition	0	0	0	1	0	0	1	0	1	0	0	0	1	4	31%
	3. Crystal structure/crystallinity	0	0	0	0	0	0	0	0	0	0	0	1	0	1	8%
	4. Particle size/size distribution	1	1	1	1	1	1	1	1	1	1	1	1	1	13	100%
	5. Purity	0	0	0	1	0	0	1	0	1	0	0	0	1	4	31%
	6. Shape	1	1	0	1	1	1	1	1	1	1	1	1	1	12	92%
	7. Surface area	1	0	0	0	0	0	0	0	0	0	0	0	0	1	8%
	8. Surface charge	0	0	0	1	1	0	1	0	1	0	0	1	1	6	46%
	9. Surface chemistry (including composition and reactivity)	0	0	0	0	0	0	0	0	0	0	0	1	0	1	8%
	10. Characterization completed in relevant experimental media	0	0	0	0	0	0	0	0	1	0	0	1	0	2	15%
	Total	**3**	**2**	**1**	**5**	**3**	**2**	**5**	**2**	**7**	**2**	**2**	**7**	**5**		**35%**

4. Study Evaluation—General Observations

The physical-chemical characterization criteria are similar between the two approaches, but the Card and Magnuson approach breaks down the criteria into more distinct categories, where chemical composition and purity, as well as shape and crystallinity, are each separate criteria for assigning a nanomaterial characterization score. Evaluating the CN studies using the criteria, we found that, in general, there was inadequate physical and chemical characterization of CNs, limiting the comparisons that can be made between studies and to other materials. However, over time, studies are improving and reporting more detailed physical and chemical characterization (Tables 2 and 3).

Animal study design, as evaluated by both sets of criteria, reveals several weaknesses that reduce the value of the studies for risk assessment. In the Card and Magnuson approach, no animal studies received a high total score, as shown in Figure 1. Most studies received a study design score of 3, indicating significant deficiencies in study design, methods, materials, and/or reporting of results for risk assessment. This was largely the result of 'mandatory minimum criteria' in the ToxRTool used to assign study scores under the Card and Magnuson approach. If one of these mandatory minimum criteria is not met, the study is automatically assigned the lowest study design score of 3. For this assessment, none of the animal studies were able to meet this mandatory minimum criterion: "Is the study design chosen appropriate for obtaining the data aimed at?" For the purpose of this evaluation, this criterion was evaluated for obtaining data relevant for risk assessment purposes, which includes exposure delivery, dose and duration of exposure, evaluation of a dose–response relationship, and inclusion of control groups [24]. The study design scores under the Card and Magnuson approach were generally lower compared to the scores assigned under the Krug and Wick approach due to 'mandatory minimum criteria' and subsequent categorization to the lowest score of 3.

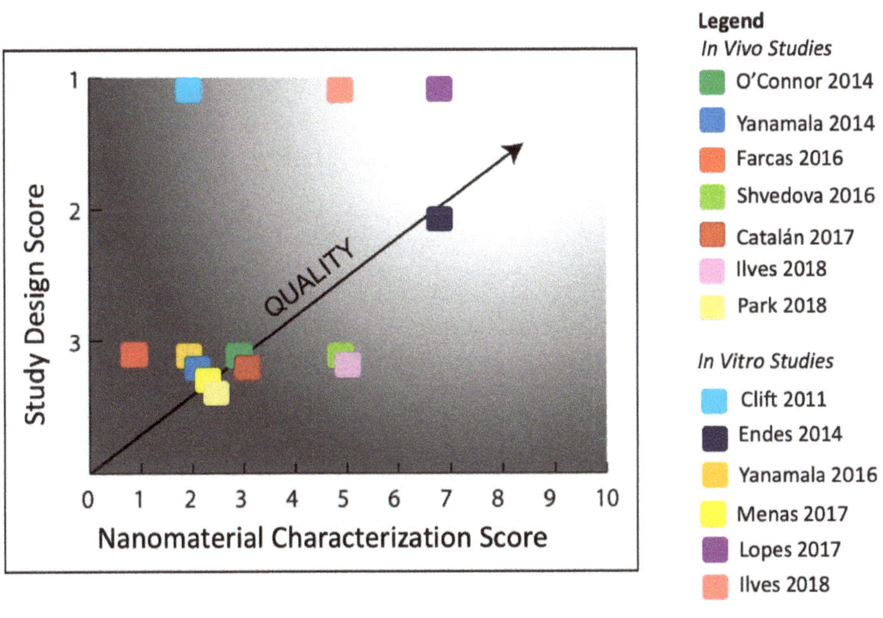

Figure 1. A schematic assessment of the overall quality of CN inhalation studies for risk assessment purposes based on its study design score and nanomaterial characterization score. The clear area represents a range of scores for which a study can be considered of high overall quality; conversely, the shaded area represents studies of low overall quality (adapted from Card and Magnuson, 2010 [11]).

Unlike Card and Magnuson, which categorizes study design scores into groups 1, 2, or 3; the Krug and Wick approach is based on ten criteria for evaluating study design with an overall 'study design score' out of 20 for animal studies. Study design scores were generally low, ranging from 1–11. Most of the studies did not adhere to some of the basic rules in the design of dose–response toxicity studies, or were not designed for that purpose. In general, studies failed to: (i) Establish a dose–response curve; (ii) ensure that the doses being administered were realistic toward human exposures; and (iii) clearly identify "overload" conditions, where doses might cause overt toxicity and be unreliable for studying toxic effects. However, similar to nanomaterial characterization, over time, studies are improving their overall design and data reporting (Tables 2 and 3).

5. Physical-Chemical Measurement and Reporting

Both the Krug and Wick and the Card and Magnuson approaches recognized that there is much uncertainty regarding the physical-chemical characterization necessary for a toxicology study. Some parameters may be more important for one nanomaterial or exposure scenario than they are for another; therefore, standardizing a list and weighting or ranking parameters was not possible. While an understanding of important parameters and the reliability of measurement has improved in nanomaterial studies since the list was developed, there are still few prescribed physical-chemical parameter lists, though some have been suggested (e.g., ECHA 2016 [25], Arts et al. 2015 [26], Oomen et al. 2015 [27]). Definitively linking physical-chemical parameters to biological activity is situation-specific. Attempts to group materials are ongoing [28], which guide the development of such nanomaterial- and situation-specific lists, though such recommendations are slow to be adopted [8].

Similar types of physical-chemical information are generally being provided in both in vitro and in vivo studies (Table 2,3). The Card and Magnuson evaluation shows that particle size/size distribution and shape is evaluated by most studies (100% and 92%, respectively). Surface charge (46%), chemical composition (31%), and purity (31%) were all measured by some studies, while agglomeration/aggregation (15%), crystallinity (8%), surface area (8%), surface chemistry (8%), and characterization in relevant media (15%) were only reported in one or two of the reviewed studies. Size and size distribution are standard reporting metrics for nanomaterial studies and are often measured with electron microscopy or atomic force microscopy [4]. Surface charge (typically measured as zeta-potential) may be an important metric to measure for insoluble nanomaterials, as the charge may affect the interaction of the nanomaterials with proteins and membranes, which could alter the risk of inflammation and lung injury [29,30]. The chemical composition and surface chemistry of CNs vary—for example, CNCs often have sulfate groups, whereas CNFs may not, affecting a myriad of physical-chemical properties [4]. Purity is considered a priority measurement for inhalation studies to distinguish test material from endotoxin or contaminant effects, as endotoxins are often a major inflammatory agent in dust [31,32]. Purity measurements may be particularly important for CNs, as they are derived from wood and plant sources that naturally contain microbes, and contamination with microbes or metals may also occur during manufacture and processing [4]. Agglomeration and aggregation properties modify available surface area and affect the uptake, translocation, and clearance in the lung [33]. Crystallinity is an important measurement for nanomaterials that have different crystal structures (e.g., rutile and anatase titanium dioxide) related to catalytic properties [34]; although, there are no indications that this is relevant for CNs. However, the degree of crystallinity is related to the stiffness of the materials. Stiffness is an important measure for fibrous materials, in which fibers of certain stiffness and length (such as glass and asbestos) can lead to chronic inflammation and eventually to more serious outcomes [35]. The relationship between surface area and pulmonary effects appears to be relevant for some nanomaterials but not others (e.g., Schmid and Stoeger 2016 [36], Warheit et al. 2006 [37]), so reporting these data will help distinguish whether there are any 'nano-specific' effects, or if an effect can be more straightforwardly attributed to scaling of the surface-area-to-volume ratio. The final criterion, characterization in relevant media, should be an essential component for all physical-chemical measurements. Most of the studies measured physical-chemical properties of their

CNs in water, rather than the dispersion media or biologically relevant fluids; therefore, these data may not be representative of the nanomaterial in the exposure media, and in the lung.

A lack of physical-chemical characterization hinders the ability to use these data for risk assessment and to confidently apply the findings of these studies to other materials. The challenge of finding appropriate test methods, as well as an absence of standardized criteria, contributes to the generally poor scores for physical-chemical criteria. Some researchers may not publish their characterization data because they acknowledge that some measurements may not be reliable; for example, dynamic light scattering (DLS) has become the standard method to measure hydrodynamic size, or agglomeration and aggregation state. However, this method is designed for spherical particles; for CNs, which tend to be rod-like or fibrillar, DLS likely does not provide accurate readings, though in some cases DLS can be useful for measuring aggregation or colloidal stability [4]. Characterization methods and best practices continue to be developed, with most based on existing methods modified for CNs [4].

6. Study Design Considerations

6.1. *In vitro versus in vivo Studies*

In vitro studies tend to score higher on study design than in vivo methods (Table 2), with the exception of 'at least two different tests for each biological end point', which is at least partially met in every in vivo study under the Krug and Wick evaluation [10]. In addition, in vitro studies more often report the dose in more than one unit; concentration (μg/mL) and as the deposited dose per cell (μg/cell or μg/cm^2), whereas in vivo studies tend to only report the dose as mass per mouse (μg/mouse). Over time, animal study designs have improved. Studies conducted from 2014–2016 received study design scores of 1–3, while the most recent studies in 2017 and 2018 received scores from 7–11 under the Krug and Wick approach [10]. This is attributed to improvements, including the use of positive, negative, and vehicle controls (including reference materials such as conventional cellulose and carbon nanotubes) [16] and longer study times more appropriate for assessing effects beyond a subacute period.

6.2. *Exposure Technique*

Most of the animal studies used a technique called pharyngeal aspiration to deliver high bolus (i.e., all at once) doses of CN to mouse lungs. This method does not accurately mimic the inhalation of CN that would occur in an occupational setting because it occurs in a moment rather than over time [24]. High-dose delivery of PSLT nanoparticles may elicit different inflammatory responses compared to low-dose delivery [38]. Even when standardized for similar lung loads, bolus delivery methods, such as intratracheal instillation, elicit an elevated inflammatory response compared to inhalation exposure [39]. High-dose bolus delivery studies may be useful for hazard ranking of materials, but these types of studies do not allow estimates of no- or lowest observed effect levels, from which to establish exposure limit values, and, therefore, are of limited utility for risk characterization [24]. Inhalation exposure is the gold standard and part of the Organisation for Economic Cooperation and Development (OECD) inhalation testing methods (e.g., OECD 412 and 413 for subacute and subchronic exposures). In inhalation studies, a test substance is aerosolized in a chamber, and the animal breathes in the substance in a more natural manner. The OECD does endorse the use of pharyngeal aspiration as a simple and inexpensive method to rapidly screen and rank substances, including fibers and nanomaterials [40] but recommends that results be supported by an inhalation study [41].

6.3. *Exposure Dose and Duration*

Most of the studies examined only high doses of CNs that are not representative of realistic levels of exposure. A previous assessment of one of these papers suggests that simulating these high doses by inhalation would require unrealistic workplace exposure concentrations in the g/m^3

range [24]. The U.S. National Institute for Occupational Safety and Health (NIOSH) has conducted exposure assessments in several U.S. governmental, academic, and industrial pilot production facilities and generally found very low total exposure levels, well below 100 μg/m^3 [42]. After assessment of four production facilities, the maximum estimated concentration of detected airborne cellulose was more than ten times below the U.S. Occupational Safety and Health Administration's 5 mg/m^3 permissible exposure limit for the respirable fraction of cellulose dust, which occurred during the milling and cutting of a CNC polymer composite (NIOSH Nanotechnology Field Studies team, personal communication, 19 June 2018). Measurement of airborne CN in full-scale commercial manufacturing facilities will be critical to establish realistic exposures prior to commencing resource-intensive long-term inhalation studies.

Instillation or aspiration of excessive doses, or exposure to high dose-rates of nanomaterials, may overwhelm the integrity of the pulmonary surfactant and permanently compromise lung mechanics [43]. As such, we do not know whether the effects observed in the studies might occur at more likely exposure levels which would be at much lower concentrations over longer periods of time. Further, many studies observed effects only 24 h after exposure, before any short-term effects from the initial exposure could resolve, so it is not possible to determine if the observed lung inflammation was transitory or persistent. More recent publications (e.g., Ilves et al. 2018 [16] and Park et al. 2018 [23]) evaluated longer-term responses, at 14- and 28-days post exposure. These studies demonstrate that for short-term CNF exposures, the initial inflammatory effects may subside by 28 days (similar to other PSLT dusts).

6.4. Lack of Dose–Response

For risk assessment purposes, a fundamental principle in the design of studies is to create a dose–response curve: Testing several concentrations of a material, including concentrations low enough that no effects are observed, all the way to high concentrations where adverse effects may be anticipated. Demonstrating a dose–response relationship associates the observed effects with material exposure, indicates at what concentrations a material might cause an effect such as lung inflammation, and allows evaluation of whether effects are likely to occur at realistic exposure levels. The in vivo inhalation studies were generally not performed according to international standard protocols. Standardized inhalation studies such as OECD 403 (acute), 412 (28-day) and 413 (90-day) typically require that at least three concentration levels be tested to allow for robust statistical analysis. Many of the animal studies published to date on CN dust inhalation tested just one or two doses and were not designed to evaluate and demonstrate a dose–response relationship, limiting their utility for risk assessment.

6.5. Control Groups

All dusts have the potential to irritate the lung when inhaled; therefore, a key question is whether CNs behave any differently or with greater potency compared to PSLT dusts or known inflammatory agents. This is assessed by including negative or positive control groups under similar conditions. Several of the reviewed studies of CNs are not relevant to the question because they did not include a conventional material, such as conventional cellulose, as a control. Ilves et al. (2018) was the first animal study to include reference material controls and demonstrated that, for short-term exposures at least, lung responses to CNF were similar to conventional cellulose in the study [16]. Studies that included positive control groups exposed to asbestos and MWCNT, demonstrated that CNs do not behave like these fibers, which caused a more potent and more persistent inflammatory response.

7. What the Studies Tell Us about the Risks of Inhaling Cellulose Nanomaterial in Dust in the Workplace

There have been some advances in knowledge of the effects of short-term inhalation exposures of CN. The findings from Ilves et al. (2018) and Park et al. (2018) suggest that exposure to CNs result in transient inflammation similar to that caused by other PSLT dusts, such as cellulose, after short-term exposures; these effects are markedly different from those caused by fibers with known toxicity such as certain types of MWCNT or asbestos described by the fiber paradigm [16,23]. As summarized in Park et al. (2018), "Asbestos was the only material not properly handled by the immune system. Nanocellulose agents did not, as evidenced here, behave as "asbestos-like" materials" [23]. However, the studies conducted to date are limited in terms of their utility for risk assessment purposes because of their designs. Some uncertainty remains regarding: (i) Whether there are differential effects from exposure to different forms of CN (e.g., CNC versus CNF; or different surface functionalizations), (ii) the effects from low-dose, long-term CN inhalation exposure that more accurately reflects workplace conditions; and (iii) whether occupational exposure limits for conventional cellulose and other PSLT apply to CNs or should be modified. The available studies provide only a limited weight of evidence for risk assessment purposes but study quality is improving with time.

7.1. Cellular (In Vitro) Assays

Cellular studies allow researchers to investigate biological mechanisms in a highly controlled system, or to make comparisons between materials. To date, six studies (Table 1), have examined the effects of CN exposure on lung cells. These studies use cell lines derived from the lining of the lung, or immune cells normally found in lung tissue, to evaluate the effects of CN exposure, including endpoints such as viability and cytokine secretion. These models give an indication of whether the immune system might be triggered upon CN exposure; however, they do not capture the true complexity of an immune response in animals or people such as inflammation. In cellular studies with CN conducted up to 2017, the results varied and were often contradictory, with some studies indicating cellular toxicity and the potential for inflammation, and others not observing these effects. These discrepancies could be due to the cell culture model system [17], different exposure times and doses, exposure methods, purity, surface chemistry, and so forth. This limits the conclusions that can be drawn for risk assessment and highlights the importance of carefully designing and reporting study details.

In vitro systems are useful in providing supporting evidence for risk assessment purposes, including information on mechanisms of toxicity, and relative responses that allow grouping, screening, and prioritization [44,45]. Currently, animal models provide a more thorough understanding of how inhalation of CNs may interact with complex biological systems such as the lung; although, alternative testing strategies (non-animal tests) are rapidly being developed and validated to support risk assessment and reduce the number of animal studies [45].

7.2. Short-Term Animal Studies

Five of the seven animal studies reviewed evaluated exposure to CNC, and three of the studies evaluated CNF. All studies examined short-term exposures to CN and reported on potential inflammation in the lung by examining various tissue, cellular, and molecular endpoints. Other reported effects include reproductive effects [12], sex differences [13], and genotoxicity [14]. However, as outlined above, the conclusions drawn from these studies for risk assessment purposes are limited.

A key finding from the Ilves et al. (2018) and Park et al. (2018) studies, which also tested MWCNT and asbestos fibers, is that there is no strong evidence CNF and CNC belong to the harmful class of fibers described by the fiber paradigm [16,23]. The fiber paradigm is a structure:toxicity model which outlines specific physical and chemical properties of fibers associated with harmful effects (e.g., inflammation, fibrosis, and increased tumor incidence) when breathed into the lung; the effects occur when macrophages fail to phagocytize or engulf fibers, an event known as frustrated phagocytosis. In particular, the fiber paradigm applies to fibers thin enough to be respirable into the lung (<3 μm); long (>5 μm), rigid enough that they cause frustrated phagocytosis and cannot be cleared from the lung, and persistent enough that they remain in the lung [46,47]. While CNC and CNF do have typical widths <3 μm, CNCs are much shorter, generally on the scale of 100 s of nm in length, and, therefore, are not likely to result in frustrated phagocytosis [17]. CNFs are typically >1 μm in length but are not rigid and generally exist as complex, tangled networks of fibers [4,48]. Studies have found evidence that conventional cellulose, CNC and CNF are biopersistent in the lung [16,49,50]. However, in Ilves et al. (2018), in vivo exposure to CNF was not associated with a persistent inflammatory response, as is common with materials defined by the fiber paradigm [16]. The authors hypothesize that the gel-like state that CNF forms in water or the formation of a protein corona might render CNF biocompatible in the lung. Thus, although CNC and CNF widths are generally below 100 nm, other physical properties suggest that these materials do not likely adhere to the paradigm, much like other organic dusts [31,51]. Further measurements of rigidity, persistence, and inflammatory response over longer periods of time are needed to validate this hypothesis for different forms of CNs.

Short-term pulmonary exposures to CNC and CNF at high levels do result in an initial inflammatory response; although, the immune response is less strong and markedly different from that of asbestos and carbon nanotubes [16,23]. The immune response to CNF is substantially reduced after 28 days and is similar in response and duration to conventional cellulose exposure [16]. This suggests CNFs induce short-term inflammatory effects when breathed into the lung, similar to other PSLT dusts, including cellulose, which subside over time. A longer term in vivo study comparing CNC exposure to conventional cellulose has not yet been reported.

These are important findings that contribute to our understanding of the potential risks from inhalation of dried CNs in occupational environments. The results suggest that CNF causes an acute inflammatory reaction in the lung that resolves, similar to other PSLT dusts such as cellulose after short-term exposure. Further, both CNC and CNF elicit reactions in the lung after short-term exposures that are markedly different from fibers with known toxicity such as certain MWCNT or asbestos. However, there is still some uncertainty about the effects of breathing CN in dust and the mode of action, especially over the long term. Surface chemistry may be an important factor to consider, and CNF may be biopersistent—how this may affect hazard over the long term remains to be determined. In particular, the lack of long-term studies at realistic exposure levels and via inhalation rather than as a bolus dose leaves significant uncertainty about health effects despite the growing database.

8. Future Research and Recommendations

Recent publications are improving the knowledge base, due to improved study designs that further our knowledge about the risks associated with inhaling CNs. These data suggest that CNF behaves similarly to other PSLT dusts such as cellulose for short-term exposures in the lung and that both CNC and CNF are markedly different from fibers with known toxicity, such as certain MWCNT, that meet the fiber paradigm. However, as outlined above, some data gaps still exist. Due to inadequate physical-chemical characterization, unrepresentative exposure methods, high dose and short duration testing, a lack of dose–response analysis, and the general absence of control groups, there is still uncertainty that limits the conclusions that can be drawn on the safety of CN in dry form in the workplace. For these reasons, the safest course of action is still to take a precautionary approach and prevent or minimize the potential for inhalation exposure. This includes adopting measures that are traditionally employed to avoid or mitigate workplace exposures to poorly soluble or insoluble dusts such as working with solutions and gel forms of CN when possible or isolating work with dusts to avoid breathing them. When not possible, measures such as enclosing mixing processes or wearing suitable personal protective equipment may be appropriate. The current data gaps and limitations for risk assessment are not specific to cellulose nanomaterials; these recommendations are applicable to most nanomaterials. Several organizations have published guidance on approaches to safely working with nanomaterials, including NIOSH in the United States and the World Health Organization.

Additional tools and studies needed to allow assessment of the potential health risks from occupational inhalation exposure include:

1. Better techniques to detect CNs in different media (e.g., air, water, biological fluids). It is challenging to measure occupational exposure to CNs due to their composition as organic carbohydrate materials. It can be difficult to identify and measure CNs, often present at very low levels, and to distinguish them from background sources of particles in the atmosphere. Detection and characterization of CN in the lung is required to understand biopersistence and clearance kinetics as they relate to any persistent negative responses, such as chronic inflammation.
2. More exposure assessments in industrial facilities. Several (unpublished) investigations indicate exposure levels are low in CN production environments. While techniques are not necessarily representative of industrial environments, studies by NIOSH in pilot CN facilities have not measured high levels of particles, and other studies have shown that common engineering control equipment such as fume hoods can effectively remove a large proportion of airborne CN. In one instance, particle counts were lower during the production of paper with CNF when compared to the control paper made without it [52]. Measurement of ambient levels of CNs typical of occupational exposures is critical to establishing appropriate dosing for long-term inhalation studies.
3. A long-term inhalation study mimicking realistic exposures. This includes using realistic exposure models (inhalation experiments instead of pharyngeal aspiration) that examine a range of realistic doses (show a dose–response) for different lengths of time, including a timepoint long enough to assess whether there is impairment when exposure occurs over time and whether recovery occurs, to distinguish transient from persistent effects. Control materials need to be included that compare CNs to conventional forms of cellulose to determine if there are any unique hazards associated with CNs or if they behave similarly and at a similar potency to other PSLT dusts. The use of positive and negative controls will also allow comparisons across materials and studies. These types of studies can be expensive but are needed to assess differences between CNs and conventional cellulose and other PSLT dusts.

Author Contributions: Conceptualization, J.E.D., M.G., A.R., and J.A.S.; methodology, J.E.D., K.J.O., and J.A.S.; validation, K.J.O.; formal analysis, J.E.D., J.A.S, and K.J.O.; investigation, J.E.D.; data curation, J.E.D; writing—original draft preparation, J.E.D, J.A.S., and K.J.O.; writing—review and editing, J.E.D., K.J.O., M.G., A.R., C.A.P.-C. and J.A.S.; visualization, J.E.D., K.J.O., and C.A.P.-C.; supervision, J.E.D. and J.A.S.; project administration, J.E.D. and J.A.S.; funding acquisition, J.A.S., A.R., M.G.

Funding: This work was partially funded by a grant from the P3Nano Public-Private partnership between the U.S. Endowment for Forestry and Communities, and the U.S. Department of Agriculture U.S. Forest Service Forest Products Laboratory.

Conflicts of Interest: The funders had no role in the collection, analyses, or interpretation of data; in the writing of the manuscript, or in the decision to publish the results.

References

1. Shatkin, J.A.; Wegner, T.H.; Bilek, E.T.; Cowie, J. Market projections of cellulose nanomaterial-enabled products—Part 1: Applications. *TAPPI J.* **2014**, *13*, 9–16.
2. Shatkin, J.A.; Kim, B. Cellulose nanomaterials: Life cycle risk assessment, and environmental health and safety roadmap. *Environ. Sci. Nano* **2015**, *2*, 477–499. [CrossRef]
3. European Centre for Ecotoxicology and Toxicology of Chemicals (ECETOC). *Poorly Soluble Particles/Lung Overload*; Technical Report 122; ECETOC: Brussels, Belgium, 2013.
4. Foster, E.J.; Moon, R.J.; Agarwal, U.P.; Bortner, M.J.; Bras, J.; Camarero-Espinosa, S.; Chan, K.J.; Clift, M.J.D.; Cranston, E.D.; Eichhorn, S.J.; et al. Current characterization methods for cellulose nanomaterials. *Chem. Soc. Rev.* **2018**, *47*, 2609–2679. [CrossRef] [PubMed]
5. Robinson, R.L.M.; Lynch, I.; Peijnenburg, W.; Rumble, J.; Klaessig, F.; Marquardt, C.; Rauscher, H.; Puzyn, T.; Purian, R.; Aberg, C. How should the completeness and quality of curated nanomaterial data be evaluated? *J. Nanoscale* **2016**, *8*, 9919–9943. [CrossRef] [PubMed]
6. Warheit, D.B. Hazard and risk assessment strategies for nanoparticle exposures: How far have we come in the past 10 years? *F1000Research* **2018**, *7*. [CrossRef] [PubMed]

7. Haase, A.; Lynch, I. Quality in nanosafety—Towards a reliable nanomaterial safety assessment. *NanoImpact* **2018**, *11*, 67–68. [CrossRef]
8. Krug, H.F. The uncertainty with nanosafety: Validity and reliability of published data. *Colloids Surf. B Biointerfaces* **2018**, *172*, 113–117. [CrossRef] [PubMed]
9. Krug, H.F. Nanosafety research—Are we on the right track? *Angew. Chem. Int. Ed. Engl.* **2014**, *53*, 12304–12319. [CrossRef] [PubMed]
10. Krug, H.F.; Wick, P. Nanotoxicology: An interdisciplinary challenge. *J. Angew. Chem. Int. Ed.* **2011**, *50*, 1260–1278. [CrossRef] [PubMed]
11. Card, J.W.; Magnuson, B.A. A method to assess the quality of studies that examine the toxicity of engineered nanomaterials. *Int. J. Toxicol.* **2010**, *29*, 402–410. [CrossRef] [PubMed]
12. Yanamala, N.; Farcas, M.T.; Hatfield, M.K.; Kisin, E.R.; Kagan, V.E.; Geraci, C.L.; Shvedova, A.A. In vivo evaluation of the pulmonary toxicity of cellulose nanocrystals: A renewable and sustainable nanomaterial of the future. *ACS Sustain. Chem. Engl.* **2014**, *2*, 1691–1698. [CrossRef] [PubMed]
13. Farcas, M.T.; Kisin, E.R.; Menas, A.L.; Gutkin, D.W.; Star, A.; Reiner, R.S.; Yanamala, N.; Savolainen, K.; Shvedova, A.A. Pulmonary exposure to cellulose nanocrystals caused deleterious effects to reproductive system in male mice. *J. Toxicol. Environ. Health A* **2016**, *79*, 984–997. [CrossRef] [PubMed]
14. Shvedova, A.A.; Kisin, E.R.; Yanamala, N.; Farcas, M.T.; Menas, A.L.; Williams, A.; Fournier, P.M.; Reynolds, J.S.; Gutkin, D.W.; Star, A.; et al. Gender differences in murine pulmonary responses elicited by cellulose nanocrystals. *Part. Fibre Toxicol.* **2016**, *13*, 28. [CrossRef] [PubMed]
15. Catalan, J.; Rydman, E.; Aimonen, K.; Hannukainen, K.S.; Suhonen, S.; Vanhala, E.; Moreno, C.; Meyer, V.; Perez, D.D.; Sneck, A.; et al. Genotoxic and inflammatory effects of nanofibrillated cellulose in murine lungs. *Mutagenesis* **2017**, *32*, 23–31. [CrossRef] [PubMed]
16. Ilves, M.; Vilske, S.; Aimonen, K.; Lindberg, H.K.; Pesonen, S.; Wedin, I.; Nuopponen, M.; Vanhala, E.; Hojgaard, C.; Winther, J.R.; et al. Nanofibrillated cellulose causes acute pulmonary inflammation that subsides within a month. *Nanotoxicology* **2018**, *12*, 729–746. [CrossRef] [PubMed]
17. Clift, M.J.; Foster, E.J.; Vanhecke, D.; Studer, D.; Wick, P.; Gehr, P.; Rothen-Rutishauser, B.; Weder, C. Investigating the interaction of cellulose nanofibers derived from cotton with a sophisticated 3D human lung cell coculture. *Biomacromolecules* **2011**, *12*, 3666–3673. [CrossRef] [PubMed]
18. Endes, C.; Schmid, O.; Kinnear, C.; Mueller, S.; Camarero-Espinosa, S.; Vanhecke, D.; Foster, E.J.; Petri-Fink, A.; Rothen-Rutishauser, B.; Weder, C.; et al. An in vitro testing strategy towards mimicking the inhalation of high aspect ratio nanoparticles. *Part. Fibre Toxicol.* **2014**, *11*, 40. [CrossRef] [PubMed]
19. Yanamala, N.; Kisin, E.R.; Menas, A.L.; Farcas, M.T.; Khaliullin, T.O.; Vogel, U.B.; Shurin, G.V.; Schwegler-Berry, D.; Fournier, P.M.; Star, A.; et al. In vitro toxicity evaluation of lignin-(un)coated cellulose based nanomaterials on human A549 and THP-1 cells. *Biomacromolecules* **2016**, *17*, 3464–3473. [CrossRef] [PubMed]
20. Menas, A.L.; Yanamala, N.; Farcas, M.T.; Russo, M.; Friend, S.; Fournier, P.M.; Star, A.; Iavicoli, I.; Shurin, G.V.; Vogel, U.B.; et al. Fibrillar vs crystalline nanocellulose pulmonary epithelial cell responses: Cytotoxicity or inflammation? *Chemosphere* **2017**, *171*, 671–680. [CrossRef] [PubMed]
21. Lopes, V.R.; Sanchez-Martinez, C.; Stromme, M.; Ferraz, N. In vitro biological responses to nanofibrillated cellulose by human dermal, lung and immune cells: Surface chemistry aspect. *Part. Fibre Toxicol.* **2017**, *14*, 1. [CrossRef] [PubMed]
22. O'Connor, B.; Berry, R.; Goguen, R. Commercialization of cellulose nanocrystal (NCC™) production: A business case focusing on the importance of proactive EHS management. In *Nanotechnology Environmental Health and Safety*, 2nd ed.; Hull, M.S., Bowman, D.M., Eds.; William Andrew Publishing: Oxford, UK, 2014; pp. 225–246.
23. Park, E.J.; Khaliullin, T.O.; Shurin, M.R.; Kisin, E.R.; Yanamala, N.; Fadeel, B.; Chang, J.; Shvedova, A.A. Fibrous nanocellulose, crystalline nanocellulose, carbon nanotubes, and crocidolite asbestos elicit disparate immune responses upon pharyngeal aspiration in mice. *J. Immunotoxicol.* **2018**, *15*, 12–23. [CrossRef] [PubMed]
24. Shatkin, J.A.; Oberdörster, G. Comment on Shvedova et al. (2016), "Gender differences in murine pulmonary responses elicited by cellulose nanocrystals". *Part. Fibre Toxicol.* **2016**, *13*, 59. [CrossRef] [PubMed]
25. European Chemicals Agency (ECHA). *Usage of (Eco)Toxicological Data for Bridging Data Gaps between and Grouping of Nanoforms of the Same Substance—Elements to Consider*; ECHA: Helsinki, Finland, 2016.

26. Arts, J.H.; Hadi, M.; Irfan, M.A.; Keene, A.M.; Kreiling, R.; Lyon, D.; Maier, M.; Michel, K.; Petry, T.; Sauer, U.G.; et al. A decision-making framework for the grouping and testing of nanomaterials (DF4nanoGrouping). *Regul. Toxicol. Pharmacol.* **2015**, *71*, S1–S27. [CrossRef] [PubMed]
27. Oomen, A.G.; Bleeker, E.A.; Bos, P.M.; van Broekhuizen, F.; Gottardo, S.; Groenewold, M.; Hristozov, D.; Hund-Rinke, K.; Irfan, M.A.; Marcomini, A.; et al. Grouping and read-across approaches for risk assessment of nanomaterials. *Int. J. Environ. Res. Public Health* **2015**, *12*, 13415–13434. [CrossRef] [PubMed]
28. Landvik, N.E.; Skaug, V.; Mohr, B.; Verbeek, J.; Zienolddiny, S. Criteria for grouping of manufactured nanomaterials to facilitate hazard and risk assessment, a systematic review of expert opinions. *Regul. Toxicol. Pharmacol.* **2018**, *95*, 270–279. [CrossRef] [PubMed]
29. Braakhuis, H.M.; Park, M.V.; Gosens, I.; De Jong, W.H.; Cassee, F.R. Physicochemical characteristics of nanomaterials that affect pulmonary inflammation. *Part. Fibre Toxicol.* **2014**, *11*, 18. [CrossRef] [PubMed]
30. Cho, W.-S.; Duffin, R.; Thielbeer, F.; Bradley, M.; Megson, I.L.; MacNee, W.; Poland, C.A.; Tran, C.L.; Donaldson, K. Zeta potential and solubility to toxic ions as mechanisms of lung inflammation caused by metal/metal oxide nanoparticles. *Toxicol. Sci.* **2012**, *126*, 469–477. [CrossRef] [PubMed]
31. Jarvholm, B. Natural organic fibers—Health effects. *Int. Arch. Occup. Environ. Health* **2000**, *73*, S69–S74. [CrossRef] [PubMed]
32. Chun, D.T.W.; Chew, V.; Bartlett, K.; Gordon, T.; Jacobs, R.R.; Larsson, B.-M.; Larsson, L.; Lewis, D.M.; Liesivuori, J.; Michel, O.; et al. Preliminary report on the results of the second phase of a round-robin endotoxin assay study using cotton dust. *Appl. Occup. Environ. Hyg.* **2000**, *15*, 152–157. [CrossRef] [PubMed]
33. Braakhuis, H.M.; Oomen, A.G.; Cassee, F.R. Grouping nanomaterials to predict their potential to induce pulmonary inflammation. *Toxicol. Appl. Pharmacol.* **2016**, *299*, 3–7. [CrossRef] [PubMed]
34. Warheit, D.B.; Webb, T.R.; Reed, K.L.; Frerichs, S.; Sayes, C.M. Pulmonary toxicity study in rats with three forms of ultrafine-TiO_2 particles: Differential responses related to surface properties. *Toxicology* **2007**, *230*, 90–104. [CrossRef] [PubMed]
35. Camarero-Espinosa, S.; Endes, C.; Mueller, S.; Petri-Fink, A.; Rothen-Rutishauser, B.; Weder, C.; Clift, M.J.D.; Foster, E.J. Elucidating the potential biological impact of cellulose nanocrystals. *J. Fibers* **2016**, *4*, 21. [CrossRef]
36. Schmid, O.; Stoeger, T. Surface area is the biologically most effective dose metric for acute nanoparticle toxicity in the lung. *J. Aerosol Sci.* **2016**, *99*, 133–143. [CrossRef]
37. Warheit, D.B.; Webb, T.R.; Sayes, C.M.; Colvin, V.L.; Reed, K.L. Pulmonary instillation studies with nanoscale TiO2 rods and dots in rats: Toxicity is not dependent upon particle size and surface area. *Toxicol. Sci.* **2006**, *91*, 227–236. [CrossRef] [PubMed]
38. Baisch, B.L.; Corson, N.M.; Wade-Mercer, P.; Gelein, R.; Kennell, A.J.; Oberdorster, G.; Elder, A. Equivalent titanium dioxide nanoparticle deposition by intratracheal instillation and whole body inhalation: The effect of dose rate on acute respiratory tract inflammation. *Part Fibre Toxicol.* **2014**, *11*, 5. [CrossRef] [PubMed]
39. Creutzenberg, O.; Koch, W.; Hansen, T.; Schuchardt, S. *Comparison of Inhalation and Intratracheal Instillatin as Testing Methods for Characterisation of Granular Biopersistent Particles (GBP)*; BAuA: Dortmund, Germany, 2018.
40. Organisation for Economic Co-operation and Development (OECD). *Draft Guidance Document on Inhalation Toxicity Testing #39*; OECD: Paris, France, 2017.
41. Organisation for Economic Co-operation and Development (OECD). *Guidance on Sample Preparation and Dosimetry for the Safety Testing of Manufactured Nanomaterials*; OECD: Paris, France, 2012.
42. Dunn, K. Overview of NIOSH field studies for the assessment and control of nanocellulose materials. In Proceedings of the International Conference on Nanotechnology for Renewable Materials, Madison, WI, USA, 11–15 June 2018.
43. Organisation for Economic Co-operation and Development (OECD). *Guidance Document on Inhalation Toxicity Studies*; OECD: Paris, France, 2009.
44. Shatkin, J.A.; Ong, K.J.; Beaudrie, C.; Clippinger, A.J.; Hendren, C.O.; Haber, L.T.; Hill, M.; Holden, P.; Kennedy, A.J.; Kim, B.; et al. Advancing risk analysis for nanoscale materials: Report from an international workshop on the role of alternative testing strategies for advancement. *Risk Anal.* **2016**, *36*, 1520–1537. [CrossRef] [PubMed]
45. Organisation for Economic Co-operation and Development (OECD). *Alternative Testing Strategies in Risk Assessment of Manufactured Nanomaterials: Current State of Knowledge and Research Needs to Advance Their Use*; OECD: Paris, France, 2016.

46. Donaldson, K.; Murphy, F.; Schinwald, A.; Duffin, R.; Poland, C.A. Identifying the pulmonary hazard of high aspect ratio nanoparticles to enable their safety-by-design. *Nanomedicine* **2011**, *6*, 143–156. [CrossRef] [PubMed]
47. Lee, D.K.; Jeon, S.; Han, Y.; Kim, S.H.; Lee, S.; Yu, I.J.; Song, K.S.; Kang, A.; Yun, W.S.; Kang, S.M.; et al. Threshold rigidity values for the asbestos-like pathogenicity of high-aspect-ratio carbon nanotubes in a mouse pleural inflammation model. *ACS Nano* **2018**, *12*, 10867–10879. [CrossRef] [PubMed]
48. De France, K.J.; Hoare, T.; Cranston, E.D. Review of hydrogels and aerogels containing nanocellulose. *Chem. Mater.* **2017**, *29*, 4609–4631. [CrossRef]
49. Stefaniak, A.B.; Duling, M.G.; Lawrence, R.B.; Thomas, T.A.; LeBouf, R.F.; Wade, E.E.; Abbas Virji, M. Dermal exposure potential from textiles that contain silver nanoparticles AU. *Int. J. Occup. Environ. Health* **2014**, *20*, 220–234. [CrossRef] [PubMed]
50. Muhle, H.; Bellmann, B.; Ernst, H. Investigation of the durability of cellulose fibres in rat lungs. *Ann. Occup. Hyg.* **1997**, *41*, 184–188. [CrossRef]
51. Schulte, P.A.; Kuempel, E.D.; Drew, N.M. Characterizing risk assessments for the development of occupational exposure limits for engineered nanomaterials. *Regul. Toxicol. Pharmacol.* **2018**, *95*, 207–219. [CrossRef] [PubMed]
52. Vartiainen, J.; Pöhler, T.; Sirola, K.; Pylkkänen, L.; Alenius, H.; Hokkinen, J.; Tapper, U.; Lahtinen, P.; Kapanen, A.; Putkisto, K.; et al. Health and environmental safety aspects of friction grinding and spray drying of microfibrillated cellulose. *J. Cellul.* **2011**, *18*, 775–786. [CrossRef]

nanomaterials

MDPI

Article

New Antioxidant Multilayer Packaging with Nanoselenium to Enhance the Shelf-Life of Market Food Products

Paula Vera [1], Elena Canellas [1,2] and Cristina Nerín [1,*]

1 Analytical Chemistry Department, GUIA Group, I3A, EINA, University of Zaragoza, Mª de Luna 3, 50018 Zaragoza, Spain; pvera@unizar.es (P.V.); elenac@unizar.es (E.C.)

2 Samtack Adhesivos Industriales, C/Cerámica, nº3, Pol. Ind. Magarola Sud, 08292 Esparraguera, Barcelona, Spain

* Correspondence: cnerin@unizar.es; Tel.: +34-976-761873

Received: 31 August 2018; Accepted: 11 October 2018; Published: 16 October 2018

Abstract: A flexible multilayer with selenium nanoparticles incorporated has been used to build an antioxidant packaging. The oxidation of hazelnuts, walnuts, and potato chips was tested at laboratory scale. Hexanal released by the nuts, fatty acids oxidation study, TBARS (thiobarbituric acid reactive substances), and tasting were compared to study the oxidation of foods packaged with this antioxidant packaging. Finally, TBARS method in combination with tasting were selected due to their simplicity and accuracy. It was found that hazelnuts packaged in nanoSe active bags released around 20% less malonaldehyde (MDA) than the blanks. In the case of the walnuts, the active ones released 25% less MDA than the blanks. As for potato chips, the improvement was around 22%. Finally, an industrial study was done. Cooked ham, chicken, and a ready-to-eat vegetable mixture seasoned with butter were industrially packaged with the new antioxidant material and improvements higher than 25% were obtained.

Keywords: selenium nanoparticles; multilayer laminates; antioxidant packaging; TBARS

1. Introduction

Lipid oxidation is an important factor that limits the shelf-life of food. It reduces the nutritional value of lipids, some fat-soluble vitamins are lost because of its reaction with free radicals, some pigments disappear, and rancidity starts. This is one of the main causes of food product rejection by consumers, since it is associated with characteristic off-flavor due to the generation of volatile short-chain aldehydes and ketones, which are responsible for off-flavor, such as malonaldehydes (MDA) [1].

The development of active packaging provides an opportunity to extend the freshness of food products. Antioxidant packaging may act by absorbing the compounds that deteriorate the food, such as oxygen or free radicals. Oxidation process is a radical reaction initiated by the free radicals derived from oxygen, which are the primary free radicals. These free radicals are transferred to the lipid chains and their oxidation takes place. Previous works demonstrates that scavenging these primary free radicals is the best and the most efficient way to protect food against oxidation [2–8]. The antioxidant properties of selenium nanoparticles (SeNPs) are well known [9–12], and based on these concepts a new multilayer material containing nanoSe was built and optimized [13]. This new antioxidant polymer was optimized and studied at lab scale using oxidizable model compounds. However, tests with real food at both lab scale and industrial scale were not carried out. The scientific literature shows many approaches of antioxidant packaging materials, but most of them fail when trying to apply them at industrial level in a packaging line. Machinability of the new active materials, as well as stability, off-odor, color, and other physical characteristics, besides the antioxidant performance, are the main reasons why these new approaches are not common in the food market. Industrial scale up is always a big challenge and often the positive results from laboratory tests turn into a different panorama, where the gain of the new material is not acceptable. In this work, an antioxidant packaging based on selenium nanoparticles incorporated to a flexible multilayer was tested with food susceptible to rancidity. Then, in addition to the mentioned difficulties, Se nanoparticles had to be produced and handled at industrial level and incorporated in a homogeneous and reliable manner into the packaging material. The shape, size, and distribution in the packaging, as well as stability, had to be under control and the resulting material should be characterized and tested in real situations. The antioxidant capacity of the laminates with SeNPs incorporated have been previously demonstrated by Vera et al. at laboratory scale. It was confirmed that SeNPs are able to trap the primary free radicals derived from oxygen [13]. With this information, the new multilayer material containing nanoSe has recently been evaluated and it received a positive opinion from EFSA (European Food Safety Authority). The absence of migration of nanoparticles, together with the efficiency as an antioxidant at lab scale were the guarantee of food safety. Thus, it can be scaled up and launched into the market, as one of the few active materials in the EU market.

The first goal is to monitor the oxidation reaction of packaged food during its shelf life. One of the strategies to evaluate this is through a sensory panel. This methodology is very useful when the antioxidant packaging is designed to increase the time the consumer considers the food acceptable. Nevertheless, this methodology depends on the people that form the panel, and it is less reproducible than an instrumental method. Thus, it is important to combine this panel with analytical techniques that provide accurate measurements, such as the determination of the fatty acid composition of the food [14–16]. The method consists of derivatizing the fatty acids to their glyceryl esters, which will be analyzed either by GC-FID or GC-MS. This way, a quantitative value of each fatty acid before and after the oxidation will be obtained, and consequently the amount of fatty acids that disappear because of the oxidation will be available. Other techniques are based on the determination of the volatile compounds responsible for the unpleasant odors and in some cases propanal, pentanal, hexanal, 2-hexenal, 3-hexenal, or 2,4-heptadienal [17,18] release were also monitored to confirm rancidity. A common methodology in food is the thiobarbituric acid reactive substances (TBARS) assay. This method is based on the determination of malondialdehyde (MDA), which is a low-molecular-weight end product usually formed via the decomposition of certain primary and secondary lipid peroxidation products [19,20]. TBARS technique has been used in many oxidation studies of food [21]. To have an objective evaluation of the industrial production of this industrial nanoSe antioxidant packaging, all these methodologies were tested to demonstrate the efficiency of the new nanoSe active packaging on packaged food.

The main targets for oxidation in the lipids are the polyunsaturated fatty acids, which are vulnerable to the action of free radicals. This is due to the presence of double bonds that weaken the C-H bonds [22,23]. The hydrolysis and/or autoxidation of fats provide short-chain aldehydes and ketones responsible for rancidity, and rancidity is associated with the characteristics off-flavor and odor [1]. Nuts have a high content of unsaturated fatty acids [24]. Owing to that, walnuts and hazelnuts were the first foods selected for this study. According to the literature, walnuts have a content of monounsaturated fatty acids (MUFA) of around 9 g/100 g and a content of polyunsaturated fatty acids (PUFA) of around 48 g/100 g. Hazelnuts have a MUFA value of approximately 46 g/100 g and PUFA of approximately 8 g/100 g [25]. Therefore, they are good candidates to study their lipid oxidation and rancidity [26], and consequently nuts and walnuts were the first target of this study. Then, several foods susceptible to rancidity, packaged by different companies and packaging lines, have been analyzed and the results are shown and discussed.

2. Materials and Methods

2.1. Reagents

Dichloromethane, cyclohexane, sodium hydroxide, boron trifluoride, sodium chloride, trichloroacetic acid, thiobarbituric acid, and 1,1,3,3-tetraethoxypropane were purchased from Sigma-Aldrich (Sigma-Aldrich-Merck, Madrid, Spain). Methanol and heptane were supplied by Scharlau Chemie S.A (Scharlau, Sentmenat, Spain). Solid phase microextraction fibers were supplied by Supelco (Bellefonte, PA, USA).

2.2. Laminates

Laminates made with selenium nanoparticles were prepared. Firstly, selenium nanoparticles were synthesized using a solution-phase approach based on the reduction of selenite with ascorbic acid in the presence of 7% of concentration of 2,4,7,9-tetramethyl-5decyne-4,7-diol ethoxylate as stabilizer agent (at pH 6.5). The final concentration of selenium nanoparticles in the solution were 1000 ppm. SeNPs between 50 and 70 nm were obtained. Figure 1 shows the SEM analysis, where the size and identification of SeNPs in the solution can be seen. The detailed description of the synthesis can be read in Vera et al. [13].

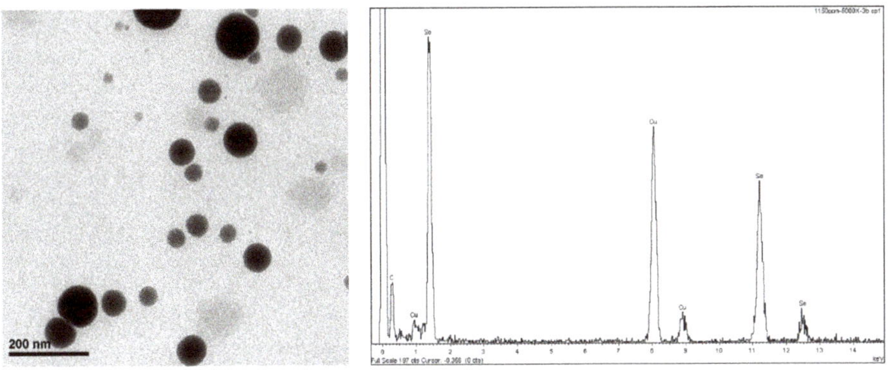

(a) (b)

Figure 1. (**a**) SEM of 1500 ppm solution of nanoSe and (**b**) EDX analysis of SeNPs. Samples on Cu grid.

This nanoselenium solution was incorporated at 10% (*w*/*w*) into a water-based adhesive dispersion. The formula of adhesive was provided under a confidential agreement and cannot be disclosed here.

Using the adhesive, the laminate structure [substrate 1-adhesive-substrate 2] was manufactured first at laboratory scale and later at industrial scale. Industrial conditions were optimized as well to provide a colorless multilayer, where nanoSe was homogeneously distributed. At laboratory scale, the laminates were prepared using an extender machine K control coater, RK printcoat instruments (RK PrintCoat Instruments Ltd, Litlington, UK) on a 20 × 30 cm of substrate 1 polyethylene terephthalate (PET, 12 μm), forming a uniform layer. The grammage of adhesive placed was 3 g/m^2, which was gravimetrically measured. Afterwards, a second 20 × 30 cm of 60 μm thickness film of low density polyethylene (LDPE) was placed on top of it, and then the laminate was pressed at 80 °C at speed number 4 in a BiO 330 model Laminator (Laminator-BIO 330, Korea)

Blank laminates were also prepared as follows: PET (12 μm)/adhesive/PE (60 μm) (laminate B) using the same formula without nanoSe.

2.3. Food Samples and Packagings

Hazelnuts, walnuts, and potato chips were bought in bulk. Fresh hazelnuts and walnuts without the peel were used. Negret hazelnut and Hartley walnut variety, additive free, were selected. Fresh fried potato chips with the following composition: Sunflower oil 80% and olive oil 20% were chosen for the study, assuming that they were prone to oxidation. In the studies made at laboratory scale, 5×2.5 cm^2 thermosealed bags were made with the laminates using a thermal sealer. An LDPE layer was inside the bag in contact with the food. Nuts and potato chips were ground to increase the surface in contact with the packaging, and to increase the oxidation effect to see differences in the oxidation of blanks and actives.

Industrial samples were provided by several companies interested in the study. Their names are confidential. Breaded Wiltshire cured ham (British stile ham) was studied. No antioxidants were added to the ham. Ready-to-eat vegetable mixture seasoned with butter consisting of sweet corn, chopped broccoli, and chopped carrot were studied. Fresh raw chicken breast was also packaged by another food company. In all cases, their own industrial packaging line and their own format (shape and size) for the selected food were used, without intervention or participation of people involved in the nanoSe project and development.

2.4. Monitoring the Hexanal

Hexanal analysis was done by headspace-solid phase microextraction (HS-SPME) coupled to a gas chromatography mass spectrometry (GC-MS). The SPME selected was Divinylbenzene/Carboxen/Polydimethylsiloxane (DVB/CAR/PDMS). The fiber was introduced into the bags made with the laminates, which contained the nuts. The extraction time was 15 min at room temperature and the desorption was done at 250 °C in the GC-MS. The equipment used was an Agilent 6890N gas chromatograph with a mass spectrometer MS 5975B detector. All of them were from Agilent Technologies (Madrid, Spain).

The capillary column used was a HP-5MS (30 m $\times$ 0.25 µm $\times$ 250 µm) from Agilent Technologies (Madrid, Spain). The oven program was as follows: 40 °C for 2 min, with a rate of 10 °C/min up to 300 °C, maintained for 2 min. The injection type was splitless and the helium flow was 1 mL/min. The mass detector was set at SCAN mode (in the range m/z 45–350) for the identification of the compounds, and SIM mode for the quantification of the compounds in the migration extracts.

2.5. Fatty Acids Oxidation Study

In the fatty acids oxidation study, a derivatization of the fatty acids was performed, where 10 g of grounded nuts were added to a recipient and then 20 mL of DCM:cyclohexane (1:1) were added. The two phases were separated. The liquid phase was transferred to another recipient and dried at room temperature. Then, 4 mL of NaOH 0.5 M were added to a 100 mg of this extract. This mixture was kept boiling for 10 min. Then, 5 mL of boron trifluoride 14% in methanol were added and the mixture was kept boiling for 2 min. Finally, 4 mL of heptane were added and the mixture was kept boiling for 1 min. The mixture was allowed to cool for 2 min. Subsequently, 15 mL of NaCl were added and the mixture was shaken with a Vortex®.

The organic phase was analyzed by GC-MS. The equipment used was a CTC Analytics CombiPal autosampler coupled to an Agilent 6890N gas chromatograph with a mass spectrometer MS 5975B detector. All of them were from Agilent Technologies (Madrid, Spain).

The capillary column used was a HP-5MS (30 m $\times$ 0.25 µm $\times$ 250 µm) from Agilent Technologies (Madrid, Spain). The oven program was as follows: 40 °C for 2 min, with a rate of 10 °C/min up to 300 °C, maintained for 2 min. The injection type was splitless and the helium flow was 1 mL/min. The mass detector was set at SCAN mode (in the range m/z 45–350) for the identification of the compounds, and SIM mode for the quantification of the compounds in the migration extracts.

2.6. TBARS (Thiobarbituric Acid Reactive Substances)

The oxidation of lipids was checked using the thiobarbituric acid reactive substances (TBARS) method. The assay was carried out following the method developed by Pfalzgraf et al. [23]. Briefly, 10 g of food were mixed with 20 mL of aqueous solution of trichloroacetic acid at a concentration of 10 μg/g, and then homogenized with an Ultra-Turrax at 18,000 rpm till uniform slurry was obtained. The supernatant was filtered using qualitative paper filter. Then, 2 mL of the filtrate were mixed with 2 mL of aqueous solution of thiobarbituric acid at a concentration 20 mM. The mixture was kept in a thermostatic bath at 97 °C during 20 min. The absorbance of the solution was measured at 532 nm against the blank sample (instead of filtered aliquot, 2 mL of trichloroacetic acid solution was used to prepare the blank sample). The results were expressed in equivalent concentration of malondialdehyde (MDA; mg of malondialdehyde/kg of meat). The malondialdehyde solution was prepared from 1,1,3,3-tetraethoxypropane dissolved in 1N aqueous hydrochloric acid for the calibration curve.

2.7. Tasting

The tasting was done by an expert panel composed of 10 tasters (8 women and 2 men). They were tasters trained for 5 years in this kind of tasting with antioxidant packaging and using different foodstuffs and different antioxidant agents in the packaging. They were also experts on active packaging materials tastings. They scored four parameters from 1 (worse) to 5 (better): appearance, aroma, taste, and rancidity. It was always a blind tasting and the data were processed by Statistical Package for the Social Sciences (SPSS) software.

3. Results

The antioxidant packaging was firstly evaluated using the methodology described by Pezo et al. [3], based on the exposure of the packaging material to an atmosphere enriched in free radicals. This method, where the free radical scavenging system is applied directly on the film, without requiring a previous extraction of the polymer, allows the evaluation of real antioxidant performance of the packaging. This method was necessary, as Vera et al. [13] demonstrated that the free radical diphenylpicrylhydrazyl (DPPH) was too big to cross the LDPE layer and arrive at the adhesive where the antioxidants, the SeNPs, are placed. Thus, the antioxidant performance can be considered as the actual one corresponding to the film. The material under study, containing SeNPs, was not previously tested on food.

As the aim of this work was to demonstrate the efficiency of this antioxidant packaging to protect the food versus oxidation processes, two series of tests were designed, the first series at laboratory scale and the second one at industrial level.

Several methodologies were applied and compared to objectively measure the oxidation of fats.

3.1. Monitorization of Hexanal Released by the Nuts

Hexanal is the main product of linoleic acid oxidation. Therefore, it is an indicator of lipid oxidation, which has been used as a good marker to monitor many foods: Potato crisps [27], Rapeseed oil [28], Iberian dry-cured loin (Muriel et al. 2007) meat and meat derivatives [29]. Nonanal and hexanal were measured in olive oil [30] and also in nuts [31].

The SPME fiber selected was Carboxen/Polydimethylsiloxane (CAR/PDMS) with 75 μm thickness, since according Pastorelli et al. [32] it had the highest sensitivity for this compound. The limit of detection for hexanal was 5 ng mL^{-1}.

In our case, hexanal did not result in a good marker. Since hexanal is a small molecule, it was able to pass through the laminate that formed the bags. Hexanal was measured after 9, 27, 63, and 105 days of storage. It was observed that the amount of hexanal was gradually decreasing throughout the study, probably due to its permeation throughout the multilayer. Therefore, the hexanal value could not be considered as reliable in this case. This also demonstrates that the presence of nanoSe in the multilayer does not enhance the barrier properties of the multilayer versus organic vapors, such as aldehydes. It is the first time that this permeation of hexanal in the presence of nanoSe in a multilayer is described.

Subsequently, this methodology for monitoring the oxidation versus time was discarded.

3.2. Fatty Acids Oxidation Study

It has been demonstrated that the fatty acid composition of oils is a strong indicative factor that could predict the oxidative state [14–16]. Owing to this, the determination of fatty acids was selected here to study their oxidation rate.

Derivatization was the selected methodology for the fatty acid analysis by GC, since the use of ester derivatives is recommended. The methyl ester derivatives are much more volatile than the corresponding fatty acids. Moreover, they are much less polar, avoiding peak tailing and peak asymmetry on GC [33].

Methyl palmitate, methyl linoleate, methyl oleate, and methyl stearate were the esters determined in this work, since palmitic acid, linoleic acid, oleic acid, and stearic acid are the most abundant fatty acids in hazelnuts and walnuts [34].

Table 1 shows the amount of these fatty acids expressed as ng/g hazelnuts and as ng/g walnuts in the bags made with laminate A (made with adhesive alone) and laminate B (made with adhesive + 10% of nanoSe). Fatty acid values were around 50% higher in the bags made with laminate B, which contained nanoSe at both day 21 and day 42, respectively. The highest differences were found for walnuts. An improvement on the amount of fatty acids from 50 to 66% at day 42 was found in the active bag. It can be seen that 11.3 ng/g of methyl oleate were found in the bag made with laminate B and 3.88 ng/g were found in the bag made with laminate A. In the case of hazelnuts, the differences on the fatty acid amount between the active bag and the blank ranged from 24 to 53% at day 42. Therefore, these results demonstrate the antioxidant effect on food and the consequent shelf life extension of the packaged nuts, since pristine fatty acids remain for much more time in nuts.

Table 1. Concentration of fatty acids expressed as ng/g of hazelnuts and walnuts obtained for the laminates A (control, without nanoselenium particles) and laminates B (with nanoselenium particles) expressed as ng/g $_{food}$.

	Hazelnuts		**Walnuts**	
	Laminate A	**Laminate B**	**Laminate A**	**Laminate B**
Methyl Palmitate				
Day 0	28.6 ± 3.4	28.6 ± 3.3	38.89 ± 3.0	38.89 ± 4.1
Day 21	8.04 ± 0.90	16.4 ± 1.51	12.1 ± 1.6	21.5 ± 1.9
Day 42	4.92 ± 0.21	9.72 ± 1.45	5.17 ± 0.42	10.3 ± 0.22
Methyl Linolelate				
Day 0	39.7 ± 7.4	39.7 ± 6.4	167 ± 10	167 ± 8
Day 21	12.9 ± 1.9	25.5 ± 3.6	35.7 ± 3.9	64.8 ± 8.4
Day 42	6.86 ± 0.91	10.9 ± 2.1	12.5 ± 1.6	21.2 ± 1.9
Methyl Oleate				
Day 0	58.4 ± 7.9	58.4 ± 8.1	66.6 ± 8.3	66.6 ± 8.0
Day 21	23.1 ± 3.1	37.3 ± 5.4	22.5 ± 1.4	39.8 ± 4.5
Day 42	14.4 ± 1.3	24.3 ± 1.5	5.22 ± 0.54	11.3 ± 1.7
Methyl Stearate				
Day 0	3.52 ± 0.33	3.52 ± 0.19	8.59 ± 0.99	8.59 ± 1.22
Day 21	3.32 ± 0.33	3.52 ± 0.35	4.57 ± 0.62	7.08 ± 0.77
Day 42	2.25 ± 0.22	2.34 ± 0.25	2.23 ± 0.29	4.45 ± 0.38

3.3. TBARS (Thiobarbituric Acid Reactive Substances)

Unsaturated fatty acids are oxidized to form odor-free, tasteless hydroperoxides. Then, they are decomposed to flavorful secondary oxidation products, which are mainly aldehydes, such as hexanal, 4-hydroxynonenal (HNE), and malondialdehyde (MDA) [35]. The most common method to determine MDA in foods is the spectrophotometric measurement of the colored adduct of MDA with 2-thiobarbituric acid (TBA). This methodology is not specific, but it has been reported to be a more accurate and sensitive parameter in assessing the oxidative deterioration than other methods [36].

Figure 2 shows the TBARS results of hazelnuts and walnuts packaged for 21 and 42 days, respectively, in both types of laminates, control and active. It can be observed that the hazelnuts packaged in active bags (made with laminate B) released around 10% less MDA than the blanks (laminate A) after 21 days of being packaged, and up to 20% when they were measured after 42 days. In the case of the walnuts study, the bags made with laminate B (active) released 5% less MDA than the blanks (laminate A) after 21 days and around 25% less after 42 days. The values of MDA were higher in walnuts than in hazelnuts. This can be attributed to the fact that walnuts contain a higher amount of PUFA [25], which are more vulnerable to oxidation. Student T-test was performed, and the results were 0.005 for hazelnuts and 0.016 for walnuts. Thus, it could be confirmed that the results with laminate A and B had significant differences with a 95% confidence level.

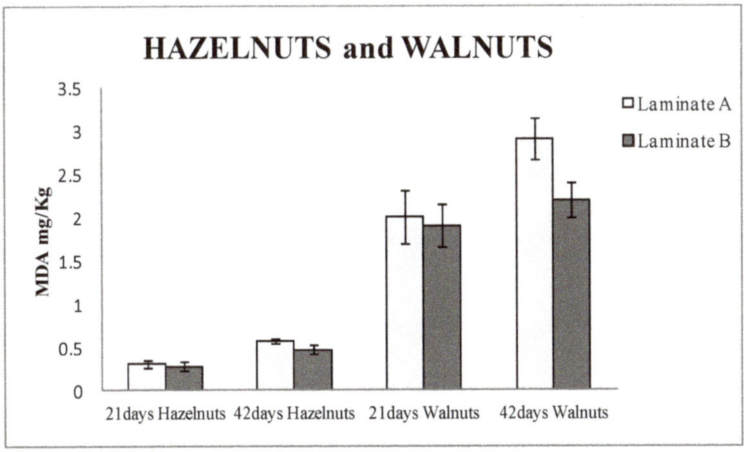

Figure 2. Thiobarbituric acid reactive substances (TBARS) results expressed as mg of malonaldehyde (MDA) per kg of hazelnuts and walnuts packaged in laminate A (control) and laminate B (actives with nanoselenium particles) for 21 and 41 days, respectively.

The results also demonstrated that the active packaging based on nanoSe prevents the oxidation of nuts and significantly extends the shelf life of the nuts.

3.4. Sensory Evaluation of Packaged Nuts

The tasting was done by an expert panel composed of 10 tasters. They scored four parameters from 1 (worse) to 5 (better): appearance, aroma, taste, and rancidity after 42 days of storage of the bags.

Figure 3a,b show the results of the tasting. It can be observed that aroma, flavor, and rancidity were scored at higher values in both hazelnuts and walnuts in laminate B (active) than in laminate A (blank). Rancidity obtained a difference of score of one point between hazelnuts packaged in active bags after 42 days of storage. Moreover, in walnuts the difference was even higher for rancidity (1.3 points of difference).

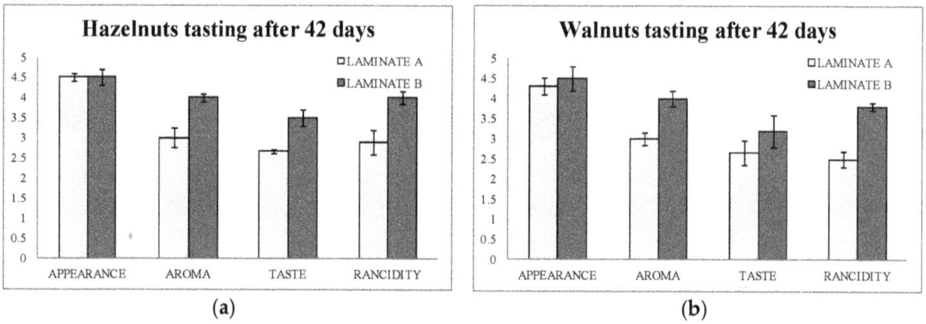

Figure 3. (**a**) Sensory evaluations of hazelnuts packaged in laminate A (control) and laminate B (actives with nanoselenium particles) after 42 days; (**b**) Sensory evaluations of walnuts packaged in laminate A (control) and laminate B (actives with nanoselenium particles) after 42 days.

This data were of great importance, since the consumer will judge the usefulness of the new active packaging. In addition to this, it can be underlined here that the tasting data matched with the analytical data found above by the two methods used.

3.5. Industrial Tests

Once different methodologies to study the oxidation were tested at laboratory scale in nuts, the active packaging was produced at industrial scale and used in other foods. TBARS method in combination with tasting (for ready-to-eat foods) were the methodologies selected for this work. The analysis of fatty acids was discarded, since TBARS offered sufficient results to compare active and blank packaged food, and TBARS was also simple and quicker.

To begin with, a laboratory scale study was done with the industrial multilayer to conduct an additional study of the active packaging with real food prior the industrial scale. The industrial active multilayer was prepared following the same steps as in the laboratory. Firstly, nanoparticles were prepared and incorporated into the adhesive. This active adhesive was then applied between the two plastic films to form the multilayer. This preliminary test was designed to save money and food coming from a large test at industrial scale in a food company. Once the antioxidant properties of the industrial multilayer were confirmed, the industrial tests with different food produce were carried out. In the preliminary study, potato chips, which are susceptible to oxidation, were selected. Once the results of this study were obtained, industrial trials were prepared with ham, vegetables seasoned with butter, and fresh chicken.

3.6. Potato Chips (Laboratory Scale Study)

Potato chips were selected for this study as they have a high amount of unsaturated fats, since they were fried in olive oil, according to the information provided by the suppliers. This kind of product is susceptible to rancidity, and it turns into a perfect target for the antioxidant packaging study.

These potato chips did not contain preservatives or antioxidants, as was mentioned in the experimental section. The time for the study was established according to the shelf-life of the product. The TBARS study and the tasting were done 21 days after the packaging.

The TBARS results (Figure 4) showed a high amount of malonaldehyde. This result was due to the high amount of fat contained in the product. A significant difference of 22% was found between the active laminates and the blanks. This data demonstrated that potato chips were protected from oxidation in this packaging, and therefore this antioxidant packaging could be a good target for its implementation in the market. Student T-test was performed, and the result was 0.045. It was confirmed that the results with laminate A and B were statistically different with a 95% confidence level.

Figure 4. TBARS results expressed as mg of MDA per kg of potato chips, ham, chicken, and ready-to-eat-vegetables packaged in laminate A (control) and laminate B (actives with nanoselenium particles) for 21 days.

The tasting results (Figure 5) showed that the detection of rancidity was almost two out of five points higher in laminate B (blank) than in laminate A (with nanoSe). Therefore, aroma and taste were consequently better in laminate A (nanoSe).

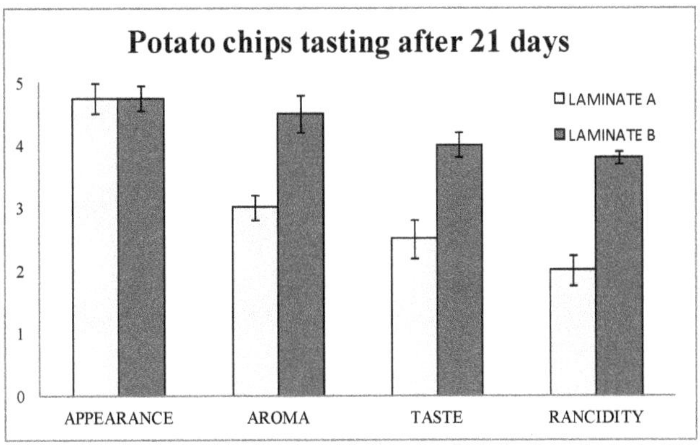

Figure 5. Sensory evaluations of potato chips packaged in laminate A (control) and laminate B (actives with nanoselenium particles) after 21 days.

3.7. Industrial Study: Cooked Ham, Chicken and Ready-to-Eat Vegetables Dressed with Butter

Cooked ham, chicken, and ready-to-eat vegetable mixture seasoned with butter were the foods studied in the industrial trial, since producers of these kind of foods expressed their interest in the antioxidant packaging for their products. A large series of each food was packaged in their industrial packaging lines, using the same procedure, packaging speed, and system as the conventional multilayer. Blank and active samples were also packaged in each case and the same procedures above mentioned and optimized, were applied to monitor the behavior of the packaged food. Half the amount of the tested food was evaluated by the food companies and half by the research team at the University. All the results arrived at the same conclusions about the active nanoSe material. Drawbacks related to the packaging lines were not found, as the material behaved the same as the conventional one, because the active agent was in the middle of the multilayer, in sandwich mode, and consequently it did not affect the performance of the plastic.

Figure 4 shows the TBARS results for all of them packaged for 21 days. In the four types of foods under study, a significantly lower amount of MDA, around 50% for ham, released in the food packaged with the nanoSe were found. Therefore, this means a considerable improvement in the quality of the market produce, since less rancid products are released. This also turns into an extended shelf-life of food produce and demonstrates that the active packaging with nanoSe is a successful option. This is the first time that an industrial active multilayer containing Se nanoparticles as an antioxidant agent was studied. This study opens the door for further active materials, where nanoparticles can be introduced in a skilled manner without affecting the behavior of the material, whilst harnessing the properties and advantages of the packaging materials. Student T-test was performed, and the result was 0.04 for ham, 0.01 for chicken, and 0.06 for vegetables, respectively. The results with laminate A and B were significantly different with a 95% confidence level, for ham and chicken. Vegetables packed with laminate A and B were not significantly different.

4. Conclusions

This works demonstrated that active packaging based on selenium nanoparticles prevents the oxidation of real food and enhances and extends its shelf-life. The presence of nanoSe in the multilayer did not increase the barrier properties for hexanal and aldehydes, but its performance as a strong free radical scavenger was demonstrated. Several industrial lines and food companies tested the material for a wide series of food, and in all cases considerable improvement of food protection versus oxidation was found. This is the first time that the efficiency of an active antioxidant material was demonstrated at industrial scale. In addition, it was demonstrated that the combination of the TBARS method with tasting (for ready-to-eat foods) were the most suitable methodology to evaluate the efficiency of the antioxidant packaging containing nanoSe. This study opens the door to new developments of active materials containing nanoparticles in which the nanoparticles are not in direct contact with the food, but act on food protection from the packaging.

Author Contributions: P.V. did the chemical analysis of fatty Acids Oxidation Study, hexanal and TBARS. E.C. prepared the laminates at lab scale and organized and processed the data from organoleptic study. C.N. designed and supervised the study both lamination, chemical, SEM and sensory analysis.

Funding: This research was funded by Gobierno de Aragón and Fondo Social Europeo, Grant T53_17R.

Acknowledgments: Thanks are given to Gobierno de Aragón and Fondo Social Europeo, Grant T53_17R for the financial help given to GUIA group.

Conflicts of Interest: The authors declare no conflict of interest.

References

1. Frankel, N.E. *Lipid Oxidation*; The Oily Press: Bridgwater, UK, 2005.
2. Pezo, D.; Salafranca, J.; Nerin, C. Design of a method for generation of gas-phase hydroxyl radicals, and use of hplc with fluorescence detection to assess the antioxidant capacity of natural essential oils. *Anal. Bioanal. Chem.* **2006**, *385*, 1241–1246. [CrossRef] [PubMed]
3. Pezo, D.; Salafranca, J.; Nerin, C. Determination of the antioxidant capacity of active food packagings by in situ gas-phase hydroxyl radical generation and high-performance liquid chromatography-fluorescence detection. *J. Chromatogr. A* **2008**, *1178*, 126–133. [CrossRef] [PubMed]
4. Lopez-de-Dicastillo, C.; Pezo, D.; Nerin, C.; Lopez-Carballo, G.; Catala, R.; Gavara, R.; Hernandez-Munoz, P. Reducing oxidation of foods through antioxidant active packaging based on ethyl vinyl alcohol and natural flavonoids. *Packag. Technol. Sci.* **2012**, *25*, 457–466. [CrossRef]
5. Colon, M.; Nerin, C. Role of catechins in the antioxidant capacity of an active film containing green tea, green coffee, and grapefruit extracts. *J. Agric. Food. Chem.* **2012**, *60*, 9842–9849. [CrossRef] [PubMed]
6. Akrami, F.; Rodriguez-Lafuente, A.; Bentayeb, K.; Pezo, D.; Ghalebi, S.R.; Nerin, C. Antioxidant and antimicrobial active paper based on Zataria (*Zataria multiflora*) and two cumin cultivars (*Cuminum cyminum*). *LWT-Food. Sci. Technol.* **2015**, *60*, 929–933. [CrossRef]

7. Carrizo, D.; Taborda, G.; Nerin, C.; Bosetti, O. Extension of shelf-life of two fatty foods using a new antioxidant multilayer packaging containing green tea extract. *Innov. Food Sci. Emerg. Technol.* **2016**, *33*, 534–541. [CrossRef]
8. Wrona, M.; Nerin, C.; Alfonso, M.J.; Caballero, M.A. Antioxidant packaging with encapsulated green tea for fresh minced meat. *Innov. Food Sci. Emerg. Technol.* **2017**, *41*, 307–313. [CrossRef]
9. Bai, Y.; Qin, B.Y.; Zhou, Y.H.; Wang, Y.D.; Wang, Z.; Zheng, W.J. Preparation and antioxidant capacity of element selenium nanoparticles sol-gel compounds. *J. Nanosci. Nanotechnol.* **2011**, *11*, 5012–5017. [CrossRef] [PubMed]
10. Torres, S.K.; Campos, V.L.; Leon, C.G.; Rodriguez-Llamazares, S.M.; Rojas, S.M.; Gonzalez, M.; Smith, C.; Mondaca, M.A. Biosynthesis of selenium nanoparticles by pantoea agglomerans and their antioxidant activity. *J. Nanopart. Res.* **2012**, *14*, 1236. [CrossRef]
11. Sarkar, B.; Bhattacharjee, S.; Daware, A.; Tribedi, P.; Krishnani, K.K.; Minhas, P.S. Selenium nanoparticles for stress-resilient fish and livestock. *Nanoscale Res. Lett.* **2015**, *10*, 371. [CrossRef] [PubMed]
12. Forootanfar, H.; Adeli-Sardou, M.; Nikkhoo, M.; Mehrabani, M.; Amir-Heidari, B.; Shahverdi, A.R.; Shakibaie, M. Antioxidant and cytotoxic effect of biologically synthesized selenium nanoparticles in comparison to selenium dioxide. *J. Trace Elem. Med. Biol.* **2014**, *28*, 75–79. [CrossRef] [PubMed]
13. Vera, P.; Echegoyen, Y.; Canellas, E.; Nerin, C.; Palomo, M.; Madrid, Y.; Camara, C. Nano selenium as antioxidant agent in a multilayer food packaging material. *Anal. Bioanal. Chem.* **2016**, *408*, 6659–6670. [CrossRef] [PubMed]
14. Yun, J.-M.; Surh, J. Fatty acid composition as a predictor for the oxidation stability of korean vegetable oils with or without induced oxidative stress. *Prev. Nutr. Food Sci.* **2012**, *17*, 158–165. [CrossRef] [PubMed]
15. Bialek, A.; Bialek, M.; Jelinska, M.; Tokarz, A. Fatty acid composition and oxidative characteristics of novel edible oils in poland. *Cyta-J. Food* **2017**, *15*, 1–8. [CrossRef]
16. Li, H.Y.; Fan, Y.W.; Li, J.; Tang, L.; Hu, J.N.; Deng, Z.Y. Evaluating and predicting the oxidative stability of vegetable oils with different fatty acid compositions. *J. Food Sci.* **2013**, *78*, H633–H641. [CrossRef] [PubMed]
17. Pastorelli, S.; Valzacchi, S.; Rodriguez, A.; Contini, C.; Simoneau, C. Hexanal as shelf-life indicator: Acelerated storage test of nuts analysed by spme. *Ital. J. Food Sci.* **2007**, *19*, 378–381.
18. Meyer, A.S.; Yi, O.S.; Pearson, D.A.; Waterhouse, A.L.; Frankel, E.N. Inhibition of human low-density lipoprotein oxidation in relation to composition of phenolic antioxidants in grapes (*Vitis vinifera*). *J. Agric. Food. Chem.* **1997**, *45*, 1638–1643. [CrossRef]
19. Janero, D.R. Malondialdehyde and thiobarbituric acid-reactivity as diagnostic indices of lipid peroxidation and peroxidative tissue injury. *Free Radical Biol. Med.* **1990**, *9*, 515–540. [CrossRef]
20. Barriuso, B.; Astiasaran, I.; Ansorena, D. A review of analytical methods measuring lipid oxidation status in foods: A. challenging task. *Eur. Food Res. Technol.* **2013**, *236*, 1–15. [CrossRef]
21. Guillen-Sans, R.; Guzman-Chozas, M. The thiobarbituric acid (TBA) reaction in foods: A. review. *Crit. Rev. Food Sci. Nutr.* **1998**, *38*, 315–330. [CrossRef] [PubMed]
22. Bekhit, A.E.A.; Hopkins, D.L.; Fahri, F.T.; Ponnampalam, E.N. Oxidative processes in muscle systems and fresh meat: Sources, markers, and remedies. *Compr. Rev. Food Sci. Food Saf.* **2013**, *12*, 565–597. [CrossRef]
23. Pfalzgraf, A.; Frigg, M.; Steinhart, H. Tocopherol contents and lipid oxidation in pork muscle and adipose tissue during storage. *J. Agric. Food. Chem.* **1995**, *43*, 1339–1342. [CrossRef]
24. De Meester, F.; Watson, R.R.; Zibadi, S. *Omega-6/3 Fatty Acids. Functions, Sustainability Strategies and Perspectives*; Humana Press, Springer: London, UK, 2013.
25. Lopez-Uriarte, P.; Bullo, M.; Casas-Agustench, P.; Babio, N.; Salas-Salvado, J. Nuts and oxidation: A systematic review. *Nutr. Rev.* **2009**, *67*, 497–508. [CrossRef] [PubMed]
26. Decker, E.; Elias, R.; McClements, D.J. *Oxidation in Foods and Beverages and Antioxidant Applications*; Woodhead Puclising Limited: Cambridge, UK, 2010; Volume 2.
27. Kaykhaii, M.; Rahmani, M. Headspace liquid phase microextraction for quantitation of hexanal in potato crisps by gas chromatography. *J. Sep. Sci.* **2007**, *30*, 573–578. [CrossRef] [PubMed]
28. Gromadzka, J.; Wardencki, W. Static headspace sampling and solid-phase microextraction for assessment of edible oils stability. *Chromatographia* **2010**, *71*, S81–S86. [CrossRef]
29. Shahidi, F.; Yun, J.; Rubin, L.J.; Wood, D.F. The hexanal content as an indicator of oxidative stability and flavor acceptability in cooked ground pork. *Can. Inst. Food. Sci. Technol. J.* **1987**, *20*, 104–106. [CrossRef]

30. Vichi, S.; Pizzale, L.; Conte, L.S.; Buxaderas, S.; Lopez-Tamames, E. Solid-phase microextraction in the analysis of virgin olive oil volatile fraction: Modifications induced by oxidation and suitable markers of oxidative status. *J. Agric. Food. Chem.* **2003**, *51*, 6564–6571. [CrossRef] [PubMed]
31. Pastorelli, S.; Torri, L.; Rodriguez, A.; Valzacchi, S.; Limbo, S.; Simoneau, C. Solid-phase micro-extraction (spme-gc) and sensors as rapid methods for monitoring lipid oxidation in nuts. *Food Addit. Contam.* **2007**, *24*, 1219–1225. [CrossRef] [PubMed]
32. Pastorelli, S.; Valzacchi, S.; Rodriguez, A.; Simoneau, C. Solid-phase microextraction method for the determination of hexanal in hazelnuts as an indicator of the interaction of active packaging materials with food aroma compounds. *Food Addit. Contam.* **2006**, *23*, 1236–1241. [CrossRef] [PubMed]
33. Leo, N.L.M. *Handbook of Food Analysis: Physical Characterization and Nutrient Analysis*; Marcel Dekker, Inc.: New York, NY, USA, 2004; Volume 1.
34. Maguire, L.S.; O'Sullivan, S.M.; Galvin, K.; O'Connor, T.P.; O'Brien, N.M. Fatty acid profile, tocopherol, squalene and phytosterol content of walnuts, almonds, peanuts, hazelnuts and the macadamia nut. *Int. J. Food Sci. Nutr.* **2004**, *55*, 171–178. [CrossRef] [PubMed]
35. Frankel, E.N. Oxidation of polyunsaturated lipids and its nutritional significance. *Abs. Pap. Am. Chem. Soc.* **1994**, *208*, 51-AGFD.
36. Nuchi, C.; Guardiola, F.; Bou, R.; Bondioli, P.; Della Bella, L.; Codony, R. Assessment of the levels of degradation in fat co-and byproducts for feed uses and their relationships with some lipid composition parameters. *J. Agric. Food. Chem.* **2009**, *57*, 1952–1959. [CrossRef] [PubMed]

nanomaterials

MDPI

Article

Electrospun Oxygen Scavenging Films of Poly(3-hydroxybutyrate) Containing Palladium Nanoparticles for Active Packaging Applications

Adriane Cherpinski [1], Melike Gozutok [1], Hilal Turkoglu Sasmazel [2], Sergio Torres-Giner [1] and Jose M. Lagaron [1,*]

1 Novel Materials and Nanotechnology Group, Institute of Agrochemistry and Food Technology (IATA), Spanish Council for Scientific Research (CSIC), Calle Catedrático Agustín Escardino Benlloch 7, 46980 Paterna, Spain; adricherpinski@iata.csic.es (A.C.); melikegozutok@gmail.com (M.G.); storresginer@iata.csic.es (S.T.-G.)

2 Department of Metallurgical and Materials Engineering, Atilim University, Incek, Golbasi, 06830 Ankara, Turkey; hilal.sasmazel@atilim.edu.tr

* Correspondence: lagaron@iata.csic.es; Tel.: +34-96-3900022; Fax: +34-96-3636301

Received: 29 May 2018; Accepted: 21 June 2018; Published: 27 June 2018

Abstract: This paper reports on the development and characterization of oxygen scavenging films made of poly(3-hydroxybutyrate) (PHB) containing palladium nanoparticles (PdNPs) prepared by electrospinning followed by annealing treatment at 160 °C. The PdNPs were modified with the intention to optimize their dispersion and distribution in PHB by means of two different surfactants permitted for food contact applications, i.e., hexadecyltrimethylammonium bromide (CTAB) and tetraethyl orthosilicate (TEOS). Analysis of the morphology and characterization of the chemical, thermal, mechanical, and water and limonene vapor barrier properties and the oxygen scavenging capacity of the various PHB materials were carried out. From the results, it was seen that a better dispersion and distribution was obtained using CTAB as the dispersing aid. As a result, the PHB/PdNP nanocomposites containing CTAB provided also the best oxygen scavenging performance. These films offer a significant potential as new active coating or interlayer systems for application in the design of novel active food packaging structures.

Keywords: polyhydroxyalkanoates; palladium nanoparticles; packaging; electrospinning

1. Introduction

An important increasing quantity of plastic waste is being generated yearly for which the precise needed time for its biodegradation is certainly unknown. This environmental awareness has driven the development and improvement of new biodegradable polymers, especially for single-use plastic items [1]. In this sense, polyhydroxyalkanoates (PHAs) are well-known biopolymers that can be produced microbially by a variety of microorganisms as an energy storage mechanism. They exhibit similar performance in terms of mechanical, thermal, and barrier properties than petroleum-derived polymers and, thus, they can potentially replace conventional thermoplastics (e.g., polyolefins) in a wide range of applications [2]. In particular, barrier properties are of fundamental importance for food packaging applications. For instance, there are many food products that are very sensitive to oxidation and, to overcome this issue, packages with reduced oxygen permeability are desirable. Additionally, the water resistance is also important, particularly for plastic materials intended for direct contact with high moisture foodstuff as well as materials to be applied in high humidity conditions during storage and/or transport [3].

Poly(3-hydroxybutyrate) (PHB) is currently the most common representative of PHAs and this biopolymer has been proposed for short-term food applications [4]. However, PHB is a brittle polymer, as its enzymatic polymerization leads to the formation of macromolecules with highly ordered stereochemical structures, resulting in a large degree of crystallinity [5]. One of the great advantages of PHB over many other biodegradable polymers is its biodegradability under aerobic as well as anaerobic conditions [6]. For this reason, PHB and their blends with other biopolymers, for instance polylactide (PLA), have been extensively investigated for food packaging applications [7–9]. One of the potential application fields of these materials is the development of films for packaging applications. As an example, Zhang et al. [10] studied PHB/PLA blends in different ratios and concluded that by blending PLA with 25 wt. % of PHB some interactions between both biopolymer matrices can be achieved. Furthermore, their results also showed improved mechanical properties.

Palladium nanoparticles (PdNPs) are well known by their ability to dissociate hydrogen molecules to single atoms. This fact is further enhanced due to its nano-sized form and resultant high surface-to-volume ratio [11]. It has been demonstrated that the oxygen scavenging activity of palladium-based oxygen scavenging films is strongly dependent on the coating substrate as well as on the palladium deposition thickness. Optimization of these parameters can result in active scavenging films where the residual headspace oxygen of packaged foods can be scavenged very quickly [10]. There is a drive to find ways to incorporate active packaging technologies directly into the package walls. In spite of the advantages that they offer in maintaining quality and extending shelf life, such systems are still little used. The reason stems from the additional cost involved, the potential toxicity of the added scavenger in the food contact layer, and even more so because of the lack of sufficient technical information on their performance and the lack of understanding of how to apply them effectively [12].

Electrospinning is a fiber production method that employs high electric forces to draw charged threads of polymer solutions or melts up to fiber diameters below 100 nm. It is a low startup cost process in which a wide variety of both polymer and non-polymer materials have been used to form mats composed of nanofibers with a high surface area-to-volume ratio [13]. The electrospinning process has a wide variety of applications such as medical, filtration, tissue engineering, food engineering, packaging, etc. [14–16]. Until now, this processing technology remained to a laboratory scale. However, recent developments in instrumentation and process aid design have allowed this process to be scaled to achieve the production volumes required in certain industrial commodity applications such as fortified foods and active packaging [17].

In active packaging, nanotechnology has a significant potential because nanostructures display a high surface-to-volume ratio and specific surface properties. Considering the high surface energy of nanoparticles, which tend to agglomerate and to prevent this aggregation, either polar polymers or surfactants can be used as protective agents and stabilizers of the nanoparticles. This is extremely necessary to obtain mono-dispersed uniform particles and to be, thereafter, used in various application purposes [18–20]. The objective of the present study was to prepare and characterize, for the first time, PHB films by the electrospinning process incorporating PdNPs. In order to improve the dispersion of the PdNPs in the PHB matrix, different surfactants were tested.

2. Materials and Methods

2.1. Materials

Bacterial aliphatic homopolyester PHB was supplied by Biomer (Krailling, Germany) as P226F. According to the manufacturer, this is certified both as compostable and food contact grade, presenting a density of 1.25 g/cm^3 and a melt flow rate (MFR) of 10 g/10 min at 180 °C and 5 kg. The weight-average molecular weight (M_W) estimated by the manufacturer was 500 kDa and the polydispersity index (PDI) was 2.

2,2,2-trifluoroethanol (TFE) with 99% purity and D-limonene with 98% purity were both purchased from Sigma-Aldrich S.A. (Madrid, Spain). The two tested surfactants, hexadecyltrimethylammonium bromide (CTAB), with 99% purity, and tetraethyl orthosilicate (TEOS), with 98% purity, as well as palladium (Pd) nano-powder, <25 nm particle size measured by transmission electronic microscopy (TEM) and $\geq$99.5% trace metals basis, were also purchased from Sigma-Aldrich S.A. All products were used as received without further purification.

2.2. *Electrospinning*

A PHB solution for electrospinning was prepared by dissolving the biopolymer at 10 wt. %. The PdNPs were added at 1 wt. % in relation to PHB to the solution. To improve the dispersion of the PdNPs, CTAB and TEOS were also added as dispersing aids at 0.5 wt. % with the PHB and PdNPs mixture.

Electrospinning was performed using a Fluidnatek® LE50 benchtop line from Bioinicia S.L. (Valencia, Spain) with a variable high-voltage 0–30 kV power supply. This device was equipped with a motorized injector that was scanning towards a metallic collector, aiming to obtain a homogeneous electrospun deposition. The different biopolymers solutions were transferred to a 30-mL plastic syringe and the syringe was connected through polytetrafluoroethylene (PTFE) tubes to a stainless-steel needle (Ø = 0.9 mm) whereas the needle tip was connected to the power supply. The solution was electrospun for 2 h under a steady flow-rate of 6 mL/h using a motorized injector, scanning horizontally towards a metallic grid. The distance between the injector and collector was optimal at 15 cm and the voltage was set at 15 kV. The biopolymer solutions were electrospun in a controlled environmental chamber at 23 °C and 40% relative humidity (RH), for a given processing time and in optimal conditions to achieve steady fiber formation. The processing and solution characterization parameters for the optimal electrospinning of this PHA grade were determined and optimized in a previous work [21].

Finally, the obtained electrospun mats were subjected to an annealing step below the biopolymer's melting point without applying pressure using a hydraulic press 4122-model from Carver, Inc. (Wabash, IN, USA). The annealing was found optimal at 160 °C, without pressure, for 5 $\pm$ 1 s, based on our previous study [21]. The resultant films were air cooled at room temperature. Prior to thermal treatment, the electrospun mats were equilibrated in a desiccator at 25 °C and 0% RH by using silica gel for at least 1 week.

2.3. *Characterization*

2.3.1. Film Thickness

Film thickness was measured with a digital micrometer series S00014, having $\pm$0.001 mm accuracy, from Mitutoyo Corporation (Kawasaki, Japan) at three random positions. The post-processed samples had a thickness of typically 55 $\pm$ 4 μm.

2.3.2. Scanning Electron Microscopy

A S-4800 microscope from Hitachi (Tokyo, Japan) was used to observe by scanning electron microscopy (SEM) the morphology of the electrospun PHB fibers and the film cross-sections and surfaces. Cross-sections of the samples were prepared by cryo-fracture of the electrospun PHB films using liquid nitrogen. Then, they were fixed to beveled holders using conductive double-sided adhesive tape, sputtered with a mixture of gold-palladium under vacuum, and observed using an accelerating voltage of 5 kV. Image J Launcher v 1.41 software was used to determine the average fiber diameter and standard deviation by measuring the diameter of at least 100 fibers.

2.3.3. Transmission Electronic Microscopy

The morphology and distribution of the PdNPs was studied using a JEOL 1010 from JEOL USA, Inc. (Peabody, MA, USA) by TEM using an accelerating voltage of 80 kV.

2.3.4. Differential Scanning Calorimetry

Thermal properties of the neat electrospun PHB fibers and films and of the multilayer systems were evaluated by differential scanning calorimetry (DSC) using a Perkin-Elmer DSC 8000 (Waltham, MA, USA) thermal analysis system under nitrogen atmosphere. The measurement was carried out on ~3 mg of each sample using a two-step program from 0 to 200 °C followed by a subsequent cooling down to −50 °C, both at a heating rate of 10 °C/min. The DSC equipment was previously calibrated with indium as a standard and the slope of the thermograms was corrected by subtracting similar scans of an empty pan. Tests were done, at least, in triplicate.

2.3.5. Thermogravimetric Analysis

Thermogravimetric analysis (TGA) was performed in a TG-STDA Mettler-Toledo model TGA/STDA851e/LF/1600 analyzer from Mettler-Toledo, LLC (Columbus, OH, USA). The samples, with an initial weight typically about 15 mg, were heated from 50 to 500 °C at a heating rate of 10 °C/min under nitrogen/air flow.

2.3.6. Infrared Spectroscopy

Fourier transform infrared spectroscopy (FTIR) spectra were collected coupling the attenuated total reflection (ATR) accessory Golden Gate of Specac, Ltd. (Orpington, UK) to Bruker Tensor 37 FTIR equipment (Rheinstetten, Germany). Single spectra were collected in the wavelength range from 4000 to 600 cm^{-1} by averaging 20 scans at a resolution of 4 cm^{-1}.

2.3.7. Mechanical Tests

Dumbbell-shaped specimens were die-cut from the electrospun films and conditioned to ambient conditions, i.e., 25 °C and 50% RH, for 24 h prior to tensile testing. Tests were carried out at room temperature in a universal mechanical testing machine AGS-X 500N from Shimadzu Corp. (Kyoto, Japan) in accordance with ASTM D638 (Type IV) standard. This was equipped with a 1-kN load cell and the cross-head speed was set at 10 mm/min. A minimum of six specimens was measured for each sample and the average results with standard deviation were reported.

2.3.8. Water Vapor Permeability

The water vapor permeability (WVP) of the samples was determined, in triplicate, using the gravimetric method ASTM E96-95. To this end, 5 mL of distilled water was placed inside a Payne permeability cup (Ø = 3.5 cm) from Elcometer Sprl (Hermalle-sous-Argenteau, Belgium) to expose the film to 100% RH on one side. The liquid was not in contact with the film. Once the films were secured with silicon rings, they were placed within a desiccator at 0% RH cabinet at 25 °C. The dryness of the cabinet was held constant using dried silica gel. Cups with aluminum films were used as control samples to estimate solvent loss through the sealing. The cups were weighted periodically using an analytical balance with a ±0.0001 g accuracy. Water vapor permeation was calculated from the steady-state permeation slopes obtained from the regression analysis of weight loss data vs. time, and the weight loss was calculated as the total loss minus the loss through the sealing.

2.3.9. D-Limonene Permeability

Permeability to limonene vapor was measured as described above for WVP. For this, 5 mL of D-limonene was placed inside the Payne permeability cups. The cups containing the films were placed at controlled environmental conditions, i.e., 23 °C and 40% RH. Limonene vapor permeation rates were estimated from the steady-state permeation slopes and the weight loss was calculated as the total cell loss minus the loss through the sealing. The samples were measured in triplicate and the limonene permeability (LP) values were calculated taking into account the average film thickness in each case.

2.3.10. Oxygen Scavenging Activity

Round-bottom Schlenk flasks from VidraFoc S.A. (Barcelona, Spain) with a PTFE stopcock and a headspace volume of 50 cm^3 were used for the oxygen scavenging measurements. The flasks contained a valve for gas flushing in and a O_2-sensitive sensor spot (PSt3, detection limit 15 ppb, 0–100% oxygen) from PreSens (Regensburg, Germany) was glued onto the inner side of the flasks for measuring oxygen depletion. The electrospun fibers and films were cut (5×5 cm^2) and placed into the flasks. The flask was subsequently flushed for 30 s at 1 bar with a gas mixture containing 1 vol. % oxygen, 4 vol. % hydrogen, and 95 vol. % nitrogen, which was provided by Abelló Linde, S.A. (Barcelona, Spain). The oxygen concentration in the cell was monitored by a non-destructive measurement method using the OXY-4 mini (PreSens) multi-channel fiber optic oxygen meter for simultaneous read-out of up to 4 oxygen sensors, used with sensors based on a 2 mm optical fiber. Oxygen concentrations over time were measured by linking the light-emitting (600–660 nm) optical fibers to the flasks inner sensing spots. The sensor emits a certain amount of luminescence depending on the oxygen concentration in the cell that was calibrated to yield concentration by the equipment. All measurements were carried out at 23 °C and at 50% and 100% RH.

2.4. Statistical Analysis

The test data were evaluated through analysis of variance (ANOVA) using STATGRAPHICS Centurion XVI v 16.1.03 from StatPoint Technologies, Inc. (Warrenton, VA, USA). Fisher's least significant difference (LSD) was used at the 95% confidence level ($p < 0.05$). Mean values and standard deviations were also calculated.

3. Results and Discussion

3.1. Morphology and Optical Properties

3.1.1. Optical Appearance

Figure 1 shows the contact transparency pictures of the electrospun PHB fibers, Figure 1a–c, as well as of their respective annealed films, Figure 1d–f. From these pictures, it can be observed that all the electrospun fiber mats were completely opaque due to the ultrathin size of the fibers that generate a significant level of porosity and hence refract the light very strongly [21]. On the other hand, the annealed films presented an improved contact transparency, specially the sample with CTAB. Due to the presence of the PdNPs, the films developed an expected dark color.

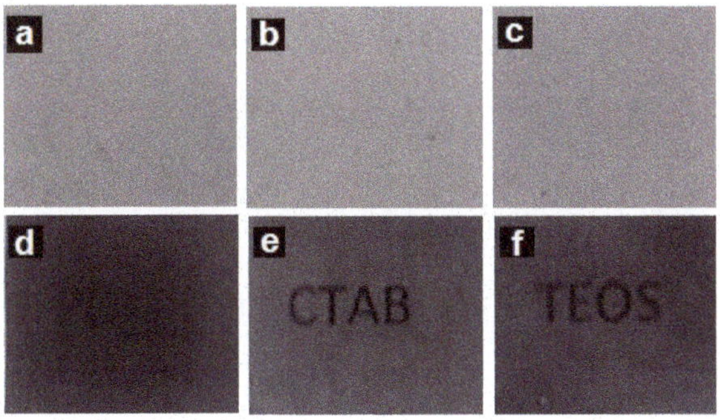

Figure 1. Contact transparency pictures of the electrospun poly(3-hydroxybutyrate) (PHB) fibers containing palladium nanoparticles (PdNPs) and their respective annealed films: (**a**) PHB/PdNP fibers; (**b**) PHB/PdNP/hexadecyltrimethylammonium bromide (CTAB) fibers; (**c**) PHB/PdNP/tetraethyl orthosilicate (TEOS) fibers; (**d**) PHB/PdNP film, (**e**) PHB/PdNP/CTAB film, (**f**) PHB/PdNP/TEOS film.

3.1.2. Morphology of Electrospun PHB Materials

The morphology of the electrospun fibers and their annealed films were studied by SEM. Representative images of all the electrospun samples are gathered in Figure 2. The images were taken from the surface and cross-sections of the obtained fibers and films. As shown in Figure 2a, the diameter of the neat electrospun PHB fibers was distributed primarily in the range of 200–600 nm, presenting a smooth and bead-free morphology. In particular, the mean diameter was found to be at 350 ± 147 nm. One can observe that the presence of the PdNPs led to a fraction with increased fiber diameter and also resulted in a spindle-type beads formation. This effect can be observed in Figure 2d,g, corresponding to the electrospun PHB/PdNP and CTAB-containing PHB/PdNP fibers, respectively. This observation may suggest that, in some fibrilar regions, partial agglomeration of the PdNPs may occur. Indeed, some degree of agglomeration is a common phenomenon in composites due to the large surface area and high total surface energy associated with nanoparticles incorporated into polymer matrices that makes them amenable to clustering [22]. The nanofibers diameter was previously reported to decrease when the surfactant concentration increased in the electrospinning solution [23]. Interestingly, the TEOS-containing PHB/PdNP fibers, shown in Figure 2j, presented a significant fraction of the fibers with reduced fiber diameter. This effect has been previously described for other electrospun materials and it has been particularly attributed to both the expected decrease in surface tension and an increase in conductivity, which in turn produce an increase in the stretching forces in the jet and consequently decreases the fiber diameter [24–26].

From the SEM images of the fibers cross-sections, one can observe that Figure 2b,e, corresponding to the neat PHB and PHB/PdNP fibers, respectively, showed similar morphologies. In particular, both electrospun mats presented cross-sections with relatively high porosity and low compaction. Alternatively, the cross-section of the surfactant-containing PHB/PdNP fibers, included in Figure 2h,k, were seen to be more compacted since the adhesion among the fibers in the layered structure was higher. For all layers, it was also possible to perceive some particles aggregation that may not be necessarily related to the presence of the PdNPs but, more probably, to additives such as the nucleating agent boron nitride, originally included in the biopolymer by the manufacturer.

In the SEM images of the films cross-sections, shown in Figure 2c,f,i,l, it can be observed that the electrospun fibers fused and interconnected among each other after the annealing treatment at 160 °C, successfully leading to the packing of the fiber mat into a continuous film. Among the here-prepared films, the neat PHB film showed a more uniform, smooth, and homogeneous surface. After the addition of the PdNPs and surfactants, the films became more heterogeneous, rougher, and also presented some cavities. This morphology change can be then related to the presence of the PdNPs, which more likely interfered in the fibers coalescence process.

3.1.3. Dispersion of PdNPs

In order to provide a more resolved information about the dispersion of the PdNPs into the PHB biopolymer matrix, TEM was performed on the nanocomposite fibers and films. The distribution of the PdNPs inside the electrospun fibers are illustrated in the TEM images included in Figure 3a,d,g. From the TEM image included in Figure 3a, one can clearly discern that the PdNPs were mainly agglomerated in certain regions of the submicron PHB fibers. One can also observe that the addition of both surfactants, i.e., CTAB and TEOS, as respectively shown in Figure 3d,g, successfully improved the PdNPs dispersion in the PHB fibers. Dispersion and distribution was, however, seen higher in the CTAB-containing sample.

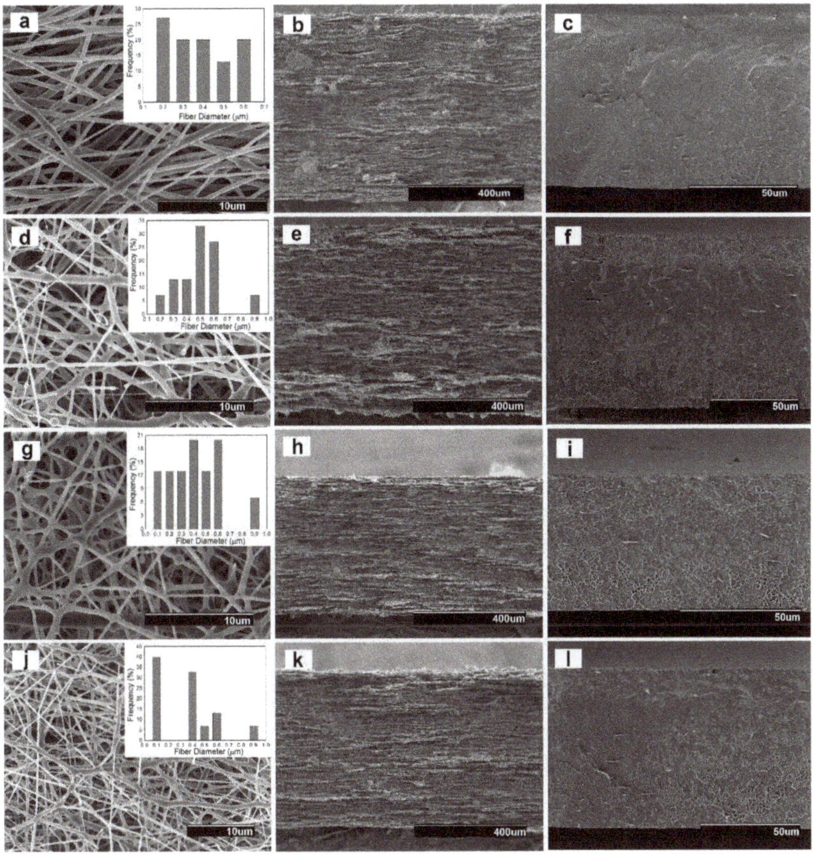

Figure 2. Scanning electron microscopy (SEM) images taken on the surface views and cross-sections of the electrospun poly(3-hydroxybutyrate) (PHB) fibers containing palladium nanoparticles (PdNPs) and their respective annealed films: (**a**) Surface view of the neat PHB fibers; (**b**) Cross-section of the neat PHB fibers; (**c**) Cross-section of the neat PHB film; (**d**) Surface view of the PHB/PdNP fibers; (**e**) Cross-section of the PHB/PdNP fibers; (**f**) Cross-section of the PHB/PdNP film; (**g**) Surface view of the PHB/PdNP/hexadecyltrimethylammonium bromide (CTAB) fibers; (**h**) Cross-section of the PHB/PdNP/CTAB fibers; (**i**) Cross-section of the PHB/PdNP/CTAB film; (**j**) Surface view of the PHB/PdNP/tetraethyl orthosilicate (TEOS) fibers; (**k**) Cross-section of the PHB/PdNP/TEOS fibers; (**l**) Cross-section of the PHB/PdNP/TEOS film.

In relation to the annealed films, the PHB/PdNP film without surfactant, shown in Figure 3b,c, still presented a clear aggregation of the particles. As opposite, Figure 3e,f, for the CTAB-containing film sample, and Figure 3h,i, for the TEOS-containing film sample, clearly showed that the PdNPs were evenly and relatively well distributed in the PHB films without forming agglomerates. In the case of the TEOS-containing film sample, the nanoparticle dispersion was less uniform when compared to the sample prepared with CTAB surfactant due to the absence of large grey dark areas. In both PHB films, very small nanoparticles of approximately 5 ± 2 nm can be seen, being homogeneously dispersed along the biopolymer matrix. The present results are in agreement with the results showed by Shaukat et al. [27], where PdNPs were incorporated into polyamide 6 (PA6)/clay nanocomposites and the nanoparticles were largely separated from each other and oriented in all possible directions in the polymer matrix. However, due to the absence of surfactant, some particles still agglomerated into clusters of bigger sizes that increased at the PA6 interfaces with the nanoclays.

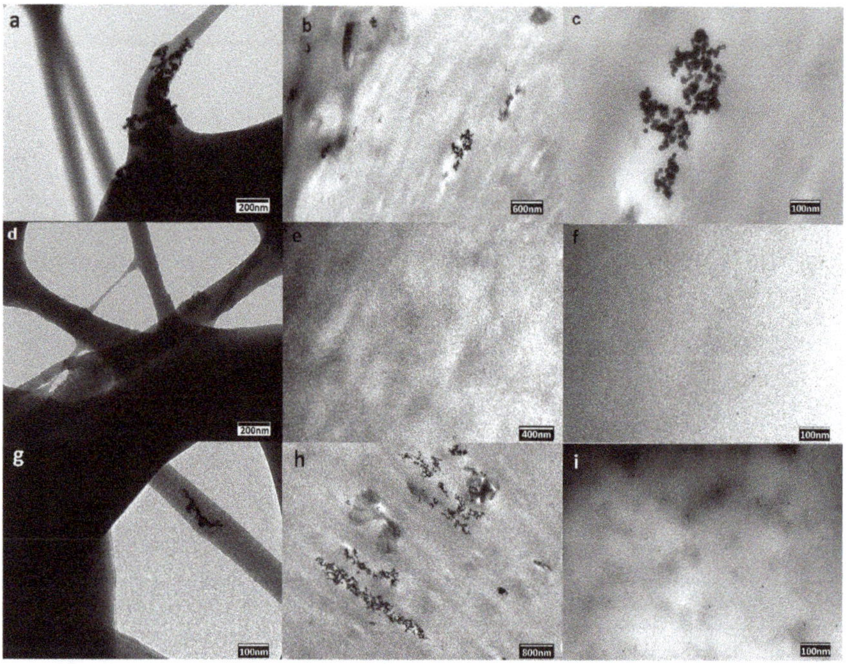

Figure 3. Transmission electron microscopy (TEM) images of the electrospun poly(3-hydroxybutyrate) (PHB) fibers containing palladium nanoparticles (PdNPs) and their respective annealed films: (**a**) PHB/PdNP fibers; (**b**,**c**) PHB/PdNP film; (**d**) PHB/PdNP/ hexadecyltrimethylammonium bromide (CTAB) fibers; (**e**,**f**) PHB/PdNP/CTAB film; (**g**) PHB/PdNP/tetraethyl orthosilicate (TEOS) fibers; (**h**,**i**) PHB/PdNP/TEOS film.

3.2. *Thermal Properties*

3.2.1. Melting Profile

The thermal properties of the electrospun PHB fibers and films containing the PdNPs nanoparticles were firstly investigated by DSC analysis. Table 1 gathers the melting temperatures (T_{m1} and T_{m2}) and enthalpies (ΔH_m) obtained from the first heating run. Likewise, the crystallization temperature (T_c) was also obtained from the cooling run. One can observe that when thermal annealing was applied to the fibers, the melting profile of the PHB materials, both temperature and enthalpy, slightly decreased. In particular, the T_m values of the neat PHB-based fibers varied in the 168–170 °C range, melting in a single peak, whereas these values were around 3 °C lower in their counterpart film samples. In addition, the incorporation of the PdNPs into PHB induced a slight decrease in both the melting temperature and enthalpy as well as resulted in the formation of multiple melting peaks. This observation may suggest that the nanoparticles interfered with the crystallization process of the homopolyester, producing more imperfect crystals composed of thinner lamellae that melted over a wider range at lower temperatures and with lower enthalpies [28]. This effect was more intense in the case of the surfactant-containing film samples, which suggests that the PdNPs were better dispersed then highly influencing the packing process of the PHB chains during cooling. In relation to the crystallization from the melt, the PHB film samples incorporating the PdNPs also showed slightly lower values of T_c than the fibers, which confirms that the nanoparticles provided an anti-nucleating effect on PHB during the film formation. In any case, given that PHB is known to exhibit cold crystallization during the dynamic DSC runs, it becomes difficult to discuss with certainty potential changes in crystallinity and crystalline morphology [21,26].

Table 1. Thermal properties obtained from the differential scanning calorimetry (DSC) curves in terms of melting temperature (T_m), normalized melting enthalpy (ΔH_m), crystallization temperature (T_c), and normalized crystallization enthalpy (ΔH_c) for the various poly(3-hydroxybutyrate) (PHB) and palladium nanoparticles (PdNPs) fibers and films with and without hexadecyltrimethylammonium bromide (CTAB) and tetraethyl orthosilicate (TEOS).

Sample	T_{m1} (°C)	T_{m2} (°C)	ΔH_m (J/g)	T_c (J/g)	ΔH_c (J/g)
PHB Fibers	-	169.1 ± 0.9^{a}	64.1 ± 1.1^{a}	110.2 ± 0.9^{a}	$59.3 \pm 2.0^{b,c}$
PHB Film	-	168.4 ± 1.3^{a}	71.8 ± 1.3^{e}	110.5 ± 1.2^{a}	$61.1 \pm 0.4^{a,b}$
PHB/PdNP Fibers	$156.4 \pm 2.1^{a,b}$	$168.3 \pm 1.1^{a,b}$	$59.4 \pm 0.9^{c,d}$	$108.4 \pm 0.5^{a,b}$	63.1 ± 1.5^{a}
PHB/PdNP Film	155.0 ± 1.2^{a}	$168.4 \pm 1.2^{a,b}$	58.5 ± 0.8^{d}	109.5 ± 2.1^{a}	$56.3 \pm 1.0^{c,d}$
PHB/PdNP/CTAB Fibers	$156.7 \pm 0.5^{a,b}$	$167.7 \pm 0.9^{a,b}$	62.2 ± 1.4^{b}	$109.1 \pm 0.9^{a,b}$	$60.3 \pm 0.8^{a,b}$
PHB/PdNP/CTAB Film	158.6 ± 1.4^{c}	165.1 ± 1.4^{b}	60.3 ± 1.5^{c}	106.8 ± 0.3^{b}	55.1 ± 0.7^{d}
PHB/PdNP/TEOS Fibers	155.2 ± 0.7^{a}	169.9 ± 0.5^{a}	59.7 ± 0.9^{d}	110.8 ± 1.0^{a}	$60.1 \pm 1.1^{a,b}$
PHB/PdNP/TEOS Film	159.4 ± 1.9^{c}	165.5 ± 1.2^{b}	$62.9 \pm 0.8^{a,b}$	110.0 ± 2.3^{a}	51.4 ± 1.8^{e}

a–e: Different superscripts within the same column indicate significant differences among the samples ($p < 0.05$).

3.2.2. Thermal Stability

Figure 4 shows the evolution of the mass loss as a function of temperature for the PHB samples, including the curves of the first derivative analysis (blue lines). As shown in Table 2, it can be seen that the neat PHB initiated degradation at 207 °C while the biopolymer fully decomposed in two steps, seen at 262 °C and 348 °C, providing a residual mass of approximately 3%. Thermal degradation of the PHB films containing the PdNPs and the surfactants also occurred in two stages with a slight increase in thermal stability. In particular, the onset degradation took place in the 220–230 °C range that continued to the first degradation stage at 270–275 °C and, at a slower rate, to over 360–380 °C, remaining a residue of 3–4% of the initial mass of the sample. In this sense, Díez-Pascual et al. [29] demonstrated that the incorporation of zinc oxide (ZnO) improved the heat resistance of PHB, which was ascribed to the barrier effect of the nanoparticles that effectively hindered the diffusion of decomposition products during thermal degradation. Therefore, the PdNPs probably also functioned as a thermal barrier, absorbing heat, thus also resulting in an enhanced thermal stability.

It is also worthy to note the slight increase observed in the thermal stability of the PHB/PdNP films with the incorporation of both surfactants, especially in the case of CTAB. This suggests that the surfactant increased the matrix–nanoparticle interaction, inducing a positive delay in thermal degradation. Taking into account that the ammonium salts degrade at the temperature range of 150–460 °C, the surfactants could successfully provide a bonding effect between both components of the nanocomposite and, eventually, enhance the whole thermal stability of PHAs.

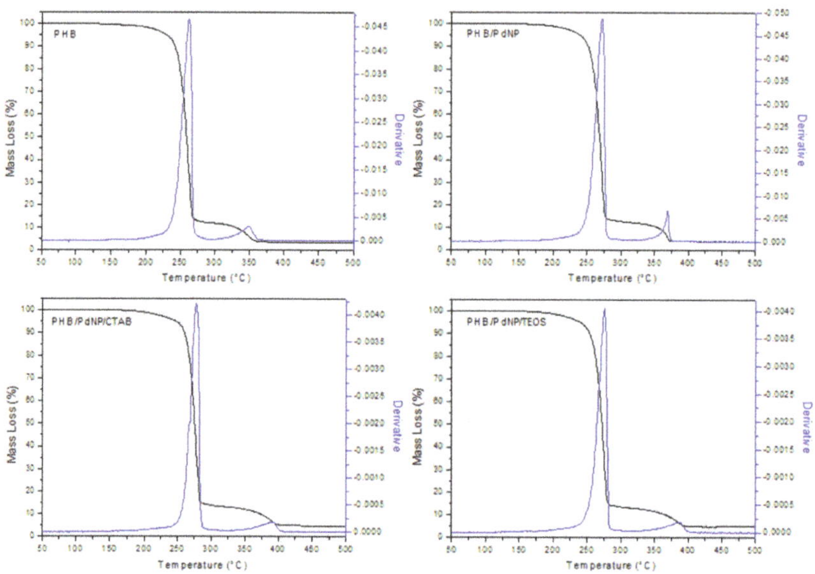

Figure 4. Thermogravimetric analysis (TGA) curves of the electrospun poly(3-hydroxybutyrate) (PHB) and palladium nanoparticles (PdNPs) films with and without hexadecyltrimethylammonium bromide (CTAB) and tetraethyl orthosilicate (TEOS) surfactants.

Table 2. Mean values of thermal stability obtained from the thermogravimetric analysis (TGA) curves of the electrospun poly(3-hydroxybutyrate) (PHB) and palladium nanoparticles (PdNPs) films with and without hexadecyltrimethylammonium bromide (CTAB) and tetraethyl orthosilicate (TEOS) in terms of degradation temperature at 5% of mass loss ($T_{5\%}$), degradation temperature (T_{deg}), and residual mass at 500 °C (R_{500}).

Film Sample	$T_{5\%}$ (°C)	T_{deg1} (°C)	T_{deg2} (°C)	R_{500} (%)
PHB	207.0 ± 4.4	262.0 ± 3.8	348.0 ± 6.4	3.08 ± 0.04
PHB/PdNP	224.1 ± 3.6	271.3 ± 2.6	368.2 ± 3.8	3.17 ± 0.03
PHB/PdNP/CTAB	220.0 ± 8.1	278.1 ± 3.4	392.2 ± 5.4	3.81 ± 0.04
PHB/PdNP/TEOS	230.2 ± 2.5	275.2 ± 4.1	386.1 ± 2.3	4.44 ± 0.04

3.3. FTIR Analysis

FTIR is a powerful tool, which is very sensitive to the molecular environment, to investigate the structural changes that occur in the material during any chemical or physical process. In Figure 5, the FTIR spectra are presented for the above-mentioned PHB-based fibers and films. From the given spectra of the PHB materials, the band centered at 1722 cm^{-1} is an indicative of the C=O stretching vibration for the biopolyester molecule. The absorption bands in the 1200–1300 cm^{-1} range were previously related to the presence of the C–O–C stretching vibrations whereas the band centered at approximately 980 cm^{-1} has been ascribed to stretching bands of the carbon–carbon single bond (C–C) in PHAs [21,30]. In the spectra of the pure surfactants, it is interesting to be noted that the bands at 2873 cm^{-1} and 2977 cm^{-1} correspond to the anti-symmetric C–H stretching of CTAB [19] while the bands located at 1168 cm^{-1}, 1103 cm^{-1}, and 1080 cm^{-1} have been assigned to the CH_3 rocking and C–O asymmetric stretching of TEOS, respectively [31]. For the surfactant-containing PHB films, the above surfactant bands or others could not be detected most likely due to the low concentration in which these additives are present in the composite.

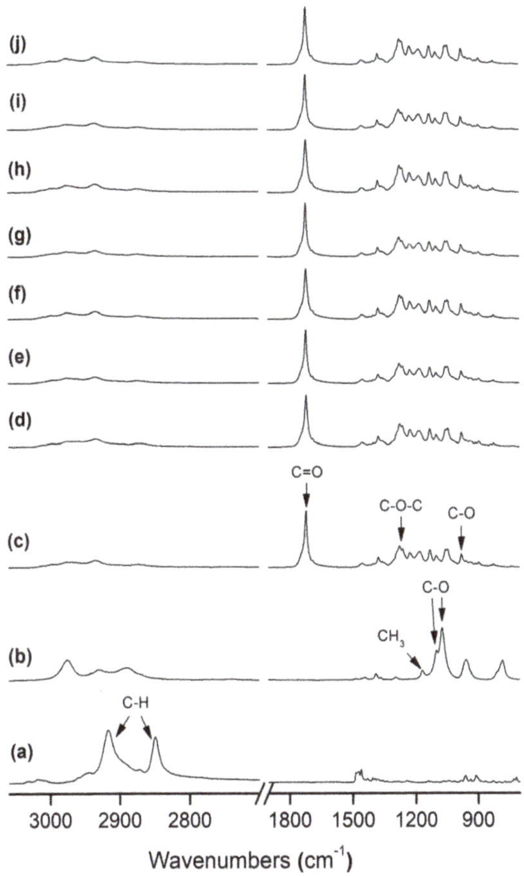

Figure 5. Fourier transform infrared (FTIR) spectra of: (**a**) Hexadecyltrimethylammonium bromide (CTAB); (**b**) Tetraethyl orthosilicate (TEOS); (**c**) Poly(3-hydroxybutyrate) (PHB) fibers; (**d**) PHB film; (**e**) PHB/palladium nanoparticles (PdNP) fibers; (**f**) PHB/PdNP film; (**g**) PHB/PdNP/CTAB fibers; (**h**) PHB/PdNP/CTAB film; (**i**) PHB/PdNP/TEOS fibers; (**j**) PHB/PdNP/TEOS film. Arrows indicate the chemical bonds and/or groups discussed in the text.

Table 3 gathers the band area ratio A1230:A1453 and the 1722 cm^{-1} carbonyl band width at half maximum, which have been recently associated with crystallinity content in electrospun PHB materials [21]. Both a 1722 cm^{-1} band broadening and reduction in the A1230:A1453 band area ratio have been previously connected with a reduction in molecular order and, hence, in crystallinity. Figure 5 and also Table 3 indicate that the band at 1722 cm^{-1} tended to broaden somehow in the case of the film samples as compared to their respective fibers, which can be indicative of a phenomenon of molecular disorder due to thermal treatment in the material. The band ratio seems to be less sensitive, as previously discussed [21], than the carbonyl band. These results may be in close agreement with those previously reported by Pachekoski et al. [32], in which it was indicated that band widening with annealing correlates with the changes in molecular backbone stability and, hence, crystallinity in PHB. Similarly, Mottin et al. [33] also reported a change in crystallinity in PHB by annealing process.

Table 3. Full width at half maximum (FWHM) of the carbonyl band centered at ~1722 cm^{-1} and the band area ratio A1230:A1453 for the electrospun poly(3-hydroxybutyrate) (PHB) and palladium nanoparticles (PdNPs) films with and without hexadecyltrimethylammonium bromide (CTAB) and tetraethyl orthosilicate (TEOS).

Sample	$FWHM_{1722}$ (cm^{-1})	A_{1230}:A_{1453}
PHB Fibers	16.20	4.28
PHB Film	16.35	4.03
PHB/PdNP Fibers	15.10	4.24
PHB/PdNP Film	15.60	4.86
PHB/PdNP/CTAB Fibers	14.44	3.58
PHB/PdNP/CTAB Film	14.23	4.28
PHB/PdNP/TEOS Fibers	15.89	4.25
PHB/PdNP/TEOS Film	16.00	4.28

3.4. Mechanical Properties

Table 4 presents the mechanical properties, obtained from the tensile tests, of the electrospun PHB films. The incorporation of the PdNPs into PHB caused an increase in both the modulus of elasticity and tensile strength, therefore increasing the elastic deformation and stiffness of the PHB film. The enhancement in mechanical resistance attained in the nanocomposite films can be attributed to the combination of fairly good nanoparticle dispersion and strong interfacial adhesion between both phases through interactions via H-bonding of PHB [29]. In relation to elongation at break, all films presented values of around 3%, which confirms the intrinsic brittleness of PHB. In any case, the presence of the PdNPs did not alter the film ductility and toughness characteristics of PHB while the effect of the surfactants addition on their mechanical performance was also not statistically significant.

Table 4. Mechanical properties in terms of elastic modulus (E), tensile strength at break (σ_b), elongation at break (%ε_b), and toughness (T) of the electrospun poly(3-hydroxybutyrate) (PHB) and palladium nanoparticles (PdNPs) films with and without hexadecyltrimethylammonium bromide (CTAB) and tetraethyl orthosilicate (TEOS).

Sample	E (MPa)	σ_b (MPa)	ε_b (%)	T (mJ/m^3)
PHB film *	1104 ± 74^{a}	17.8 ± 1.8^{a}	2.9 ± 1.0^{a}	0.3 ± 0.1^{a}
PHB/PdNP	1255 ± 15^{b}	22.5 ± 4.2^{b}	3.1 ± 1.0^{a}	0.4 ± 0.1^{a}
PHB/PdNP/CTAB	1262 ± 14^{b}	23.3 ± 1.4^{c}	3.0 ± 0.1^{a}	0.4 ± 0.1^{a}
PHB/PdNP/TEOS	1288 ± 230^{c}	21.7 ± 4.1^{b}	2.7 ± 1.0^{a}	0.3 ± 0.2^{a}

a–c: Different superscripts within the same column indicate significant differences among the samples ($p < 0.05$). * Obtained in previous work [21].

3.5. Barrier Properties

3.5.1. Water Vapor Permeability

Measuring the loss or gain in water content is a common method to estimate the WVP of film samples. The WVP values of the electrospun PHB films are gathered in Figure 6. It can be observed that the neat PHB film showed a higher barrier performance to water vapor than their respective nanocomposites with the PdNPs. In particular, while the neat PHB film showed a WVP value of 5.2×10^{-15} kg·m·m^{-2}·Pa^{-1}·s^{-1}, this value was 1.2×10^{-14} kg·m·m^{-2}·Pa^{-1}·s^{-1} for the PHB/PdNP film. The observed permeability increase in the nanocomposite films can be related to the existence of not bonded interfacial regions acting as preferential paths, especially in the vicinity of agglomerates. These preferential pathways could accelerate the diffusion of gas molecules, thus increasing the diffusion coefficient [34]. Interestingly, the WVP values of the PHB/PdNP films was lower in the case of the CTAB-containing film, i.e., 8.0×10^{-15} kg·m·m^{-2}·Pa^{-1}·s^{-1}, and considerably higher for the film with TEOS, i.e., 6.6×10^{-14} kg·m·m^{-2}·Pa^{-1}·s^{-1}. This result suggests that the dispersion in the PHB/PdNP/CTAB film was higher and, then, the sizes of such unbonded interfacial regions were lower. In addition, the high WPV value observed for the PHB/PdNP/TEOS film suggests that this film sample could be also plasticized by the surfactant, increasing the free volume of the film and favoring the diffusion of water vapor molecules through the film sample [35,36].

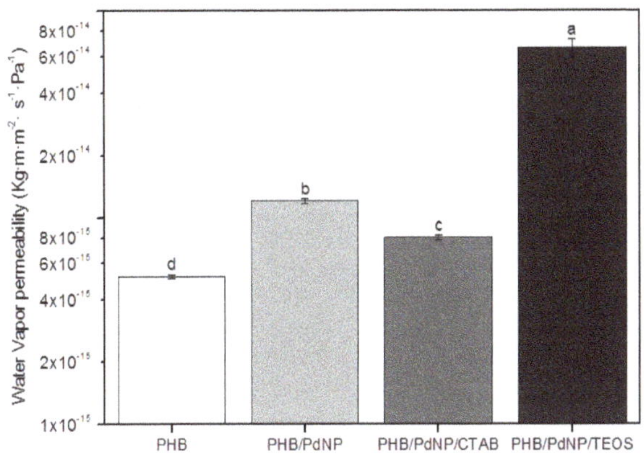

Figure 6. Values of water vapor permeability (WVP) of the electrospun poly(3-hydroxybutyrate) (PHB) and palladium nanoparticles (PdNPs) films with and without hexadecyltrimethylammonium bromide (CTAB) and tetraethyl orthosilicate (TEOS) surfactants. Different letters indicate significant differences among the samples ($p < 0.05$).

3.5.2. D-Limonene Permeability

To test the barrier performance for volatile compounds such as aromas, D-limonene is commonly used. Figure 7 shows the values of LP, where one can observe that the neat PHB film presented the lowest permeability for D-limonene with a value of 3.2×10^{-15} kg·m·m^{-2}·Pa^{-1}·s^{-1}. In the case of the nanocomposite films, the LP values increased from 4.7×10^{-15} kg·m·m^{-2}·Pa^{-1}·s^{-1}, for the PHB/PdNP film, to 9.6×10^{-15} and 9.0×10^{-15} kg·m·m^{-2}·Pa^{-1}·s^{-1}, for the CTAB- and TEOS-containing PHB/PdNP films, respectively. As discussed above, the presence of the PdNPs and their agglomerates may result in the creation of preferential paths for sorption and diffusion of the aroma molecules hence resulting in a reduced barrier performance. The here-obtained results are showing opposite behavior as the ones reported earlier by Busolo and co-workers [37] who dispersed silver nanoparticles (nAg) in PLA, yielding nanocomposites with enhanced barrier properties. Similarly, Rhim et al. [38] reported agar/nAg composites where they confirmed a substantial improvement in barrier properties of the composite. In the case of the surfactants-containing PHB/PdNP films, it should be also taken into account that permeability of D-limonene in PHB is mainly controlled by a solubilization mechanism due to the capacity of PHAs to uptake large amounts of this organic compound [39]. This supports the fact that plasticized PHB materials present increased values of aroma permeability.

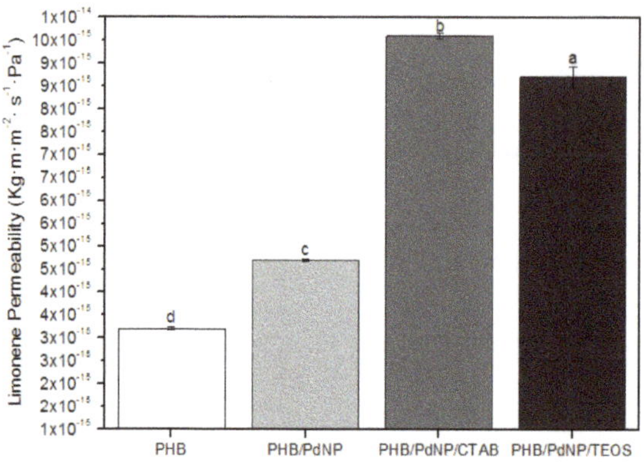

Figure 7. Values of D-limonene permeability (LP) of the electrospun poly(3-hydroxybutyrate) (PHB) and palladium nanoparticles (PdNPs) films with and without hexadecyltrimethylammonium bromide (CTAB) and tetraethyl orthosilicate (TEOS) surfactants. Different letters indicate significant differences among the samples ($p < 0.05$).

3.5.3. Oxygen Scavenging Activity

The oxygen scavenging activity of the here-prepared electrospun fibers and films containing the PdNPs was determined by measuring the oxygen scavenging rate (OSR). In relation to the electrospun fibers, Figure 8 shows the decay or depletion of the oxygen concentration as a function of time, for a span time of 800 min, at both 50% and 100% RH. From observation of the graph it can be seen that, while the neat PHB fibers were unable to reduce the amount of oxygen in the cell, comparatively, the free PdNPs in powder form were able to reduce all available headspace oxygen in an extremely short time. The incorporation of the PdNPs into the PHB fibers by electrospinning generated mat samples with intermediate oxygen scavenging activity. However, as it can also be observed in the graph, the performance of the developed nanocomposite fibers was strongly dependent on the RH conditions. All electrospun mats presented a significantly lower oxygen scavenging activity at 50% than at 100% RH. For instance, at the end of the experiment carried out at 50% RH, oxygen depletion varied from almost 40%, for the PHB/PdNP/CTAB fibers, to only ca. 10%, for the PHB/PdNP/TEOS fibers. However, at 100% RH, the electrospun PHB/PdNP mat reached a reduction in the oxygen volume of approximately 85% while both surfactant-containing PHB/PdNP fibers were able to fully consume the whole amount of oxygen after 800 min. It is also worthy to mention that the depletion rate was faster in the case of the PHB/PdNP/CTAB mat, which further confirmed the higher dispersion achieved for the nanoparticles with this surfactant. In relation to the effect of humidity, it is known that moisture favors the catalytic activity of the PdNPs, which can be mainly related to the fact that water can be associatively adsorbed directly on the PdNP surface and thereby interact with the adsorbed hydrogen and oxygen [10]. A possible mechanism is that the adsorbed atomic oxygen and hydrogen forms an OH intermediate that reacts with an adsorbed hydrogen atom or another OH molecule. Another possible explanation is the reaction of adsorbed oxygen with gas-phase hydrogen or with some kind of dihydrogen species weakly adsorbed on the surface. Finally, a concerted reaction of two adsorbed hydrogen atoms and an adsorbed oxygen atom has been also considered [40].

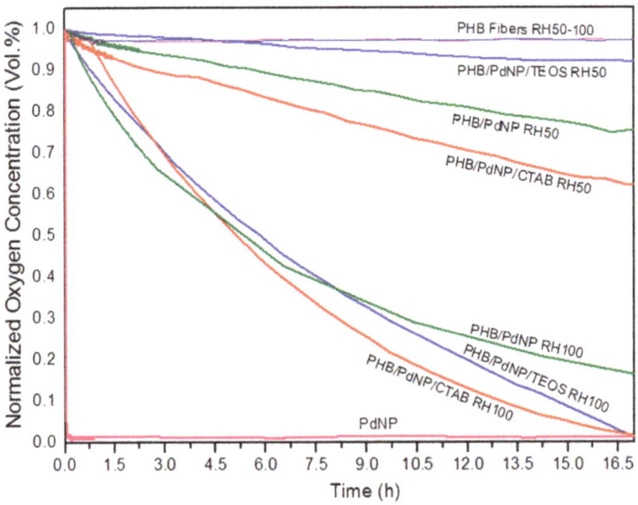

Figure 8. Oxygen depletion of the electrospun poly(3-hydroxybutyrate) (PHB) and palladium nanoparticles (PdNPs) fibers with and without hexadecyltrimethylammonium bromide (CTAB) and tetraethyl orthosilicate (TEOS) surfactants. Values were measured at 50% and 100% relative humidity (RH).

In Figure 9, the oxygen volume depletion obtained with the electrospun films at 100% RH are shown. Comparison between Figures 8 and 9 revealed that the oxygen scavenging effect of the films was considerably lower than that of the same material in the fiber form. This reduction in the OSR is related to the higher surface-to-volume ratio of the electrospun fibers than the films, since the fibers mats present an extremely high porosity. In any case, all PHB/PdNP films still presented significant oxygen scavenging capacity and, among the samples tested, the CTAB-containing films showed the highest performance as expected in view of all the above observations. This result can be explained by the better dispersion of the PdNPs achieved in the PHB matrix using CTAB. In agreement with the data reported here, Ahalawat and co-workers [41] evaluated the simultaneous effects of cationic surfactants on the textural and structural properties of silica nanoparticles. It was observed that the silica nanoparticles displayed better dispersion and lower size than those prepared with other two cationic surfactants using an aqueous TEOS precursor solution with CTAB. In relation to this, it has also been reported that an SiOx matrix between a coating of palladium and the substrate presents higher values of OSR than the palladium coated directly onto PET films [42]. These findings were later confirmed on PLA [10].

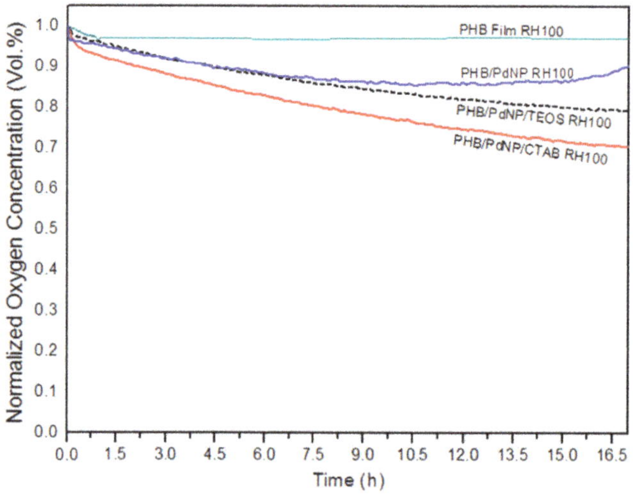

Figure 9. Oxygen depletion of the electrospun poly(3-hydroxybutyrate) (PHB) and palladium nanoparticles (PdNPs) films with and without hexadecyltrimethylammonium bromide (CTAB) and tetraethyl orthosilicate (TEOS) surfactants. Values were measured at 100% relative humidity (RH).

4. Conclusions

In this study, PdNPs were mixed with CTAB and TEOS surfactants to have better dispersion in electrospun PHB fibers. The resultant fibers were annealed at 160 °C to form continuous PHB films of direct application interest. Morphological analysis carried out by SEM and TEM showed that a better dispersion was achieved for the electrospun PHB/PdNP/CTAB film. DSC indicated that the presence of the PdNPs reduced both the melting point and the degree of crystallinity of PHB, thus acting as an anti-nucleating agent, which was further confirmed by FTIR analysis. WVP and LP measurements indicated that the nanocomposite films, including those modified with surfactants, presented lower barrier performance than the neat PHB film. These results were ascribed to the reduced crystallinity degree and to existence of unbonded interfacial regions and/or voids between the biopolymer matrix and the inorganic nanoparticles that may serve as preferential ways for the diffusion of gas molecules. Finally, the oxygen scavenging activity of the PHB materials was evaluated at different RHs. Although the electrospun films presented lower capacity to absorb oxygen that their counterpart fibers, these still presented significant activity at 100% RH. The here-developed and -characterized electrospun PHB films are suitable potential candidates as coatings or interlayer systems for active food packaging applications and the followed methodology represents a new route to prepare these films due to the relative high dispersion achieved of the PdNPs.

Author Contributions: Conceptualization was devised by J.M.L.; Methodology, Validation, and Formal Analysis was carried out by A.C., M.G., H.T.S., S.T.-G., and J.M.L. Investigation, Resources, Data Curation, Writing-Original Draft Preparation and Writing-Review & Editing was performed by A.C., M.G., H.T.S., S.T.-G., and J.M.L.; Supervision J.M.L; Project Administration, J.M.L.

Funding: This research has received funding from the Spanish Ministry of Economy and Competitiveness (MINECO, project AGL2015-63855-C2-1-R) and the EU H2020 project YPACK (reference number 773872).

Acknowledgments: A.C. and S.T.-G. would like to thank the Brazilian Council for Scientific and Technological Development (CNPq) and MINECO for her predoctoral grant (205955/2014-2) and his Juan de la Cierva contract (IJCI-2016-29675), respectively.

Conflicts of Interest: The authors declare no conflicts of interest.

References

1. Puglia, D.; Fortunati, E.; D'amico, D.A.; Manfredi, L.B.; Cyras, V.P.; Kenny, J. Influence of organically modified clays on the properties and disintegrability in compost of solution cast poly (3-hydroxybutyrate) films. *Polym. Degrad. Stab.* **2014**, *99*, 127–135. [CrossRef]
2. Ma, P.; Xu, P.; Chen, M.; Dong, W.; Cai, X.; Schmit, P.; Spoelstra, A.; Lemstra, P. Structure–property relationships of reactively compatibilized phb/eva/starch blends. *Carbohydr. Polym.* **2014**, *108*, 299–306. [CrossRef] [PubMed]
3. Molinaro, S.; Romero, M.C.; Boaro, M.; Sensidoni, A.; Lagazio, C.; Morris, M.; Kerry, J. Effect of nanoclay-type and pla optical purity on the characteristics of pla-based nanocomposite films. *J. Food Eng.* **2013**, *117*, 113–123. [CrossRef]
4. Imre, B.; Pukánszky, B. Compatibilization in bio-based and biodegradable polymer blends. *Eur. Polym. J.* **2013**, *49*, 1215–1233. [CrossRef]
5. Bittmann, B.; Bouza, R.; Barral, L.; Diez, J.; Ramirez, C. Poly (3-hydroxybutyrate-*co*-3-hydroxyvalerate)/clay nanocomposites for replacement of mineral oil based materials. *Polym. Compos.* **2013**, *34*, 1033–1040. [CrossRef]
6. Castro-Mayorga, J.; Fabra, M.; Lagaron, J.M. Stabilized nanosilver based antimicrobial poly (3-hydroxybutyrate-*co*-3-hydroxyvalerate) nanocomposites of interest in active food packaging. *Innov. Food Sci. Emerg. Technol.* **2016**, *33*, 524–533. [CrossRef]
7. Bartczak, Z.; Galeski, A.; Kowalczuk, M.; Sobota, M.; Malinowski, R. Tough blends of poly(lactide) and amorphous poly([R,S]-3-hydroxy butyrate)–morphology and properties. *Eur. Polym. J.* **2013**, *49*, 3630–3641. [CrossRef]
8. Furukawa, T.; Sato, H.; Murakami, R.; Zhang, J.; Duan, Y.-X.; Noda, I.; Ochiai, S.; Ozaki, Y. Structure, dispersibility, and crystallinity of poly (hydroxybutyrate)/poly (L-lactic acid) blends studied by ft-ir microspectroscopy and differential scanning calorimetry. *Macromolecules* **2005**, *38*, 6445–6454. [CrossRef]
9. Zhang, M.; Thomas, N.L. Blending polylactic acid with polyhydroxybutyrate: The effect on thermal, mechanical, and biodegradation properties. *Adv. Polym. Technol.* **2011**, *30*, 67–79. [CrossRef]
10. Yildirim, S.; Röcker, B.; Rüegg, N.; Lohwasser, W. Development of palladium-based oxygen scavenger: Optimization of substrate and palladium layer thickness. *Packag. Technol. Sci.* **2015**, *28*, 710–718. [CrossRef]
11. Cernohorsky, O.; Zdansky, K.; Zavadil, J.; Kacerovsky, P.; Piksova, K. Palladium nanoparticles on inp for hydrogen detection. *Nanoscale Res. Lett.* **2011**, *6*, 410. [CrossRef] [PubMed]
12. Damaj, Z.; Joly, C.; Guillon, E. Toward new polymeric oxygen scavenging systems: Formation of poly (vinyl alcohol) oxygen scavenger film. *Packag. Technol. Sci.* **2015**, *28*, 293–302. [CrossRef]
13. Wendorff, J.H.; Agarwal, S.; Greiner, A. *Electrospinning: Materials, Processing, and Applications*; John Wiley & Sons, Inc.: Hoboken, NJ, USA, 2012.
14. Uyar, T.; Kny, E. *Electrospun Materials for Tissue Engineering and Biomedical Applications: Research, Design and Commercialization*; Woodhead Publishing: New York, NY, USA, 2017; pp. 1–428.
15. Torres-Giner, S.; Pérez-Masiá, R.; Lagaron, J.M. A review on electrospun polymer nanostructures as advanced bioactive platforms. *Polym. Eng. Sci.* **2016**, *56*, 500–527. [CrossRef]
16. Torres-Giner, S. Electrospun nanofibers for food packaging applications. In *Multifunctional and Nanoreinforced Polymers for Food Packaging*; Woodhead Publishing: New York, NY, USA, 2011; pp. 108–125.
17. Echegoyen, Y.; Fabra, M.J.; Castro-Mayorga, J.L.; Cherpinski, A.; Lagaron, J.M. High throughput electro-hydrodynamic processing in food encapsulation and food packaging applications: Viewpoint. *Trends Food Sci. Technol.* **2017**, *60*, 71–79. [CrossRef]
18. Dainelli, D.; Gontard, N.; Spyropoulos, D.; Zondervan-van den Beuken, E.; Tobback, P. Active and intelligent food packaging: Legal aspects and safety concerns. *Trends Food Sci. Technol.* **2008**, *19*, S103–S112. [CrossRef]
19. Kundu, S. A new route for the formation of au nanowires and application of shape-selective au nanoparticles in sers studies. *J. Mater. Chem. C* **2013**, *1*, 831–842. [CrossRef]
20. Mayer, A.; Antonietti, M. Investigation of polymer-protected noble metal nanoparticles by transmission electron microscopy: Control of particle morphology and shape. *Colloid Polym. Sci.* **1998**, *276*, 769–779. [CrossRef]

21. Cherpinski, A.; Torres-Giner, S.; Cabedo, L.; Lagaron, J.M. Post-processing optimization of electrospun submicron poly(3-hydroxybutyrate) fibers to obtain continuous films of interest in food packaging applications. *Food Addit. Contam. Part A* **2017**, *34*, 1817–1830. [CrossRef] [PubMed]
22. Tekmen, C.; Tsunekawa, Y.; Nakanishi, H. Electrospinning of carbon nanofiber supported fe/co/ni ternary alloy nanoparticles. *J. Mater. Process. Technol.* **2010**, *210*, 451–455. [CrossRef]
23. Sainudeen, S.S.; Asok, L.B.; Varghese, A.; Nair, A.S.; Krishnan, G. Surfactant-driven direct synthesis of a hierarchical hollow mgo nanofiber–nanoparticle composite by electrospinning. *RSC Adv.* **2017**, *7*, 35160–35168. [CrossRef]
24. Sakai, S.; Kawakami, K.; Taya, M. Controlling the diameters of silica nanofibers obtained by sol–gel/electrospinning methods. *J. Chem. Eng. Jpn.* **2012**, *45*, 436–440. [CrossRef]
25. Castro-Mayorga, J.L.; Fabra, M.J.; Cabedo, L.; Lagaron, J.M. On the use of the electrospinning coating technique to produce antimicrobial polyhydroxyalkanoate materials containing in situ-stabilized silver nanoparticles. *Nanomaterials* **2016**, *7*, 4. [CrossRef] [PubMed]
26. Martínez-Abad, A.; Sanchez, G.; Lagaron, J.M.; Ocio, M.J. Influence of speciation in the release profiles and antimicrobial performance of electrospun ethylene vinyl alcohol copolymer (evoh) fibers containing ionic silver ions and silver nanoparticles. *Colloid Polym. Sci.* **2013**, *291*, 1381–1392. [CrossRef]
27. Shaukat, M.S.; Zulfiqar, S.; Sarwar, M.I. Incorporation of palladium nanoparticles into aromatic polyamide/clay nanocomposites through facile dry route. *Polym. Sci. Ser. B* **2015**, *57*, 380–386. [CrossRef]
28. Yeo, S.; Tan, W.; Bakar, M.A.; Ismail, J. Silver sulfide/poly (3-hydroxybutyrate) nanocomposites: Thermal stability and kinetic analysis of thermal degradation. *Polym. Degrad. Stab.* **2010**, *95*, 1299–1304. [CrossRef]
29. Díez-Pascual, A.M.; Díez-Vicente, A.L. Poly (3-hydroxybutyrate)/ZnO bionanocomposites with improved mechanical, barrier and antibacterial properties. *Int. J. Mol. Sci.* **2014**, *15*, 10950–10973. [CrossRef] [PubMed]
30. Torres-Giner, S.; Montanes, N.; Boronat, T.; Quiles-Carrillo, L.; Balart, R. Melt grafting of sepiolite nanoclay onto poly(3-hydroxybutyrate-*co*-4-hydroxybutyrate) by reactive extrusion with multi-functional epoxy-based styrene-acrylic oligomer. *Eur. Polym. J.* **2016**, *84*, 693–707. [CrossRef]
31. Primeau, N.; Vautey, C.; Langlet, M. The effect of thermal annealing on aerosol-gel deposited SiO_2 films: A ftir deconvolution study. *Thin Solid Films* **1997**, *310*, 47–56. [CrossRef]
32. Pachekoski, W.M.; Dalmolin, C.; Agnelli, J.A.M. The influence of the industrial processing on the degradation of poly(hidroxybutyrate)—PHB. *Mat. Res.* **2013**, *16*, 237–332. [CrossRef]
33. Mottin, A.C.; Ayres, E.; Oréfice, R.L.; Câmara, J.J.D. What changes in poly(3-hydroxybutyrate) (PHB) when processed as electrospun nanofibers or thermo-compression molded film? *Mater. Res.* **2016**, *19*, 57–66. [CrossRef]
34. Choudalakis, G.; Gotsis, A.D. Permeability of polymer/clay nanocomposites: A review. *Eur. Polym. J.* **2009**, *45*, 967–984. [CrossRef]
35. Terada, M.; Marchessault, R.H. Determination of solubility parameters for poly(3-hydroxyalkanoates). *Int. J. Biol. Macromol.* **1999**, *25*, 207–215. [CrossRef]
36. Tan, B.; Thomas, N.L. A review of the water barrier properties of polymer/clay and polymer/graphene nanocomposites. *J. Membr. Sci.* **2016**, *514*, 595–612. [CrossRef]
37. Busolo, M.A.; Fernandez, P.; Ocio, M.J.; Lagaron, J.M. Novel silver-based nanoclay as an antimicrobial in polylactic acid food packaging coatings. *Food Addit. Contam. Part A* **2010**, *27*, 1617–1626. [CrossRef] [PubMed]
38. Rhim, J.; Wang, L.; Hong, S. Preparation and characterization of agar/silver nanoparticles composite films with antimicrobial activity. *Food Hydrocoll.* **2013**, *33*, 327–335. [CrossRef]
39. Sanchez-Garcia, M.D.; Gimenez, E.; Lagaron, J.M. Morphology and barrier properties of solvent cast composites of thermoplastic biopolymers and purified cellulose fibers. *Carbohydr. Polym.* **2008**, *71*, 235–244. [CrossRef]
40. Nyberg, C.; Tengstål, C. Adsorption and reaction of water, oxygen, and hydrogen on pd (100): Identification of adsorbed hydroxyl and implications for the catalytic H_2–O_2 reaction. *J. Chem. Phys.* **1984**, *80*, 3463–3468. [CrossRef]

41. Singh, L.P.; Bhattacharyya, S.K.; Mishra, G.; Ahalawat, S. Functional role of cationic surfactant to control the nano size of silica powder. *Appl. Nanosci.* **2011**, *1*, 117–122. [CrossRef]
42. Nestorson, A.; Neoh, K.G.; Kang, E.T.; Järnström, L.; Leufvén, A. Enzyme immobilization in latex dispersion coatings for active food packaging. *Packag. Technol. Sci.* **2008**, *21*, 193–205. [CrossRef]

nanomaterials

Article

Oxygen-Scavenging Multilayered Biopapers Containing Palladium Nanoparticles Obtained by the Electrospinning Coating Technique

Adriane Cherpinski [1], Piotr K. Szewczyk [2], Adam Gruszczyński [2], Urszula Stachewicz [2] and Jose M. Lagaron [1,*]

[1] Novel Materials and Nanotechnology Group, Institute of Agrochemistry and Food Technology (IATA), Spanish Council for Scientific Research (CSIC), Calle Catedrático Agustín Escardino Benlloch 7, 46980 Paterna, Spain; adricherpinski@iata.csic.es

[2] AGH University of Science and Technology, International Centre of Electron Microscopy for Materials Science, Faculty of Metals Engineering and Industrial Computer Science, Al. A. Mickiewicza 30, 30-059 Kraków, Poland; pszew@agh.edu.pl (P.K.S.); gruszcz@agh.edu.pl (A.G.); ustachew@agh.edu.pl (U.S.)

* Correspondence: lagaron@iata.csic.es; Tel.: + 34-96-3900-022; Fax: +34-96-3636-301

Received: 11 January 2019; Accepted: 11 February 2019; Published: 14 February 2019

Abstract: The main goal of this study was to obtain, for the first time, highly efficient water barrier and oxygen-scavenging multilayered electrospun biopaper coatings of biodegradable polymers over conventional cellulose paper, using the electrospinning coating technique. In order to do so, poly(3-hydroxybutyrate) (PHB) and polycaprolactone (PCL) polymer-containing palladium nanoparticles (PdNPs) were electrospun over paper, and the morphology, thermal properties, water vapor barrier, and oxygen absorption properties of nanocomposites and multilayers were investigated. In order to reduce the porosity, and to enhance the barrier properties and interlayer adhesion, the biopapers were annealed after electrospinning. A previous study showed that electrospun PHB-containing PdNP did show significant oxygen scavenging capacity, but this was strongly reduced after annealing, a process that is necessary to form a continuous film with the water barrier. The results in the current work indicate that the PdNP were better dispersed and distributed in the PCL matrix, as suggested by focus ion beam-scanning electron microscopy (FIB-SEM) experiments, and that the Pd enhanced, to some extent, the onset of PCL degradation. More importantly, the PCL/PdNP nanobiopaper exhibited much higher oxygen scavenging capacity than the homologous PHB/PdNP, due to most likely, the higher oxygen permeability of the PCL polymer and the somewhat higher dispersion of the Pd. The passive and active multilayered biopapers developed here may be of significant relevance to put forward the next generation of fully biodegradable barrier papers of interest in, for instance, food packaging.

Keywords: polyhydroxyalkanoates; polycaprolactone; biopapers; palladium nanoparticles; oxygen scavengers; electrospinning; fiber based packaging

---

1. Introduction

Active technologies, when applied to packaging refer to the incorporation of certain additives into the packaging structure. These additives may be loose as sachets within the design, attached to the inside part or, more recently, dispersed as an additive within the packaging materials, in order to maintain or even extend product quality and shelf-life [1].

Permeated or head space oxygen in packaged foods, beverages, and pharmaceuticals can promote a range of oxidative degradation reactions and support microbial growth, ultimately impacting on product quality and shelf-life. Oxygen-scavenging active packaging systems have therefore been explored, to control headspace oxygen content [2].

The application of oxygen scavengers is one of the most important active packaging technologies, which aim to remove any residual oxygen that is present in the food packaging. In some cases, the residual levels of oxygen in the package can be reduced to < 0.01 vol %, and actively controlled, which is not possible with other packaging systems [3].

Oxygen scavengers are by far the most commercially important sub-categories of active packaging, and the market has been growing steadily over the last few years. Almost all oxygen scavenger sachets used commercially are based on the principle of iron oxidation. On the other hand, oxygen-scavenging film is a more promising emerging packaging technology, because it contains the active material within the film, and consumers are not in favor of having foreign objects such as sachets in the lining of their product packaging. With oxygen scavenger films, the consumer cannot physically see the oxygen scavenger materials, yet are able to experience its benefits [4–6].

The incorporation of scavengers into packaging films is a better way of resolving sachet-related problems. Scavengers may either be imbedded into a solid, dispersed in the plastic, or introduced into various layers of the package, including adhesive, lacquer, or enamel layers. Multi-layer oxygen scavengers more effectively absorb oxygen than single-layer scavenging systems [7].

Several new oxygen-scavenging technologies have been developed over the last decade, incorporating active substances and metals directly into packaging films or containers [8]. However, only a few of them have been successfully implemented in real food systems, due to, for instance, in the case of metals that function by chemical reduction, low reaction capacities and the need for triggering mechanisms, among other factors. Consequently, real application studies demonstrating the benefits of alternative oxygen-scavenging systems to particular food products are rather rare [9].

Recently, Hutter et al. [10] developed an oxygen scavenging film based on a catalytic system with palladium (CSP), which is able to reduce residual headspace oxygen very quickly. Palladium, in very low dosages, catalyzes the oxidation of hydrogen into water, and thus can remove the residual oxygen in the headspace of a modified atmosphere package containing hydrogen. Catalytic systems based on palladium have also been reported to have other interesting applications, such as the construction of complex molecules [11–14].

The main difficulties of this approach are the dispersion of the scavenger in the matrix, the accessibility of the scavenger to oxygen, and the necessity of an activation system for the oxygen absorption reaction. Without an activation system, the oxygen-scavenging capacity of the active film would be consumed during storage, before the packaging is used [15].

In addition, consumer trends for better quality, fresh-like, and convenient food products have intensified over the last decade. Therefore, a variety of active packaging technologies have been developed to provide better quality, wholesome, and safe foods, and also to limit package-related environmental pollution and disposal problems.

Recently, the environmental impact of persistent plastic packaging wastes is raising general global concern, since disposal methods are limited. Biopolymers have been considered as a potential environmentally-friendly substitute for the use of non-biodegradable and non-renewable plastic packaging materials [16].

Polycaprolactone (PCL) is petroleum-based, but it can be degraded by microorganisms, and the polyhydroxyalkanoate (PHA) homopolymer called poly(hydroxybutyrate) (PHB) is produced from biomass or renewable resources, and it is readily biodegradable [17]. The aim of this emerging and developing field is to change the nature of polymer products and to minimize the environmental impact. Various approaches are currently being investigated for possible polymers that may be utilized to design adequate environmentally friendly packaging [18].

Electrospinning is a feasible, efficient, and convenient technique for obtaining biopolymer-active nanofibers of interest in many application fields, such as active packaging, and since recently, it has also been scaled up for mass production [19–23]. Many factors influence fiber morphology and diameter, including solid concentrations, types of solvent, surface tension, additivation, solution viscosity, polymer molecular weight, flow rate, injector design, spinneret diameter, solution conductivity, injector to collector distance, and applied voltage. Of the many parameters discussed, concentration/solution viscosity, surface tension, and conductivity are probably the most important factors affecting the final fiber morphology and diameter [24–26]. A previous study [15] dealt with the development of a monolayer of oxygen-scavenging electrospun PHB containing palladium nanoparticles (PdNP). This monolayer demonstrated oxygen scavenging, but after annealing of the fibers to reduce porosity and to generate a water barrier, the material reduced the oxygen scavenging capacity to a significant extent.

The present work focuses, for the first time, on the preparation of significantly enhanced oxygen-scavenging bilayered coatings of PHB and PCL electrospun fibers, so-called biopapers, containing PdNPs, so-called nanobiopapers, deposited on a cellulose paper, to derive an optimized passive and active coating of interest in biodegradable fiber-based packaging.

2. Materials and Methods

2.1. Materials

The microbial homobiopolyester PHB, P226F grade, was obtained from Biomer (Krailling Germany). This grade is certified as both compostable and food contact. It has a density of 1.25 g/cm^3 and a melt flow rate (MFR) of 10 g/10 min when tested at 180 °C using a 5 kg load.

Polycaprolactone (PCL) (Mw: Mn 80 kDa), 2,2,2-trifluoroethanol (TFE) with 99% purity and D-limonene with 98% purity were purchased from Sigma-Aldrich S.A. (Madrid, Spain). Hexadecyltrimethylammonium bromide (CTAB) with 99% purity and palladium (Pd) nano-powder, <25 nm particle size measured by transmission electronic microscopy (TEM) and ≥99.5% trace metals basis, were also purchased from Sigma-Aldrich S.A. Chloroform (≥99%) and 1-butanol (99.5%) were purchased from AppliChem. All products were used as received without further purification. CTAB was selected as surfactant for PHB, because it is currently permitted for food contact applications by FDA and EFSA.

The conventional cellulose fiber-based packaging substrate was prepared using commercial bleached Kraft eucalyptus pulp as raw material, which was kindly provided by Ence-Celulosas y Energia S.A. (Madrid, Spain). Briefly, the pulp was disintegrated in a pulp disintegrator for 1 h at 3000 rpm to achieve a consistency of 1.5%. Paper sheets of 700 × 16 mm^2 with a final grammage of 75 g/m^2 were fabricated in an isotropic Rapid-Köthen sheet former and conditioned at 23 °C and 50% of relative humidity (RH) according to ISO standard 187. The grammage and thickness were evaluated following ISO standards 536 and 534, respectively. Further details can be found in previous research [22,27].

2.2. Preparation of the Films

Before electrospinning, the PHB solution was prepared by dissolving 10 wt % in TFE under magnetic stirring conditions at 50 °C. The PHB/PdNP suspension was prepared by adding CTAB (0.25 wt. % in the fibers) surfactant and PdNP (1 wt. % in the fibers) to the PHB solution.

PCL was prepared by dissolving 10 wt % in butanol:chloroform (25:75) under magnetic stirring conditions at room temperature. To prepare the solution of PCL/PdNP, PdNP were added (1 wt % in the fibers) in the previous solution and dissolved while magnetically stirring.

The electrospinning device used was a high throughput Fluidnatek® LE-500, used in lab mode with temperature and relative humidity control pilot plant equipment from Bioinicia S.L. (Valencia, Spain), a variable 0–60 kV dual polarizer high-voltage power supply, and a scanning injector, to obtain a homogeneous deposition of fibers. To obtain the electrospun PHB layers, the biopolymer solution was transferred to a 30 mL plastic syringe and coupled by a Teflon tube to a stainless-steel needle (∅ = 0.9 mm) that was connected to the power supply. PHB and PCL solutions were electrospun at 25 °C and 30% RH on a flat metallic collector, for 2 and 1 h under a steady flow-rate of 6 mL/h and 2 mL/h, respectively, using a motorized injector, scanning vertically toward a metallic grid collector.

A distance between the injector and collector was both optimal at 15 cm, and the applied voltage was 16 kV and 12 kV.

The electrospun PHB and PCL coatings were subjected to an annealing post-processing step below the polymers' melting points, at temperatures of 160 °C and 50 °C, respectively using a hydraulic press 4122-model from Carver, Inc. (Indiana, IN, USA). This post-processing thermal treatment was applied for 5 s without pressure, to ensure the coalescence of the fibers mat into a continuous film. The conditions were selected based on the research conducted in previous works [21,28,29].

2.3. *Characterization of the Films*

2.3.1. Film Thickness

Prior to characterization, the whole thickness of all of the structures was measured by using a digital micrometer (Series S00014, having ± 0.001 mm accuracy, from Mitutoyo Corporation (Kawasaki, Japan)). Measurements were performed at three random positions, and values were averaged. All samples were stored before evaluation, in desiccators containing dried silica gel at 25 °C.

2.3.2. Focus Ion Beam Scanning Electron Microscopy (FIB-SEM)

Electrospun samples were prepared for microscopy evaluation by fixing nanofibers deposited on aluminum foil with carbon tape, and they were gold coated with a 5 nm layer, using a rotary-pump sputter coater (Q150RS, Quorum Technologies, UK). The samples were imaged with a scanning electron microscope (SEM), using an accelerating voltage of 3 kV, 0.15 nA current and a working distance of 5 mm. 3D tomography of nanofibers with nanoparticles was achieved by using a dual beam system (NEON CrossBeam 40EsB, Zeiss, Germany) integrating a SEM with a focused ion beam (FIB). The sample stage was tilted at 54° so that the sample surface was perpendicular to the FIB direction [30], as demonstrated in previous research [31]. The 12–14 nm thick cross-sectional slices were milled by using FIB from the nanofiber sample at 30 kV and a beam current of 0.5 nA. [32,33]. The collected SEM images during FIB sectioning were filtered and reconstructed in 3D using Avizo Fire (version 6.3—FEI Edition, U.S.A.). To obtain the 3D reconstructions of the PCL fibers, 50 images and 108 images, respectively, for fiber 1 and 2 were used, maintaining the voxel size (5 × 5 × 14 nm). In the case of the PHB fibers, 50 images and a voxel size of 4 × 4 × 12 nm were used. Additionally, Avizo Fire was used to calculate the Pd nanoparticle concentration in the investigated pieces of electrospun fibers from the 3D reconstructions already obtained.

2.3.3. Scanning Electron Microscopy

An S-4800 SEM microscope from Hitachi (Tokyo, Japan) was further used to observe the morphology of the electrospun PHB films, and their cross-sections and surfaces. Cross-sections of the samples were prepared by cryo-fracture of the electrospun PHB films in liquid nitrogen. Then, they were fixed to beveled holders by using conductive double-sided adhesive tape, sputtered with a mixture of gold-palladium under a vacuum, and observed using an accelerating voltage of 5 kV.

2.3.4. Transmission Electronic Microscopy

The morphology and distribution of Pd nanoparticles were studied in electrospun fibers directly deposited onto clamping holders, and in the case of the films, on ultrathin microtomed sections as described in reference [15], using a Jeol 1010 (Hitachi, Tokyo, Japan) transmission electronic microscope, at an accelerating voltage of 80 kV.

2.3.5. Differential Scanning Calorimetry (DSC)

Thermal properties of neat PCL and Pd containing PCL electrospun fibers and films were evaluated by differential scanning calorimetry (DSC) using a Perkin-Elmer DSC 8000 (Waltham, MA, USA) thermal analysis system under a nitrogen atmosphere. The analysis was carried out on ~3 mg of each sample at a heating rate of 10 °C/min, from −25 °C to 125 °C, with subsequent cooling to −25°C. The DSC equipment was calibrated with indium as a standard, and the slope of the thermograms was corrected by subtracting similar scans of an empty pan. Tests were done at least in triplicate.

2.3.6. Thermogravimetric Analysis (TGA)

The TGA was performed in a TG-STDA Mettler Toledo model TGA/STDA851e/LF/1600 analyzer. The samples with an initial weight of typically about 15 mg were heated from 50 to 1300 °C at a heating rate of 10 °C/min under nitrogen/air flow.

2.3.7. Infrared Spectroscopy

Fourier transform infrared spectroscopy (FTIR) spectra were collected coupling the attenuated total reflection (ATR) accessory Golden Gate of Specac, Ltd. (Orpington, UK) to Bruker Tensor 37 FTIR equipment (Rheinstetten, Germany). Single spectra were collected in the wavelength range from 4000 to 600 cm^{-1} by averaging 20 scans at a resolution of 4 cm^{-1}.

2.3.8. Water Vapor Permeance

The water vapor permeance was determined by using the ASTM 2011 gravimetric method. To this end, 5 mL of distilled water was placed inside a Payne permeability cup (Ø = 3.5 cm) from Elcometer Sprl (Hermalle-sous-Argenteau, Belgium). The films were placed in the cups so that on one side, they were exposed to 100% RH on the coated side, avoiding direct contact with water. The cups containing the films were then secured with silicon rings and stored in a desiccator at 0% RH using dried silica gel, at 25 °C. Identical cups with aluminum films were used as control samples to estimate water loss through the sealing. The cups were weighed periodically using an analytical balance of ±0.0001 g accuracy. The water vapor transmission rate (WVTR), also called water vapor permeance when corrected for permeant partial pressure, was determined from the steady-state permeation slope obtained from the regression analysis of weight loss data per unit area versus time, in which the weight loss was calculated as the total cell loss minus the loss through the sealing. Measurements were performed in triplicate.

2.3.9. Measurement of Oxygen Scavenging Activity

Round-bottom flasks (to Schlenk) from VidraFoc S.A. (Barcelona, Spain) with a polytetrafluoroethylene (PTFE) stopcock and a headspace volume of 50 cm^3 was used for the oxygen scavenging measurements. The flasks contained a valve for flushing gas in, and an O_2-sensitive sensor spot (PSt3, detection limit 15 ppb, 0–100% oxygen) from PreSens (Regensburg, Germany) was glued onto the inner side of the flasks for the oxygen depletion measurements. Electrospun fibers and multilayers containing electrospun fibers with same sample areas were cut (5 × 5 cm^2) and placed into the flasks. The flask was subsequently flushed for 30 s at 1 bar with a gas mixture containing 1 vol % oxygen, 4 vol % hydrogen, and 95 vol % nitrogen, which was provided by Abelló Linde, S.A. (Barcelona, Spain). The oxygen concentration in the cell was monitored by a non-destructive measurement method, using the OXY-4 mini (PreSens) multi-channel fiber optic oxygen meter for simultaneous read-outs of up to four oxygen sensors, and used with sensors based on a 2 mm optical fiber. Oxygen concentrations over time were measured by linking the light-emitting (600–660 nm) optical fibers to the flasks' inner sensing spots. The sensor emits a certain amount of luminescence, depending on the oxygen concentration in the cell that is calibrated to yield the concentration by the equipment. All measurements were carried out at 23 °C and 50% RH, simulating typical ambient conditions.

2.4. Statistical Analysis

The test data were evaluated through analysis of variance (ANOVA) using STATGRAPHICS Centurion XVI v 16.1.03 from StatPoint Technologies, Inc. (Warrenton, VA, USA). Fisher's least significant difference (LSD) was used at the 95% confidence level ($p < 0.05$). Mean values and standard deviations were also calculated.

3. Results and Discussion

3.1. Morphology of the Electrospun PCL Fibers and Films

As it can be seen from the observation of Figure 1a,b, a narrow distribution of fiber diameter with an average at 2.75 ± 0.4 μm in PCL fibers, and 2.25 ± 0.7 μm in PCL/PdNP fibers was observed. The surface of the formed fibers was seen to be smooth and without beaded regions. The diameters of the fibers produced by electrospinning primarily depended on the spinning parameters, the most crucial being the solution concentration [34]. The smaller average diameter of the PCL/PdNP fibers can be attributed to an expected increase in conductivity, in agreement with previous works making use of metallic nanoparticles. Thus, in the case of nanoparticles of ZnO, the authors hypothesized that the solution was seen to have a larger charge capacity, and then to be driven by a stronger electric force along the fibers; therefore, smaller fiber diameters were obtained [35].

The SEM images of the films' cross-sections, shown in Figure 1c,d, indicated the presence of compact structures that resulted from the annealing post-processing step, which was in good agreement with previous works [20–22]).

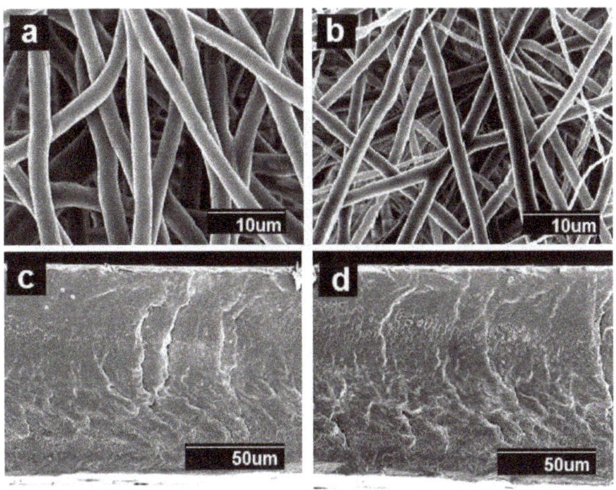

Figure 1. Scanning electron microscopy (SEM) images of the surface view and the cross-section of the polycaprolactone (PCL) fibers, with and without palladium nanoparticles (PdNPs), and their respective annealed films: (**a**) Surface view of the neat PCL fibers; (**b**) Surface view of the PCL/PdNP fibers; (**c**) Cross-section of the neat PCL film; (**d**) Cross-section of the PCL/PdNP film.

Additionally, FIB-SEM was used to cross-section the internal structure of the electrospun fibers. The detailed examples of the FIB-SEM images of PCL /PdNP fibers collected during FIB sectioning are presented in Figure 2, showing that Pd particles are incorporated in the two fibers. The cross-sectional images enabled the visualization of the PCL/PdNP fibers in 3D, as shown in Figure 3. The Pd nanoparticles were seen to be distributed along the PCL fibers, forming small agglomerates at different parts within the fiber cross-sections. Finally, the concentration of particles was also estimated per the given volume of the piece of fiber analyzed, and for the PCL fiber 1, this was 1.1%, and for PCL fiber 2, this was 0.9%. These observations suggest that the PdNP are better dispersed and distributed across the PCL fibers, and that the agglomerates may account for most of the Pd in the nanocomposites.

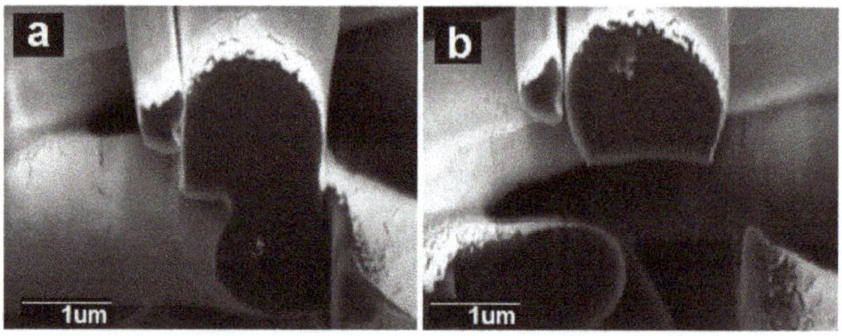

Figure 2. The cross-sectional SEM image of two PCL fibers with Pd nanoparticles after (FIB-SEM) sectioning. (**a**) PCL fiber 1 at the bottom with visible bright nanoparticles and (**b**) PCL fiber 2 at the top of the image with visible bright nanoparticles.

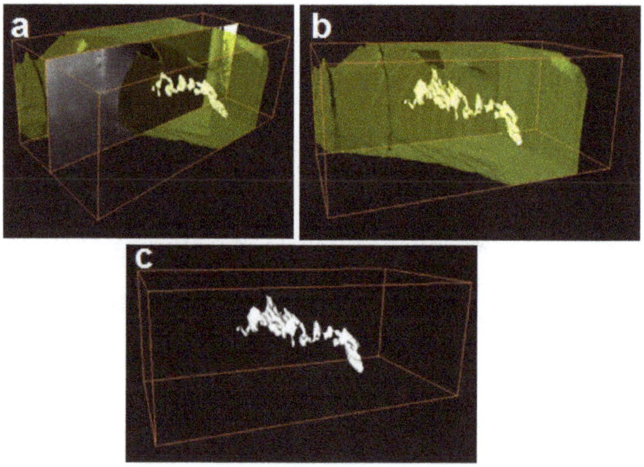

Figure 3. (**a**) The side view of 3D reconstructions of PCL fiber 1 (in semi-transparent yellow) and Pd nanoparticles (in white), including the SEM image (obtained from FIB sectioning) in the middle of the reconstruction, (**b**) 3D reconstruction of PCL fiber 1 with Pd nanoparticles inside, (**c**) 3D reconstruction of particles only inside of PCL fiber 1. The binding box size for this reconstruction had the following dimensions $3.105 \times 1.185 \times 1.526$ µm.

In spite of the very revealing FIB-SEM results, the SEM technique is thought to be inadequate for resolving highly dispersed Pd nanoparticles within the polymer matrix [36]. In order to check for this, additional TEM experiments were conducted on the samples.

The additional TEM experiments displayed in Figure 4 indicate that the Pd nanoparticles, in agreement with the FIB-SEM experiments, exhibit a significant degree of aggregation within the fiber. Due to attractive forces (Van der Waals and others), particles tend to agglomerate, even in suspension, unless stabilized by equivalent repulsive forces such as surface charge or steric effects. Thus, the smaller the particle size, the greater the relative attractive forces per unit mass. This means that it becomes progressively more difficult to disperse nanoscale materials as the size decreases [37]. In any case, TEM also revealed the presence of some highly dispersed and distributed nanoparticles within the cross-section of the biopaper film. The smallest particles were seen to have diameters of ca. 6 ± 2 nm, and they seemed evenly distributed throughout the fibers/film. Our prior studies of PHB/PdNPs electrospun fibers, also showed a similar dispersion of Pd nanoparticles, with some clear agglomeration zones within the fibers [20].

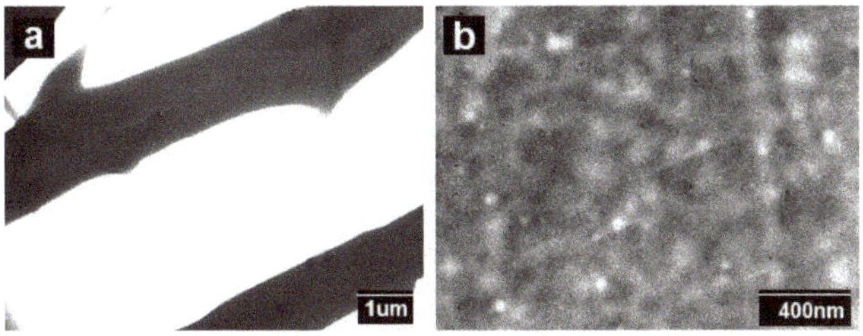

Figure 4. Scanning electron microscopy (TEM) images taken (**a**) directly on electrospun polycaprolactone (PCL) fibers containing palladium nanoparticles (PdNPs) and on (**b**) microtomed sections of their corresponding annealed film.

3.2. FTIR Analysis of the PCL Electrospun Fibers and Films

The FTIR spectra of the electrospun neat PCL fibers and film, and the PCL/PdNP fibers and film are shown in Figure 5.

The PCL spectrum displays the characteristic peaks of C=O stretching vibrations at 1726 cm^{-1}, CH2 bending modes at 1361, 1397, and 1473 cm^{-1}, and CH2 asymmetric stretching at 2942 and symmetric stretching at 2862 cm^{-1}. The C-O-C stretching vibrations yield peaks at 1042, 1107 and 1233 cm^{-1}. The bands at 1160 and 1290 cm^{-1} are assigned to C-O and C-C stretching in the amorphous and crystalline phases, respectively [38–40].

The overall PCL spectrum, including the main bands ascribed to PCL, such as the peaks at 2949 and 2865 cm^{-1} from methylene (CH2) groups, and the strong carbonyl (C=O) peak centered at 1720 cm^{-1}, were not seen to be affected either by incorporating PdNP, nor by the post-processing step, suggesting a lack of changes across the polymer molecular backbone.

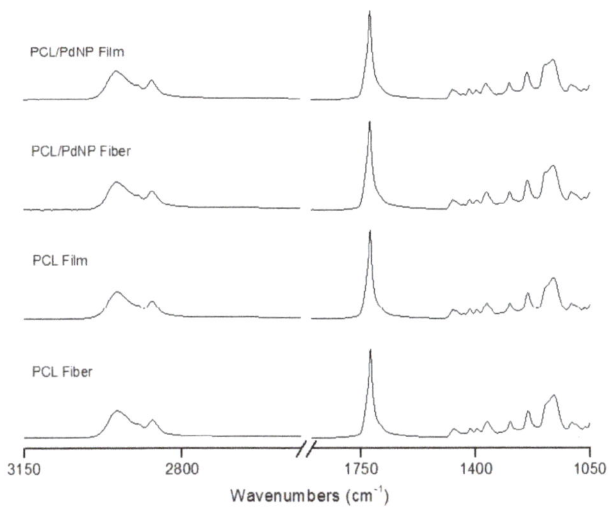

Figure 5. ATR-FTIR spectra of the electrospun PCL and the PCL/PdNP fiber mats and annealed films.

3.3. Thermal Properties of the PCL Electrospun Fibers

Table 1 shows the thermal properties, melting and crystallization points and enthalpies, in the first heating run and the subsequent crystallization run from the melt for the PCL and PCL/PdNP fibers mats. With the exception of the melting point, the rest of the thermal features were very similar for the neat PCL and nanocomposite fibers. The melting temperature for pure PCL is typically reported at 60 °C, and the glass transition temperature is −60 °C. As discussed, in the melting point, there was an increase of ca. 5 °C in comparison with the mean PCL, which must be explained by the addition of PdNPs to the polymer. This may due to the interaction of the polymer chains with the surface of the particles, which can change the chain kinetics in the region immediately surrounding the nanoparticles [41]. Similar results were reported by Bajsić et al. [42], where the melting point of the PCL/TiO_2 composites was found to increase slightly with an increasing load of TiO_2 micro- and nanoparticles.

Table 1. Thermal properties obtained by DSC in terms of melting temperature (T_m), normalized melting enthalpy (ΔH_m), crystallization temperature (T_c), and crystallization enthalpies (ΔH_c) for PCL and PCL/PdNP fibers.

Sample	T_m (°C)	ΔH_m (J/g)	T_c (J/g)	ΔH_c (J/g)
PCL Fibers	59.7 ± 1.2 [a]	33.8 ± 2.0 [b]	32.6 ± 0.9 [b]	41.4 ± 2.1 [b]
PCL/PdNP Fibers	64.7 ± 0.7 [b]	27.8 ± 1.5 [a]	31.2 ± 1.2 [a]	38.2 ± 0.9 [a]

Thermogravimetric analysis (TGA) was carried out to evaluate the degradation temperature of the PCL and PCL/PdNP fibers, including the curves of the first derivative analysis (blue lines) (see Figure 6), and the results are summarized in Table 2.

From Figure 6, it can be observed that PCL and PCL/PdNP initiated degradation at 342 and 355 °C, respectively, exhibiting two transition peaks: The first transition peaks were at 388 and 391 °C, and the second transition peaks were at 449 and 447 °C, respectively. The residual material of PCL and PCL/PdNP had a slight difference of ca. 0.8%, which is ascribed to the Pd that is present in the sample.

The data in the work reported here indicate that adding 1 wt % PdNP resulted in a slightly higher degree of thermal stability for the composite. Previous studies showed that other metallic nanofillers can impact the degradation temperature of PCL in different ways. Thus, Wang et al. demonstrated that the thermal stability of PCL was depressed by the incorporation of Fe_3O_4/GO nanoparticles, most likely due to the filler acting as a catalyst for polymer degradation [43]. Castro-Mayorga et al. also observed that the degradation temperature of poly(3-hydroxybutyrate-co-3-hydroxyvalerate) (PHBVs containing ZnO nanoparticles showed lower degradation temperatures than that of pure PHBV3. This was attributed to the high thermal conductivity and catalytic properties of the ZnO nanoparticles [19]. Other studies suggest that a temperature drop can also be explained by the fact that nanoparticles can weaken to some extent the interactive force of polymer inter-chains, and hence assist the thermal decomposition of the nanocomposite [44]. However, in the current work, it was observed that the nanofiller induced a somewhat improved degree of thermal stabilization, which may be ascribed to a better adhesion between the nanoparticles and the polymer matrix, resulting in both a hindered diffusion of volatile decomposition products, and/or the sorption of these over the filler surface, for the nanocomposites. Thus, the effect of a filler in thermal stability is in fact thought to depend on the type, content, interfacial interaction, and the degree of dispersion and distribution of this into the polymer matrix [45]. Thus, Ag [45], Fe, and Zn nanoparticles [46,47] have also been previously reported to enhance the thermal stability of PHA, supporting the current results for PCL.

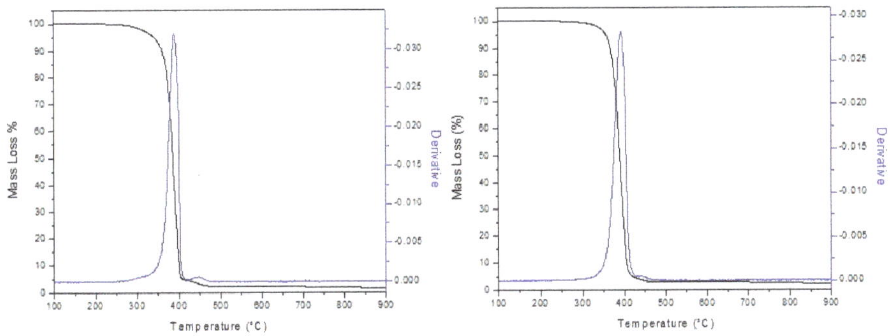

Figure 6. Thermogravimetric analysis (TGA) curves of the electrospun PCL (**left**) and PCL-containing palladium nanoparticles (PdNPs) fibers (**right**).

Table 2. Values of thermal stability obtained from the thermogravimetric analysis (TGA) curves of the electrospun PCL and PCL/PdNP electrospun fibers in terms of degradation temperature at 5% of mass loss (T5%), maximum degradation temperature of the two degradation peaks (Td1, Td2), and residual mass at 900 °C (R900).

Sample	T5% (°C)	Td1 (°C)	Td2 (°C)	R900 (%)
PCL	342.5 ± 4.1	388.0 ± 5.3	449.3 ± 4.2	1.3 ± 0.05
PCL-PdNP	355.1 ± 4.5	391.7 ± 3.6	447.8 ± 5.3	2.1 ± 0.03

3.4. PHB Electrospun Fiber Morphology

Similarly to the FIB-SEM investigation of PCL fibers, the sectioning of PHB mats was also carried out, and it is displayed in Figure 7. The SEM images in Figure 7 indicate that the PHB fibers morphology is similar to the one reported in earlier work, with the fiber cross-section ranging between 200 and 400 nm, and showing a smooth and beads-free fiber morphology [20–22].

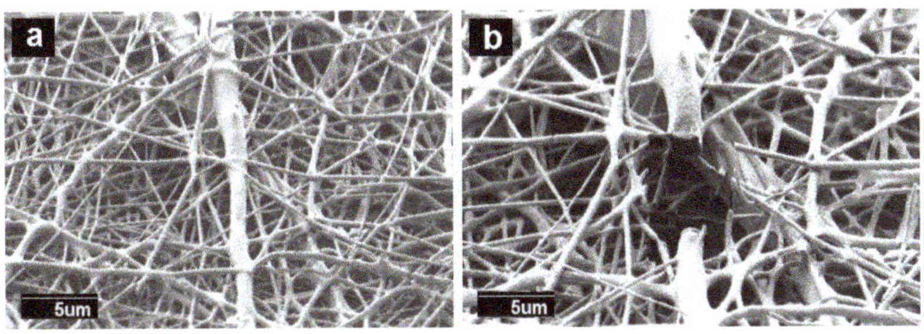

Figure 7. SEM images of the PHB sample. (**a**) Overview of PHB nanofibers, (**b**) PHB fibers after FIB sectioning.

The cross-section of the individual PHB fiber shown in Figure 8 revealed that the PdNP agglomerated even more strongly than seen in PCL at the core of fiber, even with a surfactant, which was also visualized with the 3D reconstructions present in Figure 9. The 3D tomography allowed us to verify the presence of PdNP in the individual PHB fiber section scanned, similarly to the PCL data shown in Figure 3. The estimation of the Pd nanoparticle concentration of the fiber section analyzed yielded a concentration of 3%, suggesting that Pd is not as well-distributed as in PCL. This is a relevant finding that suggests that even when a surfactant was added to the PHB to improve filler distribution, as suggested by our earlier work [20], the distribution was still seen lower than for PCL without a surfactant. The reason for this could be explained by the different chemistry, but also by the fact that

PHB is known to be a more rigid polymer that crystallizes to a greater extent into thick spherulites than PCL, which has more available free volume in the amorphous phase for dispersion.

Figure 10 shows the TEM analysis of the fibers and the corresponding film, which are in agreement with the FIB-SEM results, in which some Pd agglomerations could be observed but also the presence of some Pd nanoparticles dispersed and distributed across the material matrix. These results are also in agreement with previous findings for PHB [20].

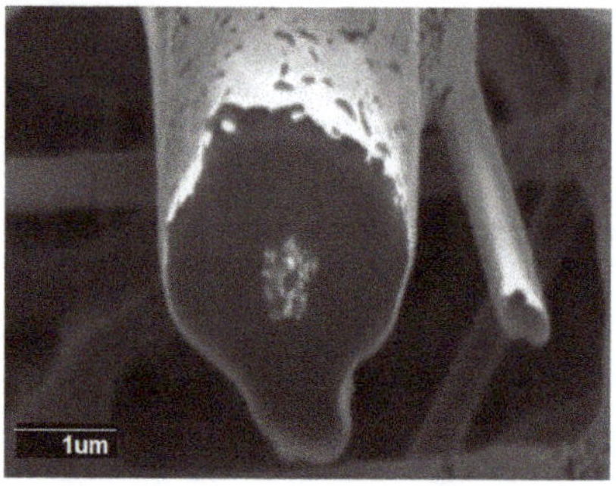

Figure 8. The cross-sectional SEM image of two PHB fibers with Pd nanoparticles after FIB sectioning.

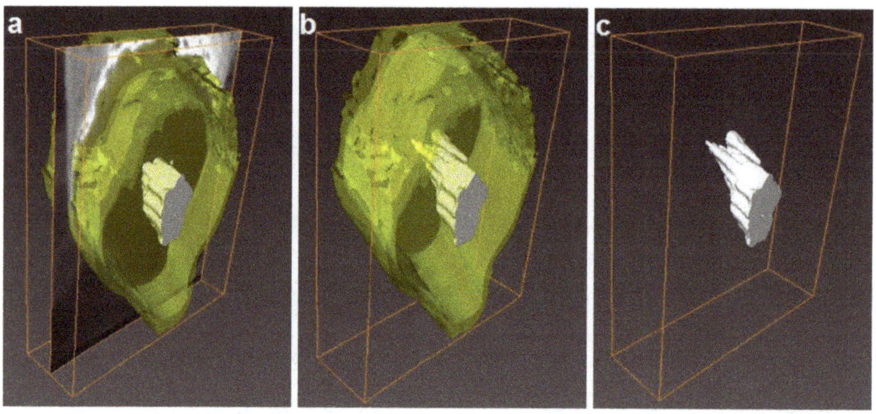

Figure 9. (**a**) The side view of the 3D reconstruction of the PHB fiber (in semi-transparent yellow) and Pd nanoparticles (in white), including the SEM image (obtained from FIB sectioning) in the middle of the reconstruction, (**b**) 3D reconstruction of the PHB fiber with Pd nanoparticles inside (**c**) 3D reconstruction of particles only from the inside of PHB fiber 2. The binding box size for this reconstruction had the following dimensions: 1.996 × 2.464 × 0.612 nm.

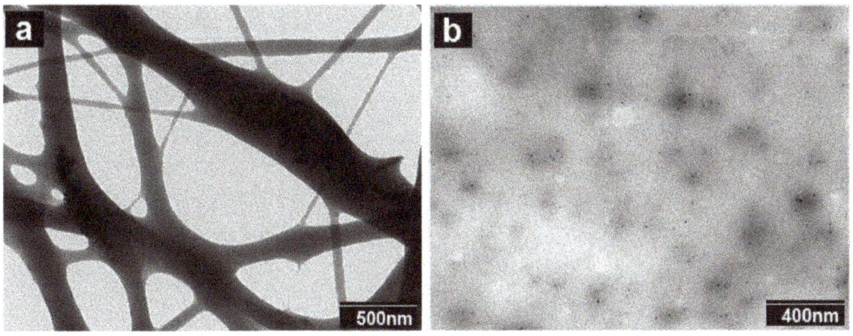

Figure 10. TEM images of (**a**) the PHB/PdNP fibers and (**b**) the PHB/PdNP film.

3.5. Morphology of the Multilayers

Figure 11 shows the SEM images of the multilayer structures obtained, which contain a paper substrate, and electrospun PHB/PCL fibers containing PdNP and their annealed films. In all cases, the amount of PCL deposited (1 hr coating time) was lower than the amount of PHB (2 hr coating time) because PHB is a better water barrier than PCL. In a previous paper, PHB containing Pd nanoparticles showed a significant decrease in oxygen absorption after annealing; hence, in this work we intended to explore the feasibility of using some PCL in the coating, in the hope of enhancing the oxygen scavenging effect of Pd. As a result, the PCL layer has a lower thickness, and it was set as the top layer. After annealing, there is an additional packing of the fibers, resulting in an even lower layer thickness, as can be seen in Figure 11.

The thickness of the paper layer was 117 μm and, of course, the fiber mean cross section of the paper was found to be higher, 18.37 ± 2.45 μm, compared to the electrospun fibers. Figure 11b shows that the surface of the electrospun neat PCL fibers exhibit significant fiber interconnections suggesting either remnant solvent induced coalescence and/or coarser fibers due to a drop in electric field as a result of deposition over other insulating materials as compared to direct deposition over the metallic collector (compare with Figure 1a).

The multilayers presented in Figure 11c–f indicate the coalescence of fibers during the annealing step as expected, leading to a much less porous continuous film. Even though the samples had similar morphologies, it seems that sample (c) showed somewhat greater porosity, and sample (**f**) showed the least porosity. This was surely the effect of annealing in both coatings, leading to a higher level of packing structure.

In agreement with previous studies carried out on coatings with electrospun fibers over paper or polymer substrates, it is seen that the adhesion achieved after annealing between layers was very strong, due to the high surface-to-volume ratio of the electrospun fibers [22,48].

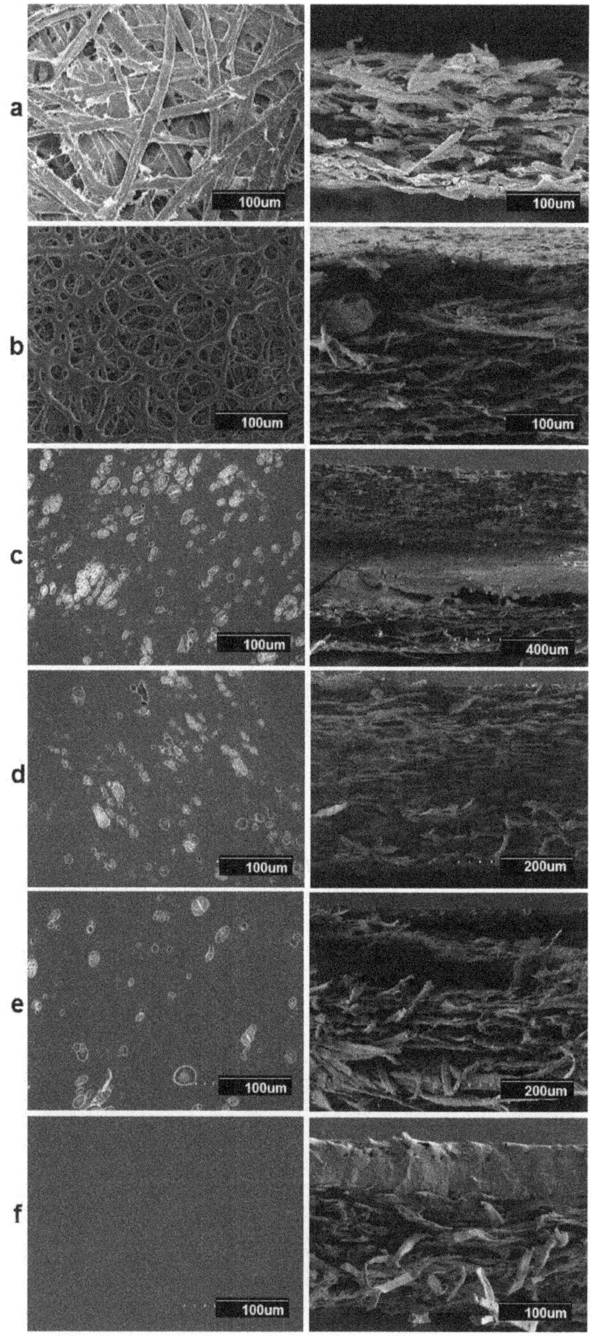

Figure 11. SEM images of the top view, and the cross section of (a) paper; (**b**) paper/PHB fibers/PCL fibers; (**c**) paper/PHB fibers/PCL-PdNP film; (**d**) Paper/PHB-PdNP fibers/PCL film; (**e**) Paper/PHB-PdNP fiber/PCL-PdNP film; (**f**) paper/PHB film/PCL-PdNP film.

3.6. Passive and Active Barrier Properties of the Multilayers

3.6.1. Water Vapor Passive Permeance

In general, the barrier properties of materials depend on the solubility and diffusion of the permeants, and hence, they depend on the permeant size, shape, and polarity; but also on the crystallinity, degree of cross-linking, and polymer chain segmental motion of the polymer matrix, among other factors [49,50].

The diffusion coefficient of water in an amorphous or semi-crystalline polymer is related to the particular molecular dynamics or segmental motions within the amorphous regions of the polymer. In addition, in semi-crystalline materials, a low crystallinity index and the formation of crystals of inferior quality confer a high degree of mobility to the macromolecular chains, resulting in lower barrier performance [51]. It is known that PHA is a better barrier material than PCL [52]. However, for fiber-based materials such as the paper/PHB fibers/PCL fibers multilayer generated here, the barrier performance was expected to be as low as paper, due to the existing porosity between the adjacent fibers.

The water permeance data of the multilayer samples are gathered in Figure 12. From this figure, it can be clearly seen that the samples that contained fibers and that had porosity at the surface did not significantly enhance the barrier of the multilayers compared to neat paper. However, it can also be seen that when the two electrospun coatings underwent annealing, this led to a porosity reduction and stronger adhesion, and the water barrier performance was significantly increased.

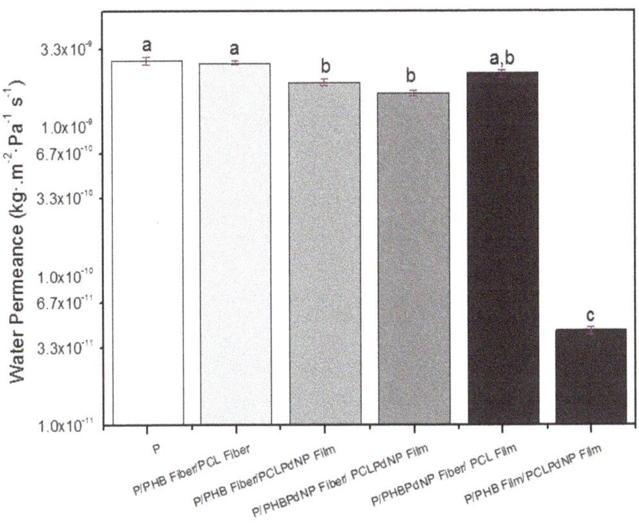

Figure 12. Water vapor permeance (WVP) of paper and Paper/PHB/PCL multilayers with and without palladium nanoparticles (PdNPs). Different letters indicate significant differences among samples ($p < 0.05$).

The higher water barrier of the sample paper/PHB film/PCLPdNP film is then explained by a reduction in sample porosity, and also by the expected improvement in the crystalline morphology that occurs in the sample after annealing, which is known be both impermeable to the diffusion of sorbed water molecules, and impose restraints to the mobility of the amorphous phase [53]. The barrier data gathered for this sample (4.3×10–11 kg/m^2·s^{-1}·Pa^{-1}) is in the same order of magnitude as the results obtained in a previous work for PHB (9.6×10–11 kg/m^2·s^{-1}·Pa^{-1}) [20].

3.6.2. Active Oxygen Scavenging Performance

Figure 13 shows the oxygen scavenging rate (OSR) of the PCL and PHB fibers, and the prepared paper based multilayers. The oxygen absorption of the fibers and resulting multilayers was investigated at 23 °C, with an initial oxygen concentration of 1.0% in the headspace of the measuring flasks, and at an RH of 50%. As reported earlier, due to the higher barrier of the material against water and oxygen compared to PCL, the PHB-PdNP fibers are not very efficient as oxygen scavengers at low or intermediate relative humidity, and even at high relative humidity as a film. However, PCL fibers and films are extremely quick at removing oxygen from the head space, suggesting that PCL is more adequate for hosting PdNP for oxygen scavenging purposes. The higher fractional free volume of the PCL polymer allows for moisture and oxygen to reach the catalyst quickly, and hence oxygen removal is more efficient. The greater reduction in the OSR of the fibers is, of course, related to the high surface-to-volume ratio of the electrospun fibers as compared to the annealed films. The dissociation rate of the hydrogen molecules into hydrogen atoms over the Pd surface in the films depends on the available surface area that is presented by the PdNP within the film.

As mentioned above, another important factor to be considered is that oxygen depletion is dependent on the RH conditions applied, and oxygen scavenging decreases when the RH increases [15,54]. In this work, we intended to conduct testing at medium relative humidities to simulate an intermedium case study. Figure 13 clearly indicates that, as expected, neither the paper nor the neat PCL polymer had any oxygen scavenging capacity.

Figure 13 also indicates that the best performing multilayer materials, in terms of OSR were, as expected, paper/PHB fibers/PCL-PdNP film and paper/PHB-PdNP fibers/PCL-PdNP film. In the cases where the intermediate layer is a PHB film or where the PdNP are not at the surface, the performance is reduced, since oxygen and moisture will possess slower kinetics of diffusion.

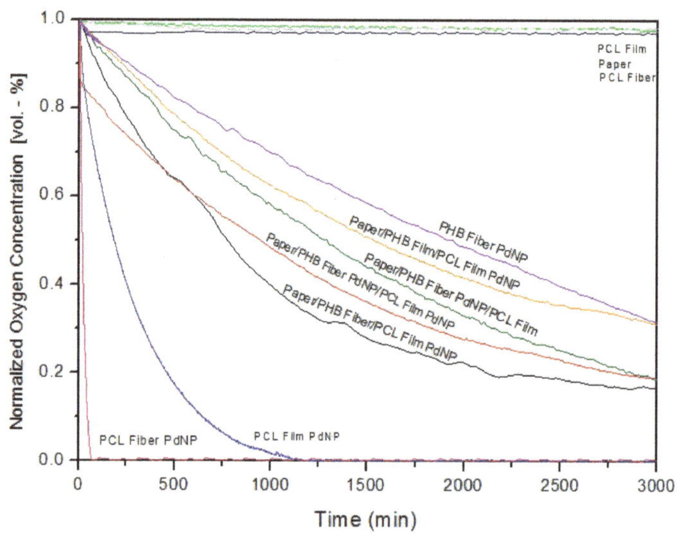

Figure 13. Oxygen depletion of PCL and PHB fibers, and paper/PHB/PCL multilayers and films with and without palladium nanoparticles (PdNPs). Values were measured at 50% relative humidity (RH).

4. Conclusions

Electrospun nanobiopaper bilayer coatings of water barrier PCL and PHB biopolymers containing PdNP were here developed for the first time over conventional cellulose paper. Some of the materials were post-processed by annealing, to achieve both better adhesion among layers, and to form a continuous structure to improve the barrier properties to water. SEM and FIB-SEM results indicated that the PdNP agglomerated in certain areas of the fiber cross-section, but TEM results also indicated that some of the PdNP were dispersed and distributed within the biopolymer matrix. Better distribution of the PdNP was inferred for the PCL matrix. FIB-SEM 3D reconstruction was a very powerful tool for visualizing composites beyond the conventional SEM results, and it seen to exhibit good correspondence with TEM results. However, TEM was further able to resolve better at the nanoscale, and it showed that some of the PdNP were highly dispersed and distributed within the fibers. The water barrier was enhanced, as expected, after annealing of the fiber-based materials, but the annealing process also decreased the oxygen absorption capacity. A previous work indicated that with electrospun PHB fibers containing PdNP, the oxygen scavenging capacity of the films reduced to a significant extent after annealing, compared to the non-heated highly porous electrospun fibers. In this more advanced study on the topic, the oxygen scavenging of the PdNP was largely enhanced, even after annealing, by incorporating these within PCL, a more oxygen permeable material that is still biodegradable.

The fully biodegradable fiber-based multilayered materials developed here show their tremendous potential for becoming the next generation of barrier papers, with demonstrated water barrier and oxygen scavenging capacity that are of interest in, for instance, food packaging applications.

Author Contributions: Conceptualization was devised by J.M.L.; Methodology, Validation, and Formal Analysis was carried out by A.C., J.M.L., P.K.S., A.G.; U.S. Investigation, Resources, Data Curation, Writing—Original Draft Preparation and Writing—Review & Editing was performed by A.C., J.M.L., P.K.S., A.G., U.S. Supervision, J.M.L.; Project Administration, J.M.L.

Funding: This research was funded by the Spanish Ministry of Economy and Competitiveness (MINECO) project AGL2015-63855-C2-1-R, the H2020 EU YPACK (reference number 773872) and ResUrbis (Reference number 730349).

Acknowledgments: A. Cherpinski would like to thank the Brazilian Council for Scientific and Technological Development (CNPq) of the Brazilian Government for her predoctoral grant (205955/2014-2). U. Stachewicz thanks the National Science Centre in Poland for the Sonata Bis 5 project, No 2015/18/E/ST5/00230 allowing this study, and a PhD scholarship for P.K. Szewczyk. The 3D imaging was supported by the infrastructure at the International Centre of Electron Microscopy for Materials Science (IC-EM) at AGH University of Science and Technology.

Conflicts of Interest: The authors declare no conflict of interest.

References

1. Kerry, J.P.; O'Grady, M.N.; Hogan, S.A. Past, Current and Potential Utilisation of Active and Intelligent Packaging Systems for Meat and Muscle-Based Products: A Review. *Meat Sci.* **2006**, *74*, 113–130. [CrossRef]
2. Wong, D.E.; Andler, S.M.; Lincoln, C.; Goddard, J.M.; Talbert, J.N. Oxygen Scavenging Polymer Coating Prepared by Hydrophobic Modification of Glucose Oxidase. *J. Coat. Technol. Res.* **2017**, *14*, 489–495. [CrossRef]
3. Yildirim, S.; Röcker, B.; Rüegg, N.; Lohwasser, W. Development of Palladium-Based Oxygen Scavenger: Optimization of Substrate and Palladium Layer Thickness. *Packag. Technol. Sci.* **2015**, *28*, 710–718. [CrossRef]
4. Brody, A.L.; Strupinsky, E.P.; Kline, L.R. *Active Packaging for Food Applications*; CRC Press: Boca Raton, FL, USA, 2001.
5. Dobrucka, R.; Cierpiszewski, R. Active and Intelligent Packaging Food-Research and Development—A Review. *Pol. J. Food Nutr. Sci.* **2014**, *64*, 7–15. [CrossRef]
6. Kaufman, J.; Lacoste, A.; Schulok, J.; Shehady, E.; Yam, K. An Overview of Oxygen Scavenging Packaging and Applications. 2000. Available online: https://www.bakeryonline.com/doc/an-overview-of-oxygen-scavenging-packaging-an-0002 (accessed on 8 October 2017).
7. Ozdemir, M.; Floros, J.D. Active Food Packaging Technologies. *Crit. Rev. Food Sci. Nutr.* **2004**, *44*, 185–193. [CrossRef] [PubMed]
8. Solovyov, S.E. Oxygen Scavengers. In *Kirk-Othmer Encyclopedia of Chemical Technology*; John Wiley and Sons: New York, NY, USA, 2000; pp. 1–31.

9. Yildirim, S.; Röcker, B.; Pettersen, M.K.; Nilsen-Nygaard, J.; Ayhan, Z.; Rutkaite, R.; Radusin, T.; Suminska, P.; Marcos, B.; Coma, V. Active Packaging Applications for Food. *Compr. Rev. Food Sci. Food Saf.* **2018**, *17*, 165–199. [CrossRef]
10. Hutter, S.; Rüegg, N.; Yildirim, S. Use of Palladium Based Oxygen Scavenger to Prevent Discoloration of Ham. *Food Packag. Shelf Life* **2016**, *8*, 56–62. [CrossRef]
11. Li, M.; Posevins, D.; Gustafson, K.P.J.; Tai, C.; Shchukarev, A.; Qiu, Y.; Bäckvall, J. Diastereoselective Cyclobutenol Synthesis: A Heterogeneous Palladium-Catalyzed Oxidative Carbocyclization-Borylation of Enallenols. *Chem. Eur. J.* **2019**, *25*, 210–215. [CrossRef] [PubMed]
12. Li, M.; Inge, A.K.; Posevins, D.; Gustafson, K.P.J.; Qiu, Y.; Backvall, J. Chemodivergent and Diastereoselective Synthesis of Γ-Lactones and Γ-Lactams: A Heterogeneous Palladium-Catalyzed Oxidative Tandem Process. *J. Am. Chem. Soc.* **2018**, *140*, 14604–14608. [CrossRef] [PubMed]
13. Zhu, C.; Yang, B.; Backvall, J. Highly Selective Cascade C–C Bond Formation Via Palladium-Catalyzed Oxidative Carbonylation–Carbocyclization–Carbonylation–Alkynylation of Enallenes. *J. Am. Chem. Soc.* **2015**, *137*, 11868–11871. [CrossRef]
14. Zhu, C.; Yang, B.; Jiang, T.; Bäckvall, J. Olefin-Directed Palladium-Catalyzed Regio-and Stereoselective Oxidative Arylation of Allenes. *Angew. Chem. Int. Ed.* **2015**, *54*, 9066–9069. [CrossRef] [PubMed]
15. Mahieu, A.; Terrié, C.; Youssef, B. Thermoplastic Starch Films and Thermoplastic Starch/Polycaprolactone Blends with Oxygen-Scavenging Properties: Influence of Water Content. *Ind. Crop Prod.* **2015**, *72*, 192–199. [CrossRef]
16. Rhim, J.; Park, H.; Ha, C. Bio-Nanocomposites for Food Packaging Applications. *Prog. Polym. Sci.* **2013**, *38*, 1629–1652. [CrossRef]
17. Tokiwa, Y.; Calabia, B.P.; Ugwu, C.U.; Aiba, S. Biodegradability of Plastics. *Int. J. Mol. Sci.* **2009**, *10*, 3722–3742. [CrossRef] [PubMed]
18. Pan, Y.; Farmahini-Farahani, M.; O'Hearn, P.; Xiao, H.; Ocampo, H. An Overview of Bio-Based Polymers for Packaging Materials. *J. Bioresour. Bioprod.* **2016**, *1*, 106–113.
19. Castro-Mayorga, J.L.; Martínez-Abad, A.; Fabra, M.J.; Olivera, C.; Reis, M.; Lagarón, J.M. Stabilization of Antimicrobial Silver Nanoparticles by a Polyhydroxyalkanoate Obtained from Mixed Bacterial Culture. *Int. J. Biol. Macromol.* **2014**, *71*, 103–110. [CrossRef] [PubMed]
20. Cherpinski, A.; Gozutok, M.; Sasmazel, H.; Torres-Giner, S.; Lagaron, J. Electrospun Oxygen Scavenging Films of Poly(3-Hydroxybutyrate) Containing Palladium Nanoparticles for Active Packaging Applications. *Nanomaterials* **2018**, *8*, 469. [CrossRef]
21. Cherpinski, A.; Torres-Giner, S.; Cabedo, L.; Lagaron, J.M. Post-Processing Optimization of Electrospun Sub-Micron Poly(3-Hydroxybutyrate) Fibers to Obtain Continuous Films of Interest in Food Packaging Applications. *Food Addit. Contam. Part A* **2017**, *34*, 1817–1830. [CrossRef]
22. Cherpinski, A.; Torres-Giner, S.; Cabedo, L.; Méndez, J.A.; Lagaron, J.M. Multilayer Structures Based on Annealed Electrospun Biopolymer Coatings of Interest in Water and Aroma Barrier Fiber-Based Food Packaging Applications. *J. Appl. Polym. Sci.* **2018**, *135*, 45501. [CrossRef]
23. Hu, M.; Li, C.; Li, X.; Zhou, M.; Sun, J.; Sheng, F.; Shi, S.; Lu, L. Zinc Oxide/Silver Bimetallic Nanoencapsulated in Pvp/Pcl Nanofibres for Improved Antibacterial Activity. *Artif. Cells Nanomed. Biotechnol.* **2018**, *46*, 1248–1257. [CrossRef]
24. Deitzel, J.M.; Kleinmeyer, J.; Harris, D.E.A.; Tan, N.C.B. The Effect of Processing Variables on the Morphology of Electrospun Nanofibers and Textiles. *Polymer* **2001**, *42*, 261–272. [CrossRef]
25. Theron, S.A.; Zussman, E.; Yarin, A.L. Experimental Investigation of the Governing Parameters in the Electrospinning of Polymer Solutions. *Polymer* **2004**, *45*, 2017–2030. [CrossRef]
26. Zander, N.E. Hierarchically Structured Electrospun Fibers. *Polymers* **2013**, *5*, 19–44. [CrossRef]
27. Tarrés, Q.; Delgado-Aguilar, M.; Pèlach, M.A.; González, I.; Boufi, S.; Mutjé, P. Remarkable Increase of Paper Strength by Combining Enzymatic Cellulose Nanofibers in Bulk and Tempo-Oxidized Nanofibers as Coating. *Cellulose* **2016**, *23*, 3939–3950. [CrossRef]
28. Figueroa-Lopez, K.J.; Castro-Mayorga, J.L.; Andrade-Mahecha, M.M.; Cabedo, L.; Lagaron, J.M. Antibacterial and Barrier Properties of Gelatin Coated by Electrospun Polycaprolactone Ultrathin Fibers Containing Black Pepper Oleoresin of Interest in Active Food Biopackaging Applications. *Nanomaterials* **2018**, *8*, 199. [CrossRef] [PubMed]

29. Lasprilla-Botero, J.; STorres-Giner, E.; Pardo-Figuerez, M.; Álvarez-Láinez, M.; Lagaron, J.M. Superhydrophobic Bilayer Coating Based on Annealed Electrospun Ultrathin Poly (E-Caprolactone) Fibers and Electrosprayed Nanostructured Silica Microparticles for Easy Emptying Packaging Applications. *Coatings* **2018**, *8*, 173. [CrossRef]
30. Stachewicz, U.; Modaresifar, F.; Bailey, R.J.; Peijs, T.; Barber, A.H. Manufacture of Void-Free Electrospun Polymer Nanofiber Composites with Optimized Mechanical Properties. *ACS Appl. Mater. Interfaces* **2012**, *4*, 2577–2582. [CrossRef] [PubMed]
31. Stachewicz, U.; Qiao, T.; Rawlinson, S.C.F.; Almeida, F.V.; Li, W.; Cattell, M.; Barber, A.H. 3d Imaging of Cell Interactions with Electrospun Plga Nanofiber Membranes for Bone Regeneration. *Acta Biomater.* **2015**, *27*, 88–100. [CrossRef]
32. Bailey, R.J.; Geurts, R.; Stokes, D.J.; de Jong, F.; Barber, A.H. Evaluating Focused Ion Beam Induced Damage in Soft Materials. *Micron* **2013**, *50*, 51–56. [CrossRef]
33. Stachewicz, U.; Bailey, R.J.; Zhang, H.; Stone, C.A.; Willis, C.R.; Barber, A.H. Wetting Hierarchy in Oleophobic 3d Electrospun Nanofiber Networks. *ACS Appl. Mater. Interfaces* **2015**, *7*, 16645–16652. [CrossRef]
34. Garg, K.; Bowlin, G.L. Electrospinning Jets and Nanofibrous Structures. *Biomicrofluidics* **2011**, *5*, 013403. [CrossRef] [PubMed]
35. Yu, W.; Lan, C.; Wang, S.; Fang, P.; Sun, Y. Influence of Zinc Oxide Nanoparticles on the Crystallization Behavior of Electrospun Poly(3-Hydroxybutyrate-Co-3-Hydroxyvalerate) Nanofibers. *Polymer* **2010**, *51*, 2403–2409. [CrossRef]
36. Kane, M. Investigation and Characterization of the Dispersion of Nanoparticles in a Polymer Matrix by Scattering Techniques. Ph.D. Thesis, University of Florida, Gainesville, FL, USA, 2007.
37. Powers, K.W.; Palazuelos, M.; Moudgil, B.M.; Roberts, S.M. Characterization of the Size, Shape, and State of Dispersion of Nanoparticles for Toxicological Studies. *Nanotoxicology* **2007**, *1*, 42–51. [CrossRef]
38. Abdelrazek, E.M.; Hezma, A.M.; El-khodary, A.; Elzayat, A.M. Spectroscopic Studies and Thermal Properties of Pcl/Pmma Biopolymer Blend. *Egypt. J. Basic Appl. Sci.* **2016**, *3*, 10–15. [CrossRef]
39. Liverani, L.; Lacina, J.; Roether, J.A.; Boccardi, E.; Killian, M.S.; Schmuki, P.; Schubert, D.W.; Boccaccini, A.R. Incorporation of Bioactive Glass Nanoparticles in Electrospun Pcl/Chitosan Fibers by Using Benign Solvents. *Bioact. Mater.* **2018**, *3*, 55–63. [CrossRef] [PubMed]
40. Mohamed, A.; Finkenstadt, V.L.; Gordon, S.H.; Biresaw, G.; Palmquist, D.E.; Rayas-Duarte, P. Thermal Properties of Pcl/Gluten Bioblends Characterized by Tga, Dsc, Sem, and Infrared-Pas. *J. Appl. Polym. Sci.* **2008**, *110*, 3256–3266. [CrossRef]
41. Aliah, N.N.; Ansari, M.N.M. Thermal Analysis on Characterization of Polycaprolactone (Pcl) E Chitosan Scaffold for Tissue Engineering. *Int. J. Sci. Res. Eng. Technol.* **2017**, *6*, 2278–2882.
42. Bajsić, E.G.; Bulatović, V.O.; Slouf, M.; Šitum, A. Characterization of Biodegradable Polycaprolactone Containing Titanium Dioxide Micro and Nanoparticles. *Int. J. Chem. Nucl. Met. Mater. Eng.* **2014**, *8*, 572–576.
43. Wang, G.; Yang, S.; Wei, Z.; Dong, X.; Wang, H.; Qi, M. Facile Preparation of Poly(E-Caprolactone)/ Fe3o4@Graphene Oxide Superparamagnetic Nanocomposites. *Polym. Bull.* **2013**, *70*, 2359–2371. [CrossRef]
44. Hong, R.Y.; Qian, J.Z.; Cao, J.X. Synthesis and Characterization of Pmma Grafted Zno Nanoparticles. *Powder Technol.* **2006**, *163*, 160–168. [CrossRef]
45. Castro-Mayorga, J.L.; Fabra, M.J.; Lagaron, J.M. Stabilized Nanosilver Based Antimicrobial Poly (3-Hydroxybutyrate-Co-3-Hydroxyvalerate) Nanocomposites of Interest in Active Food Packaging. *Innov. Food Sci. Emerg. Technol.* **2016**, *33*, 524–533. [CrossRef]
46. Díez-Pascual, A.M.; Díez-Vicente, A.L. Poly (3-Hydroxybutyrate)/Zno Bionanocomposites with Improved Mechanical, Barrier and Antibacterial Properties. *Int. J. Mol. Sci.* **2014**, *15*, 10950–10973. [CrossRef]
47. Panaitescu, D.M.; Ionita, E.R.; Nicolae, C.; Gabor, A.R.; Ionita, M.D.; Trusca, R.; Lixandru, B.; Codita, I.; Dinescu, G. Poly (3-Hydroxybutyrate) Modified by Nanocellulose and Plasma Treatment for Packaging Applications. *Polymers* **2018**, *10*, 1249. [CrossRef]
48. Cherpinski, A.; Torres-Giner, S.; Vartiainen, J.; Peresin, M.S.; Lahtinen, P.; Lagaron, J.M. Improving the Water Resistance of Nanocellulose-Based Films with Polyhydroxyalkanoates Processed by the Electrospinning Coating Technique. *Cellulose* **2018**, *25*, 1291–1307. [CrossRef]
49. Siracusa, V. Food Packaging Permeability Behaviour: A Report. *Int. J. Polym. Sci.* **2012**. [CrossRef]
50. Zeman, S.; Kubík, L. Permeability of Polymeric Packaging Materials. *Tech. Sci.* **2007**, *10*, 26–34. [CrossRef]

51. Cavalcante, M.; Rodrigues, E.; Tavares, M.I. Avaliação Da Cristalinidade De Blendas De Polihidróxibutirato E Policaprolactona. In Proceedings of the 13th Congresso Brasileiro de Polímeros, Natal-RN, Brazil, 18 October 2015.
52. Sanchez-Garcia, M.D.; Gimenez, E.; Lagaron, J.M. Morphology and Barrier Properties of Nanobiocomposites of Poly (3-Hydroxybutyrate) and Layered Silicates. *J. Appl. Polym. Sci.* **2008**, *108*, 2787–2801. [CrossRef]
53. Shogren, R. Water Vapor Permeability of Biodegradable Polymers. *J. Environ. Polym. Degrad.* **1997**, *5*, 91–95. [CrossRef]
54. Gibis, D.; Rieblinger, K. Oxygen Scavenging Films for Food Application. *Procedia Food Sci.* **2011**, *1*, 229–234. [CrossRef]

nanomaterials

MDPI

Article

Antioxidant Bilayers Based on PHBV and Plasticized Electrospun PLA-PHB Fibers Encapsulating Catechin

Marina P. Arrieta [1,2,*], Alberto Díez García [1], Daniel López [1], Stefano Fiori [3] and Laura Peponi [1,*]

1 Institute of Polymer Science and Technology, ICTP-CSIC, Juan de la Cierva 3, 28006 Madrid, Spain; albertodiezgarcia1982@yahoo.es (A.D.G.); daniel.l.g@csic.es (D.L.)

2 Departamento de Química Orgánica I, Facultad de Ciencias Químicas, Universidad Complutense de Madrid, Av. Complutense s/n, 28040 Madrid, Spain

3 Condensia Química S.A., R&D Department, C/La Cierva 8, 08184 Barcelona, Spain; s.fiori@condensia.com

* Correspondence: marrie06@ucm.es (M.P.A.); lpeponi@ictp.csic.es (L.P.)

Received: 31 January 2019; Accepted: 20 February 2019; Published: 3 March 2019

Abstract: The main objective of this work was to develop bio-based and biodegradable bilayer systems with antioxidant properties. The outer layer was based on a compression-molded poly(3-hydroxybutyrate-co-3-hydroxyvalerate) (PHBV)-based material while antioxidant electrospun fibers based on poly(lactic acid) (PLA) and poly(3-hydroxybutyrate) (PHB) blends formed the inner active layer. In particular, PLA was blended with 25 wt% of PHB to increase the crystallinity of the fibers and reduce the fiber defects. Moreover, in order to increase the stretchability and to facilitate the electrospinning process of the fiber mats, 15 wt% of oligomeric lactic acid was added as a plasticizer. This system was further loaded with 1 wt% and 3 wt% of catechin, a natural flavonoid with antioxidant activity, to obtain antioxidant-active mats for active food packaging applications. The obtained bilayer systems showed effective catechin release capacity into a fatty food simulant. While the released catechin showed antioxidant effectiveness. Finally, bilayer films showed appropriate disintegration in compost conditions in around three months. Thus, showing their potential as bio-based and biodegradable active packaging for fatty food products.

Keywords: electrospinning; antioxidant active packaging; bio-based polymers; biodegradable polymers; poly(3-hydroxybutyrate-co-3-hydroxyvalerate) (PHBV); poly(lactic acid) (PLA); poly(3-hydroxybutyrate) (PHB); oligomeric lactic acid (OLA); catechin

---

1. Introduction

There is a growing attitude towards the incorporation of active agents into food packaging materials instead of directly into the food, thus allowing a controlled release of the active component from the packaging to the foodstuff to maintain and even to enhance the food quality and safety [1–3]. Particularly, antioxidant packaging seeks to prevent the oxidation of food components (i.e., lipids and proteins), which lead to the deterioration of physical characteristics of food, such as flavor and color [4]. In fact, scavenging the primary free radicals of the oxidation process is the most efficient way to protect food against oxidation [1]. Catechin has been widely used for the development of antioxidant packaging materials mainly because it shows good scavenging activity [3,5]. Catechin is a natural flavonoid that can be obtained from several species of plants, in particular, from green tea and grapes [3,6,7]. The scientific literature reports many approaches for the development of antioxidant packaging systems, but most of them fail when they are trying to be developed into a packaging line at an industrial level [1]. In this sense, the electrospinning technique has gained considerable interest during the last year for the development of active layers for multilayer packaging systems [4,8,9]. Nowadays, the electrospinning process is a simple, extremely flexible, and low-cost process for fiber at an industrial level that can find several applications in the food packaging industry (i.e., active and intelligent packaging systems) [10,11]. It can produce multifunctional thin polymeric materials with different functionalities in the form of non-woven fibers from polymeric solutions subjected to high electric fields and at room temperature [2,12]. Food packaging is required to contain food products and protect them from the surroundings avoiding contamination, humidity, and oxidation processes [13,14]. Although electrospun fiber mats are not resistant enough to be directly used as films for food packaging, they can be used in multilayer packaging approaches [4,15,16]. Moreover, the electrospinning process is currently one of the most promising encapsulation techniques, and electrospun fibers have been considered interesting carriers for many active compounds [4,17,18]. Particularly, electrospun active layers have been recently developed and proposed as a food contact layer in multilayer systems [4,8,19].

On the other side, there is a current trend in the food packaging industry to substitute the use of non-renewable and non-degradable polymers with bio-based and biodegradable polymers. The use of bio-based polymers permits the reduction of the global dependence on fossil petrol sources for polymer production, while biodegradable polymers allow composting as a simple and sustainable end-life option for the packaging material. In this regard, catechin has been widely used in combination with biopolymeric matrices for the development of active packaging and composites [5,20–23]. Furthermore, in recent decades, bio-based and biodegradable polyesters, such as poly(lactic acid) (PLA) and the family of poly(hydroxybutyrates) (PHAs) have gained industrial attention. In this sense, the crystallinity of PLA has been successfully increased by blending with 25 wt% of poly(3-hydroxybutyrate) (PHB), allowing the formation of straight and bead-less electrospun fibers in comparison with neat PLA [10]. Moreover, catechin has been recently successfully incorporated into PLA-PHB (75:25) electrospun fiber mats obtaining fibers without defects and increased mechanical resistance when it was incorporated in amounts lower than 5 wt% [7]. To overcome the inherent brittleness of both biopolymers, plasticizers are frequently added to PLA-PHB blend systems [24], such as citrate esters [5,10], poly(ethylene glycol) (PEG) [25], and oligomeric lactic acid (OLA) [26]. Moreover, the combination of electrospun mats with a thermoplastic copolyester from the PHAs family such as poly(3-hydroxybutyrate-co-3-hydroxyvalerate) (PHBV) could represent an advantageous strategy since it will govern the resulting mechanical and barrier properties of the bilayer films [27]. In fact, PHBV has gained attention in the packaging field due to its commercial availability, ease of processing by using conventional thermoplastic equipment, and due to the fact that it possesses equilibrated mechanical properties in terms of stiffness and tensile strength [15]. Moreover, PHBV is frequently used to increase the barrier performance of multilayer packaging films [8,16]. However, PHBV still presents poorer mechanical performance than traditional packaging materials, such as lower impact strength and toughness [28]. In this sense, the modification of bio-based and biodegradable polymers by a blending approach has many advantages because it allows improving a wide range of the physical properties through cost-effective and readily available processing technologies at an industrial level [14]. It has been observed that by blending PHBV with an aliphatic-co-aromatic biodegradable polyester, poly(butylene adipate-co-terephthalate) (PBAT), a good balance of stiffness and toughness is obtained and, thus, PHBV applications are broadened [29].

In this work, PLA-PHB (75:25) electrospun fibers were loaded with 1 wt% and 3 wt% of catechin (Cat) to develop an antioxidant inner layer for biodegradable bilayer packaging systems. To facilitate the electrospinning process and to increase the stretchability of the final mats, 15 wt% of oligomeric lactic acid (OLA) was added as a plasticizer. Since electrospun mats are non-woven materials, an outer layer was required to achieve structural resistance of the final packaging material. To this end, a commercial PHBV-based pre-blend (PHBV/PBAT) was selected for the preparation of the outer layer by compression molding. The structural, thermal, and mechanical performance of each layer, as well as of the final bilayer system was studied. Since these bilayers are intended for active food packaging applications, the release ability of catechin, as well as its antioxidant response, was studied in a fatty food simulant. Finally, the disintegration under composting conditions at a laboratory-scale level of such bilayer systems was assayed to demonstrate their sustainable end-life option.

2. Materials and Methods

2.1. Materials

Poly(lactic acid) (PLA 3051, Mn = 110,000 Da, 3 wt% D-isomer) was supplied by NatureWorks (USA), poly(3-hydroxybutyrate) (PHB, under the trade name P226, Mw = 426,000 Da) was supplied by Biomer (Krailling, Germany) and poly(3-hydroxybutyrate-co-3-hydroxyvalerate) (PHBV, under the trade name ENMAT 5010P) was supplied by Tianan Biologic Materials Company, Ltd. (Ningbo, China) as a compound of PLA/PBAT in 45/55 proportion [30]. Oligomeric lactic acid (OLA 00A/8, Mn = 957 g mol^{-1}) was synthesized according to a previously reported process [31] and kindly supplied by Condensia Química S.A (Barcelona, Spain). Catechin (Cat, 98% purity, anhydrous grade) was purchased from Sigma-Aldrich (Madrid, Spain). Chloroform (CL, 99.6% purity, boiling point 60 °C) and dimethylformamide (DMF, 99.5% purity, boiling point 153 °C) and 2,2-diphenyl-1-picrylhydrazyl (DPPH) 95% free radical were supplied by Sigma Aldrich (Madrid, Spain).

2.2. Bilayer Systems Preparation

The inner layer was prepared by means of an electrospinning technique following previously optimized conditions for plasticized electrospun PLA-PHB-based materials [7,10]. Briefly, PLA pellets were previously dried at 80 °C overnight, PHB pellets, OLA, and catechin powder were dried at 40 °C for 4 h. Polymer solutions were prepared at 8 wt% in a mixture of cloroform:dimethylformamide (CL:DMF = 80:20 [32]) and further electrospun (polymer and solvent flow rate = 1.0 mL·h^{-1}, positive and negative voltages = 10.8 kV and −10.8 kV and working distance = 14 cm) in a coaxial Electrospinner (Y flow 2.2.D-XXX, Nanotechnology Solutions). The electrospun fibers were randomly collected during 4 h and the obtained mats were vacuumed for 48 h to eliminate any potential residual solvents. Each formulation was prepared by blending PLA-PHB in 75:25 proportion on the basis of our previous results [10] and plasticized with 15 wt% of OLA. The plasticized PLA-PHB systems were further reinforced with 1 wt% and 3 wt% of Cat. To improve the Cat particles dispersion the solutions were sonicated during 10 min before being processed by electrospinning [7]. The obtained mat formulations and the proportion of each component are summarized in Table 1.

Table 1. Electrospun PLA-PHB fibers and their composites.

Formulations	PLA (wt%)	PHB (wt%)	OLA (wt%)	Cat (wt%)	Fibers Diameter (nm)	Dinamic Viscosity (η) (Pa.s)
PLA-PHB	75	25	-	-	215 ± 67	0.12 ± 0.01
PLA-PHB-OLA	63.75	21.25	15	-	260 ± 78	0.06 ± 0.01
PLA-PHB-Cat1	74.25	24.75	-	1	405 ± 143	0.13 ± 0.03
PLA-PHB-OLA-Cat1	63.0	21.0	15	1	228 ± 57	0.06 ± 0.02
PLA-PHB-Cat3	72.75	24.25	-	3	400 ± 116	0.14 ± 0.01
PLA-PHB-OLA-Cat3	61.5	20.5	15	3	206 ± 57	0.07 ± 0.01

The outer PHBV-based layer was processed into films by compression molding at 180 °C in a hot press (Dr. COLLIN 200 × 200) by using a film mold (50 × 50 mm^2). PHBV was previously dried at 40 °C for 4 h. PHBV pellets were kept between the plates at atmospheric pressure for 1 min until melting and they were further submitted to pressure cycle: 5 kPa for 1 min, 10 kPa for 1 min, and then quenched to room temperature at 5 kPa for 1 min.

Finally, both layers were compression-molded to obtain a continuous bilayer film following already reported processes for the development of bilayer systems based on a PHBV outer layer with an electrospun inner layer [8]. In brief, a post-annealing process was applied by placing the inner electrospun layer onto the compression-molded PHBV and assembled in a hot press at 150 °C for 1 min and cooled down to room temperature in 2 min at 5 kPa. The obtained bilayer film formulations were labeled as the electrospun inner layer with the prefix PHBV: PHBV/PLA-PHB,

PHBV/PLA-PHB-OLA, PHBV/PLA-PHB-Cat1, PHBV/PLA-PHB-OLA-Cat1, PHBV/PLA-PHB-Cat3, and PHBV/PLA-PHB-OLA-Cat3.

2.3. Characterization Techniques

The dynamic viscosity of the electrospun polymeric solutions was determined using an AR-G2 TA Instruments rheometer parallel plate geometry (40 mm in diameter). Rotational tests were conducted using a stepped shear rate from 0.01 to 1500 s^{-1} at 20 °C.

The morphology of the obtained electrospun fibers in each mat and/or bilayer systems, as well as the cryo-fractured cross-sections of bilayer systems, were studied using a PHILIPS XL30 Scanning Electron Microscope (SEM). Samples were previously sputtered with a gold/palladium layer. The fiber diameters were statistically calculated from the SEM images with ImageJ software.

Isothermal and dynamic thermogravimetric analysis (TGA) tests were conducted by means of a TA Instruments, TGA Q500 thermal analyzer. For isothermal TGA analysis, electrospun mats were heated at 150 °C during 30 min under air conditions. The electrospun mats, as well as the bilayer systems, were heated under TGA dynamic mode from 30 °C to 700 °C at 10 °C min^{-1} under a nitrogen atmosphere (flow rate 50 mL min^{-1}). Film sample masses were between 5–7 mg. Initial degradation temperatures (T_0) were determined at 5% of mass loss and the maximum degradation temperatures (T_{max}) were calculated from the first derivative of the TGA curves (DTG).

The mechanical properties of monolayer and bilayer films were evaluated with the use of tensile test measurements conducted at room temperature by an Instron dynamometer (model 3366) equipped with a 100-N load cell, at a crosshead speed of 10 mm·min^{-1} and initial length of 30 mm. Dogbone-style film samples were used and at least five specimens were tested for each formulation. Release studies were performed in triplicate by the determination of catechin-specific migration tests into solutions of 50% v/v ethanol (food simulant D1) since these materials are intended for fatty food-packaging applications [33]. Thus, pre-weighed bilayer film samples were immersed in the food simulant D1 and were kept at 40 °C for 10 days. The released amount of catechin into the food simulant after 10 contact days was determined by the measurement of UV absorbance at 280 nm attributed to the B ring of catechin moiety, by means of a UV-Vis Perkin Elmer (Lambda 35, Waltham, MA, USA) UV-VIS spectrophotometer.

The antioxidant effectiveness of the released catechin was measured according to the DPPH-method [34] by determining the absorbance at 517 nm of the released catechin in the food simulant D1 at 10 days by means of a UV-Vis Perkin Elmer (Lambda 35, Waltham, MA, USA) spectrophotometer. The antioxidant activity was determined according to Equation (1):

$$(\%) = \left(\frac{A_{control} - A_{sample}}{A_{control}}\right) \times 100\% \quad (1)$$

where I (%) is the percentage of inhibition. $A_{control}$ the absorbance of DPPH at 517 nm in ethanolic solution and A_{sample} is the absorbance of DPPH at 517 nm after 15 min in contact with the food simulant containing the released catechin. The % of inhibition was expressed as the equivalent of gallic acid (GA) concentration (mg kg^{-1}) by using a calibrated curve of gallic acid concentration versus I (%).

The bilayer films were disintegrated in composting conditions at a laboratory-scale level following the ISO 20200 standard [35]. Samples (15 mm × 15 mm) were buried at a depth of 4–6 cm in appropriate reactors containing a solid synthetic wet waste (10% of compost (Mantillo, Spain), 30% rabbit food, 10% starch, 5% sugar, 1% urea, 4% corn oil, and 40% sawdust) and approximately 50 wt% of water content. The reactors were incubated under aerobic conditions at 58 °C. Film samples were recovered at different disintegration times (6, 23, 37, 51, 65, and 90 days). The disintegration degree at different days of incubation under compositing conditions was calculated by normalizing the sample weight to the initial weight.

3. Results

It is widely known that to prepare solvent-based electrospun biopolymeric fibers, the polymeric matrix should be homogeneously dissolved in a proper solvent [2,10]. In this context, good solubility of PLA and PHB matrices has been observed in CF:DMF in a proportion of 80:20 [7,10,36], and it has been ascribed to a similarity in their chemical structure which leads to similar solubility parameters (δ) [25] considering their group contribution according to the Small's cohesive energies [37]: δ_{PLA} = 19.5–20.5 $MPa^{1/2}$ [38] and δ_{PHB} = 18.5–20.1 $MPa^{1/2}$ [39], while solvents also show similar solubility parameters: δ_{CF} = 19 $MPa^{1/2}$ and δ_{DMF} = 24.9 $MPa^{1/2}$ [32]. CF (boiling point 60°C) has been considered as an effective solvent for PLA-PHB blends; while DMF with lower evaporation rate (boiling point 153 °C) usually produces better electrospun PLA-PHB fibers [10]. Thus, allowing the solvent evaporation during electrospun fibers processing [7,10,32,36]. The good miscibility of PLA with OLA [40] has also been directly related with the high similarity in their chemical structure and solubility parameters (δ_{PLA} = 19.5–20.5 $MPa^{1/2}$ [38] and δ_{OLA} = 17.7 $MPa^{1/2}$ [41]). Catechin also shows the solubility parameter in the same order of magnitude (δ_{Cat}= 11.9 $MPa^{1/2}$ [42]) and if it is used as additive for PLA-PHB electrospun fibers in an amount less than 5 wt%, bead-free electrospun fibers can be obtained, while it induces a decrease of the average fiber diameter [7]. Therefore, the combination of PLA-PHB polymeric matrices in the proportion 75:25, the plasticization of the PLA-PHB blends in 15 wt% with OLA, as well as the use of catechin as antioxidant additive, in the amount of less than 5 wt% (i.e., 1 and 3 wt%) should be miscible.

The viscosity of the polymeric solutions was reduced with the presence of OLA plasticizer, see Table 1, but without the formation of droplets showing enough viscosity to be spinnable. Meanwhile, the viscosity of the polymeric solutions increased with the addition of catechin with a consequent higher average fiber diameter, see Table 1.

The morphological aspects, as well as the average fiber diameter of the electrospun mats, were studied by SEM, see Figure 1. A PLA-PHB mat exhibits randomly oriented uniform, straight, and bead-less electrospun fibers in agreement with already reported electrospun PLA-PHB (75:25) mats [10,36]. The plasticization of the PLA-PHB system with OLA lead to less straight fibers with a coarser surface and a slightly increased average fiber diameter, see Table 1. Cat increased the fiber diameter of PLA-PHB mats, see Figure 1c,e. Nevertheless, in plasticized systems Cat slightly reduced the fiber diameter, see Table 1. This behavior has been ascribed to the better interaction due to the presence of plasticizer which allows a homogenous distribution of Cat into the polymeric matrix which better interacts with all components in the system (PLA, PHB, and OLA) by means of hydrogen bonding [7]. Catechin-loaded plasticized electrospun PLA-PHB-OLA fibers show some spindle-like defects (beads), probably due to the reduced viscosity of plasticized electrospun solutions.

Figure 1g,h shows the cross-section of the bilayer PHBV/PLA-PHB and PHBV/PLA-PHB-OLA films respectively, as an example. The two layers with good adhesion between them can be clearly distinguished, the electrospun inner layer was very thin (thickness of about 20–30 μm) in comparison to the overall thickness of the bilayer system (200 ± 20 μm). The absence of phase separations between two polymeric layers has been ascribed to the thermal treatment between them [9]. However, less adhesion can be observed in the case of PHBV/PLA-PHB, see Figure 1g. Meanwhile, the presence of plasticizer in PLA-PHB electrospun fibers favors the adhesion of the mat to the PHBV surface, see Figure 1h. Moreover, in order to characterize the morphology of the electrospun fibers obtained after the hot-pressed assembly process used for bilayer preparation, the inner layer of the bilayer systems was observed by SEM. As an example, in Figure 1i,j the surface of the inner layer of PHBV/PLA-PHB-OLA of bilayer system is shown. It can be observed that the electrospun fiber structure was mainly maintained after heat compression treatment. However, fibers in direct contact with the hot plates in the compression molding press resulted in higher diameters and they showed less straight and coarser fibers. Thus, the overall average fiber diameter increased from 260 ± 78 nm in PLA-PHB electrospun fiber mats to 342 ± 120 nm in PHBV/PLA-PHB-OLA, showing more scattered values.

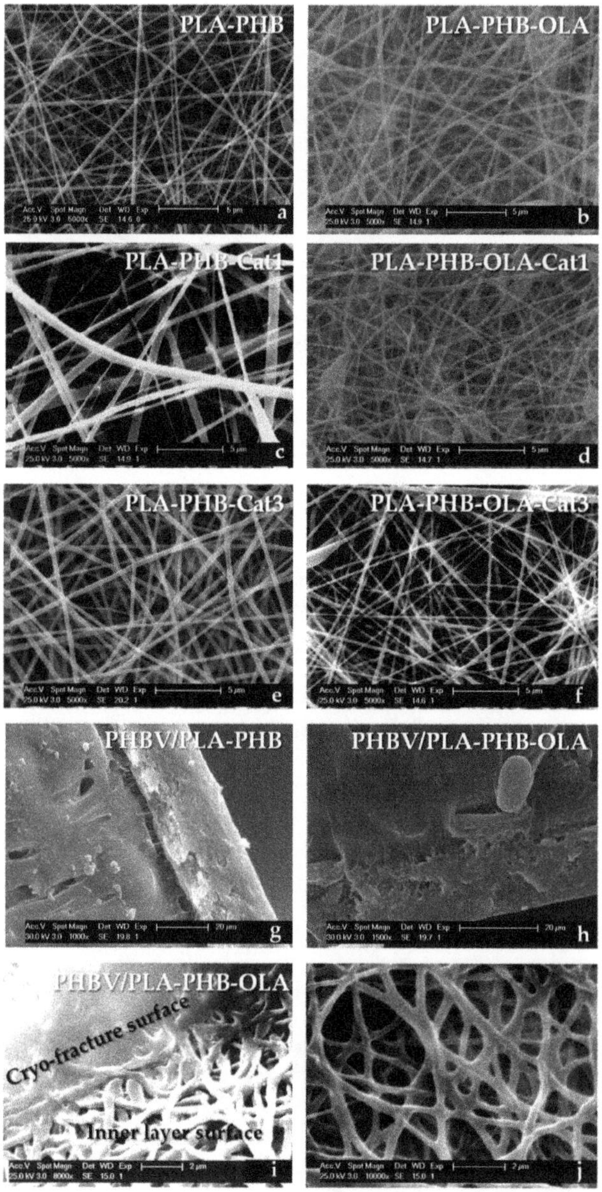

Figure 1. Scanning electron microscope (SEM) observations of electrospun mats: (**a**) PLA-PHB, (**b**) PLA-PHB-OLA, (**c**) PLA-PHB-Cat1, (**d**) PLA-PHB-OLA-Cat1, (**e**) PLA-PHB-Cat3, and (**f**) PLA-PHB-OLA-Cat3; SEM observations of the cry-fracture section of (**g**) PHBV/PLA-PHB and (**h**) PHBV/PLA-PHB-OLA as well as (**i**,**j**) inner surface of PHBV/PLA-PHB-OLA.

The effect of Cat and OLA on the thermal properties of the electrospun PLA-PHB blends was investigated by thermogravimetric isothermal and dynamic measurements. The isothermal TGA analysis, Figure 2, was carried out at 150 °C in order to ensure sufficient thermal stability of the electrospun active layer for assembly with the PHBV-based outer layer by compression molding for industrial purposes. Under isothermal conditions, no significant differences were observed between the PHBV-based pellet and the corresponding thermally processed PHBV-based films, suggesting that no thermal degradation has occurred during processing. Electrospun PLA-PHB mat showed slightly less thermal stability than the PHBV-based layer, which was slightly improved with 3 wt% of catechin addition (PLA-PHB-Cat3). The plasticized electrospun PLA-PHB mat (PLA-PHB-OLA) showed the lowest thermal stability. Meanwhile, plasticized systems loaded with catechin showed somewhat higher thermal stability since catechin protects the polymeric matrix from thermal degradation. It should be highlighted that the two layers are assembled into the final bilayer system by compression molding at 150 °C during 1 min and cooled down in 2 min. Thus, considering the actual assembly time of 1 min at 150 °C, the electrospun mats lost less than 0.5% of the mass.

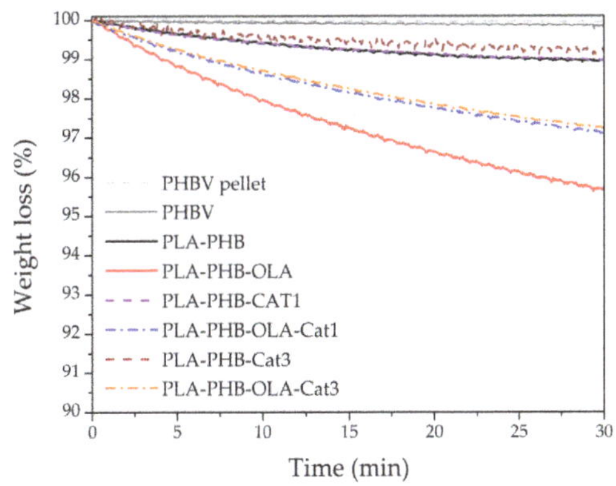

Figure 2. Isotheral thermogravimetric analysis (TGA) at 150 °C of electrospun mats.

The main thermal parameters obtained from the dynamic TGA and DTG curves, see Figure 3, are summarized in Table 2. Electrospun PLA-PHB and plasticized electrospun PLA-PHB mats present a two-step thermal degradation process in which PHB shows its maximum degradation temperature at around 280 °C (T_{max1}) and PLA at around 340 °C (T_{max2}). The presence of OLA plasticizer decreased the thermal stability of the electrospun PLA-PHB mats, shifting the onset degradation temperature to around 30 °C. Nevertheless, OLA improved the dispersion of catechin, since those materials with catechin at 3 wt% and OLA (PLA-PHB-OLA-Cat3) showed improved thermal stability with respect to the un-plasticized systems. It should be mentioned that the onset degradation of electrospun inner layers occurred between 197 °C and 258 °C, which is a temperature higher than that required for the compression-molded assembly process of bilayer systems, that is 150 °C, as it was already discussed for isothermal TGA results.

Catechin addition in 1 wt% increased the T_{max1} of PHB and T_{max2} of PLA (around 10 °C) in electrospun PLA-PHB-OLA-Cat1 with respect to the PLA-PHB-OLA counterpart, showing an effective stabilizing effect for both polymeric matrices. A higher amount of catechin (3 wt%) was not able to stabilize either the PHB or PLA matrix, as indicated by a significant decrease of the T_{max1} and T_{max2}. This unexpected result has been related to the fact that the optimum stabilization effect of phenolic compounds on the polymer matrices is characterized by an optimum amount, while higher amounts over the optimal point do not provide a higher stabilization effect [43,44].

The PHBV sample showed a two-step degradation behavior since it is a compound of PHBV/PBAT, see Figure 3c,d and Table 3. The first step corresponds to the degradation of PHBV, while the second is the degradation of PBAT [30]. In bilayer systems, it was observed that, in general, all mats slightly reduce the thermal stability of PHBV, as it can be seen by the decrease of the onset of thermal degradation, as well as the maximum degradation temperatures. Nevertheless, those systems with higher amounts of catechin in the inner layer showed an increase of the T_0 and the T_{max1}

in bilayer films, see Table 3, with respect to their unloaded counterparts. The antioxidant ability of catechin protects the PHBV-based polymer compound from thermal degradation at the first stage [5]. Meanwhile, the T_{max2} of PBAT was shifted to lower values.

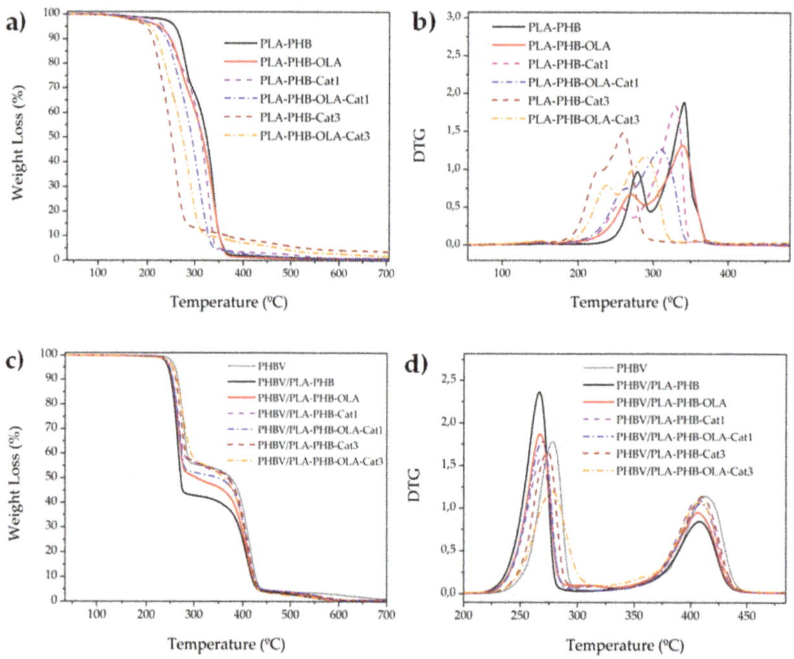

Figure 3. Dynamic TGA measurements of electrospun mats: (**a**) TGA and (**b**) DTG, as well as of bilayer systems: (**c**) TGA and (**d**) DTG.

Table 2. TGA and DTG results of the electrospun fiber mats.

Electrospun Mats	T_0 (°C)	T_{max1} (°C)	T_{max2} (°C)
PLA-PHB	258.1	278.7	341.0
PLA-PHB-OLA	225.0	270.2	339.2
PLA-PHB-Cat1	232.9	253.8	329.7
PLA-PHB-OLA-Cat1	220.3	263.2	310.4
PLA-PHB-Cat3	197.5	270.8	344.7
PLA-PHB-OLA-Cat3	205.9	250.8	282.5

Figure 4 shows the mechanical properties of the monolayer (left) and bilayer systems (right). In monolayer systems, it is possible to observe that catechin acts as a filler, reinforcing the PLA-PHB matrix and leading to an increase of the Young's modulus, as shown in Figure 4a, and tensile strength, see Figure 4c. The reinforcing effect of catechin particles in PLA-based materials has already been observed, particularly in low amounts when it is well dispersed into the polymeric matrix and is attributed to the strong interaction of catechin hydroxyl groups with PLA carbonyl groups [5,7,43]. In this work, this reinforcing effect was particularly marked when catechin was at 1 wt%, due to the good dispersion achieved at this proportion, as well as to the presence of fibers without defects (such as beads). On the other side, the presence of OLA increased the elongation at break, showing the success of the plasticizing effect, see Figure 4e. [40]. Regarding the bilayer systems, the fibers increased the Young's modulus of PHBV-based monolayer, see Figure 4b, confirming the good adhesion in the bilayer structure. Meanwhile, the tensile strength, see Figure 4d, and the elongation at break, see Figure 4f, of the PHBV-based material were not significantly altered by the addition of the electrospun active layer.

Table 3. TGA and DTG results of bilayer systems.

Electrospun Mats	T_0 (°C)	T_{max1} (°C)	T_{max2} (°C)
PHVB	260.0	278.2	413.1
PHBV/PLA-PHB	244.9	266.6	407.8
PHBV/PLA-PHB-OLA	247.5	266.9	406.3
PHBV/PLA-PHB-Cat1	248.5	267.8	406.7
PHBV/PLA-PHB-OLA-Cat1	248.5	268.8	408.9
PHBV/PLA-PHB-Cat3	254.7	273.6	410.1
PHBV/PLA-PHB-OLA-Cat3	255.2	277.3	408.2

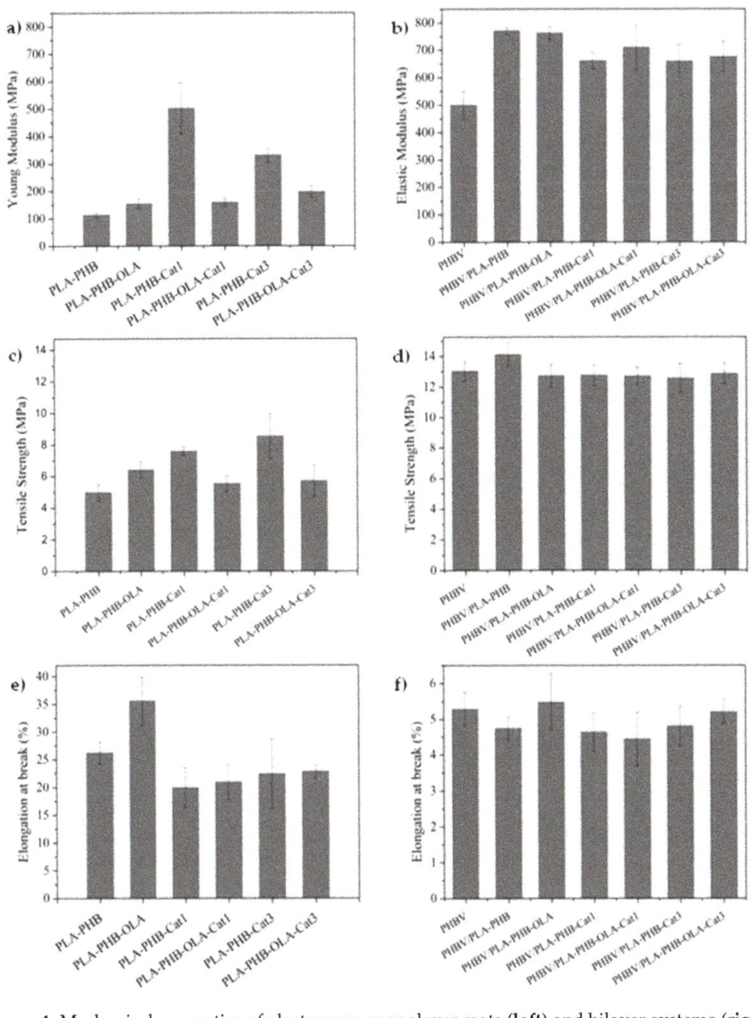

Figure 4. Mechanical properties of electrospun monolayer mats (**left**) and bilayer systems (**right**).

The release of catechin from the bilayer systems was studied in fatty food simulant D1 and it was expressed as the amount of catechin and epicatechin released from the inner layer to the food simulant D1 at 1, 6, 10, and 20 days, see Figure 5a. Catechin presents high solubility in ethanol (50 g L^{-1}), this is the reason why it is able to interact with fatty food simulant and be released from the polymeric matrix. The incorporation of OLA showed a significant increase in the catechin release capacity from electrospun PLA-PHB electrospun fibers. This behavior has been ascribed to the increased polymer chain mobility due to the plasticizing effect which favors the active compound release [3,5,18,26].

In order to confirm the effectiveness of the developed bilayer structures as antioxidant bilayer packaging systems, the reduction of stable free radical DPPH caused by catechin presence in the food simulant D1 was studied, see Figure 5b. A higher antioxidant effect was observed in the plasticized systems with a higher amount of catechin (3 wt%). This can be explained as plasticizer presence improved the release capacity of catechin from the polymeric matrix to the foodstuff and, thus, OLA leads to materials with higher antioxidant effectiveness than unplasticized ones. As expected, the behavior of the antioxidant activity as a function of time is comparable with the catechin release.

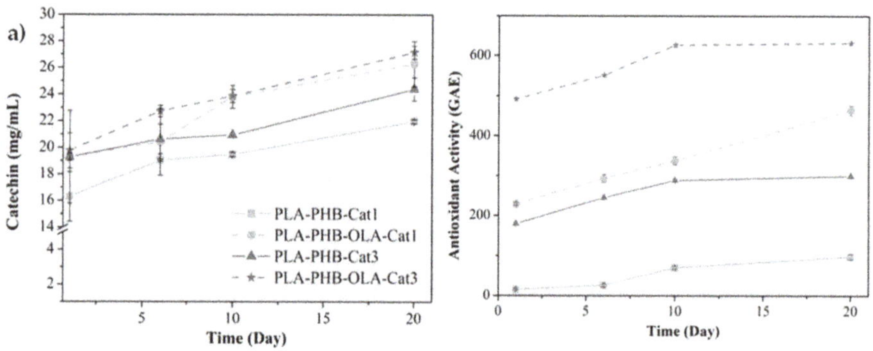

Figure 5. (**a**) Catechin release from bilayer materials to the food simulant and (**b**) antioxidant activity expressed as gallic acid concentration measured by DPPH radical scavengers.

Finally, the disintegrability under composting conditions was conducted to corroborate the biodisintegrable character of the developed bilayer materials, see Figure 6. Although the biodegradable character was governed by the outer layer of PHBV, which is thicker than the electrospun layer, some insights regarding the influence of the electrospun layer can be obtained. After 23 days of disintegration, the bilayer systems started to become breakable, see Figure 6a, suggesting that the hydrolysis process starts in the electrospun materials and continues in the PHBV matrix. Thus, the electrospun layer somewhat speeds up the disintegration process. Catechin slightly delays the disintegration process, while OLA speeds it up. The disintegration process particularly increases when both additives were incorporated into the system. It is worth noting that all formulations were totally disintegrated under composting conditions in less than three months (90% of disintegration according to ISO 20200, Figure 6b), showing their inherent biodegradable character by thermophilic bacteria.

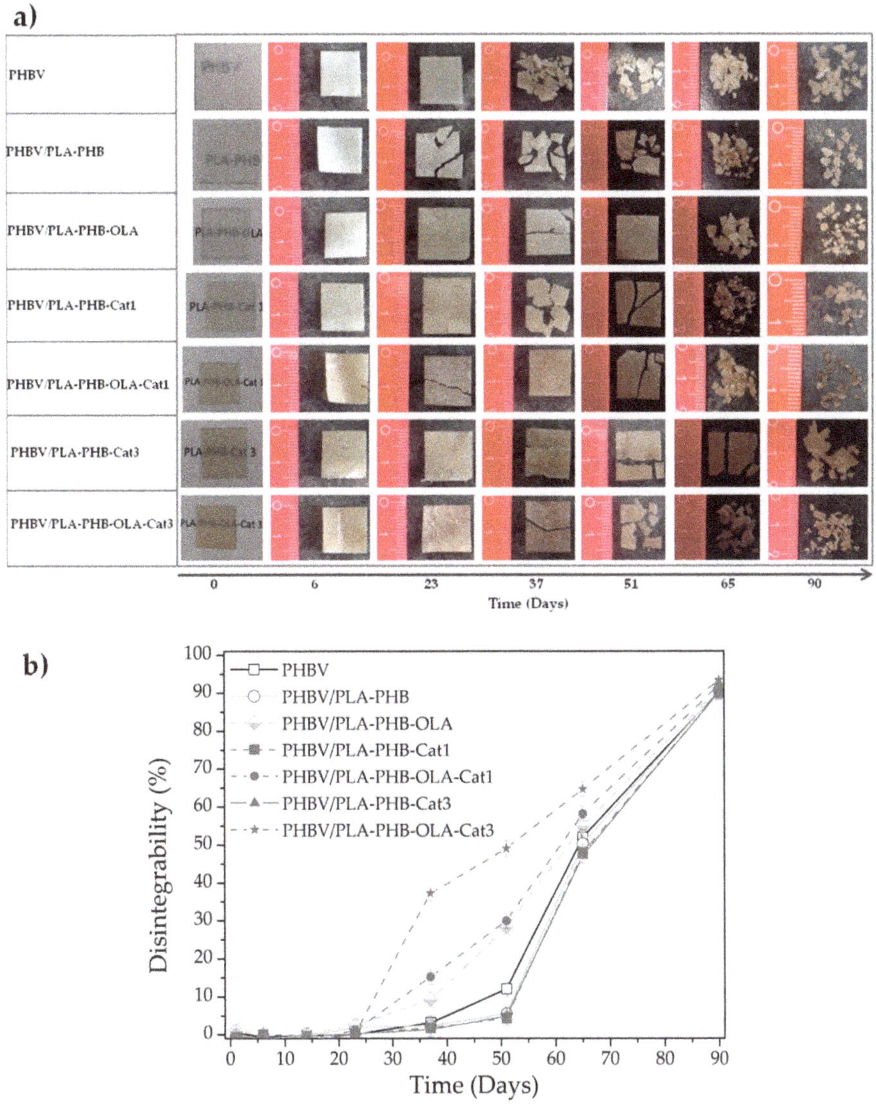

Figure 6. Disintegration under composting conditions: (**a**) Visual appearance of bilayer films before and after different incubation days under composting conditions, and (**b**) disintegration degree of bilayer films as a function of time under composting conditions.

4. Conclusions

Biodegradable antioxidant bilayer films were successfully developed by using an outer layer based on a compression-molded PHVB/PBAT-based material and an inner antioxidant layer based on plasticized PLA-PHB electrospun fibers loaded with catechin. Plasticized electrospun PLA-PHB fibers encapsulating catechin were successfully obtained by means of the electrospinning technique. The addition of 1 wt% of Cat was enough to improve the mechanical properties of the electrospun mats, while 3 wt% produced some structural defects reducing the mechanical performance of the final materials. The obtained mats were used as inner layers for PHBV-based bilayer systems, where a PHBV/PBAT-based material provided the mechanical resistance to the final packaging film. The bilayer systems presented antioxidant activity in a fatty food simulant showing their potential as active food packaging materials. Meanwhile, they were totally disintegrated under composting conditions highlighting their potential application in the sustainable food packaging industry.

Author Contributions: M.P.A. and L.P. conceived and designed the paper structure. M.P.A. and A.D.G. performed the experiments, measurements and data analysis. M.P.A. and L.P. wrote the manuscript. S.F. synthesized the O.L.A. plasticizer and co-wrote the manuscript. D.L. supervises the rheological measurements and co-wrote the manuscript. L.P. and D.L. performed the project administration.

Funding: This research was funded by the Ministerio de Ciencia, Innovación y Universidades (MICINN, Spain), the Agencia Nacional de Investigación (AEI, Spain) and el Fondo Europeo de Desarrollo Regional (FEDER, EU), project number MAT2017-88123-P. M.P. Arrieta thanks Universidad Complutense de Madrid for her postdoctoral contract (Ayuda posdoctoral de formación en docencia e investigación en los Departamentos de la UCM). L. Peponi thanks MICINN for her "Ramon y Cajal" contract RYC-2014-15595.

Conflicts of Interest: The authors declare no conflict of interest.

References

1. Vera, P.; Canellas, E.; Nerín, C. New antioxidant multilayer packaging with nanoselenium to enhance the shelf-life of market food products. *Nanomaterials* **2018**, *8*, 837. [CrossRef] [PubMed]
2. Mujica-Garcia, A.; Sonseca, A.; Arrieta, M.P.; Yusef, M.; López, D.; Gimenez, E.; Kenny, J.M.; Peponi, L. Electrospun fibers based on biopolymers. In *Advanced Surface Engineering Materials*; Tiwari, A., Wang, R., Wei, B., Eds.; Scrivener Publishing LLC: Salem, MA, USA, 2016; pp. 385–438.
3. Castro-López, M.M.; López de Dicastillo, C.; López-Vilariño, J.M.; González-Rodríguez, M.A.V. Improving the capacity of polypropylene to be used in antioxidant active films: Incorporation of plasticizer and natural antioxidants. *J. Agric. Food Chem.* **2013**, *61*, 8462–8470.
4. Fabra, M.J.; López-Rubio, A.; Lagaron, J.M. Use of the electrohydrodynamic process to develop active/bioactive bilayer films for food packaging applications. *Food Hydrocoll.* **2016**, *55*, 11–18. [CrossRef]
5. Arrieta, M.P.; Castro-López, M.D.M.; Rayón, E.; Barral-Losada, L.F.; López-Vilariño, J.M.; López, J.; González-Rodríguez, M.V. Plasticized poly(lactic acid)-poly(hydroxybutyrate) (PLA-PHB) blends incorporated with catechin intended for active food-packaging applications. *J. Agric. Food Chem.* **2014**, *62*, 10170–10180. [CrossRef] [PubMed]
6. Teixeira, A.; Baenas, N.; Dominguez-Perles, R.; Barros, A.; Rosa, E.; Moreno, D.; Garcia-Viguera, C. Natural bioactive compounds from winery by-products as health promoters: A review. *Int. J. Mol. Sci.* **2014**, *15*, 15638–15678. [CrossRef] [PubMed]
7. Arrieta, M.P.; López, J.; López, D.; Kenny, J.M.; Peponi, L. Effect of chitosan and catechin addition on the structural, thermal, mechanical and disintegration properties of plasticized electrospun PLA-PHB biocomposites. *Polym. Degrad. Stab.* **2016**, *132*, 145–156. [CrossRef]
8. Castro Mayorga, J.L.; Fabra Rovira, M.J.; Cabedo Mas, L.; Sánchez Moragas, G.; Lagarón Cabello, J.M. Antimicrobial nanocomposites and electrospun coatings based on poly(3-hydroxybutyrate-co-3-hydroxyvalerate) and copper oxide nanoparticles for active packaging and coating applications. *J. Appl. Polym. Sci.* **2018**, *135*, 45673. [CrossRef]
9. Lasprilla-Botero, J.; Torres-Giner, S.; Pardo-Figuerez, M.; Álvarez-Láinez, M.; Lagaron, J.M. Superhydrophobic bilayer coating based on annealed electrospun ultrathin poly(ε-caprolactone) fibers and electrosprayed nanostructured silica microparticles for easy emptying packaging applications. *Coatings* **2018**, *8*, 173. [CrossRef]
10. Arrieta, M.P.; López, J.; López, D.; Kenny, J.M.; Peponi, L. Development of flexible materials based on plasticized electrospun PLA-PHB blends: Structural, thermal, mechanical and disintegration properties. *Eur. Polym. J.* **2015**, *73*, 433–446. [CrossRef]
11. Torres-Giner, S.; Echegoyen, Y.; Teruel-Juanes, R.; Badia, J.; Ribes-Greus, A.; Lagaron, J. Electrospun poly (ethylene-co-vinyl alcohol)/graphene nanoplatelets composites of interest in intelligent food packaging applications. *Nanomaterials* **2018**, *8*, 745. [CrossRef] [PubMed]
12. Torres-Giner, S.; Gimenez, E.; Lagarón, J.M. Characterization of the morphology and thermal properties of zein prolamine nanostructures obtained by electrospinning. *Food Hydrocoll.* **2008**, *22*, 601–614. [CrossRef]

13. Lagaron, J.M.; Lopez-Rubio, A. Nanotechnology for bioplastics: Opportunities, challenges and strategies. *Trends Food Sci. Technol.* **2011**, *22*, 611–617. [CrossRef]
14. Arrieta, M.P.; Fortunati, E.; Burgos, N.; Peltzer, M.A.; López, J.; Peponi, L. Nanocellulose-based polymeric blends for food packaging applications. In *Multifunctional Polymeric Nanocomposites Based on Cellulosic Reinforcements*; Puglia, D., Fortunati, E., Kenny, J.M., Eds.; William Andrew Publishing: Norwich, NY, USA, 2016; pp. 205–252.
15. Cherpinski, A.; Torres-Giner, S.; Cabedo, L.; Méndez, J.A.; Lagaron, J.M. Multilayer structures based on annealed electrospun biopolymer coatings of interest in water and aroma barrier fiber-based food packaging applications. *J. Appl. Polym. Sci.* **2018**, *135*, 45501. [CrossRef]
16. Cerqueira, M.A.; Fabra, M.J.; Castro-Mayorga, J.L.; Bourbon, A.I.; Pastrana, L.M.; Vicente, A.A.; Lagaron, J.M. Use of electrospinning to develop antimicrobial biodegradable multilayer systems: Encapsulation of cinnamaldehyde and their physicochemical characterization. *Food Bioprocess Technol.* **2016**, *9*, 1874–1884. [CrossRef]
17. López de Dicastillo, C.; Patiño, C.; Galotto, M.J.; Palma, J.L.; Alburquenque, D.; Escrig, J. Novel antimicrobial titanium dioxide nanotubes obtained through a combination of atomic layer deposition and electrospinning technologies. *Nanomaterials* **2018**, *8*, 128. [CrossRef] [PubMed]
18. Arrieta, M.P.; López de Dicastillo, C.; Garrido, L.; Roa, K.; Galotto, M.J. Electrospun pva fibers loaded with antioxidant fillers extracted from durvillaea antarctica algae and their effect on plasticized PLA bionanocomposites. *Eur. Polym. J.* **2018**, *103*, 145–157. [CrossRef]
19. Castro-Mayorga, J.; Fabra, M.; Cabedo, L.; Lagaron, J. On the use of the electrospinning coating technique to produce antimicrobial polyhydroxyalkanoate materials containing in situ-stabilized silver nanoparticles. *Nanomaterials* **2017**, *7*, 4. [CrossRef] [PubMed]
20. Arrieta, M.P.; Sessini, V.; Peponi, L. Biodegradable poly(ester-urethane) incorporated with catechin with shape memory and antioxidant activity for food packaging. *Eur. Polym. J.* **2017**, *94*, 111–124. [CrossRef]
21. Sessini, V.; Arrieta, M.P.; Fernández-Torres, A.; Peponi, L. Humidity-activated shape memory effect on plasticized starch-based biomaterials. *Carbohydr. Polym.* **2018**, *179*, 93–99. [CrossRef] [PubMed]
22. Sessini, V.; Arrieta, M.P.; Kenny, J.M.; Peponi, L. Processing of edible films based on nanoreinforced gelatinized starch. *Polym. Degrad. Stab.* **2016**, *132*, 157–168. [CrossRef]
23. Han, J.; Shin, S.H.; Park, K.M.; Kim, K.M. Characterization of physical, mechanical, and antioxidant properties of soy protein-based bioplastic films containing carboxymethylcellulose and catechin. *Food Sci. Biotechnol.* **2015**, *24*, 939–945. [CrossRef]
24. Arrieta, M.P.; Samper, M.D.; Aldas, M.; López, J. On the use of PLA-PHB blends for sustainable food packaging applications. *Materials* **2017**, *10*, 1008. [CrossRef] [PubMed]
25. Arrieta, M.P.; Samper, M.D.; López, J.; Jiménez, A. Combined effect of poly(hydroxybutyrate) and plasticizers on polylactic acid properties for film intended for food packaging. *J. Polym. Environ.* **2014**, *22*, 460–470. [CrossRef]
26. Burgos, N.; Armentano, I.; Fortunati, E.; Dominici, F.; Luzi, F.; Fiori, S.; Cristofaro, F.; Visai, L.; Jiménez, A.; Kenny, J.M. Functional properties of plasticized bio-based poly(lactic acid)_poly(hydroxybutyrate) (PLA_PHB) films for active food packaging. *Food Bioprocess Technol.* **2017**, *10*, 770–780. [CrossRef]
27. Requena, R.; Jiménez, A.; Vargas, M.; Chiralt, A. Poly[(3-hydroxybutyrate)-*co*-(3-hydroxyvalerate)] active bilayer films obtained by compression moulding and applying essential oils at the interface. *Polym. Int.* **2016**, *65*, 883–891. [CrossRef]
28. Javadi, A.; Kramschuster, A.J.; Pilla, S.; Lee, J.; Gong, S.; Turng, L.S. Processing and characterization of microcellular PHBV/PBAT blends. *Polym. Eng. Sci.* **2010**, *50*, 1440–1448. [CrossRef]
29. Javadi, A.; Srithep, Y.; Lee, J.; Pilla, S.; Clemons, C.; Gong, S.; Turng, L.-S. Processing and characterization of solid and microcellular PHBV/PBAT blend and its RWF/nanoclay composites. *Compos. Part A Appl. Sci. Manuf.* **2010**, *41*, 982–990. [CrossRef]
30. Nagarajan, V.; Mohanty, A.K.; Misra, M. Sustainable green composites: Value addition to agricultural residues and perennial grasses. *ACS Sustain. Chem. Eng.* **2013**, *1*, 325–333. [CrossRef]
31. Fiori, S.; Ara, P. Method for Plasticizing Lactic Acid Polymers. Patent WO 2009/092825, 25 January 2008.
32. Mujica-Garcia, A.; Navarro-Baena, I.; Kenny, J.M.; Peponi, L. Influence of the processing parameters on the electrospinning of biopolymeric fibers. *J. Renew. Mater.* **2014**, *2*, 23–34. [CrossRef]

33. European Commission. No. 10/2011 of 14 january 2011 on plastic materials and articles intended to come into contact with food. *Off. J. Eur. Union* **2011**, *L12*, 1–89.
34. López de Dicastillo, C.; Castro-López, M.M.; Lasagabaster, A.; Vilariño, J.M.L.; Rodríguez, M.V.G. Interaction and release of catechin from anhydride maleic-grafted polypropylene films. *ACS Appl. Mater Interfaces.* **2013**, *5*, 3281–3289.
35. UNE-EN ISO. *Determination of the Degree of Disintegration of Plastic Materials under Simulated Composting Conditions in a Laboratory-Scale Test*; UNE-EN ISO-20200:2015; AENOR: Madrid, España, 2016.
36. Arrieta, M.P.; López, J.; López, D.; Kenny, J.M.; Peponi, L. Biodegradable electrospun bionanocomposite fibers based on plasticized PLA–PHB blends reinforced with cellulose nanocrystals. *Ind. Crop. Prod.* **2016**, *93*, 290–301. [CrossRef]
37. Weast, R.C.; Astle, M.J.; Beyer, W.H. *Crc Handbook of Chemistry and Physics*; CRC Press: Boca Raton, FL, USA, 1989; Volume 1990.
38. Auras, R.; Harte, B.; Selke, S. An overview of polylactides as packaging materials. *Macromol. Biosci.* **2004**, *4*, 835–864. [CrossRef] [PubMed]
39. Calvao, P.S.; Chenal, J.M.; Gauthier, C.; Demarquette, N.R.; Bogner, A.; Cavaille, J.Y. Understanding the mechanical and biodegradation behaviour of poly (hydroxybutyrate)/rubber blends in relation to their morphology. *Polym. Int.* **2012**, *61*, 434–441. [CrossRef]
40. Burgos, N.; Martino, V.P.; Jiménez, A. Characterization and ageing study of poly(lactic acid) films plasticized with oligomeric lactic acid. *Polym. Degrad. Stab.* **2013**, *98*, 651–658. [CrossRef]
41. Burgos, N.; Tolaguera, D.; Fiori, S.; Jiménez, A. Synthesis and characterization of lactic acid oligomers: Evaluation of performance as poly(lactic acid) plasticizers. *J. Polym. Environ.* **2014**, *22*, 227–235. [CrossRef]
42. Kang, J.H.; Chung, S.T.; Row, K.H. Comparison of experimental and calculated solubilities of (+)-catechin in green tea. *J. Ind. Eng. Chem.* **2002**, *8*, 354–358.
43. Arrieta, M.P.; Peponi, L. Polyurethane based on PLA and PCL incorporated with catechin: Structural, thermal and mechanical characterization. *Eur. Polym. J.* **2017**, *89*, 174–184. [CrossRef]
44. Samper, M.; Fages, E.; Fenollar, O.; Boronat, T.; Balart, R. The potential of flavonoids as natural antioxidants and UV light stabilizers for polypropylene. *J. Appl. Polym. Sci.* **2013**, *129*, 1707–1716. [CrossRef]

nanomaterials

MDPI

Article

Physicochemical, Antioxidant and Antimicrobial Properties of Electrospun Poly(ε-caprolactone) Films Containing a Solid Dispersion of Sage (*Salvia officinalis* L.) Extract

Ana Salević [1], Cristina Prieto [2,3], Luis Cabedo [4], Viktor Nedović [1] and Jose Maria Lagaron [2,*]

[1] Department of Food Technology and Biochemistry, Faculty of Agriculture, University of Belgrade, Nemanjina 6, 11080 Zemun, Belgrade, Serbia; ana.salevic@agrif.bg.ac.rs (A.S.); vnedovic@agrif.bg.ac.rs (V.N.)

[2] Novel Materials and Nanotechnology Group, Institute of Agrochemistry and Food Technology (IATA), Spanish Council for Scientific Research (CSIC), Calle Catedrático Agustín Escardino Benlloch 7, 46980 Paterna, Valencia, Spain; cprieto@iata.csic.es

[3] Bioinicia R&D Department, Bioinicia S.L., Calle Algepser 65, 46980 Paterna, Spain

[4] Polymers and Advanced Materials Group (PIMA), Jaume I University (UJI), Avenida de Vicent Sos Baynat s/n, 12071 Castellón, Spain; lcabedo@uji.es

* Correspondence: lagaron@iata.csic.es; Tel.: +34-963-900-022

Received: 22 January 2019; Accepted: 12 February 2019; Published: 15 February 2019

Abstract: In this study, novel active films made of poly(ε-caprolactone) (PCL) containing a solid dispersion of sage extract (SE) were developed by means of the electrospinning technique and subsequent annealing treatment. Initially, the antioxidant and antimicrobial potential of SE was confirmed. Thereafter, the effect of SE incorporation at different loading contents (5%, 10%, and 20%) on the physicochemical and functional properties of the films was evaluated. The films were characterized in terms of morphology, transparency, water contact angle, thermal stability, tensile properties, water vapor, and aroma barrier performances, as well as antioxidant and antimicrobial activities. Thin, hydrophobic films with good contact transparency were produced by annealing of the ultrathin electrospun fibers. Interestingly, the effect of SE addition on tensile properties and thermal stability of the films was negligible. In general, the water vapor and aroma permeability of the PCL-based films increased by adding SE to the polymer. Nevertheless, a strong 2,2-diphenyl-1-picrylhydrazyl (DPPH˙) free radical scavenging ability, and a strong activity against foodborne pathogens *Staphylococcus aureus* and *Escherichia coli* were achieved by SE incorporation into PCL matrix. Overall, the obtained results suggest great potential of the here-developed PCL-based films containing SE in active food packaging applications with the role of preventing oxidation processes and microbial growth.

Keywords: poly(ε-caprolactone); sage; electrospinning; nanofibers; active packaging; antioxidant activity; antimicrobial activity

1. Introduction

Oxidation processes and microbial growth are common causes of food deterioration which result in color changes, texture modifications, development of off-flavor, and loss of nutritional value and quality of foodstuffs [1,2]. In this regard, spices and herbs have been traditionally added to food, not just as flavoring and healing agents, but also as preservatives [3]. Nowadays, many spices and herbs are recognized as sources of bioactive compounds which are able to stabilize free radicals and prevent oxidation processes and/or act as bacteriostatic or bactericidal agents [2,3]. In addition, there is a growing interest from consumers and the food industry for the use of natural active compounds in food preservation due to their synergy, potency, and presumed low side effects when compared to synthetic additives [4]. However, a major drawback for direct food application of extracted active compounds represents their susceptibility to adverse external factors, chemical instability, and interactions with food constituents [5]. Another limiting factor is a rapid actuation and diffusion of active compounds within the bulk of food [6,7]. A promising approach to overcome these problems is the incorporation of active compounds in a polymeric matrix by encapsulation, to provide stability, keep functionality, and increase effectiveness during time, creating a physical barrier between actives and their environment [8].

Many efforts done to prevent food deterioration and achieve higher effectiveness of active compounds have triggered innovations in food packaging. This current trend is reflected in the development of active food packaging with improved functionality. In this context, the passive role of traditional packaging in protecting and marketing of a food product has evolved into a novel function as a carrier of active compounds [9,10]. This novel concept is based on the incorporation of various active agents into a packaging material with the aim to maintain or enhance quality and safety, extend shelf life of a packaged product, and to reduce the packaging related environmental pollution [10,11]. Nowadays, the use of natural active compounds as functional ingredients in active food packaging is highlighted.

Sage (*Salvia officinalis* L.) is a herb widely used in cookery due to its seasoning and flavoring properties as well as in traditional medicine to treat dyspepsia, excessive sweating, age-related cognitive disorders, and throat and skin inflammations [12]. Moreover, sage is one of the herbs with a great potential for use as a functional ingredient for the development of active food packaging due to its well-known antioxidant [13,14], antibacterial [3], and antifungal [14] effects. These beneficial activities are positively related to phenolic compounds [3,14], such as phenolic diterpenoids (carnosic acid, carnosol, rosmanol), phenolic acids (caffeic acid, rosmarinic acid, ferulic acid) and flavonoids (luteolin derivatives, apigenin derivatives), among others [13–15]. Essential oils and extracts of *Salvia officinalis* are generally recognized as safe according to the U.S. Food and Drug Administration [16].

While formulating active food packaging, an environmental issue related to an increasing quantity of disposed plastic packaging with unknown biodegradation time should be taken into account. This problem has set a strong challenge towards replacement of non-biodegradable polymers by biodegradable ones, especially for single-use plastic items [17]. In this sense, poly(ε-caprolactone) (PCL), a commercially available and biodegradable aliphatic polyester, could represent an alternative. It is a hydrophobic, semi-crystalline polymer with a low melting point and is miscible with many other polymers [18,19]. The rheological and viscoelastic properties allow an easy processability of PCL [18]. This material is very interesting for packaging applications [19], as well as being used as a carrier of active compounds and development of active materials. For instance, Martínez-Abad et al. studied the potential of PCL for preparation of antibacterial solvent casted films containing cinnamaldehyde and allyl isothiocyanate [20]. Also, PCL electrospun fibers loaded with nettle extract and embedded to whey protein isolate were successfully applied as a bioactive coating to inhibit bacterial growth and extend quality of fresh fish fillets [21].

The design of active packaging materials represents a very dynamic field and a real challenge [10]. In this regard, the electrospinning technique supposes an innovative nanofabrication approach for the development of active food packaging coating and interlayer materials. This approach employs a high-voltage electric field imposed on a polymer solution to create ultrathin mats composed of polymeric fibers with diameters in micro, submicro, and nano range [22–24]. Because of a high trapping efficiency and no need for high processing temperatures, the electrospinning technique is very suitable for entrapping active compounds, such as antioxidants [24] and antimicrobials [25] within a fibrous polymer matrix [22]. Fiber-based systems have gained a lot of attention as a way to improve active functionality and achieve an optimal effect during the food storage [6]. Also, characteristics such as cost-effectiveness, continuous fabricating capability, and a facile operating process make the electrospinning technique an excellent candidate for the development of active packaging materials [22].

There are some studies on the incorporation of sage extracts into whey protein isolate [26] and chitosan matrix [2] by means of solvent casting technique. The study on the whey protein isolate-based film has proven that there is great potential of sage extract to be used as the active constituent of food packaging for antioxidant protection of meat products. Nevertheless, there was no information on film properties [26]. On the other hand, the chitosan-based film loaded with sage extract has been evaluated with respect to its physical properties, regardless of its active functionality [2]. Despite those research works, according to our knowledge there is no commercial application of sage extract in commercial food packaging. When compared to casted films, electrospun films would provide more desirable properties, such as homogeneous dispersion of film constituents and better solvent resistance [27]. However, to the best of our knowledge, the use of the electrospinning technique to develop a PCL-sage extract system has not been carried out so far.

The main aim of this study was to develop and characterize active PCL-based films loaded with sage extract. Formulations containing different sage extract loadings were prepared, employing the electrospinning technique followed by annealing treatment. A comprehensive investigation was performed to assess the effect of the sage extract addition on morphology, transparency, hydrophobicity, thermal stability, tensile properties, water vapor and aroma barrier performance, and, most importantly, on antioxidant and antimicrobial activities of the films.

2. Materials and Methods

2.1. Materials

Sage (*Salvia officinalis* L.) was supplied by the Institute for Medicinal Plants Research "Dr. Josif Pančić" (Belgrade, Serbia). Poly(ε-caprolactone) (Mn=80,000), 2,2-diphenyl-1-picrylhydrazyl (DPPH), (±)-6-hydroxy-2,5,7,8-tetramethylchroman-2-carboxylic acid (Trolox) and D-limonene were obtained from Sigma Aldrich (St. Louis, MO, USA). Ethanol, chloroform, butanol, and methanol were purchased from Panreac Quimica SLU (Barcelona, Spain). Müller Hinton Broth and Agar (MHB and MHA) were provided by Merck KGaA (Darmstadt, Germany). Resazurin sodium salt was procured from MP Biomedicals, LLC (Solon, OH, USA). Phosphate buffered saline (PBS) was purchased from Amresco, LLC (Cleveland, OH, USA). All chemicals were used as received without any further purification.

2.2. Preparation of Sage Extract

Sage extract (SE) was prepared by maceration. An aqueous solution of ethanol (50% v/v) was used as extraction medium. Drug to solvent ratio was 1:20. The extraction process was carried out on an orbital shaker (Stuart SSL1, Staffordshire, UK) with continuous agitation fixed at 200 rpm for 90 min at room temperature. The obtained extract was filtered through a medical gauze and the solvent was evaporated by casting in a fume hood (Flores Valles, Madrid, Spain). Sage extract powder was stored under refrigeration conditions prior to use.

2.3. Characterization of Sage Extract

2.3.1. Antioxidant Activity

The antioxidant capacity of SE was estimated using the DPPH$^{\cdot}$ free radical scavenging assay [28], with some modifications. SE (10 mg) was dissolved in 1 mL of methanol. Thereafter, 100 μL of the appropriate diluted solution was mixed with 1.9 mL of a DPPH methanolic solution (0.094 mM). A control sample was prepared by adding 100 μL of methanol to 1.9 mL of the DPPH solution. The free radical$^{\cdot}$ scavenging ability was determined by measuring the absorbance at 517 nm after incubation in dark at room temperature for 30 min. An UV/Vis spectrophotometer (model 4000, Dinko instruments, Barcelona, Spain) was used. Percentage of DPPH˙ inhibition was calculated following Equation (1):

$$I(\%) = \frac{Ac - As}{Ac} \times 100 \tag{1}$$

where Ac is the absorbance of the control and As is the absorbance of the sample. Trolox (0–1000 μM) was used as a standard to prepare calibration curve. The antioxidant activity was determined from the calibration curve and expressed as mM Trolox equivalents (TE)/g SE.

2.3.2. Antimicrobial Activity

The antibacterial properties of SE and subsequently of the films were ascertained in triplicate against *Staphylococcus aureus* (ATCC 6538P) and *Escherichia coli* (ATCC 25922). The bacterial strains were obtained from the Spanish Type Culture Collection (CECT, Valencia, Spain) and stored in PBS with 10% MHB and 10% glycerol at −80 °C. To prepare fresh inoculum, a loopful of the bacteria was cultivated in MHB at optimal growth conditions overnight and an aliquot was again transferred to MHB and grown at 37 °C to the mid-exponential phase of growth. Suspensions containing approximately 5×10^5 CFU/mL were used for antimicrobial activity assays. Previously to each assay, the samples were sterilized by UV radiation for 30 min in a Biostar cabinet (Telstar S.A., Madrid, Spain).

The antibacterial potential of SE was evaluated by the broth microdilution method [29]. 90 μL of the bacterial suspension was added into a 96-well microtiter plate (Thermo Fischer Scientific, Roskilde, Denmark) containing 10 μL of two-fold serially diluted extract (concentration range from 0.31 to 80 mg/mL). Wells containing only MHB and wells with the bacterial suspension in MHB were used as positive and negative controls, respectively. The plates were incubated at 37 °C for 24 h. Afterward, 10 μL of resazurin, a metabolic indicator, was added into each well and incubated at 37 °C for 3 h. Minimum inhibitory concentration (MIC) was considered as the lowest extract concentration that inhibited bacterial growth according to a resazurin color. The contents from the wells containing dilutions designated as MIC were sub-cultured on MHA. Minimum bactericidal concentration (MBC) was established as the lowest extract concentration for which no bacterial growth was observed after incubation at 37 °C for 24 h.

2.4. Preparation of Poly(ε-caprolactone) Based Films

2.4.1. Preparation of Solutions for Electrospinning

Plain PCL solution (10% *w*/*w*, designated as PCL) was prepared by dissolving PCL pellets in a solvent comprising chloroform and butanol (chloroform:butanol = 3:1, *v*/*v*) under magnetic stirring at room temperature. Three PCL-based active systems containing different contents of the sage extract (5%, 10%, and 20% *w*/*w* with respect to the polymer content) were formulated. These formulations were designated as PCL-SE5, PCL-SE10, and PCL-SE20, respectively. The solutions were prepared by dissolving the required amount of the sage extract in the chloroform-butanol mixture and stirring overnight. Subsequently, the solutions were centrifuged at 10,000 rpm for 10 min. Supernatants were collected, filtered through 0.22 μm polytetrafluoroethylene (PTFE) filters, and filled up to the initial weight. Afterwards, polymer was added to the solutions and stirred until it was completely dissolved.

2.4.2. Characterization of the Solutions

The viscosity was measured using a rotational viscometer Visco Basic Plus L (Fungilab S.A., Sane Feliu de Llobregat, Spain). The surface tension was determined applying the Wilhemy plate method an Easy Dyne K20 tensiometer (Krüss GmbH, Hamburg, Germany). The conductivity was measured using a conductivity meter (HI98192 portable meter HANNA Instruments, Gothenburg, Sweden). The measurements were made in triplicate at room temperature.

2.4.3. Electro-Hydrodynamic Processing

The preparation of the fibrous mats was carried out using a high throughput Fluidnatek LE-500 pilot line electrospinning equipment with temperature and relative humidity control in the lab mode with a single needle injector (Bioinicia S.L., Valencia, Spain). The solutions were drawn in a 20 mL plastic syringe that was placed on a syringe pump and connected by PTFE tube to a stainless steel needle (20 Gauge). A positive electrode of a high voltage power supply was coupled to the needle. The solutions were electrospun at the constant flow rate of 3 mL/h and the voltage of 19 kV for 2 h. The fibers were homogenously deposited onto an aluminum foil sheet placed on a metallic collector using a scanning injector. The distance between the needle tip and the collector was 15 cm. The process was performed at 25 °C and 30% RH.

Subsequently, an annealing step was applied in order to obtain ultrathin, transparent, and continuous films by fibers coalescence. This process was carried out using a hydraulic press (4122 model, Carver Inc., Wabash, IN, USA) at temperatures 55 °C for 25 s without pressure. The annealing, carried out below the polymer melting point at 60 °C, was optimized for PCL electrospun fibers in a previous work [25]. The films were stored before physical characterization in a desiccator containing dried silica gel at 0% RH and 25 °C.

2.5. Characterization of The PCL Based Films

2.5.1. Film Thickness

Before further analysis, the film thickness was measured at five random points of each sample by a digital micrometer with an accuracy of 0.001 mm (S00014, Mitutoyo Corporation, Kawasaki, Japan).

2.5.2. Morphology

Scanning electron microscopy (SEM) analysis of the electrospun fibers and the cryofractured annealed films was performed employing a Hitachi S-4800 microscope (Tokyo, Japan). The films were cryofractured using liquid nitrogen. The samples were attached to beveled holders using a double-sided adhesive tape, coated with a gold-palladium mixture under vacuum, and examined using an accelerating voltage of 5 kV. The average fiber diameter was determined using ImageJ program (National Institutes of Health, Bethesda, MD, USA).

2.5.3. Transparency

The light transmission spectrum was determined by measuring the light absorption in the wavelength range of 200–800 nm using the spectrophotometer (model UV/Vis 4000, Dinko instruments, Barcelona, Spain). Triplicates of the film specimens of 20 mm × 50 mm were fixed to a test cell perpendicularly to the light beam. The transparency (T) was calculated following the Equation (2):

$$T\left(mm^{-1}\right) = \frac{A_{600}}{L} \quad (2)$$

where A_{600} is the absorbance measured at 600 nm and L is the film thickness (mm).

2.5.4. Water Contact Angle

The contact angle of water on the film surface was measured using an optical tensiometer (Theta Lite, Staffordshire, UK). Five droplets (5 μL) of ultrapure water were placed on different positions of three species (20 mm × 50 mm) of each sample and the mean values of the contact angle were calculated.

2.5.5. Thermal Analysis

Thermogravimetric analysis (TGA) was performed by a TG-STDA thermobalance (TGA/STDA851e/LF/1600 model, Mettler-Toledo, LLC, Columbus, OH, USA). The samples (~13 mg) were heated from 25 to 700 °C at a heating rate of 5 °C/min under a dynamic nitrogen atmosphere (flow rate 50 mL/min). The measurements were done in five replicates.

2.5.6. Mechanical Properties

Tensile tests were carried out, employing an universal testing machine (AGS-X 500 N model, Shimadzu, Kyoto, Japan) according to the ASTM Standard D638 [30]. The samples were conditioned at 25 °C and 50% RH for 24 h and cut in dumbbell shaped specimens (5 mm × 25 mm). The tests were performed at a crosshead rate of 10 mm/min on five replicates of each film. The stress–strain curves were prepared on the basis of the force–distance data and used to determine elastic modulus, tensile strength, elongation at break, and toughness.

2.5.7. Water Vapor Permeability (WVP)

WVP was measured gravimetrically in triplicate, following the ASTM E96-95 method [31]. Payne permeability cups of 3.5 cm diameter (Elcometer Sprl, Hermelle-sous-Argenteau, Belgium) were filled with 5 mL of distilled water. The films were sealed with silicon rings in the cups and exposed to 100% RH on one side without direct contact with water. The cups were placed in a desiccator at 0% RH and 25 °C and weighted periodically (±0.0001 g) until the steady state was reached. The cups containing aluminum foil with water and PCL-based films without water in the cups were used as control samples to estimate losses of water through the sealing and of volatile compounds, respectively. Water vapor permeation rate (WVPR) was determined from the permeation slopes obtained from the regression analysis of weight loss data versus time. The weight loss was compensated by the very marginal losses through the sealing and volatiles. WVPR was multiplied by the film thickness to determine WVP.

2.5.8. D-Limonene Permeability (LP)

LP was determined in triplicate, as previously described for WVP. The Payne permeability cups containing 5 mL of D-limonene and the sealed films were stored at 40% RH and 25 °C. Limonene permeation rate (LPR) was determined from the permeation slopes. LPR was multiplied by the film thickness to determine LP.

2.5.9. Antioxidant Activity

Dynamics of DPPH˙ free radical inhibition by the films was monitored. 1.9 mL of the DPPH˙ solution was added into a vial containing 1 mg of film sample. Control sample contained 1.9 mL of the DPPH˙ solution without film. Inhibition rate was determined by measuring the absorbance at 517 nm after incubation of the vials at 100 rpm in dark at room temperature for 0.5, 6, 12, and 24 h. The results were obtained following the Equation (1) and expressed as percentage of DPPH˙ inhibition. All measurements were performed in triplicate.

2.5.10. Antimicrobial Activity

The antibacterial activity of the films was assessed using a modified Japanese Industrial Standard (JIS) Z2801 [32]. The bacterial suspension was spread between the film samples (dimensions of 2 cm × 2 cm). The inoculated film samples were incubated at 95% RH and 25 °C for 24 h. Thereafter, the bacteria were recovered with PBS and plated onto MHA. The number of viable cells was determined after incubation at 37 °C for 24 h. The antibacterial activity of the films was defined by surface reduction of $\log_{10}$ CFU of the test culture during incubation. Surface reduction (R) was calculated according to the Equation (3):

$$R = \left[\log\left(\frac{B}{A}\right) - \log\left(\frac{C}{A}\right)\right] = \log\left(\frac{B}{C}\right) \quad (3)$$

where A is the average of the viable bacterial counts on the control sample immediately after inoculation, B is the average of the viable bacterial counts on the control sample after 24 h and C is the average of the viable bacterial counts on the test sample after 24 h. According to the value of the surface reduction, the antibacterial activity of the films was evaluated with the following assessment: nonsignificant ($R < 0.5$), slight ($0.5 \leq R < 1$), significant ($1 \leq R < 3$), and strong ($R \geq 3$).

2.5.11. Statistical Analysis

The results were expressed as mean ± standard deviation. The data were subjected to one-way analysis of variance (ANOVA) using Statgraphics Centurion XV software (StatPoint, Inc., Warrenton, VA, USA). Tukey's HSD test, at 95% confidence level, was performed to determine the influence of SE addition on the properties of the PCL-based films.

3. Results and Discussion

3.1. Sage Extract Characterization

3.1.1. Antioxidant Activity

The antioxidant potential of SE was evaluated in terms of DPPH$^{\cdot}$ free radicals scavenging ability. SE exhibited the free radical inhibition of 50.56% at concentration of 0.5 mg SE/mL, which corresponds to an antioxidant activity of 1.07 mmol TE/g SE. The here-obtained extract presented similar free radical inhibition as some herb extracts well-known for a strong antioxidant activity, such as *Camellia sinensis* L. and *Rosmarinus officinalis* L., which showed 49.47% and 55.32% scavenging effects, respectively, at a concentration of 0.5 mg/mL [33]. Potent antioxidant activity of sage extracts was reported in many studies in which phenolic compounds, i.e., abietane diterpenoids (carnosol and carnosic acid) and caffeic acid derivates (rosmarinic acid, chlorogenic acid, caffeic acid) were marked as the most effective constituents with the free radical scavenging ability [15,34,35]. The antioxidant activity of phenolic compounds is due to hydroxyl groups positioned along the aromatic phenolic ring, which act as hydrogen or electron donors enabling termination of free radical chain reactions, as well as from the aromatic ring which is able to stabilize and delocalize the unpaired electron [36,37]. In accordance with the presented results, SE may be used as a functional constituent of novel packaging materials to prevent oxidation reactions.

3.1.2. Antimicrobial Activity

Antimicrobial potential of SE was scrutinized in terms of MIC and MBC values against food-borne pathogens: Gram-positive *S. aureus* and G-negative *E. coli*. The obtained results have shown higher sensitiveness of *S. aureus* to SE with MIC and MBC values of 0.31 and 20 mg/mL, respectively. On the other hand, higher concentrations of SE presented the inhibitory (2.50 mg/mL) and biocidal (40 mg/mL) effects against *E. coli*. According to the literature, the antibacterial activity of SE is highly correlated with phenolic compounds [3]. The mechanism of biocidal action of plant compounds is based on degradation of cell wall, damage to cytoplasmic membrane and membrane proteins, leakage of content out of the cell, and coagulation of the cytoplasm [38]. The noticed higher susceptibility of *S. aureus* as a Gram-positive bacteria is in agreement with the previous study and related to the above-mentioned mechanism of biocidal action. In particular, this behavior originated from differences between cell wall structure of Gram-positive and Gram-negative bacteria. Namely, Gram-negative bacteria possess a more complex cell wall and an outer membrane which acts as a barrier to the penetration of the antimicrobial compounds providing higher resistance [3]. Generally, the obtained results suggest that SE has a potential to be used as a functional ingredient for development of materials with not just antioxidant, but also antibacterial activity.

3.2. Solution Characterization

Electrospinning behavior is greatly influenced by the solution properties, typically viscosity, surface tension, and conductivity [6]. Plain PCL solution, as a control sample, and the PCL-based solutions containing SE at three different contents (5%, 10%, and 20% SE) were prepared. As it can be seen in Table 1, the solution parameters are affected by SE addition. A greater reduction in viscosity was observed when the SE content rose. This reduction can be related to the presence of low molecular weight compounds in SE and possible effects on the configuration of polymer chains in the solutions. Furthermore, the addition of a higher amount of SE (10% and 20% SE) significantly increased conductivity. This effect can be attributed to changes in viscosity and mobility of charged species [39]. Regarding the surface tension values, a significant surfactant effect was noticed only for the highest content of SE (20%).

Table 1. Properties of the poly(ε-caprolactone)-based solutions, diameter of the electrospun fibers and thickness of the annealed films.

Formulation	Viscosity (cP)	Conductivity (μS/cm)	Surface Tension (mN/m)	Fiber Diameter (μm)	Film Thickness (mm)
PCL	1908.1 ± 41.2^{a}	0.02 ± 0.00^{a}	27.2 ± 0.2^{a}	4.95 ± 0.29	0.09 ± 0.01^{a}
PCL-SE5	1695.6 ± 36.6^{b}	0.02 ± 0.00^{a}	27.4 ± 0.1^{a}	3.80 ± 0.28	0.10 ± 0.01^{a}
PCL-SE10	$1579.2 \pm 33.7^{b,c}$	0.10 ± 0.00^{b}	27.5 ± 0.1^{a}	3.65 ± 0.21	0.08 ± 0.01^{b}
PCL-SE20	1565.9 ± 35.6^{c}	0.09 ± 0.00^{b}	24.7 ± 0.1^{b}	3.31 ± 0.21	0.08 ± 0.01^{b}

Data are expressed as mean ± standard deviation. Different letters within the same column indicate significant differences among samples ($p < 0.05$).

Physical properties of the solutions were suitable for producing continuous fibers under stable electrospinning process without dripping of the solutions or formation of beaded areas (processing parameters are described in Section 2.4.3). Ultrathin films were made by subsequent exposure of the mats to the annealing post-processing treatment.

3.3. Film Characterization

3.3.1. Morphology

SEM micrographs of the electrospun fibers before the annealing treatment and their corresponding diameter distribution histograms are shown in Figure 1 (A1-D1, A2-D2, vertically). The electrospun mats presented a fibrous, bead-free morphology. The fibers were regular, smooth, and without aggregates, suggesting that a homogeneous solid dispersion of SE is achieved in the fibers (Figure 1, A1-D1, vertically). As shown in histograms (Figure 1, A2-D2, vertically), SE addition affected the fibers' diameter. It was observed that the increase in SE content led to reduced mean values of fiber diameter (Table 1). This effect is related to the changes in the solution properties induced by SE addition. Likewise, higher conductivity and lower viscosity and surface tension altered the polymer chain entanglements, generating an increase in the stretching forces in the jet which resulted in the decreased electrospun fiber diameters [6,40]. Similarly, decreased diameter of PCL fibers was reported when other active compounds, such as black pepper oleoresins [25], carvacrol [41], or cefazolin [42] were incorporated within the polymeric matrix.

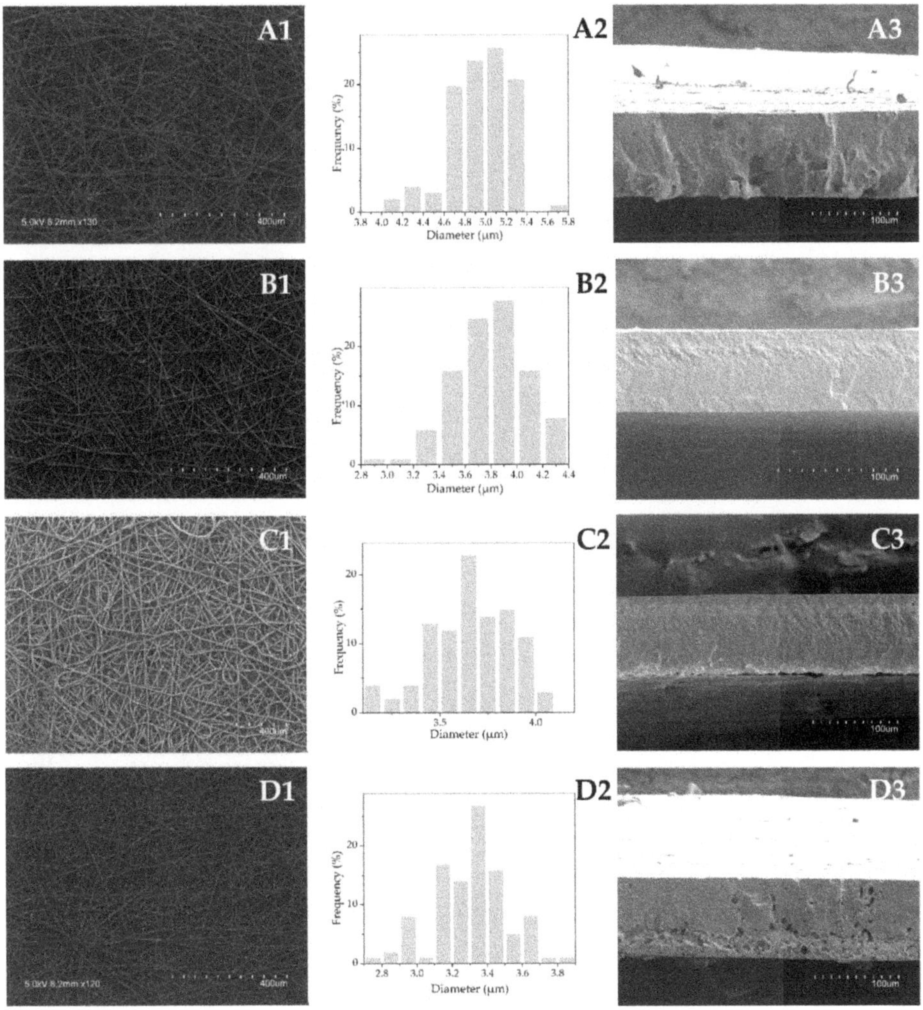

Figure 1. Scanning electron microscopy (SEM) micrographs of the electrospun fibers (A1-D1, vertically), diameter distribution histograms (A2–D2, vertically) and cross-sections of the annealed films (A3–D3, vertically) of the plain poly(ε-caprolactone) PCL (A1–3) and the formulations containing 5% (B1–3), 10% (C1–3), and 20% (D1–3) of SE.

Figure 1 (A3–D3, vertically) presents the cross-sections of the cryo-fractured annealed films. As can be observed, the electrospun fibrous mats were packed into continuous films due to fibers' coalescence during the annealing treatment. Furthermore, different structures were revealed depending on the SE content. The plain PCL film exhibited a somewhat rougher appearance as compared to the samples containing SE at levels 5% and 10%, which appeared somewhat more compact, smooth, and homogeneous. Higher roughness for the PCL film may be due to the higher fiber diameter. On the other hand, the incorporation of 20% SE led to a more porous morphology. This may be due to the generation of volatiles arising from the higher SE concentration during the annealing process.

Film thickness was found to be lower for the samples with the higher SE content due to most likely lower fiber diameter as a result of a more efficient fiber packing after the annealing treatment (see Table 1).

3.3.2. Optical Properties

The developed films were highly transparent, as shown in Figure 2. Film transparency is a desirable property for packaging applications, since the packaging should enable visual assessment of its content. The active films exhibited a yellowish color, which was more intense with increasing the active ingredient content. However, the SE addition did not significantly ($p > 0.05$) alter the transparency, as compared to the control. In particular, the plain PCL film presented a transparency value of 12.96 mm^{-1}, while the transparency of the films containing 5%, 10%, and 20% SE was 13.31, 14.24, and 16.16 mm^{-1}, respectively. The slight decrease in transparency of the film loaded with the highest extract content (20%) is related to the more heterogeneous morphology and light scattering [43].

Figure 2. Contact transparency image of the plain poly(ε-caprolactone) PCL film (A) and the PCL-based films containing: 5% (B), 10% (C), and 20% (D) SE.

Light, especially in the UV range, triggers photo-oxidation processes, which leads to rapid quality loss or deterioration of packaged food products [44]. Evaluation of the UV light transmission capacity implied that, in general, the developed films very effectively blocking UV light. At 300 nm the plain PCL film presented a good light barrier, with a transmission value 1.84%. The incorporation of SE within the PCL matrix significantly helped to decrease penetration of the UV light to very low levels. The light transmission rate at 300 nm was 1.07%, 1.00%, and 0.63% for the films containing 5%, 10%, and 20% SE, respectively. This barrier property makes the studied systems suitable for protection of the products susceptible to photo-oxidation.

3.3.3. Water Contact Angle

Contact angle between a drop of water and the film surface is an indicator of the surface hydrophilicity ($\theta < 65°$) and hydrophobicity ($\theta > 65°$) [45]. The here developed systems present hydrophobic surfaces with poor water affinity ($\theta > 65°$). The plain PCL film presented a contact angle of 76.6° which is in line with the literature ($\theta \sim 74°$) [25]. Interestingly, the film hydrophobicity was increased by increasing the extract content. In particular, the incorporation of 5%, 10% and 20% of SE resulted in the contact angle values of 73.6°, 92.3°, and 100.8°, respectively. This behavior has been ascribed before to changes in surface topology, since an increase in surface roughness and heterogeneity leads to higher values of contact angle [46,47]. Thus, air could be trapped within these micro or submicron size interfiber valleys making air pockets which lead to an increase in water contact angle [48,49]. It is expected that the thinner the fiber and the higher the heterogeneity along the fiber due to increasing SE content, the higher could be the contact angle, as observed. The trend towards greater hydrophobicity of materials containing natural extracts was also reported for starch films with incorporated yerba mate extract, which was explained by observed roughness when the extract was incorporated [50]. The observed water resistance is a highly desirable property for potential food packaging applications.

3.3.4. Thermogravimetric Analysis

Thermal stability of the free SE and the prepared films was determined by thermogravimetric analysis (TGA). TGA curves of the mass loss as a function of temperature (blue lines) and the first derivative analysis (orange lines) are presented in Figure 3. Degradation of the free SE occurred in several phases, starting from around 106 °C at 5% of weight loss, due to most likely moisture and volatiles evaporation. The maximum degradation rate with a mass loss of about 76% was reached at 452 °C, while residual mass at 600 °C was at about 9%. As it can be seen from the Figure 3, thermal degradation properties of the plain PCL film are not altered by the SE incorporation. The PCL-based films exhibited similar thermal degradation patterns, regardless of the SE content. Generally, the thermal decomposition process of the here-obtained films took place within the range between 350 and 480 °C, largely coinciding with the main degradation of the SE. This thermal degradation range corresponds to the one reported for PCL nanofibrous mat [51]. The maximum degradation rate with a mass loss of about 60% was observed at 398, 399, 399, and 396 °C for the plain PCL film and the films containing 5%, 10%, and 20% SE, respectively. It can be concluded from TGA results that the incorporation of SE into the PCL matrix did not detrimentally affect the thermal stability of the composites when compared to the control sample (without SE).

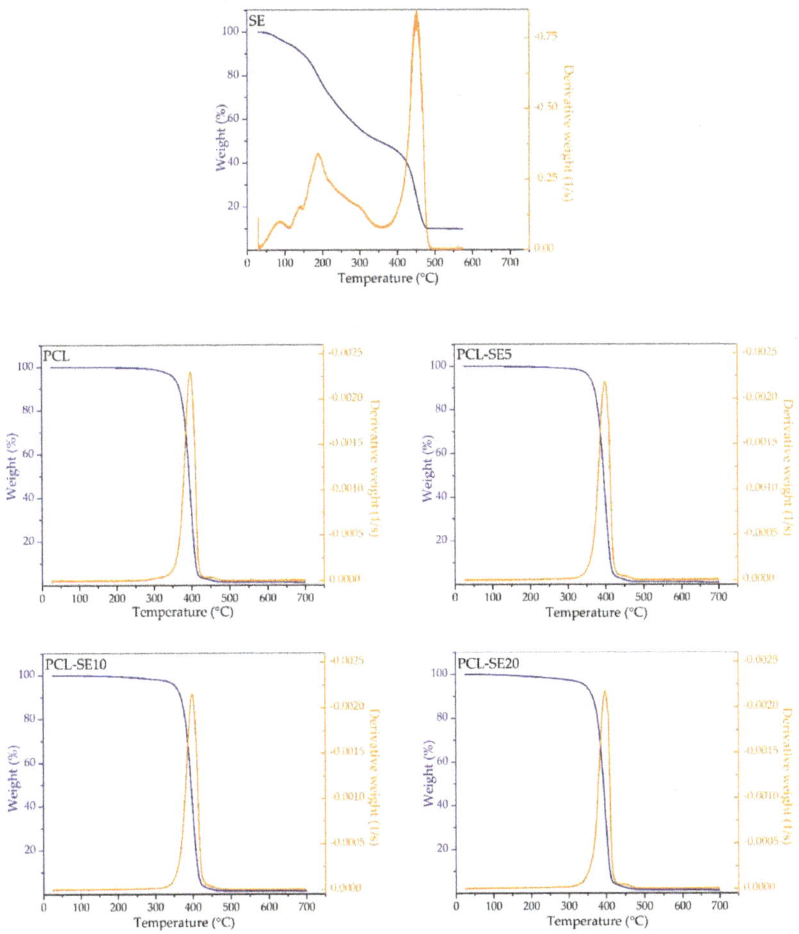

Figure 3. Thermogravimetric curves of the free sage extract (SE) and the poly(ε-caprolactone) (PCL)-based films.

3.3.5. Mechanical Properties

The tensile properties (elastic modulus, tensile strength, elongation at break, and toughness) of the developed systems are presented in Table 2. The incorporation of SE induced a slight, not statistically significant ($p > 0.05$) decrease in elastic modulus, and an increase in tensile strength, elongation at break, and toughness compared to the PCL film. The observations suggest that the SE, if anything, acts as a very slight plasticizer to PCL. This tensile behavior is in line with the afore-observed thermal behavior. The negligible effect of SE on the mechanical properties of the films indicates that there are not significant interactions between the polymer matrix and the SE compounds [52]. Similarly, mechanical properties of a PCL–gelatin–PCL multilayer system were not significantly affected by addition of black pepper oleoresins into the PCL layers [25]. As compared to a commercial packaging material, the here developed films were more resistant to fracture, but less ductile than LDPE. As reported, the pure LDPE film with a thickness of ca. 44 μm presents a tensile strength and an elongation at break values around 5.88 MPa and 112.39%, respectively [53].

3.3.6. Water Vapor and D-Limonene Permeability

The barrier properties of the materials are relevant for their application, but also to understand the relationship between composition, structure, processing, and properties. The barrier performance of the PCL-based films in terms of water vapor and D-limonene permeability is gathered in Figure 4.

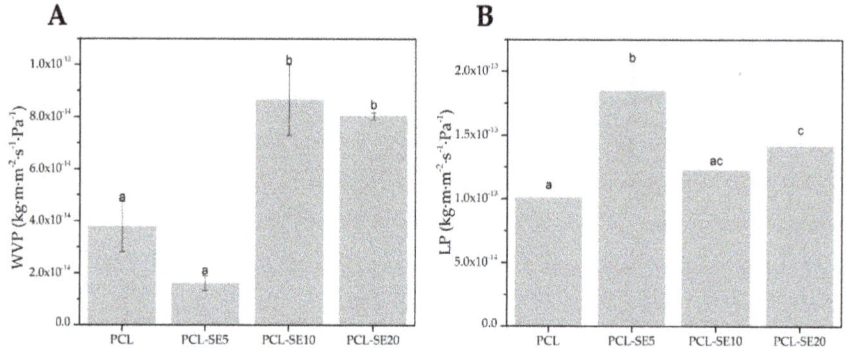

Figure 4. Water vapor (WVP) and D-limonene permeability (LP) of the poly(ε-caprolactone) (PCL)-based films. Different letters within the same column indicate significant differences among samples ($p < 0.05$).

Table 2. Mechanical properties in terms of elastic modulus (E), tensile strength (σ_b), elongation at break (ε_b) and toughness (T) of the poly(ε-caprolactone) (PLC)-based films.

Formulation	E (MPa)	σ_b (MPa)	ε_b (%)	T (mJ/m^3)
PCL	420.77 ± 61.45 ^{a}	24.32 ± 3.79 ^{a}	4.60 ± 0.34 ^{a}	1.41 ± 0.19 ^{a}
PCL-SE5	425.40 ± 40.85 ^{a}	26.11 ± 2.76 ^{a}	4.90 ± 0.48 ^{a}	1.51 ± 0.09 ^{a}
PCL-SE10	410.78 ± 24.70 ^{a}	31.98 ± 4.42 ^{a}	5.44 ± 0.78 ^{a}	1.62 ± 0.02 ^{a}
PCL-SE20	359.25 ± 63.43 ^{a}	26.43 ± 5.19 ^{a}	5.28 ± 0.57 ^{a}	1.52 ± 0.15 ^{a}

Data are expressed as mean ± standard deviation. Different letters within the same column indicate significant differences among samples ($p < 0.05$).

From Figure 4, it can be observed that the plain PCL film presents a higher ($p < 0.05$) water vapor barrier as compared to its counterparts loaded with high extract content (10% and 20%) (Figure 4A). However, the sample with 5% SE loading shows a higher water barrier than neat PCL. This particular sample was seen to have a smoother, less porous morphology than pure PCL. Thus, the reason for the overall changes in permeability to water may be related to sample porosity, but also to SE content. SE may have higher affinity for water than PCL, but at low contents it results in finer fibers that pack better, and at the higher content it produces some porosity due to volatiles leaving the sample during the annealing process. When compared to cellophane as a commercial material widely used in the packaging industry, the plain PCL film and the film containing 5% of SE presented lower water vapor permeability, while the values of the films loaded with 10% and 20% SE were in the same order of magnitude as the value reported for cellophane (6.90×10^{-14} kg·m·m^{-2} s^{-1} Pa^{-1}) [54].

D-limonene is commonly used as a standard compound to assess the aroma barrier of packaging materials. The SE incorporation was seen to increase the permeability of limonene through the PCL film for all samples (Figure 4B). D-limonene, an apolar permeant, is known to strongly plasticize PCL [49,55]. Therefore, the higher permeability seen for the samples with SE suggests that the comparatively low molar mass of SE facilitates even more the diffusion of the permeant through the materials. The LP of the developed PCL-based films are within the same order of magnitude as the LP of widely used neat poly(ethylene terephthalate) (PET) film produced by compression molding (1.17×10^{-13} kg·m·m^{-2} s^{-1} Pa^{-1}) [56].

3.3.7. Antioxidant Activity

Figure 5 shows the antioxidant activity of the PCL-based films during 24 h of contact with DPPH˙ free radicals solution. According to the results, the unloaded PCL film presented a certain level of antioxidant activity (6.12%) after 24 h. However, the antioxidant activity of the plain PCL film was very weak and might be attributed to a DPPH˙ absorption by the film during the contact [51]. The antioxidant activity of the films was achieved when SE was incorporated into the PCL matrix, as expected according to the free radical neutralizing effect of the unloaded SE. Namely, the increase in the loaded SE content led to a significantly stronger ($p < 0.05$) free radical scavenging activity of the films (Figure 5). Also, it may be noted that the DPPH˙ free radicals scavenging ability of the studied systems was more pronounced with an increase in the contact time. This behavior can be explained by release of active compounds from the films into the free radicals solution which allows rapid inhibition or quenching of DPPH˙ radicals [57]. In particular, the antioxidant activity after 0.5 h of the contact between the free radicals solution and the films loaded with 5%, 10%, and 20% SE was increased from 13.32%, 22.64%, and 53.88%, respectively to 28.89%, 47.17%, and 85.27%, respectively after 24 h. This is in the agreement with a study on chitosan films containing caraway essential oil and beeswax [58] which also reported an increase in antioxidant activity of the films during incubation time. The obtained results point out the efficiency of the PCL-based films containing SE in free radicals neutralization, suggesting their potential role in prevention of oxidation processes in food products.

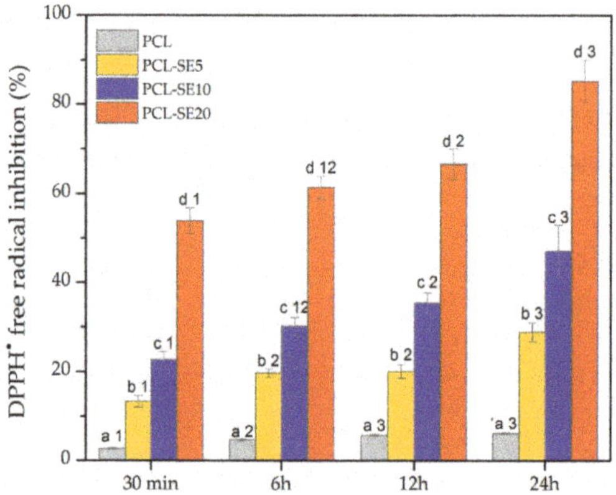

Figure 5. Antioxidant activity, expressed as DPPH˙ free radicals scavenging ability of the PCL-based films. Different letters indicate significant difference among different films at certain incubation time ($p < 0.05$). Different numbers indicate significant difference of the same sample at different incubation time ($p < 0.05$).

3.3.8. Antimicrobial Activity

The antibacterial efficiency of the PCL-based films against *S. aureus* and *E. coli* is presented in Table 3. The plain PCL film did not exhibit any antibacterial activity (R < 0.5) as expected. Inhibition effects on the bacterial growth were successfully induced by the incorporation of SE into the PCL matrix. Interestingly, complete biocidal effect on *S. aureus* (no viable counts) was detected after 24 h of exposure to any of the films containing SE. On the other hand, the results indicated a higher resistance of *E. coli* towards the films. This behavior is in concordance with the results obtained for the unloaded extract and attributed to the more complex cell wall of *E. coli* as a Gram negative bacteria compared to the cell wall of *S. aureus* as a Gram positive bacteria. Thus, the inhibition effect on *E. coli* was dependent on the SE content in the films. In particular, *E. coli* was slightly inhibited ($0.5 \leq R < 1$) when in contact with the films containing 5% and 10% SE, while strong inhibition (R > 3) was achieved when in contact with the film loaded with 20% SE. Therefore, the strong antibacterial surface against foodborne pathogens was obtained by incorporating 20% SE into the PCL matrix, which reveals its remarkable potential for antimicrobial food packaging applications.

Table 3. Antimicrobial activity against *S. aureus* and *E. coli* of the PCL-based films.

Formulation	*S. aureus*		*E. coli*	
	Bacterial Counts log CFU/mL	Surface Reduction R	Bacterial Counts log CFU/mL	Surface Reduction R
Control	6.10 ± 0.09	-	7.09 ± 0.12	-
PCL	5.77 ± 0.15	0.33	6.89 ± 0.09	0.20
PCL-SE5	no viable counts	biocidal effect	6.36 ± 0.13	0.73
PCL-SE10	no viable counts	biocidal effect	6.18 ± 0.16	0.91
PCL-SE20	no viable counts	biocidal effect	2.40 ± 0.52	4.69

CFU—Colony Forming Units; R—Surface Reduction.

4. Conclusions

In the present study, SE solid dispersions with proven antioxidant and antimicrobial activities were successfully incorporated at different loading contents (5%, 10%, and 20%) within the PCL matrix by the electrospinning technique with the aim to develop antioxidant and antimicrobial packaging materials. The electrospun fibrous mats were post-processed by means of the annealing treatment to produce continuous, transparent, active films. The obtained results gave an insight into the effect of SE addition on the physicochemical and functional properties of the PCL-based film. Thus, the properties of the PCL-based solutions and morphology of the formulated films were affected by the SE content. Namely, thinner electrospun fibers were produced when higher SE contents were loaded. Generally, the here obtained systems presented a good contact transparency. UV-light barrier and water resistance, i.e., hydrophobicity, were generally enhanced when SE was loaded within the PCL matrix. The presence of SE had a negligible effect on thermal stability and tensile parameters of the films. On the other hand, SE led to a general decrease in barrier properties to water and D-limonene. Finally, the SE incorporation triggered remarkable DPPH$^{\cdot}$ free radical scavenging ability and antimicrobial action against *Staphylococcus aureus* and *Escherichia coli*. As the SE content increased, the films presented more effective antiradical ability and activity against *Escherichia coli*. The evaluated characteristics of the PCL-based films containing SE suggest that the here-developed systems are potential candidates as active materials in food packaging applications, which would delay oxidation processes and prevent microbiological contamination.

Author Contributions: Conceptualization was devised by V.N. and J.M.L.; methodology, validation and formal analysis were performed by A.S., C.P. and J.M.L.; investigation, resources, data curation, writing-original draft preparation, writing-review & editing were carried out by A.S., C.P., L.C., V.N. and J.M.L.; supervision: J.M.L. and C.P.; project administration: J.M.L.; funding acquisition: V.N. and J.M.L.

Funding: This research was funded by COST Action FP1405 (STSM reference number 39168), the EU H2020 ResUrbis project (Reference number 730349) and the Ministry of Education, Science and Technological Development (Republic of Serbia) project number III46010.

Acknowledgments: A.S. is a recipient of a contract from the Serbian Ministry of Education, Science and Technological Development (Project No III46010) and the STSM grant. The CSIC-UJI Joint Unit in "Polymer Technology" is also acknowledged.

Conflicts of Interest: The authors declare no conflict of interest. The funders had no role in the design of the study; in the collection, analyses, or interpretation of data; in the writing of the manuscript, or in the decision to publish the results.

References

1. Talón, E.; Trifkovic, K.T.; Nedovic, V.A.; Bugarski, B.M.; Vargas, M.; Chiralt, A.; González-Martínez, C. Antioxidant edible films based on chitosan and starch containing polyphenols from thyme extracts. *Carbohydr. Polym.* **2017**, *157*, 1153–1161. [CrossRef] [PubMed]
2. Souza, V.G.L.; Fernando, A.L.; Pires, J.R.A.; Rodrigues, P.F.; Lopes, A.A.S.; Fernandes, F.M.B. Physical properties of chitosan films incorporated with natural antioxidants. *Ind. Crop. Prod.* **2017**, *107*, 565–572. [CrossRef]
3. Shan, B.; Cai, Y.Z.; Brooks, J.D.; Corke, H. The in vitro antibacterial activity of dietary spice and medicinal herb extracts. *Int. J. Food Microbiol.* **2007**, *117*, 112–119. [CrossRef] [PubMed]
4. Carocho, M.; Morales, P.; Ferreira, I.C.F.R. Natural food additives: Quo vadis? *Trends Food Sci. Technol.* **2015**, *45*, 284–295. [CrossRef]
5. Belščak-Cvitanović, A.; Đorđević, V.; Karlović, S.; Pavlović, V.; Komes, D.; Ježek, D.; Bugarski, B.; Nedović, V. Protein-reinforced and chitosan-pectin coated alginate microparticles for delivery of flavan-3-ol antioxidants and caffeine from green tea extract. *Food Hydrocoll.* **2015**, *51*, 361–374. [CrossRef]
6. Castro-Mayorga, J.; Fabra, M.; Cabedo, L.; Lagaron, J. On the Use of the Electrospinning Coating Technique to Produce Antimicrobial Polyhydroxyalkanoate Materials Containing In Situ-Stabilized Silver Nanoparticles. *Nanomaterials* **2016**, *7*, 4. [CrossRef] [PubMed]
7. Talón, E.; Trifkovic, K.T.; Vargas, M.; Chiralt, A.; González-Martínez, C. Release of polyphenols from starch-chitosan based films containing thyme extract. *Carbohydr. Polym.* **2017**, *175*, 122–130. [CrossRef] [PubMed]
8. Nedovic, V.; Kalusevic, A.; Manojlovic, V.; Levic, S.; Bugarski, B. An overview of encapsulation technologies for food applications. *Procedia Food Sci.* **2011**, *1*, 1806–1815. [CrossRef]
9. Salmieri, S.; Lacroix, M. Physicochemical properties of alginate/polycaprolactone-based films containing essential oils. *J. Agric. Food Chem.* **2006**, *54*, 10205–10214. [CrossRef]

10. Dainelli, D.; Gontard, N.; Spyropoulos, D.; Zondervan-van den Beuken, E.; Tobback, P. Active and intelligent food packaging: Legal aspects and safety concerns. *Trends Food Sci. Technol.* **2008**, *19*. [CrossRef]
11. Ozdemir, M.; Floros, J.D. Active food packaging technologies. *Crit. Rev. Food Sci. Nutr.* **2004**, *44*, 185–193. [CrossRef] [PubMed]
12. Ghorbani, A.; Esmaeilizadeh, M. Pharmacological properties of *Salvia officinalis* and its components. *J. Tradit. Complement. Med.* **2017**, *7*. [CrossRef] [PubMed]
13. Roby, M.H.H.; Sarhan, M.A.; Selim, K.A.H.; Khalel, K.I. Evaluation of antioxidant activity, total phenols and phenolic compounds in thyme (*Thymus vulgaris* L.), sage (*Salvia officinalis* L.), and marjoram (*Origanum majorana* L.) extracts. *Ind. Crop. Prod.* **2013**, *43*, 827–831. [CrossRef]
14. Martins, N.; Barros, L.; Santos-Buelga, C.; Henriques, M.; Silva, S.; Ferreira, I.C.F.R. Evaluation of bioactive properties and phenolic compounds in different extracts prepared from *Salvia officinalis* L. *Food Chem.* **2014**, *170*, 378–385. [CrossRef] [PubMed]
15. Cuvelier, M.E.; Richard, H.; Berset, C. Antioxidative activity and phenolic composition of pilot-plant and commercial extracts of sage and rosemary. *Jaocsj. Am. Oil Chem. Soc.* **1996**, *73*, 645–652. [CrossRef]
16. U.S. Food and Drug Administration. Food for Human Consumption, Part 182—Substances Generally Recognized as Safe. *Fed. Regist.* **2016**, *81*, 54960–55055.
17. Cherpinski, A.; Gozutok, M.; Sasmazel, H.; Torres-Giner, S.; Lagaron, J.; Cherpinski, A.; Gozutok, M.; Sasmazel, H.T.; Torres-Giner, S.; Lagaron, J.M. Electrospun Oxygen Scavenging Films of Poly(3-hydroxybutyrate) Containing Palladium Nanoparticles for Active Packaging Applications. *Nanomaterials* **2018**, *8*, 469. [CrossRef] [PubMed]
18. Woodruff, M.A.; Hutmacher, D.W. The return of a forgotten polymer—Polycaprolactone in the 21st century. *Prog. Polym. Sci.* **2010**, *35*, 1217–1256. [CrossRef]
19. Labet, M.; Thielemans, W. Synthesis of polycaprolactone: A review. *Chem. Soc. Rev.* **2009**, *38*, 3484–3504. [CrossRef] [PubMed]
20. Martínez-Abad, A.; Sánchez, G.; Fuster, V.; Lagaron, J.M.; Ocio, M.J. Antibacterial performance of solvent cast polycaprolactone (PCL) films containing essential oils. *Food Control* **2013**, *34*, 214–220. [CrossRef]
21. Erbay, E.A.; Dağtekin, B.B.; Türe, M.; Yeşilsu, A.F.; Torres-Giner, S. Quality improvement of rainbow trout fillets by whey protein isolate coatings containing electrospun poly(ε-caprolactone) nanofibers with Urtica dioica L. extract during storage. *LWT* **2017**, *78*, 340–351. [CrossRef]
22. Echegoyen, Y.; Fabra, M.J.; Castro-Mayorga, J.L.; Cherpinski, A.; Lagaron, J.M. High throughput electro-hydrodynamic processing in food encapsulation and food packaging applications: Viewpoint. *Trends Food Sci. Technol.* **2017**, *60*, 71–79. [CrossRef]
23. Ghorani, B.; Tucker, N. Fundamentals of electrospinning as a novel delivery vehicle for bioactive compounds in food nanotechnology. *Food Hydrocoll.* **2015**, *51*, 227–240. [CrossRef]
24. Fabra, M.J.; López-Rubio, A.; Sentandreu, E.; Lagaron, J.M. Development of multilayer corn starch-based food packaging structures containing β-carotene by means of the electro-hydrodynamic processing. *Starch/Staerke* **2016**, *68*, 603–610. [CrossRef]
25. Figueroa-Lopez, K.; Castro-Mayorga, J.; Andrade-Mahecha, M.; Cabedo, L.; Lagaron, J. Antibacterial and Barrier Properties of Gelatin Coated by Electrospun Polycaprolactone Ultrathin Fibers Containing Black Pepper Oleoresin of Interest in Active Food Biopackaging Applications. *Nanomaterials* **2018**, *8*, 199. [CrossRef] [PubMed]
26. Akcan, T.; Estévez, M.; Serdaroğlu, M. Antioxidant protection of cooked meatballs during frozen storage by whey protein edible films with phytochemicals from *Laurus nobilis* L. and *Salvia officinalis*. *LWT-Food Sci. Technol.* **2017**, *77*, 323–331. [CrossRef]
27. Deng, L.; Kang, X.; Liu, Y.; Feng, F.; Zhang, H. Characterization of gelatin/zein films fabricated by electrospinning vs solvent casting. *Food Hydrocoll.* **2018**, *74*, 324–332. [CrossRef]
28. Brand-Williams, W.; Cuvelier, M.E.; Berset, C. Use of a Free Radical Method to Evaluate Antioxidant Activity. *LWT-Food Sci. Technol.* **1995**, *28*, 25–30. [CrossRef]
29. Clinical and Laboratory Standards Institute. *Methods for Dilution Antimicrobial Susceptibility Tests for Bacteria That Grow Aerobically; Approved Standard—Tenth Edition*; Clinical and Laboratory Standards Institute: Wayne, PA, USA, 2015.
30. ASTM D638-10. *Standard Test Method for Tensile Properties of Plastics*; ASTM International: West Conshohocken, PA, USA, 2010.

31. ASTM E96-95. *Standard Test Methods for Water Vapour Transmission of Materials*; ASTM International: West Conshohocken, PA, USA, 1995.
32. Japanese Industrial Standard Committee. *JIS Z 2801:2000 Antimicrobial Products—Test for Antimicrobial Activity and Efficacy*; Japanese Standards Association: Tokyo, Japan, 2000.
33. Chen, H.Y.; Lin, Y.C.; Hsieh, C.L. Evaluation of antioxidant activity of aqueous extract of some selected nutraceutical herbs. *Food Chem.* **2007**, *104*, 1418–1424. [CrossRef]
34. Farhat Ben, M.; Landoulsi, A.; Chaouch-hamada, R.; Sotomayor, J.A.; Jordán, M.J. Characterization and quantification of phenolic compounds and antioxidant properties of Salvia species growing in different habitats. *Ind. Crop. Prod.* **2013**, *49*, 904–914. [CrossRef]
35. Lu, Y.; Foo, L.Y. Antioxidant activities of polyphenols from sage (*Salvia officinalis*). *Food Chem.* **2001**, *75*, 197–202. [CrossRef]
36. Gorjanović, S.; Komes, D.; Pastor, F.T.; Belščak-Cvitanović, A.; Pezo, L.; Hečimović, I.; Sužnjević, D.S. Antioxidant Capacity of Teas and Herbal Infusions: Polarographic Assessment. *J. Agric. Food Chem.* **2012**, *60*, 9573–9580. [CrossRef]
37. Dey, T.B.; Chakraborty, S.; Jain, K.K.; Sharma, A.; Kuhad, R.C. Antioxidant phenolics and their microbial production by submerged and solid state fermentation process: A review. *Trends Food Sci. Technol.* **2016**, *53*, 60–74. [CrossRef]
38. Cetin-Karaca, H.; Newman, M.C. Antimicrobial efficacy of plant phenolic compounds against Salmonella and Escherichia Coli. *Food Biosci.* **2015**, *11*, 8–16. [CrossRef]
39. Tampau, A.; Gonzalez-Martinez, C.; Chiralt, A. Carvacrol encapsulation in starch or PCL based matrices by electrospinning. *J. Food Eng.* **2017**, *214*, 245–256. [CrossRef]
40. Neo, Y.P.; Ray, S.; Jin, J.; Gizdavic-Nikolaidis, M.; Nieuwoudt, M.K.; Liu, D.; Quek, S.Y. Encapsulation of food grade antioxidant in natural biopolymer by electrospinning technique: A physicochemical study based on zein–gallic acid system. *Food Chem.* **2013**, *136*, 1013–1021. [CrossRef] [PubMed]
41. Tampau, A.; González-Martínez, C.; Chiralt, A. Release kinetics and antimicrobial properties of carvacrol encapsulated in electrospun poly-(ε-caprolactone) nanofibres. Application in starch multilayer films. *Food Hydrocoll.* **2018**, *79*, 158–169. [CrossRef]
42. Radisavljevic, A.; Stojanovic, D.B.; Perisic, S.; Djokic, V.; Radojevic, V.; Rajilic-Stojanovic, M.; Uskokovic, P.S. Cefazolin-loaded polycaprolactone fibers produced via different electrospinning methods: Characterization, drug release and antibacterial effect. *Eur. J. Pharm. Sci.* **2018**, *124*, 26–36. [CrossRef] [PubMed]
43. Fabra, M.J.; López-Rubio, A.; Lagaron, J.M. Three-Layer Films Based on Wheat Gluten and Electrospun PHA. *Food Bioprocess Technol.* **2015**, *8*, 2330–2340. [CrossRef]
44. Tian, F.; Decker, E.A.; Goddard, J.M. Controlling lipid oxidation of food by active packaging technologies. *Food Funct.* **2013**, *4*, 669–680. [CrossRef]
45. Vogler, E.A. Structure and reactivity of water at biomaterial surfaces. *Adv. Colloid Interface Sci.* **1998**, *74*, 69–117. [CrossRef]
46. Erbil, H.Y.; Demirel, A.L.; Avci, Y.; Mert, O. Transformation of a simple plastic into a superhydrophobic surface. *Sci.* **2003**, *299*, 1377–1380. [CrossRef] [PubMed]
47. Han, J.H.; Krochta, J.M. Wetting properites and water vapor permeability of whey-protein-coated paper. *Trans. Asae* **1999**, *42*, 1375–1382. [CrossRef]
48. Pardo-Figuerez, M.; López-Córdoba, A.; Torres-Giner, S.; Lagaron, J. Superhydrophobic Bio-Coating Made by Co-Continuous Electrospinning and Electrospraying on Polyethylene Terephthalate Films Proposed as Easy Emptying Transparent Food Packaging. *Coatings* **2018**, *8*, 364. [CrossRef]
49. Lasprilla-Botero, J.; Torres-Giner, S.; Pardo-Figuerez, M.; Álvarez-Láinez, M.; Lagaron, J.M. Superhydrophobic Bilayer Coating Based on Annealed Electrospun Ultrathin Poly(ε-caprolactone) Fibers and Electrosprayed Nanostructured Silica Microparticles for Easy Emptying Packaging Applications. *Coatings* **2018**, *8*, 173. [CrossRef]
50. Jaramillo, C.M.; González Seligra, P.; Goyanes, S.; Bernal, C.; Famá, L. Biofilms based on cassava starch containing extract of yerba mate as antioxidant and plasticizer. *Starch/Staerke* **2015**, *67*, 780–789. [CrossRef]
51. Aytac, Z.; Uyar, T. Antioxidant activity and photostability of α-tocopherol/β-cyclodextrin inclusion complex encapsulated electrospun polycaprolactone nanofibers. *Eur. Polym. J.* **2016**, *79*, 140–149. [CrossRef]

52. Yang, X.; Cui, C.; Tong, Z.; Sabanayagam, C.R.; Jia, X. Poly(ε-caprolactone)-based copolymers bearing pendant cyclic ketals and reactive acrylates for the fabrication of photocrosslinked elastomers. *Acta Biomater.* **2013**, *9*, 8232–8244. [CrossRef]
53. Gaikwad, K.K.; Lee, S.M.; Lee, J.S.; Lee, Y.S. Development of antimicrobial polyolefin films containing lauroyl arginate and their use in the packaging of strawberries. *J. Food Meas. Charact.* **2017**, *11*, 1706–1716. [CrossRef]
54. Péroval, C.; Debeaufort, F.; Despré, D.; Voilley, A. Edible arabinoxylan-based films. 1. Effects of lipid type on water vapor permeability, film structure, and other physical characteristics. *J. Agric. Food Chem.* **2002**, *50*, 3977–3983. [CrossRef]
55. Cava, D.; Giménez, E.; Gavara, R.; Lagaron, J.M. Comparative performance and barrier properties of biodegradable thermoplastics and nanobiocomposites versus PET for food packaging applications. *J. Plast. Film Sheeting* **2006**, *22*, 265–274. [CrossRef]
56. Sanchez-Garcia, M.D.; Gimenez, E.; Lagaron, J.M. Novel PET nanocomposites of interest in food packaging applications and comparative barrier performance with biopolyester nanocomposites. *J. Plast. Film Sheeting* **2007**, *23*, 133–148. [CrossRef]
57. Benbettaïeb, N.; Tanner, C.; Cayot, P.; Karbowiak, T.; Debeaufort, F. Impact of functional properties and release kinetics on antioxidant activity of biopolymer active films and coatings. *Food Chem.* **2018**, *242*, 369–377. [CrossRef]
58. Hromiš, N.M.; Lazić, V.L.; Markov, S.L.; Vaštag, Ž.G.; Popović, S.Z.; Šuput, D.Z.; Džinić, N.R.; Veličanski, A.S.; Popović, L.M. Optimization of chitosan biofilm properties by addition of caraway essential oil and beeswax. *J. Food Eng.* **2015**, *158*, 86–93. [CrossRef]

nanomaterials

MDPI

Article

Electrospun Antimicrobial Films of Poly(3-hydroxybutyrate-*co*-3-hydroxyvalerate) Containing Eugenol Essential Oil Encapsulated in Mesoporous Silica Nanoparticles

Beatriz Melendez-Rodriguez [1], Kelly J. Figueroa-Lopez [1], Andrea Bernardos [2,3,4,5], Ramón Martínez-Máñez [2,3,4,5], Luis Cabedo [6], Sergio Torres-Giner [1] and Jose M. Lagaron [1,*]

1 Novel Materials and Nanotechnology Group, Institute of Agrochemistry and Food Technology (IATA), Spanish Council for Scientific Research (CSIC), Calle Catedrático Agustín Escardino Benlloch 7, 46980 Paterna, Spain; beatriz.melendez@iata.csic.es (B.M.-R.); kjfigueroal@iata.csic.es (K.J.F.-L.); storresginer@iata.csic.es (S.T.-G.)

2 Instituto Interuniversitario de Investigación de Reconocimiento Molecular y Desarrollo Tecnológico (IDM), Universitat Politècnica de València (UPV), Universitat de València (UV), camí de Vera s/n, 46022 Valencia, Spain; anberba@upv.es (A.B.); rmaez@qim.upv.es (R.M.-M.)

3 CIBER de Bioingeniería, Biomateriales y Nanomedicina (CIBER-BBN), Camino de Vera s/n, 46022 Valencia, Spain

4 Unidad Mixta de Investigación en Nanomedicina y Sensores, Universitat Politècnica de València (UPV), Instituto de Investigación Sanitaria La Fe, 46026 Valencia, Spain

5 Unidad Mixta UPV-CIPF de Investigación en Mecanismos de Enfermedades y Nanomedicina, Universitat Politècnica de València (UPV), Centro de Investigación Príncipe Felipe, 46012 Valencia, Spain

6 Polymers and Advanced Materials Group (PIMA), Universitat Jaume I, 12071 Castellón, Spain; lcabedo@uji.es

* Correspondence: lagaron@iata.csic.es

Received: 4 January 2019; Accepted: 2 February 2019; Published: 8 February 2019

Abstract: The main goal of this study was to develop poly(3-hydroxybutyrate-*co*-3-hydroxyvalerate) (PHBV) films with long-term antimicrobial capacity of interest in food packaging applications. To this end, eugenol was first highly efficiently encapsulated at 50 wt.-% in the pores of mesoporous silica nanoparticles by vapor adsorption. The eugenol-containing nanoparticles were then loaded in the 2.5–20 wt.-% range into PHBV by electrospinning and the resultant electrospun composite fibers were annealed at 155 °C to produce continuous films. The characterization showed that the PHBV films filled with mesoporous silica nanoparticles containing eugenol present sufficient thermal resistance and enhanced mechanical strength and barrier performance to water vapor and limonene. The antimicrobial activity of the films was also evaluated against foodborne bacteria for 15 days in open vs. closed conditions in order to simulate real packaging conditions. The electrospun PHBV films with loadings above 10 wt.-% of mesoporous silica nanoparticles containing eugenol successfully inhibited the bacterial growth, whereas the active films stored in hermetically closed systems increased their antimicrobial activity after 15 days due to the volatile portion accumulated in the system's headspace and the sustained release capacity of the films. The resultant biopolymer films are, therefore, potential candidates to be applied in active food packaging applications to provide shelf life extension and food safety.

Keywords: PHBV; MCM-41; eugenol; antimicrobial properties; active packaging

1. Introduction

Polyhydroxyalkanoates (PHAs) currently represent one of the most important alternative to petroleum-based materials in the frame of the Circular Economy [1]. PHAs, which are synthesized by a wide range of microorganisms as carbon storage material, are thermoplastic materials, biodegradable, and present similar physical properties to other plastics, e.g., polypropylene (PP) and polystyrene (PS), such as high mechanical strength and water resistance [2]. PHAs have been prompted as potential packaging applications due to their biocompatibility and physical properties [3]. However, the PHA production currently associates a high cost due to the carbon sources of the raw materials, i.e., low yield and productivity, and the down-stream process [4]. The synthesis of PHA through fermentation from industrial by-products and waste, particularly the use of mixed microbial cultures, is nowadays seen as an option to reduce the production costs [5].

Among PHAs, the most widely studied and easiest-to-produce member of this family is poly(3-hydroxybutyrate) (PHB). This isotactic homopolyester presents a relatively high melting temperature (T_m) and good stiffness due to its high crystallinity (>50%). However, the use of PHB has been limited due to several drawbacks, particularly its poor impact-strength resistance and low thermal stability. To overcome these shortcomings, the use of its copolymers, such as those made with 3-hydroxyvalerate (3HV) or 4-hydroxybutyrate (4HB) to produce poly(3-hydroxybutyrate-*co*-3-hydroxyvalerate) (PHBV) and poly(3-hydroxybutyrate-*co*-4-hydroxybutyrate) (P(3HB-*co*-4HB)), can improve these limitations and widen its processing window [6,7]. In particular, PHBV is a potential candidate to be applied for packaging of films, blow-molded bottles, paper coatings, etc. [8]. To this end, different studies have explored the use of PHBV due to its potential as a sustainable packaging material [9,10]. For instance, PHBVs have been applied in the form of films, fibers, and foams for everyday articles such as shampoo bottles and plastic beverage bottles due to its renewability, biodegradability, and high water vapor barrier [11]. In addition, the incorporation of antimicrobial and/or antioxidant substances into a PHA-based packaging material can result in high interest to improve both protection and shelf life of foodstuffs during the storage period [12–14].

Electrospinning is an innovative technology to generate ultrathin fibrous mats made of a wide range of polymer and biopolymer materials with fiber diameters ranging from several nanometers to a few microns [15]. Electrospun ultrathin fibers have prompted their use in a wide range of industrial sectors, including packaging applications [16,17]. This technique is highly suitable for the encapsulation and/or sustained delivery of active and bioactive substances at the nanoscale level due to both the high surface-to-volume ratios of the electrospun fibers and the high porosity of their mats [18,19]. In particular, electrospinning is interesting for the development of antimicrobial materials by either the use of inherently antimicrobial polymers or the nanoencapsulation of biocide substances [20]. As a result, within the frame of active packaging, different recent studies have reported the encapsulation of metal nanoparticles (MNPs) in electrospun matrices. For instance, poly(vinyl alcohol) (PVOH) and poly(N-isopropylacrylamide) (PNIPAAm) membranes containing silver nanoparticles (AgNPs) immobilized onto cellulose nanowhiskers (CNWs) presented antimicrobial activity against several Gram-negative (G-) and Gram-positive (G+) bacteria [21]. In another study, polyvinylpyrrolidone (PVP)/poly(ε-caprolactone) (PCL) nanofibers functionalized with zinc oxide nanoparticles (ZnONPs) and AgNPs, also prepared by electrospinning, showed a high antibacterial activity against *Staphylococcus aureus* (*S. aureus*) and *Escherichia coli* (*E. coli*) [22]. Similarly, electrospun chitosan/poly(ethylene oxide) (PEO) membranes containing AgNPs presented antimicrobial effect against *E. coli* [23]. Recently developed electrospun PHA materials containing AgNPs [24] and copper oxide nanoparticles (CuONPs) [25] have been also able to considerably reduce bacterial growth at very low contents. These novel NPs-containing electrospun materials offer significant potential as new antimicrobial coatings or interlayers, that is, internal layers in a multilayer system, for application in the design of active food packaging structures.

Natural antimicrobials, such as essential oils (EOs), are currently regarded as an alternative to synthetic preservatives of food because they are considered as Generally Recognized As Safe (GRAS) substances, being acceptable to consumers [26] and having the capacity to exert a multitude of biological effects [27]. For instance, eugenol, which has potential antimicrobial and antioxidant actions, has been effectively applied against foodborne pathogens [28,29]. However, EOs are frequently unstable and can be easily degraded in stressful situations such as in the presence of oxygen, temperature and light, so that they can lose their antimicrobial activity [30]. To avoid this issue, encapsulation is considered a good way to protect and preserve the effectiveness of active and bioactive substances [31]. In this sense, silica mesoporous supports (SMPSs) [32] show a great deal of potential for the storage and release of active substances [33,34]. In particular, the typical sizes of SMPSs range from microns to nanometers, presenting tailor-made pores of around 2–10 nm [35]. The particular morphology of SMPSs renders a very large specific surface area, up to 1200 m^2/g and, then, an enhanced loading capacity for the encapsulation and release of natural antimicrobials [36]. Within SMPSs, Mobil Composition of Matter (MCM), including both MCM-41 and MCM-48, are among of the most popular mesoporous molecular sieves in which their pore diameter can be nicely controlled by adjusting their synthesis conditions and/or by employing surfactants with different chain lengths in their preparation [37]. Silica mesoporous materials are thus able to encapsulate organic molecules, forming host–guest complexes with volatile molecules (e.g., EOs) to efficiently control their volatility and reactivity. So far, many studies have employed MCM to encapsulate active substances with positive results in different applications, for instance, caprylic acid against foodborne pathogens [38], EOs as antifungal [36,39,40] and antimicrobial systems [41], and poplar-type propolis in drug delivery platforms [42]. In particular, the antimicrobial and antifungal effect of the EOs-functionalized supports improved compared to the free compounds due to the EOs encapsulated inside MCM released in a controlled manner [39–41]. These previous results suggest that the immobilization of EOs onto silica supports can represent a novel strategy to develop a new generation of long-term antimicrobial systems that may not only enhance the antimicrobial activity of EOs, but also mask their characteristic odor/taste for food-related applications.

In this study, it is initially reported the preparation of nanometric MCM-41 particles loaded with eugenol, a phenylpropene and an allyl chain-substituted guaiacol that is primarily extracted from cinnamon, bay leaf, nutmeg, basil, and clove [43]. The resultant MCM-41 particles containing eugenol were thereafter incorporated, for the first time, into PHBV by electrospinning. The generated electrospun composite fibers were thermally post-treated to produce films that were characterized in terms of their morphology, thermal, mechanical, and barrier properties. Finally, the antimicrobial performance against foodborne bacteria was also determined. In a packaging context, the active tests were carried out as a function of time in open vs. close conditions in order to simulate potential real conditions.

2. Materials and Methods

2.1. Materials

Commercial PHBV was ENMAT^{TM} Y1000P, produced by Tianan Biologic Materials (Ningbo, China) and delivered in the form of pellets by NaturePlast (Ifs, France). According to the manufacturer, this biopolymer resin presents a density of 1.23 g/cm^3 and a melt flow index (MFI) of 5–10 g/10 min (190 °C, 2.16 kg). The 3HV fraction in the copolyester is 2–3 mol.-%.

Eugenol, with 99% purity, tetraethyl orthosilicate (TEOS), n-cetyltrimethylammonium bromide (CTAB), sodium hydroxide (NaOH), 2,2,2-trifluoroethanol (TFE), ≥99% purity, and D-limonene, with 98% purity, were all purchased from Sigma Aldrich S.A. (Madrid, Spain).

2.2. Synthesis and Complexation of MCM-41

2.2.1. Synthesis of MCM-41

The MCM-41 type mesoporous particles were synthesized using the following procedure [44]: 2 g of CTAB, 5.48 mmol, was first dissolved in 960 mL of deionized water. Then, 7.00 mL of NaOH, 2 M, was added to the CTAB solution, followed by adjusting the solution temperature to 95 °C. Later, 10 mL of TEOS, $5.14 \cdot 10^{-2}$ mol, was added dropwise to the surfactant solution. The mixture was allowed to stir for 3 h to produce a white precipitate. The solid product was centrifuged and washed several times with deionized water and ethanol and, thereafter, dried at 60 °C to obtain solid MCM particles. Lastly, to prepare the final porous material, i.e., the MCM-41 type particles, the as-synthetized MCM particles were calcined at 550 °C using air atmosphere for 5 h so that their template phase was removed.

2.2.2. Eugenol Complexation on MCM-41

Silica loading with eugenol was achieved via vapor adsorption by mixing 100 mg of eugenol with 100 mg of the MCM-41 type particles in a tightly closed vial [36]. The mixture was incubated in an oven at 40 °C for 24 h while being continuously shaken. The amount of eugenol loaded in the MCM-41 type support was determined by monitoring the sample weight increase before and after the loading process. Approximately 500 mg/g of the final weight corresponded to eugenol.

2.3. Electrospinning Process

Prior to electrospinning, different PHBV solutions were prepared by dissolving the biopolymer at 10 wt.-% in TFE. Then, the MCM-41 type particles, with and without eugenol, were added to the PHBV solutions at 2.5, 5, 7.5, 10, 15, and 20 wt.-%. A neat PHBV solution without MCM-41 type particles was also prepared as a control sample. All PHBV solutions were processed by electrospinning using a high-throughput Fluidnatek® LE-500 pilot-plant device with temperature and relative humidity (RH) control manufactured by Bioinicia S.L. (Valencia, Spain). The equipment was operated in the lab mode using a motorized single needle injector, scanning vertically towards a metallic fixed collector. The conditions were set at a flow-rate of 6 mL/h, 20 kV of voltage, and 15 cm of needle-to-collector distance. Each solution was electrospun for 2 h at 25 °C and 40% RH. The collected mats were stored in darkness at room temperature in a desiccator at 0% RH for one week before physical characterization.

2.4. Film Preparation

The resultant electrospun PHBV fibers mats were subjected to annealing in a 4122-model press from Carver, Inc. (Wabash, IN, USA) at 155 °C, for 5 s, without pressure. These conditions were selected based on our previous work [45]. The thermally post-processed samples had an average thickness of approximately 60 μm.

2.5. Characterization

2.5.1. Electron Microscopy

The morphologies of the MCM-41 type particles as well as the electrospun PHBV fibers and films were observed by scanning electron microscopy (SEM) using an S-4800 device from Hitachi (Tokyo, Japan). The samples were fixed to beveled holders using conductive double-sided adhesive tape and sputtered with a mixture of gold-palladium under vacuum prior to observation. An accelerating voltage of 10 kV was used. For the cross-section observations, the films were previously cryo-fractured by immersion in liquid nitrogen.

Detailed morphology of the MCM-41 particles and their distribution in the PHBV fibers was further studied by transmission electron microscopy (TEM) using a JEOL 1010 from JEOL USA, Inc. (Peabody, MA, USA) using an accelerating voltage of 100 kV. The estimation of the dimensions was performed by means of the Aperture software from Apple (Cupertino, CA, USA) using a minimum of 20 SEM or TEM micrographs in their original magnification.

2.5.2. Thermal Analysis

Thermal transitions were studied by differential scanning calorimetry (DSC) on a DSC-7 analyzer from PerkinElmer, Inc. (Waltham, MA, USA), equipped with a cooling accessory Intracooler 2 also from PerkinElmer, Inc. A heating program was applied from −30 °C to 190 °C, followed by a cooling program to −30 °C. The heating and cooling rates were both set at 10 °C/min under nitrogen atmosphere with a flow-rate of 20 mL/min. The typical sample weight was ~3 mg while an empty aluminum pan was used as reference. Calibration was performed using an indium sample. All tests were carried out, at least, in duplicate.

Thermogravimetric analysis (TGA) was performed in a TG-STDA model TGA/STDA851e/LF/1600 thermobalance from Mettler-Toledo, LLC (Columbus, OH, USA). The samples, with a weight of about 15 mg, were heated from 50 °C to 800 °C at a heating rate of 10 °C/min under a nitrogen atmosphere with a flow-rate of 50 mL/min.

2.5.3. Mechanical Tests

Tensile tests of the PHBV films were performed according to ASTM standard method D638 using an Instron 4400 universal testing machine, equipped with a 1-kN load cell, from Instron (Norwood, MA, USA). The tests were performed, at room conditions, with 115×16 mm^2 stamped dumb-bell shaped specimens using a cross-head speed of 10 mm/min. Samples were conditioned for 24 h prior to tensile assay. A minimum of six specimens was measured for each sample and the average values with standard deviation (SD) were reported.

2.5.4. Permeability Tests

The water vapor permeability (WVP) of the film samples was determined using the gravimetric method ASTM E96-95 in triplicate. For this, 5 mL of distilled water was placed inside a Payne permeability cup (diameter of 3.5 cm) from Elcometer Sprl (Hermallesous-Argenteau, Belgium). The films were not in direct contact with water but exposed to 100% RH on one side and secured with silicon rings. The samples were placed within a desiccator, filled with dried silica gel, at 0% RH and 25 °C. The control samples were cups with aluminum films to estimate the solvent loss through the sealing and samples placed in cups but without permeant to compensate for mass losses due to eugenol release. The cups were weighted periodically using an analytical balance (±0.0001 g). WVP was calculated from the regression analysis of weight loss data vs. time and the weight loss was compensated by the marginal losses through the sealing and eugenol release. The permeability was obtained by multiplying the permeance by the film thickness.

Similar as described above for WVP, limonene permeability (LP) was measured placing 5 mL of D-limonene inside the Payne permeability cups. The cups containing the films were placed at the controlled room conditions of 25 °C and 40% RH. The samples were measured in triplicate and the limonene vapor permeation rate (LPRT) values were estimated from the steady-state permeation slopes and the weight loss was compensated by the comparatively marginal loss through the sealing and by the fluctuations in mass of the films due to eugenol evaporation and potential water sorption. LP was calculated taking into account the average film thickness in each case.

2.6. Antimicrobial Assays

The antibacterial activity of the neat eugenol, the eugenol-containing MCM-41 particles, and the electrospun films with MCM-41 with eugenol was evaluated against *S. aureus* CECT240 (ATCC 6538P) and *E. coli* CECT434 (ATCC 25922). These strains were obtained from the Spanish Type Culture Collection (CECT, Valencia, Spain) and stored in phosphate buffered saline (PBS) with 10 wt.-% tryptic soy broth (TSB, Conda Laboratories, Madrid, Spain) and 10 wt.-% glycerol at −80 °C. Previous to each study, a loopful of each bacteria was transferred to 10 mL of TSB and incubated at 37 °C for 24 h. A 100-μL aliquot from the culture was again transferred to TSB and grown at 37 °C to the mid-exponential phase of growth. An approximate count of 5×10^5 colony-forming units (CFU)/mL of a culture resulted in an absorbance value of 0.20, as determined by optical density at 600 nm (UV 4000 spectrophotometer, Dinko Instruments, Barcelona, Spain).

The minimum inhibitory concentration (MIC) and minimum bactericide concentration (MBC) of eugenol against the selected foodborne bacteria was tested following the plate micro-dilution protocol, as described in the Methods for Dilution Antimicrobial. Susceptibility Tests for Bacteria That Grow Aerobically; Approved Standard Tenth. Edition (M07-A10) by the Clinical and Laboratory Standards Institute (CLSI). For this, a 96-well plate with an alpha numeric coordination system (columns 12 and rows A-H) were used, where 10 μL of the tested samples were introduced in the wells with 90 μL of the bacteria medium. In the wells corresponding to A, B, C, E, F, and G columns different concentrations of eugenol, that is, 0.312, 0.625, 1.25, 2.5, 5, 10, 20, 40, 80, 160 μL/mL, were tested, in triplicate, from rows 1 to 10. Columns D and H were used as control of eugenol in TSB without bacteria. Row 11 was taken as positive control, that is, only TSB, and row 12 was used as negative control, that is, *S. aureus* and *E. coli* in TSB. The plates were incubated at 37 °C for 24 h. Thereafter, 10 μL of resazurin, a metabolic indicator, was added to each well and incubated again at 37 °C for 2 h. Upon obtaining the resazurin change, the wells were read through color difference. The MIC value was determined as the lowest concentration of eugenol presenting growth inhibition.

The antimicrobial performance of the films was evaluated by using a modification of the Japanese Industrial Standard JIS Z2801 (ISO 22196:2007). A microorganism suspension of *S. aureus* and *E. coli* was applied onto the test films of PHBV/MCM-41 with eugenol and also PHBV/MCM-41, as negative control without eugenol, both sizing 2×2 cm^2. After incubation for 24 h at 24 °C and at a RH of at least 95%, bacteria were recovered with PBS, 10-fold serially diluted and incubated at 37 °C for 24 h in order to quantify the number of viable bacteria by conventional plate count. The antimicrobial activity was evaluated from 1 (initial day), 8, and 15 days. The antibacterial activity was taken as the test surface reduction (R) using the equation 1:

$$R = [\log(B/A) - \log(C/A)] = \log(B/C), \quad (1)$$

where A is the mean of bacterial counts of the control sample immediately after inoculation, B is the mean of bacterial counts of the control sample after 24 h, and C is the mean of bacterial counts of the test sample after 24 h. Antimicrobial activity was evaluated with the following assessment: Nonsignificant ($R < 0.5$), slight ($R \geq 0.5$ and <1), significant ($R \geq 1$ and <3), and strong ($R \geq 3$) [46].

3. Results

3.1. Morphology

Figure 1 shows the morphology of the here-obtained MCM-41 powder. Figure 1a,b present the SEM images of the MCM-41 powders with and without eugenol, respectively. One can observe that the silica particles presented a spherical shape with a mean size of around 100 nm, where the incorporation of eugenol slightly reduced their particle size. Therefore, the incorporation of eugenol did not alter the morphology of the mesoporous MCM-41 type nanoparticles. TEM was carried out in order to further ascertain the morphology of the MCM-41 particles. Figure 1c confirmed the spherical shape of the MCM-41 particle without eugenol, showing that their mean size was 96.1 ± 3.8 nm. A similar morphology can be observed in Figure 1d for the MCM-41 powder with eugenol, having a mean diameter of 88.6 ± 2.1 nm. Similar results were reported by Ribes et al. [40] in which the immobilization of eugenol and thymol on the surface of MCM-41 did not affect the integrity of the mesoporous silica particles. Also, Ruiz-Rico et al. [41] observed that the appearance of fumed silica, amorphous silica, and MCM-41 particles did not change after functionalization with thymol. Indeed, MCM-41 has been widely used as a model material in the context of porosity characterization owing to its peculiar features, such as high surface area, large pore volume, low toxicity, high chemical and thermal stability, and versatile chemical modifiable surface. It has been reported that the pore structure is organized in the form of hexagonal arrays of uniform tubular channels of controlled width [47,48]. As a result, mesoporous silica nanoparticles are excellent candidates for reference adsorbents for standardizing adsorption measurements and methods for characterization of porous solids due to their regular pore structure, high stability, and also convenient method of synthesis [49,50].

Figure 1. Scanning electron microscopy (SEM) images of: (**a**) Mobil Composition of Matter (MCM-41); (**b**) MCM-41 with eugenol. Scale markers of 1 µm. Transmission electron microscopy (TEM) images of: (**c**) MCM-41 and (**d**) MCM-41 with eugenol. Scale markers of 100 nm.

Figure 2 shows the resultant electrospun mats obtained from the neat PHBV solution and the different solutions of PHBV/MCM-41 with eugenol. One can observe that, in all cases, the electrospinning process generated a mat composed of non-woven fibers with a similar morphology. Table 1 summarizes the mean diameters of the electrospun fibers. The neat PHBV fibers without MCM-41, processed in the same conditions, presented a mean diameter of 0.89 ± 0.30 µm. It can be observed that the mean diameters of the electrospun fibers varied in the 0.6–0.7 µm range when the silica particles were incorporated. However, one can observe that the electrospun fibers with the highest particle contents, that is, 15 and 20 wt.-% MCM-41, presented certain cross-linking or fibers coalescence. This can be related to difficulties encountered during the fiber formation more likely due to a phenomenon of particle aggregation in the electrospinning process. Indeed, it is known that high nano-sized filler contents habitually lead to the formation of beaded regions in the electrospun fibers [51,52].

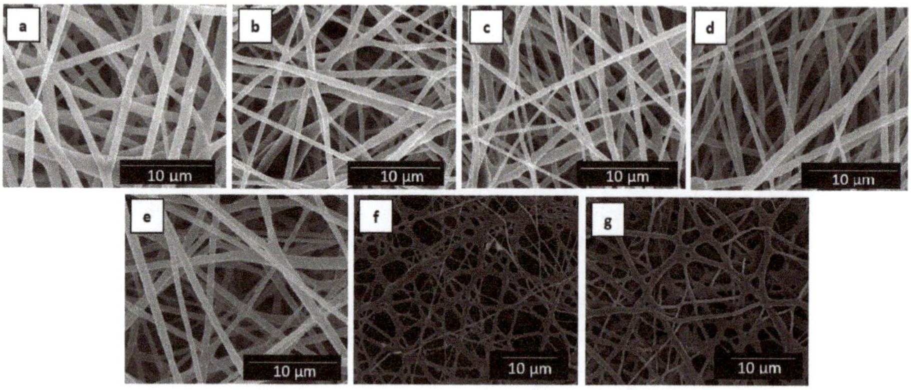

Figure 2. Scanning electron microscopy (SEM) images of the electrospun fibers of poly(3-hydroxybutyrate-*co*-3-hydroxyvalerate) (PHBV)/Mobil Composition of Matter (MCM)-41 with eugenol: (**a**) Neat PHBV; (**b**) 2.5 wt.-% MCM-41 + eugenol; (**c**) 5 wt.-% MCM-41 + eugenol; (**d**) 7.5 wt.-% MCM-41 + eugenol; (**e**) 10 wt.-% MCM-41 + eugenol; (**f**) 15 wt.-% MCM-41 + eugenol; (**g**) 20 wt.-% MCM-41 + eugenol. Scale markers of 10 µm.

Table 1. Mean diameters of the electrospun fibers of poly(3-hydroxybutyrate-*co*-3-hydroxyvalerate) (PHBV)/Mobil Composition of Matter (MCM)-41 with eugenol.

Fibers	Diameter (µm)
PHBV	0.89 ± 0.30
PHBV/2.5 wt.-% MCM-41 + eugenol	0.65 ± 0.19
PHBV/5 wt.-% MCM-41 + eugenol	0.66 ± 0.16
PHBV/7.5 wt.-% MCM-41 + eugenol	0.63 ± 0.18
PHBV/10 wt.-% MCM-41 + eugenol	0.64 ± 0.19
PHBV/15 wt.-% MCM-41 + eugenol	0.65 ± 0.19
PHBV/20 wt.-% MCM-41 + eugenol	0.67 ± 0.24

TEM was also performed in order to evaluate the distribution of the MCM-41 particles inside the electrospun fibers. The detailed morphologies of the electrospun mats of PHBV/MCM-41 with eugenol, at different particle contents, are shown in Figure 3. One can observe that at low contents, that is, from 2.5 wt.-% to 7.5 wt.-% MCM-41 with eugenol, the functionalized silica nanoparticles were relatively well distributed inside the electrospun fibers. However, for higher filler contents, the MCM-41 particles were mainly agglomerated in certain regions of the fibers. This fact supports the above-described morphology during the SEM analysis by which the silica nanoparticles interconnected the fibers in the electrospun mats. A similar morphology was recently reported, for instance, by Cherpinski et al. [53] in PHB fibers containing palladium nanoparticles (PdNPs).

The morphology of the electrospun materials was also analyzed by SEM in order to ascertain the effect of the film-forming process on the PHBV fibers. Figure 4 shows the SEM images at both the cross-section and surface of the electrospun PHBV materials containing different amounts of MCM-41 with eugenol. The surface cryo-fractures of the electrospun materials, shown in the left column, revealed the formation of a continuous film with much reduced porosity. This process has been ascribed to a process of fibers coalescence that occurs during annealing, that is, at a temperature below the polymer's T_m [54]. In the case of the electrospun films having the highest particle contents, that is, 15 and 20 wt.-% MCM-41 with eugenol, the films presented a higher porosity and also certain plastic deformation. This observation can be related to the above-described fiber morphology and, more importantly, to the presence of high loadings of eugenol that could plasticize the PHBV matrix and/or migrate during the annealing process. In the top view of the electrospun films, shown in the right column, one can clearly observe that the film sample containing 20 wt.-% MCM-41 presented higher porosity on its surface. This morphology confirms that contents above 15 wt.-% MCM-41 with eugenol are not optimal to be processed by electrospinning and thermally post-treatment at 160 °C. Similar findings were concluded when electrospun mats of PHBV with ~20 mol.-% HV were post-treated at higher temperatures than optimal, resulting in an increased porosity due to partial polymer melting and/or degradation [55].

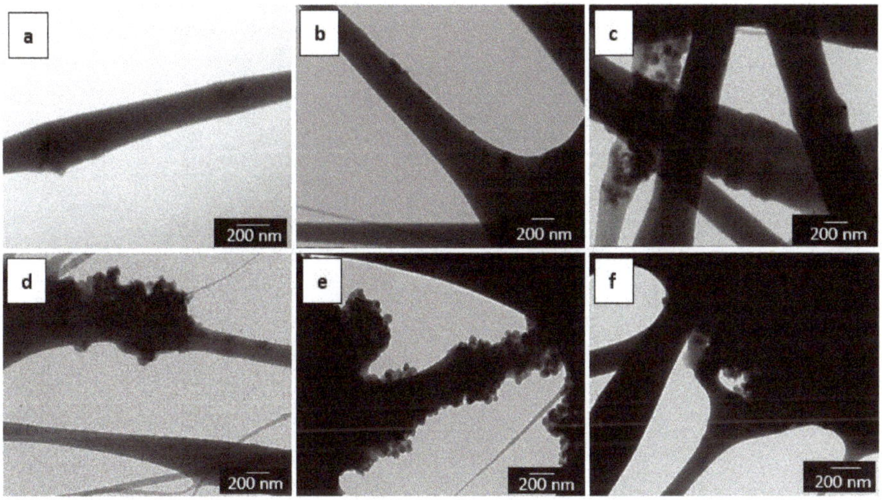

Figure 3. Transmission electron microscopy (TEM) images of the electrospun fibers of poly(3-hydroxybutyrate-*co*-3-hydroxyvalerate) (PHBV)/Mobil Composition of Matter (MCM)-41 with eugenol: (**a**) 2.5 wt.-% MCM-41 + eugenol; (**b**) 5 wt.-% MCM-41 + eugenol; (**c**) 7.5 wt.-% MCM-41 + eugenol; (**d**) 10 wt.-% MCM-41 + eugenol; (**e**) 15 wt.-% MCM-41 + eugenol; (**f**) 20 wt.-% MCM-41 + eugenol. Scale markers of 200 nm.

Figure 5 shows the visual aspect of the resulting annealed electrospun PHBV films containing MCM-41 with eugenol. Although the contact transparency of the films was similar in all the samples, the films with the highest particle contents, that is, 15 and 20 wt.-% MCM-41 with eugenol, developed a yellow color. A similar yellowing and, in some cases, browning was previously observed by Muratore et al. [56] after the incorporation of eugenol into commercial paper prepared by grafting of this EO onto cellulose at 120–180 °C. This effect was ascribed to the intrinsic eugenol color, which is a pale yellow oily liquid, as well as secondary reactions and/or by-products due to thermal oxidation and chain scission of the substrate favored by high temperatures and prolonged time. Therefore, the incorporation of up to 10 wt.-% MCM-41 with eugenol successfully allows the production of contact transparent films of PHBV.

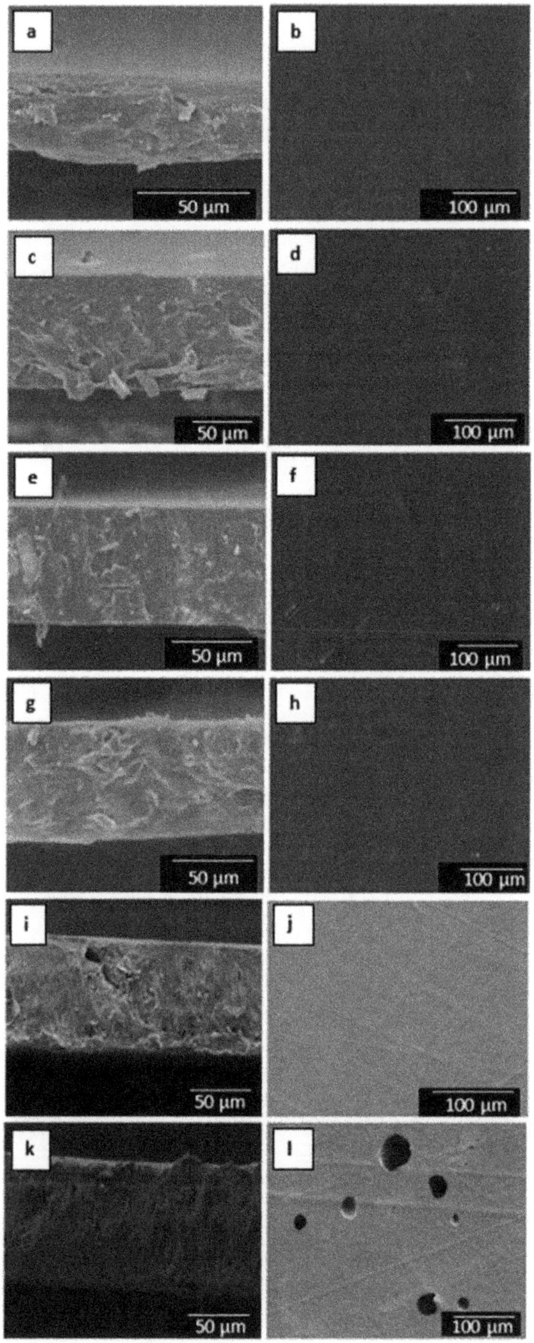

Figure 4. Scanning electron microscopy (SEM) images of the films cross-section (left) and top view (right) of poly(3-hydroxybutyrate-*co*-3-hydroxyvalerate) (PHBV)/Mobil Composition of Matter (MCM)-41 with eugenol: (**a**,**b**) 2.5 wt.-% MCM-41 + eugenol; (**c**,**d**) 5 wt.-% MCM-41 + eugenol; (**e**,**f**) 7.5 wt.-% MCM-41 + eugenol; (**g**,**h**) 10 wt.-% MCM-41 + eugenol; (**k**,**j**) 15 wt.-% MCM-41 + eugenol; (**k**,**l**) 20 wt.-% MCM-41 + eugenol. Scale markers of 50 µm and 100 µm.

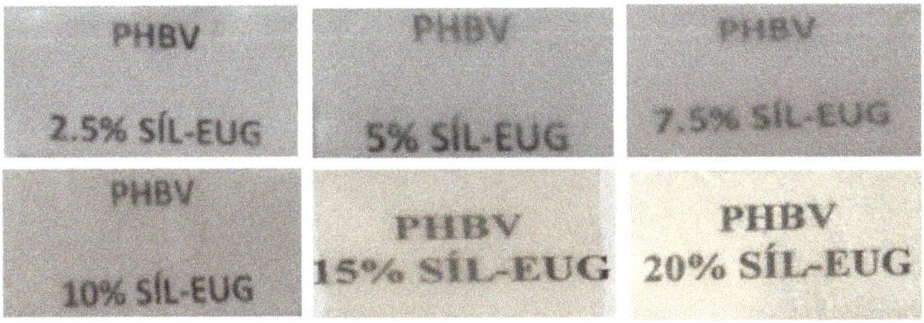

Figure 5. Visual aspect of the electrospun films of poly(3-hydroxybutyrate-*co*-3-hydroxyvalerate) (PHBV)/Mobil Composition of Matter (MCM)-41 with eugenol.

3.2. Thermal Properties

Table 2 displays the main thermal transitions, obtained by DSC during the heating and cooling steps, of the annealed electrospun neat PHBV film and the films containing MCM-41 without and with eugenol. It can be observed that the neat PHBV film presented a glass transition temperature (T_g) of 2.6 ± 0.4, while the addition of MCM-41 without eugenol had a negligible effect on T_g. Interestingly, after the incorporation of MCM-41 with eugenol, the T_g values were reduced to 1.8–0.6 °C in the PHBV film samples. Reductions of T_g are habitually associated to a plasticization process by low-molecular weight (M_W) molecules with high chemical affinity to the polymer matrix by which the free volume of the polymer is enlarged since they increase the distance between the polymer chains and then favor segmental motion [57]. Some previous studies have already reported the plasticizing effect of eugenol on different polymer matrices. For instance, Fernandes Nassar et al. [58] reported a reduction in T_g when eugenol was incorporated into soy protein isolate (SPI) films, ascribing this effect to the plasticizing role that the aroma compound played in the protein matrix. Also, Narayanan et al. [59] observed a reduction in T_g from 4 °C, for the neat PHB film, to −14 °C, for PHB films containing up to 200 µg/g of eugenol.

Whereas cold crystallization phenomenon was not observed in any of the PHBV films during heating, all the samples crystallized from the melt during cooling. In particular, the neat PHBV film showed a crystallization temperature (T_c) of 116.8 ± 0.5 °C. The presence of MCM-41 without eugenol increased the crystallization temperature of PHBV, up to reaching a maximum value of 120.5 °C for the film filled at 15 wt.-%. This result suggests that the nanoparticles provided a nucleating effect on the PHBV molecules, except for the film filled with 20 wt.-% MCM-41, possibly due to nanoparticles agglomeration as previously described during the morphological analysis. On the contrary, one can observe that the T_c values of the film samples containing MCM-41 with eugenol decreased as the filler content increased. Then, the T_c value was reduced up to a value of 113.8 ± 0.6 °C for the film filled with 20 wt.-% MCM-41 with eugenol. This restrained crystallization of PHBV can be ascribed to the above-described plasticizing effect of eugenol, which impair the packing of the polymer chains to form crystals.

During heating, the neat PHBV film melted in a single peak at 170.4 ± 0.2 °C while all the PHBV films containing MCM-41 without eugenol presented similar T_m values in the 169–171 °C range. However, the T_m values progressively reduced in the PHBV films containing MCM-41 with eugenol was as the filler content increased. Up to contents of 15 wt.-% MCM-41 with eugenol, the PHBV films presented a single melting peak in the 163–171 °C range, whereas the film filled with 20 wt.-% MCM-41 with eugenol showed two endothermic peaks, starting melting at 160.5 ± 1.5 °C. Therefore, the MCM-41 particles when loaded with eugenol were able to impair and induce some defects in the PHBV crystals, particularly at the highest tested contents. It is also worthy to note that the presence of MCM-41 without eugenol, up to fillings of 15 wt.-%, increased the values of enthalpy of melting (ΔH_m), confirming the formation of more perfect PHBV crystals with thicker lamellae by a nucleation phenomenon. As opposite, all the PHBV films with MCM-41 with eugenol presented lower values of ΔH_m, being this reduction significantly noticeable for the films filled with contents above 15 wt.-%. Therefore, the presence of MCM-41 with eugenol impaired the crystallization of PHBV due to the above-described reduction of the biopolymer segments packing. It has been similarly reported that the addition of mesoporous silica nanoparticles has a slight influence on T_g or T_m in polymer nanocomposites [60,61], therefore supporting that the here-observed suppressed effect on the melt behavior is ascribed to eugenol. In this sense, Garrido-Miranda et al. [62] showed that the T_m value of PHB/thermoplastic starch (TPS)/organically modified montmorillonite (OMMT) nanocomposites was reduced by approximately 4 °C when 3 wt.-% eugenol was incorporated, concluding that eugenol induces the formation of less perfect crystals. Woranuch et al. [63] also observed a ΔH_m reduction when eugenol-loaded chitosan nanoparticles were incorporated into thermoplastic flour (TPF) made of cassava, rice, and waxy rice through an extrusion process. The reduction observed was related to a plasticization by eugenol.

Table 2. Thermal properties in terms of glass transition temperature (T_g), crystallization temperature (T_c), melting temperature (T_m), and normalized enthalpy of melting (ΔH_m) for the electrospun films of poly(3-hydroxybutyrate-*co*-3-hydroxyvalerate) (PHBV) and PHBV/Mobil Composition of Matter (MCM)-41 without and with eugenol.

Film	T_g (°C)	T_c (°C)	T_m (°C)	ΔH_m (J/g)
PHBV	2.6 ± 0.4	116.8 ± 0.5	170.4 ± 0.2	83.2 ± 3.0
PHBV/2.5 wt.-% MCM-41	2.6 ± 0.2	117.8 ± 0.6	169.9 ± 0.1	85.7 ± 0.5
PHBV/5 wt.-% MCM-41	2.3 ± 0.5	118.1 ± 0.4	170.1 ± 0.6	86.2 ± 4.0
PHBV/7.5 wt.-% MCM-41	2.4 ± 0.4	118.2 ± 0.2	171.7 ± 2.1	87.4 ± 7.3
PHBV/10 wt.-% MCM-41	2.3 ± 0.3	118.9 ± 0.1	170.2 ± 0.6	89.7 ± 6.3
PHBV/15 wt.-% MCM-41	2.2 ± 0.2	120.5 ± 0.3	170.1 ± 1.0	100.6 ± 7.4
PHBV/20 wt.-% MCM-41	2.5 ± 0.1	116.9 ± 0.1	169.0 ± 0.1	67.9 ± 5.9
PHBV/2.5 wt.-% MCM-41 + eugenol	1.8 ± 0.6	116.7 ± 0.1	169.0 ± 1.6	77.4 ± 3.4
PHBV/5 wt.-% MCM-41 + eugenol	1.4 ± 0.8	116.4 ± 0.6	167.4 ± 0.3	74.9 ± 6.2
PHBV/7.5 wt.-% MCM-41 + eugenol	1.5 ± 0.3	118.5 ± 0.3	168.8 ± 3.6	72.3 ± 6.9
PHBV/10 wt.-% MCM-41 + eugenol	1.3 ± 0.4	114.9 ± 0.7	165.4 ± 0.4	66.8 ± 4.8
PHBV/15 wt.-% MCM-41 + eugenol	0.9 ± 0.5	113.9 ± 0.2	163.4 ± 0.3	50.3 ± 4.5
PHBV/20 wt.-% MCM-41 + eugenol	0.6 ± 0.2	113.8 ± 0.6	160.5 ± 1.5/168.6 ± 0.1	47.4 ± 2.1

Figure 6 depicts the TGA curves of MCM-41 and MCM-41 with eugenol powders, the eugenol-free oil, and the electrospun films made of neat PHBV and PHBV/MCM-41 without and with eugenol. Table 3 gathers the main relevant thermal parameters obtained from the TGA curves. As one can observe in the graph, the neat MCM-41 particles presented a mass loss of ~5% at a temperature close to 100 °C, which can be ascribed to residual humidity on the surface and/or in the pores of the nanoparticles. In addition, the neat MCM-41 particles provided a residual mass of 95.0 ± 2.3% measured at 800 °C. On the contrary, the eugenol free oil had a relatively low thermal stability, showing full decomposition at approximately 200 °C. Moreover, comparison of the TGA curves of the MCM-41 nanoparticles with and without eugenol corroborated that the eugenol loading was 49.5 ± 1.2%. This loading capacity of MCM-41 was higher than other encapsulation techniques reported for polyphenols [64].

In relation to the neat PHBV film, a low-intense first weight loss process (<1%) was observed at 100 °C due to absorbed moisture and/or volatiles leaving the samples. Trapped solvent losses were discarded by Fourier transform infrared (FTIR) spectroscopy and TGA of the neat PHBV fibers (results not shown). One can also observe that the biopolymer presented the onset of degradation, measured at the temperature at which the mass loss was 5% ($T_{5\%}$), at 259.9 ± 1.2 °C. The degradation temperature (T_{deg}) occurred at 277.3 ± 0.6 °C, degrading in a single step and producing a residual mass of 2.0 ± 0.2% at 800 °C. In addition, the weight loss process corresponding to thermal decomposition reaction of the biopolymer chain occurred sharply, approximately from 225 °C to 275 °C. The thermal degradation onset was shifted to lower temperatures when both the MCM-41 without and with eugenol, in all the composition range, was incorporated. This result suggests that the nanoparticles catalyzed thermal degradation. Interestingly, the $T_{5\%}$ and T_{deg} values were slightly improved at the lowest content of MCM-41 with eugenol, which can be related to the above-described nucleating effect and restricted mobility of the biopolymer chains by the presence of MCM-41 and eugenol. However, the thermal stability was reduced at the higher filler contents, that is, 15 wt.-% and 20 wt.-% MCM-41 with eugenol, due to the high content of both MCM-41 and eugenol. Furthermore, the residual weight at 800 °C of the PHBV/MCM-41 with eugenol films increased due to the presence of the mesoporous silica nanoparticles. In any case, the incorporation of up to 10 wt.-% of MCM-41 with eugenol had a relatively low influence on the thermal stability of the PHBV films, which can be considered a positive result since they encapsulate an active component with low thermal stability. In this sense, Requena et al. [65] reported that the incorporation of carvacrol and eugenol enhanced the thermal sensitivity of PHBV, decreasing the onset temperature, whereas the incorporation of whole essential oils (oregano and clove) slightly promoted its thermal stability. The latter effect was suggested to occur due to a strong bonding of the eugenol with the polymer network.

Table 3. Thermal properties in terms of mass loss was 5% ($T_{5\%}$), degradation temperature (T_{deg}), mass loss at T_{deg}, and a residual mass at 800 °C for Mobil Composition of Matter (MCM)-41, MCM-41 with eugenol, eugenol free oil, and electrospun films of poly(3-hydroxybutyrate-*co*-3-hydroxyvalerate) (PHBV) and PHBV/MCM-41 without and with eugenol.

Sample	$T_{5\%}$ (°C)	T_{deg} (°C)	Mass Loss (%)	Residual Mass (%)
MCM-41 powder	-	-	-	95.0 ± 2.3
MCM-41 with eugenol powder	143.7 ± 2.4	178.3 ± 0.7	16.3 ± 0.5	52.4 ± 1.9
Eugenol free oil	105.9 ± 3.2	185.5 ± 0.6	80.6 ± 1.1	0.9 ± 0.1
PHBV	259.9 ± 1.2	277.3 ± 0.6	62.0 ± 0.8	2.0 ± 0.2
PHBV/2.5 wt.-% MCM-41	250.8 ± 2.3	270.9 ± 0.2	74.4 ± 1.2	3.3 ± 0.4
PHBV/5 wt.-% MCM-41	245.3 ± 2.7	265.4 ± 1.0	72.1 ± 0.6	5.4 ± 0.8
PHBV/7.5 wt.-% MCM-41	245.3 ± 2.2	265.4 ± 0.8	70.3 ± 0.3	8.0 ± 0.4
PHBV/10 wt.-% MCM-41	240.7 ± 1.8	262.7 ± 0.5	70.8 ± 0.2	9.1 ± 1.0
PHBV/15 wt.-% MCM-41	230.0 ± 2.0	262.0 ± 0.2	70.2 ± 0.7	14.6 ± 0.9
PHBV/20 wt.-% MCM-41	215.2 ± 1.7	256.4 ± 0.6	80.5 ± 0.9	17.2 ± 0.2
PHBV/2.5 wt.-% MCM-41 + eugenol	259.8 ± 2.6	281.0 ± 2.5	71.3 ± 0.3	2.5 ± 0.4
PHBV/5 wt.-% MCM-41 + eugenol	251.7 ± 1.4	276.4 ± 1.6	71.4 ± 0.3	3.4 ± 0.7
PHBV/7.5 wt.-% MCM-41 + eugenol	248.0 ± 2.7	273.7 ± 0.9	72.8 ± 0.4	3.8 ± 0.8
PHBV/10 wt.-% MCM-41 + eugenol	247.1 ± 3.2	271.8 ± 0.7	73.9 ± 0.7	4.3 ± 1.0
PHBV/15 wt.-% MCM-41 + eugenol	215.2 ± 7.3	261.0 ± 4.2	74.6 ± 1.4	7.5 ± 3.6
PHBV/20 wt.-% MCM-41 + eugenol	205.1 ± 5.1	259.2 ± 4.4	75.6 ± 1.2	9.1 ± 3.7

3.3. Mechanical Properties

Since the resultant electrospun films may be subjected to various kinds of stress during use, the determination of the mechanical properties involves not only scientific but also technological and practical aspects. Table 4 displays the values of elastic modulus (E), tensile strength at yield (σ_y), elongation at break (ε_b), and toughness (T) of the electrospun films made of PHBV and PHBV/MCM-41 with eugenol calculated from their strain–stress curves. In general, all the electrospun films presented characteristics of a brittle material associated to the inherent low ductility of PHBV, showing ε_b and T values below 3% and 0.5 mJ/m^3, respectively. The film specimens also presented a relative high mechanical strength. In particular, the mean values of E were comprised in the of 1250–2000 MPa range while σ_y varied from approximately 18 to 30 MPa. The here-obtained mechanical properties of the PHBV films are similar to those recently reported in our group by Cherpinski et al. [54] for PHB films also prepared by electrospinning and thereafter thermally post-treated, having a E value of 1104 MPa and ε_b and T values of 2.9% and 0.3 mJ/m^3, respectively.

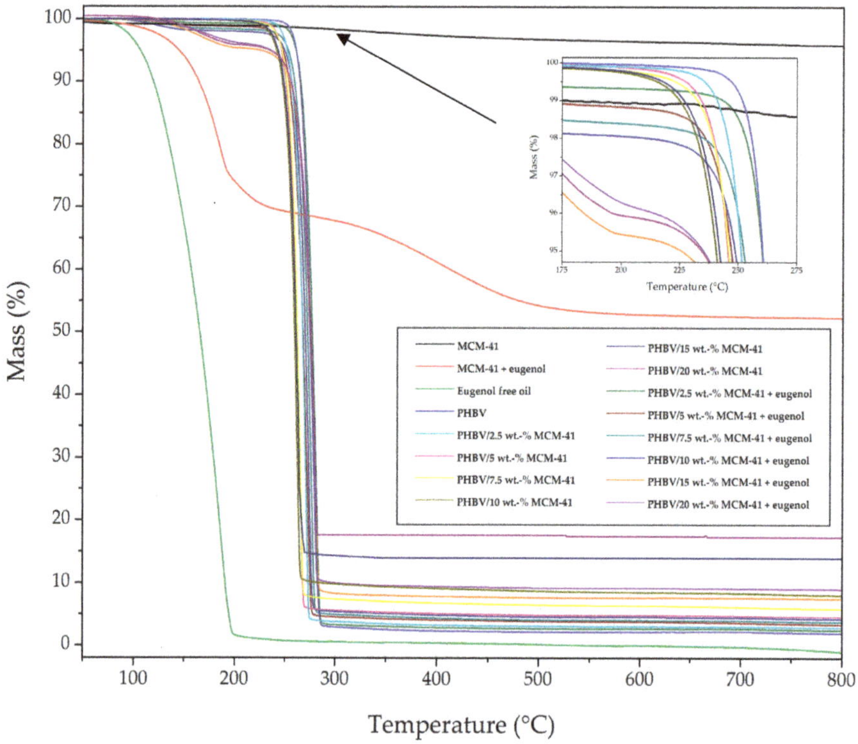

Figure 6. Thermogravimetric analysis (TGA) curves for Mobil Composition of Matter (MCM)-41, eugenol, MCM-41 with eugenol, poly(3-hydroxybutyrate-*co*-3-hydroxyvalerate) (PHBV), and PHBV/Mobil Composition of Matter (MCM)-41 without and with eugenol.

It can be observed that the incorporation of MCM-41 with eugenol increased the mechanical strength of the PHBV films while the ductility was slightly reduced. This effect can be related to the reinforcing effect of MCM-41 as a filler in the PHBV matrix, while the smaller impact in ductility may be accounted for the plasticizing effect that the released eugenol may have in the polymer matrix. This mechanical enhancement of E and σ_y indicates a good transfer of mechanical energy from the hard filler, that is, MCM-41, as well as the interaction between the biopolymer matrix and the silica nanoparticles. Considering both the low concentration of MCM-41 and the presence of eugenol, which acts as plasticizer, the mechanical reinforcement of the filler is thought to dominate the enhancement in E and σ_y. However, a comparative reduction, change in trend, in mechanical strength was observed when the content of the antimicrobial filler exceeded 10 wt.-%. This effect may be ascribed to a balance between filler agglomeration and stronger plasticizing effect of the released eugenol. High tensile strengths are generally necessary for food packaging films in order to withstand the normal stress encountered during their application, subsequent shipping, and handling [66]. Similarly, Voon et al. [67] reported that the addition of 3 wt.-% of mesoporous silica nanoparticles to bovine gelatin films improved their mechanical resistant properties, that is, σ_y, while it reduced ε_b. Others studies have also demonstrated that the incorporation of mesoporous silica nanoparticles can remarkably enhance the mechanical strength in PVOH-based materials due to the intermolecular interactions between the fillers and the polymer when prepared by in situ radical copolymerization [68,69]. As compared to commercial biopolymers for packaging applications, the here-developed electrospun films of PHBV/MCM-41 with eugenol are slightly less deformable but more elastic than thermo-compressed PHBV films, stiffer but less ductile than rigid polylactide

(PLA) films, and mechanically stronger but considerably more brittle than flexible poly(butylene adipate-*co*-terephthalate) (PBAT) [70].

Table 4. Mechanical properties in terms of elastic modulus (E), tensile strength at yield (σ_y), elongation at break (ε_b), and toughness (T) for the electrospun films of poly(3-hydroxybutyrate-*co*-3-hydroxyvalerate) (PHBV) and PHBV/Mobil Composition of Matter (MCM)-41 with eugenol.

Film	*E*	σ_y (MPa)	ε_b (%)	T (mJ/m^3)
PHBV	1252 ± 79	18.1 ± 2.1	2.4 ± 0.3	0.3 ± 0.1
PHBV/2.5 wt.-% MCM-41 + eugenol	1735 ± 60	27.0 ± 2.4	2.4 ± 0.1	0.4 ± 0.1
PHBV/5 wt.-% MCM-41 + eugenol	1976 ± 162	25.1 ± 7.8	1.7 ± 0.4	0.2 ± 0.1
PHBV/7.5 wt.-% MCM-41 + eugenol	2000 ± 365	27.7 ± 5.4	2.2 ± 0.2	0.4 ± 0.1
PHBV/10 wt.-% MCM-41 + eugenol	1802 ± 288	21.7 ± 5.8	1.8 ± 0.2	0.2 ± 0.1
PHBV/15 wt.-% MCM-41 + eugenol	1702 ± 140	29.4 ± 3.4	2.0 ± 0.3	0.3 ± 0.1
PHBV/20 wt.-% MCM-41 + eugenol	1462 ± 358	25.1 ± 5.4	2.1 ± 0.5	0.3 ± 0.2

3.4. Barrier properties

Table 5 gathers the WVP and LP values of the electrospun PHBV/MCM-41 with eugenol films. It can be observed that the incorporation of low contents of MCM-41 with eugenol, that is, 2.5 and 5 wt.-%, induced an increase in the WVP values of the electrospun PHBV films while the water vapor barrier properties were improved for contents higher than 7.5 wt.-%. A similar effect was observed in the case of LP. The resultant increase in permeability observed at low filler loadings can be related to the plasticizing effect of eugenol on the PHBV matrix outweighing the barrier effect of the MCM-41 filler, as above discussed during the thermal analysis, with a subsequent increase in the matrix free volume. At higher contents, however, the barrier improvements can be ascribed to the presence of large quantities of mesoporous silica nanoparticles. Then, MCM-41 successfully acted as barrier elements forcing the permeant molecules to travel through a longer path to permeate across according to the early theory suggested by Nielsen [71]. It is also worthy to mention the change in the barrier trend observed for the film samples containing the highest filler contents, that is, 20 wt.-% MCM-41 with eugenol. This permeability change in trend can also be ascribed to the above-mentioned higher filler agglomeration resulting in a somewhat increased porosity as observed in the morphological analysis, leading to preferential paths for diffusion.

The here-prepared electrospun films of PHBV/MCM-41 with eugenol showed higher WVP values than PHBV films with 12 mol.-% HV prepared by solvent casting, that is, 1.27×10^{-14} kg·m·m^{-2}·Pa^{-1}·s^{-1} [72] or PHB films obtained by compression-molded, that is, 1.7×10^{-15} kg·m·m^{-2}·Pa^{-1}·s^{-1} [73], which can be related to the higher mol.-% HV fraction in the here-used copolyester. Therefore, it can be considered that intermediate contents of MCM-41 with eugenol inside the fibers promoted lower free volume available for diffusion. Similar results were reported for instance by Hashemi Tabatabaei et al. [74], who showed that the incorporation of 5 wt.-% mesoporous silica microparticles decreased the WVP value from 8.9 g·m·m^{-2}·Pa^{-1}·s^{-1} to 1.6×10^{-11} g·m·m^{-2}·Pa^{-1}·s^{-1} in gelatin/k-carrageenan films. Also, Hassannia-Kolaee et al. [75] reported a reduction of up to approximately 33% in WVP for whey protein isolate (WPI)/pullulan (PUL) films containing 1, 3, and 5 wt.-% mesoporous silica nanoparticles prepared by a casting method. The barrier improvement achieved was attributed to the formation of hydrogen bonds between the polymer hydroxyl groups and the oxygen atoms of silica and also to the good dispersion of the nanoparticles in the polymer matrix. In relation to eugenol, some studies have also demonstrated that the direct incorporation of EOs, among them eugenol, in polymer films do not induce significant change in WVP, concluding that water permeability basically depends on the hydrophilic–hydrophobic ratio of the film constituents [76,77]. Other studies have reported that the addition of EOs can negatively increase the water permeability depending on the nature of the polymer matrix and the type and concentration of EO [78]. However, this impairment may be attributed to the difficulties encountered to integrate the hydrophobic EO in hydrophilic networks that might cause matrix disruptions and create void spaces at the polymer–oil interface [79]. However, in the current study, the expected plasticizing effect of the hydrophobic eugenol within the PHBV matrix is seen detrimental for the barrier performance at lower silica loadings.

Table 5. Permeability values in terms of water vapor permeability (WVP) and D-limonene permeability (LP) for the electrospun films of poly(3-hydroxybutyrate-*co*-3-hydroxyvalerate) (PHBV) and PHBV/Mobil Composition of Matter (MCM)-41 with eugenol.

Sample	WVP × 10^{-14} (kg·m·m^{-2}·Pa^{-1}·s^{-1})	LP × 10^{-14} (kg·m·m^{-2}·Pa^{-1}·s^{-1})
PHBV	5.34 ± 1.79	2.68 ± 1.82
PHBV/2.5 wt.-% MCM-41 + eugenol	8.68 ± 3.57	3.41 ± 0.97
PHBV/5 wt.-% MCM-41 + eugenol	8.84 ± 4.36	3.49 ± 1.17
PHBV/7.5 wt.-% MCM-41 + eugenol	4.25 ± 4.04	3.51 ± 0.54
PHBV/10 wt.-% MCM-41 + eugenol	2.99 ± 0.95	2.32 ± 0.68
PHBV/15 wt.-% MCM-41 + eugenol	0.25 ± 0.19	0.38 ± 0.20
PHBV/20 wt.-% MCM-41 + eugenol	4.08 ± 1.98	4.66 ± 2.91

3.5. Antimicrobial activity

S. aureus and *E. coli* are common microorganisms associated with food-related diseases. Therefore, the incorporation of active substances in the design of packaging materials can be an important technology not only to avoid food waste but also to enhance food safety [80]. For the pure eugenol in its original liquid form, the MIC and BIC values for *S. aureus* were 1.25 μL/mL and 2.5 μL/mL, respectively, and for *E. coli* these values were 2.5 μL/mL and 5 μL/mL, respectively. The MCM-41 particles with eugenol presented a MIC value against *S. aureus* and *E. coli* of 10 μg/mL and 20 μg/mL, respectively, while the BIC values were 40 μg/mL for both bacteria. The higher value observed for *E. coli* can be ascribed to the greater bacterial resistance of G- bacteria than G+ ones [81], thus a higher dose of the antimicrobial was needed to obtain the same efficacy.

The antimicrobial activity of the film samples was evaluated using the JIS Z2801. The reduction values in the open system against *S. aureus* and *E. coli* are gathered in Tables 6 and 7, respectively. Tables 8 and 9 includes the values against *S. aureus* and *E. coli*, respectively, in the closed system. As expected, it can be observed that both the unfilled PHBV film and the different PHBV films containing MCM-41 without eugenol showed no inhibition effect on the bacterial growth ($R \leq 1$). In contrast, the incorporation of MCM-41 with eugenol into the PHVB film exhibited significant antibacterial activity against both bacteria. In the open system, at the initial day, that is, for the tests carried out the same day of the film production, the bacterial reduction on the film surface gradually increased with the content of MCM-41 with eugenol. As it can be seen in Table 6 for *S. aureus*, at the lowest contents, that is, 2.5 and 5 wt.-% MCM-41 with eugenol, the films presented a slight antibacterial activity ($R \geq 1$ and < 2). For the highest tested contents, that is, 7.5 and 10 wt.-% MCM-41 with eugenol, the films generated a significant surface reduction ($R \geq 1$ and < 3). Although none of the films produced a strong reduction ($R \geq 3$), materials with values of surface reduction in the 1–2 range are usually considered as bacteriostatic [82]. Therefore, electrospun PHBV films with 10 wt.-% MCM-41 with eugenol were able to provide a bacteriostatic effect against *S. aureus*. As also shown in the table, after 15 days, the films still kept a significant antibacterial activity. In particular, the films with 7.5 and 10 wt.-% MCM-41 with eugenol still presented significant values of reduction ($R \geq 1$ and < 3) while these presented slight values ($R \geq 0.5$ and < 1) for loadings of 2.5 and 5 wt.-%. This suggests that, although part of eugenol was released from the films, MCM-41 was still able to retain over time a significant amount of EO. Regarding *E. coli*, shown in Table 7, the required concentration of MCM-41 with eugenol to generate an antimicrobial effect in the open system was 15 wt.-%. At this content, the films presented a significant value of reduction, that is, R values of 1.30 and 1.40 at days 0 and 15, respectively. This supports the above-described higher antimicrobial resistance of

E. coli, as a G- bacterium, which would need more exposure time to the active oil to render a similar antimicrobial activity.

The tested closed system was aimed to better represent the real conditions in a packaging material. In the case of *S. aureus*, which is shown in Table 8, the film with 10 wt.-% MCM-41 with eugenol was selected since this sample showed a high R value at a relatively low content of filler. One can observe that the antimicrobial activity was higher than that observed in the open system, showing R values of 1.35 and 1.64 for day 0 and 15, respectively. This confirms the high volatility of eugenol, which remained enclosed and still active in the system in comparison to the open one. In Table 9, for *E. coli*, the R values were 1.34 and 1.58 for day 0 and 15, respectively, in the closed system. Therefore, the here-achieved antimicrobial effect was somehow higher in the closed system than in the open one. This result has been recently ascribed to the volatile portion of active components accumulated in the system's headspace, which successfully contributed to decrease bacterial growth [83]. In any case, the differences in bacterial reduction in both tested packaging conditions, that is, the open and closed systems, for each type of bacteria was relatively low. This observation can be related to the use of MCM-41 that successfully performed as vehicles to control the release of eugenol and to render high antimicrobial activity.

Similar to this study, other authors have previously reported the antibacterial activity of eugenol in different biopolymer articles. For instance, PCL/gelatin electrospun membranes loaded with active peptide containing 30 wt.-% of eugenol successfully inhibited the growth of *E. coli* and *S. aureus* with inhibition rates of 71.6% and 78.6%, respectively [84]. In another study, compression-molded PHBV bilayer films were sprayed with four active components, among them eugenol, resulting in antimicrobial systems against G- and G+ bacteria such as *E. coli* and *Listeria innocua* (*L. innocua*) [65]. In this previous research, the added active agents were more effective against G- than G+, which in agreement with the present results. The benefit of loading antimicrobial agents in MCM-41 has been also studied elsewhere, both against bacteria and fungi. For instance, Park et al. [85] loaded allyl isothiocyanate, a natural antimicrobial, in MCM-41 as a novel controlled release vector against selected foodborne pathogenic microorganisms. In other studies, other volatile EOs were immobilized on the surface of mesoporous silica materials acting as antifungal agents and showing improved antimicrobial activity than the free compounds [39,40].

Table 6. Antibacterial activity against *Staphylococcus aureus* (*S. aureus*) in the open system for the electrospun films of poly(3-hydroxybutyrate-*co*-3-hydroxyvalerate) (PHBV) and PHBV/Mobil Composition of Matter (MCM)-41 with eugenol.

Films	Initial		After 15 days	
	Bacterial Counts [log (CFU/mL)]	R	Bacterial Counts [log (CFU/mL)]	R
Control day 0	5.75 ± 0.09	-	5.75 ± 0.09	-
Control 24 h	5.67 ± 0.07	-	5.68 ± 0.03	-
PHBV	5.39 ± 0.56	0.28 ± 0.52	5.29 ± 0.41	0.38 ± 0.38
PHBV/2.5 wt.-% MCM-41	4.86 ± 0.54	0.81 ± 0.58	5.06 ± 0.48	0.61 ± 0.46
PHBV/2.5 wt.-% MCM-41 + eugenol	4.33 ± 0.35	1.04 ± 0.39	4.75 ± 0.09	0.92 ± 0.12
PHBV/5 wt.-% MCM-41	5.47 ± 0.58	0.20 ± 0.65	5.51 ± 0.09	0.16 ± 0.06
PHBV/5 wt.-% MCM-41 + eugenol	4.60 ± 0.23	1.07 ± 0.23	4.69 ± 0.14	0.99 ± 0.14
PHBV/7.5 wt.-% MCM-41	5.77 ± 0.07	0.10 ± 0.01	5.44 ± 0.55	0.24 ± 0.57
PHBV/7.5 wt.-% MCM-41 + eugenol	4.55 ± 0.06	1.12 ± 0.11	4.55 ± 0.12	1.12 ± 0.15
PHBV/10 wt.-% MCM-41	5.98 ± 0.57	0.31 ± 0.06	4.70 ± 0.06	0.97 ± 0.09
PHBV/10 wt.-% MCM-41 + eugenol	4.43 ± 0.24	1.23 ± 0.20	4.55 ± 0.07	1.24 ± 0.10

Table 7. Antibacterial activity against *Escherichia coli* (*E. coli*) in the open system for the electrospun films of poly(3-hydroxybutyrate-*co*-3-hydroxyvalerate) (PHBV) and PHBV/Mobil Composition of Matter (MCM)-41 with eugenol.

Films	Initial		After 15 days	
	Bacterial Counts [log (CFU/mL)]	R	Bacterial Counts [log (CFU/mL)]	R
Control day 0	5.76 ± 0.01	-	5.71 ± 0.02	-
Control 24 h	6.81 ± 0.01	-	6.80 ± 0.02	-
PHBV	5.99 ± 0.07	0.82 ± 0.01	6.08 ± 0.03	0.72 ± 0.05
PHBV/15 wt.-% MCM-41	6.41 ± 0.01	0.40 ± 0.03	6.15 ± 0.04	0.65 ± 0.06
PHBV/15 wt.-% MCM-41 + eugenol	5.51 ± 0.02	1.30 ± 0.02	5.40 ± 0.01	1.40 ± 0.06

Table 8. Antibacterial activity against *Staphylococcus aureus* (*S. aureus*) in the closed system for the electrospun films of poly(3-hydroxybutyrate-*co*-3-hydroxyvalerate) (PHBV) and PHBV/Mobil Composition of Matter (MCM)-41 with eugenol.

Films	Initial		After 15 days	
	Bacterial Counts [log (CFU/mL)]	R	Bacterial Counts [log (CFU/mL)]	R
Control day 0	5.61 ± 0.03	-	5.65 ± 0.01	-
Control 24 h	6.82 ± 0.06	-	6.85 ± 0.01	-
PHBV	6.23 ± 0.08	0.59 ± 0.01	6.11 ± 0.03	0.74 ± 0.05
PHBV/10 wt.-% MCM-41	6.30 ± 0.01	0.52 ± 0.03	6.09 ± 0.04	0.76 ± 0.06
PHBV/10 wt.-% MCM-41 + eugenol	5.47 ± 0.01	1.35 ± 0.15	5.21 ± 0.01	1.64 ± 0.09

Table 9. Antibacterial activity against *Escherichia coli* (*E. coli*) in the closed system for the electrospun films of poly(3-hydroxybutyrate-*co*-3-hydroxyvalerate) (PHBV) and PHBV/Mobil Composition of Matter (MCM)-41 with eugenol.

Films	Initial		After 15 days	
	Bacterial Counts [log (CFU/mL)]	R	Bacterial Counts [log (CFU/mL)]	R
Control day 0	5.68 ± 0.03	-	5.66 ± 0.06	-
Control 24 h	6.83 ± 0.01	-	6.60 ± 0.01	-
PHBV	6.10 ± 0.01	0.73 ± 0.01	6.11 ± 0.03	0.49 ± 0.04
PHBV/15 wt.-% MCM-41	6.24 ± 0.01	0.59 ± 0.03	6.26 ± 0.06	0.34 ± 0.01
PHBV/15 wt.-% MCM-41 + eugenol	5.49 ± 0.03	1.34 ± 0.03	5.02 ± 0.07	1.58 ± 0.01

4. Conclusions

EOs are well known for their antimicrobial properties, being suitable as food preservatives. However, to ensure their long-term effect, which is controlled by their volatility, it may be necessary to encapsulate them in, for instance, porous materials. The present study evaluated the complexation of eugenol EO on MCM-41 to be thereafter incorporated into PHBV biopolymers by electrospinning. The resultant electrospun mats were annealed below the biopolymer melting point to generate continuous films. The thermal analysis performed on the films showed that the incorporation of MCM-41 with eugenol induced certain plasticization on PHBV as well as a reduction in crystallinity. Interestingly, the incorporation of MCM-41 with eugenol up to 10 wt.-% had a relatively low influence on the thermal stability of the PHBV films. During the mechanical analysis, it was observed that the mechanical strength of the PHBV films was increased while the ductility was only slightly reduced after the incorporation of MCM-41 with eugenol. The barrier properties were also enhanced due to the presence of the eugenol-containing nanofillers and were optimal around contents of 15 wt.-%. Finally, the antimicrobial activity against *S. aureus* and *E. coli* was studied in both an open and closed system to better represent the real conditions in packaging applications. The electrospun biopolymer films showed antibacterial activity after 15 days, being higher (as expected) in the ones that were studied in the closed system, which was ascribed to the accumulation of eugenol in the system's headspace. For all this, the films developed can be regarded as a sustainable material to be used in the form of interlayers or coatings for active food packaging applications.

Author Contributions: Conceptualization was devised by J.M.L., A.B., R.M.-M.; Methodology, J.M.L., B.M.-R. and S.T.-G.; A.B. and R.M.-M. synthesized and provided the nanoparticles and eugenol; B.M.-R. prepared the films and carry out most of the characterization; L.C. conducted the mechanical measurements; B.M.-R. and K.J.F-L carried out the antimicrobial experiments; Writing-Original Draft Preparation was performed by B.M.-R.; Writing-Review & Editing, S.T.-G.; Supervision, S.T.-G. and J.M.L.; Project Administration, J.M.L.; S.T.-G. and J.M.L. designed the work, supervised the execution and interpretation of all experiments and carried out the final version of the manuscript.

Funding: This research was supported by the Ministry of Science, Innovation, and Universities (MICIU) program numbers AGL2015-63855-C2-1-R and MAT2015-64139-C4-1-R, by the Generalitat Valenciana (GVA) PROMETEO/2018/024 program, and by the EU H2020 projects YPACK (reference number 773872) and ResUrbis (reference number 730349).

Acknowledgments: B.M.-R. and S.T.-G. acknowledge MICIU for her FPI grant (BES-2016-077972) and his Juan de la Cierva - Incorporación contract (IJCI-2016-29675), respectively. K.J.F-L. also acknowledges GVA for her Santiago Grisolia grant (GRISOLIAP/2017/101). A.B. would also like to thank GVA (POSTD/2014/016) and MICIU for her Juan de la Cierva - Incorporación contract (IJCI-2014-21534). The authors also thank the Joint Unit in Polymers Technology between IATA–CSIC and PIMA-Universitat Jaume I.

Conflicts of Interest: The authors declare no conflict of interest.

References

1. Torres-Giner, S.; Montanes, N.; Fombuena, V.; Boronat, T.; Sanchez-Nacher, L. Preparation and characterization of compression-molded green composite sheets made of poly(3-hydroxybutyrate) reinforced with long pita fibers. *Adv. Polym. Technol.* **2018**, *37*, 1305–1315. [CrossRef]
2. Reddy, C.S.K.; Ghai, R.; Rashmi; Kalia, V.C. Polyhydroxyalkanoates: An overview. *Bioresour. Technol.* **2003**, *87*, 137–146. [CrossRef]
3. Keshavarz, T.; Roy, I. Polyhydroxyalkanoates: Bioplastics with a green agenda. *Curr. Opin. Microbiol.* **2010**, *13*, 321–326. [CrossRef] [PubMed]
4. Lee, S.Y. Plastic bacteria? Progress and prospects for polyhydroxyalkanoate production in bacteria. *Trends Biotechnol.* **1996**, *14*, 431–438. [CrossRef]
5. Choi, J.I.; Lee, S.Y. Process analysis and economic evaluation for poly(3-hydroxybutyrate) production by fermentation. *Bioprocess Eng.* **1997**, *17*, 335–342. [CrossRef]
6. Díez-Pascual, A.M.; Díez-Vicente, A.L. ZnO-reinforced poly(3-hydroxybutyrate-co-3-hydroxyvalerate) bionanocomposites with antimicrobial function for food packaging. *ACS Appl. Mater. Interfaces* **2014**, *6*, 9822–9834. [CrossRef] [PubMed]
7. Torres-Giner, S.; Montanes, N.; Boronat, T.; Quiles-Carrillo, L.; Balart, R. Melt grafting of sepiolite nanoclay onto poly(3-hydroxybutyrate-co-4-hydroxybutyrate) by reactive extrusion with multi-functional epoxy-based styrene-acrylic oligomer. *Eur. Polym. J.* **2016**, *84*, 693–707. [CrossRef]
8. Khosravi-Darani, K.; Bucci, D.Z. Application of poly(hydroxyalkanoate) in food packaging: Improvements by nanotechnology. *Chem. Biochem. Eng. Q.* **2015**, *29*, 275–285. [CrossRef]
9. Kulkarni, S.O.; Kanekar, P.P.; Jog, J.P.; Patil, P.A.; Nilegaonkar, S.S.; Sarnaik, S.S.; Kshirsagar, P.R. Characterisation of copolymer, poly (hydroxybutyrate-co-hydroxyvalerate) (PHB-co-PHV) produced by halomonas campisalis (MCM B-1027), its biodegradability and potential application. *Bioresour. Technol.* **2011**, *102*, 6625–6628. [CrossRef]
10. Keskin, G.; Kızıl, G.; Bechelany, M.; Pochat-Bohatier, C.; Öner, M. Potential of polyhydroxyalkanoate (PHA) polymers family as substitutes of petroleum based polymers for packaging applications and solutions brought by their composites to form barrier materials. *Pure Appl. Chem.* **2017**, *89*, 1841–1848. [CrossRef]
11. Philip, S.; Keshavarz, T.; Roy, I. Polyhydroxyalkanoates: Biodegradable polymers with a range of applications. *J. Chem. Technol. Biotechnol.* **2007**, *82*, 233–247. [CrossRef]
12. Robertson, G.L. *Food Packaging: Principles and Practice*, 3rd ed.; Taylor & Francis: Boca Raton, FL, USA, 2012.

13. Requena, R.; Vargas, M.; Chiralt, A. Release kinetics of carvacrol and eugenol from poly(hydroxybutyrate-co-hydroxyvalerate) (PHBV) films for food packaging applications. *Eur. Polym. J.* **2017**, *92*, 185–193. [CrossRef]
14. Torres-Giner, S.; Hilliou, L.; Melendez-Rodriguez, B.; Figueroa-Lopez, K.J.; Madalena, D.; Cabedo, L.; Covas, J.A.; Vicente, A.A.; Lagaron, J.M. Melt processability, characterization, and antibacterial activity of compression-molded green composite sheets made of poly(3-hydroxybutyrate-co-3-hydroxyvalerate) reinforced with coconut fibers impregnated with oregano essential oil. *Food Packag. Shelf Life* **2018**, *17*, 39–49. [CrossRef]
15. Li, D.; Xia, Y. Electrospinning of Nanofibers: Reinventing the Wheel? *Adv. Mater.* **2004**, *16*, 1151–1170. [CrossRef]
16. Torres-Giner, S. Electrospun nanofibers for food packaging applications. In *Multifunctional and Nanoreinforced Polymers for Food Packaging*; Lagaron, J.M., Ed.; Woodhead Publishing Ltd.: Cambridge, UK, 2011; pp. 108–125.
17. Torres-Giner, S.; Busolo, M.; Cherpinski, A.; Lagaron, J.M. Electrospinning in the packaging industry. In *Electrospinning: From Basic Research to Commercialization*; Kny, E., Ghosal, K., Thomas, S., Eds.; The Royal Society of Chemistry: Cambridge, UK, 2018; pp. 238–260.
18. Torres-Giner, S.; Pérez-Masiá, R.; Lagaron Jose, M. A review on electrospun polymer nanostructures as advanced bioactive platforms. *Polym. Eng. Sci.* **2016**, *56*, 500–527. [CrossRef]
19. Torres-Giner, S.; Wilkanowicz, S.; Melendez-Rodriguez, B.; Lagaron, J.M. Nanoencapsulation of *aloe vera* in synthetic and naturally occurring polymers by electrohydrodynamic processing of interest in food technology and bioactive packaging. *J. Agric. Food Chem.* **2017**, *65*, 4439–4448. [CrossRef] [PubMed]
20. Torres-Giner, S. Novel antimicrobials obtained by electrospinning methods. In *Antimicrobial Polymers*; John Wiley & Sons, Inc.: Hoboken, NJ, USA, 2011; pp. 261–285.
21. Spagnol, C.; Fragal, E.H.; Pereira, A.G.B.; Nakamura, C.V.; Muniz, E.C.; Follmann, H.D.M.; Silva, R.; Rubira, A.F. Cellulose nanowhiskers decorated with silver nanoparticles as an additive to antibacterial polymers membranes fabricated by electrospinning. *J. Colloid Interface Sci.* **2018**, *531*, 705–715. [CrossRef] [PubMed]
22. Hu, M.; Li, C.; Li, X.; Zhou, M.; Sun, J.; Sheng, F.; Shi, S.; Lu, L. Zinc oxide/silver bimetallic nanoencapsulated in PVP/PCL nanofibres for improved antibacterial activity. *Artif. Cells Nanomed. Biotechnol.* **2018**, *46*, 1248–1257. [CrossRef] [PubMed]
23. An, J.; Zhang, H.; Zhang, J.; Zhao, Y.; Yuan, X. Preparation and antibacterial activity of electrospun chitosan/poly(ethylene oxide) membranes containing silver nanoparticles. *Colloid Polym. Sci.* **2009**, *287*, 1425–1434. [CrossRef]
24. Castro-Mayorga, J.L.; Fabra, M.J.; Cabedo, L.; Lagaron, J.M. On the use of the electrospinning coating technique to produce antimicrobial polyhydroxyalkanoate materials containing in situ-stabilized silver nanoparticles. *Nanomaterials* **2017**, *7*, 4. [CrossRef]
25. Castro Mayorga, J.L.; Fabra Rovira, M.J.; Cabedo Mas, L.; Sánchez Moragas, G.; Lagarón Cabello, J.M. Antimicrobial nanocomposites and electrospun coatings based on poly(3-hydroxybutyrate-co-3-hydroxyvalerate) and copper oxide nanoparticles for active packaging and coating applications. *J. Appl. Polym. Sci.* **2018**, *135*, 45673. [CrossRef]
26. Burt, S. Essential oils: Their antibacterial properties and potential applications in foods—A review. *Int. J. Food Microbiol.* **2004**, *94*, 223–253. [CrossRef] [PubMed]
27. Lang, G.; Buchbauer, G. A review on recent research results (2008–2010) on essential oils as antimicrobials and antifungals. A review. *Flavour Fragr. J.* **2012**, *27*, 13–39. [CrossRef]
28. Da Silva, F.F.M.; Monte, F.J.Q.; de Lemos, T.L.G.; do Nascimento, P.G.G.; de Medeiros Costa, A.K.; de Paiva, L.M.M. Eugenol derivatives: Synthesis, characterization, and evaluation of antibacterial and antioxidant activities. *Chem. Cent. J.* **2018**, *12*, 34. [CrossRef] [PubMed]
29. Wieczyńska, J.; Cavoski, I. Antimicrobial, antioxidant and sensory features of eugenol, carvacrol and trans-anethole in active packaging for organic ready-to-eat iceberg lettuce. *Food Chem.* **2018**, *259*, 251–260. [CrossRef] [PubMed]
30. Majeed, H.; Bian, Y.-Y.; Ali, B.; Jamil, A.; Majeed, U.; Khan, Q.F.; Iqbal, K.J.; Shoemaker, C.F.; Fang, Z. Essential oil encapsulations: Uses, procedures, and trends. *RSC Adv.* **2015**, *5*, 58449–58463. [CrossRef]
31. Kailasapathy, K. Encapsulation technologies for functional foods and nutraceutical product development. *CAB Rev. Perspect. Agric. Vet. Sci. Nutr. Nat. Resour.* **2009**, *4*, 1–19. [CrossRef]

32. Kresge, C.T.; Leonowicz, M.E.; Roth, W.J.; Vartuli, J.C.; Beck, J.S. Ordered mesoporous molecular sieves synthesized by a liquid-crystal template mechanism. *Nature* **1992**, *359*, 710–712. [CrossRef]
33. Vallet-Regi, M.; Rámila, A.; Del Real, R.P.; Pérez-Pariente, J. A new property of MCM-41: Drug delivery system. *Chem. Mater.* **2001**, *13*, 308–311. [CrossRef]
34. He, D.; He, X.; Wang, K.; Zou, Z.; Yang, X.; Li, X. Remote-controlled drug release from graphene oxide-capped mesoporous silica to cancer cells by photoinduced pH-jump activation. *Langmuir* **2014**, *30*, 7182–7189. [CrossRef]
35. Muñoz, B.; Rámila, A.; Pérez-Pariente, J.; Díaz, I.; Vallet-Regí, M. MCM-41 organic modification as drug delivery rate regulator. *Chem. Mater.* **2003**, *15*, 500–503. [CrossRef]
36. Bernardos, A.; Marina, T.; Žáček, P.; Pérez-Esteve, É.; Martínez-Mañez, R.; Lhotka, M.; Kouřimská, L.; Pulkrábek, J.; Klouček, P. Antifungal effect of essential oil components against aspergillus niger when loaded into silica mesoporous supports. *J. Sci. Food Agric.* **2015**, *95*, 2824–2831. [CrossRef] [PubMed]
37. Fan, J.; Yu, C.; Gao, F.; Lei, J.; Tian, B.; Wang, L.; Luo, Q.; Tu, B.; Zhou, W.; Zhao, D. Cubic mesoporous silica with large controllable entrance sizes and advanced adsorption properties. *Angew. Chem. Int. Ed.* **2003**, *42*, 3146–3150. [CrossRef]
38. Ruiz-Rico, M.; Fuentes, C.; Pérez-Esteve, É.; Jiménez-Belenguer, A.I.; Quiles, A.; Marcos, M.D.; Martínez-Máñez, R.; Barat, J.M. Bactericidal activity of caprylic acid entrapped in mesoporous silica nanoparticles. *Food Control* **2015**, *56*, 77–85. [CrossRef]
39. Janatova, A.; Bernardos, A.; Smid, J.; Frankova, A.; Lhotka, M.; Kourimská, L.; Pulkrabek, J.; Kloucek, P. Long-term antifungal activity of volatile essential oil components released from mesoporous silica materials. *Ind. Crop. Prod.* **2015**, *67*, 216–220. [CrossRef]
40. Ribes, S.; Ruiz-Rico, M.; Pérez-Esteve, É.; Fuentes, A.; Talens, P.; Martínez-Máñez, R.; Barat, J.M. Eugenol and thymol immobilised on mesoporous silica-based material as an innovative antifungal system: Application in strawberry jam. *Food Control* **2017**, *81*, 181–188. [CrossRef]
41. Ruiz-Rico, M.; Pérez-Esteve, É.; Bernardos, A.; Sancenón, F.; Martínez-Máñez, R.; Marcos, M.D.; Barat, J.M. Enhanced antimicrobial activity of essential oil components immobilized on silica particles. *Food Chem.* **2017**, *233*, 228–236. [CrossRef] [PubMed]
42. Popova, M.; Lazarova, H.; Trusheva, B.; Popova, M.; Bankova, V.; Mihály, J.; Najdenski, H.; Tsvetkova, I.; Szegedi, Á. Nanostructured silver silica materials as potential propolis carriers. *Microporous Mesoporous Mater.* **2018**, *263*, 28–33. [CrossRef]
43. Chatterjee, D.; Bhattacharjee, P. Comparative evaluation of the antioxidant efficacy of encapsulated and un-encapsulated eugenol-rich clove extracts in soybean oil: Shelf-life and frying stability of soybean oil. *J. Food Eng.* **2013**, *117*, 545–550. [CrossRef]
44. Estela, C.; Pilar, C.; Dolores, M.M.; Ramón, M.M.; Félix, S.; Juan, S. Selective chromofluorogenic sensing of heparin by using functionalised silica nanoparticles containing binding sites and a signalling reporter. *Chem. Eur. J.* **2009**, *15*, 1816–1820.
45. Cherpinski, A.; Torres-Giner, S.; Vartiainen, J.; Peresin, M.S.; Lahtinen, P.; Lagaron, J.M. Improving the water resistance of nanocellulose-based films with polyhydroxyalkanoates processed by the electrospinning coating technique. *Cellulose* **2018**, *25*, 1291–1307. [CrossRef]
46. Torres-Giner, S.; Torres, A.; Ferrándiz, M.; Fombuena, V.; Balart, R. Antimicrobial activity of metal cation-exchanged zeolites and their evaluation on injection-molded pieces of bio-based high-density polyethylene. *J. Food Saf.* **2017**, *37*, 1–12. [CrossRef]
47. Beck, J.S.; Vartuli, J.C.; Roth, W.J.; Leonowicz, M.E.; Kresge, C.T.; Schmitt, K.D.; Chu, C.T.W.; Olson, D.H.; Sheppard, E.W.; McCullen, S.B.; et al. A new family of mesoporous molecular sieves prepared with liquid crystal templates. *J. Am. Chem. Soc.* **1992**, *114*, 10834–10843. [CrossRef]
48. Alfredsson, V.; Keung, M.; Monnier, A.; Stucky, G.D.; Unger, K.K.; Schüth, F. High-resolution transmission electron microscopy of mesoporous MCM-41 type materials. *J. Chem. Soc. Chem. Commun.* **1994**, 921–922. [CrossRef]
49. Ravikovitch, P.I.; O'Domhnaill, S.C.; Neimark, A.V.; Schiith, F.; Unger, K.K. Capillary hysteresis in nanopores: Theoretical and experimental studies of nitrogen adsorption on MCM-41. *Langmuir* **1995**, *11*, 4765–4772. [CrossRef]

50. Sayed, E.; Karavasili, C.; Ruparelia, K.; Haj-Ahmad, R.; Charalambopoulou, G.; Steriotis, T.; Giasafaki, D.; Cox, P.; Singh, N.; Giassafaki, L.-P.N.; et al. Electrosprayed mesoporous particles for improved aqueous solubility of a poorly water soluble anticancer agent: In vitro and ex vivo evaluation. *J. Control. Release* **2018**, *278*, 142–155. [CrossRef] [PubMed]
51. Torres-Giner, S.; Lagaron, J.M. Zein-based ultrathin fibers containing ceramic nanofillers obtained by electrospinning. I. Morphology and thermal properties. *J. Appl. Polym. Sci.* **2010**, *118*, 778–789. [CrossRef]
52. Torres-Giner, S.; Martinez-Abad, A.; Lagaron, J.M. Zein-based ultrathin fibers containing ceramic nanofillers obtained by electrospinning. II. Mechanical properties, gas barrier, and sustained release capacity of biocide thymol in multilayer polylactide films. *J. Appl. Polym. Sci.* **2014**, *131*, 9270–9276. [CrossRef]
53. Cherpinski, A.; Gozutok, M.; Sasmazel, H.; Torres-Giner, S.; Lagaron, J. Electrospun oxygen scavenging films of poly(3-hydroxybutyrate) containing palladium nanoparticles for active packaging applications. *Nanomaterials* **2018**, *8*, 469. [CrossRef]
54. Cherpinski, A.; Torres-Giner, S.; Cabedo, L.; Lagaron, J.M. Post-processing optimization of electrospun submicron poly(3-hydroxybutyrate) fibers to obtain continuous films of interest in food packaging applications. *Food Addit. Contam. Part A* **2017**, *34*, 1817–1830. [CrossRef]
55. Melendez-Rodriguez, B.; Castro-Mayorga, J.L.; Reis, M.A.M.; Sammon, C.; Cabedo, L.; Torres-Giner, S.; Lagaron, J.M. Preparation and characterization of electrospun food biopackaging films of poly(3-hydroxybutyrate-*co*-3-hydroxyvalerate) derived from fruit pulp biowaste. *Front. Sustain. Food Syst.* **2018**, *2*, 38. [CrossRef]
56. Muratore, F.; Martini, R.E.; Barbosa, S.E. Bioactive paper by eugenol grafting onto cellulose. Effect of reaction variables. *Food Packag. Shelf Life* **2018**, *15*, 159–168. [CrossRef]
57. Torres-Giner, S.; Montanes, N.; Fenollar, O.; García-Sanoguera, D.; Balart, R. Development and optimization of renewable vinyl plastisol/wood flour composites exposed to ultraviolet radiation. *Mater. Des.* **2016**, *108*, 648–658. [CrossRef]
58. Fernandes Nassar, S.; Dombre, C.; Gastaldi, E.; Touchaleaume, F.; Chalier, P. Soy protein isolate nanocomposite film enriched with eugenol, an antimicrobial agent: Interactions and properties. *J. Appl. Polym. Sci.* **2017**, *135*, 45941. [CrossRef]
59. Narayanan, A.; Neera; Mallesha; Ramana, K.V. Synergized antimicrobial activity of eugenol incorporated polyhydroxybutyrate films against food spoilage microorganisms in conjunction with pediocin. *Appl. Biochem. Biotechnol.* **2013**, *170*, 1379–1388. [CrossRef] [PubMed]
60. Ju, C.; Kim, T.; Kang, H. Renewable, eugenol—Modified polystyrene layer for liquid crystal orientation. *Polymers* **2018**, *10*, 201. [CrossRef]
61. Loganathan, S.; Jacob, J.; Valapa, R.B.; Thomas, S. Influence of linear and branched amine functionalization in mesoporous silica on the thermal, mechanical and barrier properties of sustainable poly(lactic acid) biocomposite films. *Polymer* **2018**, *148*, 149–157. [CrossRef]
62. Garrido-Miranda, K.A.; Rivas, B.L.; Pérez-Rivera, M.A.; Sanfuentes, E.A.; Peña-Farfal, C. Antioxidant and antifungal effects of eugenol incorporated in bionanocomposites of poly(3-hydroxybutyrate)-thermoplastic starch. *LWT* **2018**, *98*, 260–267. [CrossRef]
63. Woranuch, S.; Yoksan, R. Eugenol-loaded chitosan nanoparticles: II. Application in bio-based plastics for active packaging. *Carbohydr. Polym.* **2013**, *96*, 586–592. [CrossRef] [PubMed]
64. Fang, Z.; Bhandari, B. Encapsulation of polyphenols—A review. *Trends Food Sci. Technol.* **2010**, *21*, 510–523. [CrossRef]
65. Requena, R.; Jiménez, A.; Vargas, M.; Chiralt, A. Poly[(3-hydroxybutyrate)-co-(3-hydroxyvalerate)] active bilayer films obtained by compression moulding and applying essential oils at the interface. *Polym. Int.* **2016**, *65*, 883–891. [CrossRef]
66. Rivero, S.; García, M.A.; Pinotti, A. Composite and bi-layer films based on gelatin and chitosan. *J. Food Eng.* **2009**, *90*, 531–539. [CrossRef]
67. Voon, H.C.; Bhat, R.; Easa, A.M.; Liong, M.T.; Karim, A.A. Effect of addition of halloysite nanoclay and SiO_2 nanoparticles on barrier and mechanical properties of bovine gelatin films. *Food Bioprocess Technol.* **2012**, *5*, 1766–1774. [CrossRef]
68. Jia, X.; Li, Y.; Cheng, Q.; Zhang, S.; Zhang, B. Preparation and properties of poly(vinyl alcohol)/silica nanocomposites derived from copolymerization of vinyl silica nanoparticles and vinyl acetate. *Eur. Polym. J.* **2007**, *43*, 1123–1131. [CrossRef]

69. Tang, S.; Zou, P.; Xiong, H.; Tang, H. Effect of nano-SiO_2 on the performance of starch/polyvinyl alcohol blend films. *Carbohydr. Polym.* **2008**, *72*, 521–526. [CrossRef]
70. Quiles-Carrillo, L.; Montanes, N.; Lagaron, J.M.; Balart, R.; Torres-Giner, S. In situ compatibilization of biopolymer ternary blends by reactive extrusion with low-functionality epoxy-based styrene–acrylic oligomer. *J. Polym. Environ.* **2019**, *27*, 84–96. [CrossRef]
71. Nielsen, L.E. Models for the permeability of filled polymer systems. *J. Macromol. Sci. Part A Chem.* **1967**, *1*, 929–942. [CrossRef]
72. Sanchez-Garcia, M.D.; Gimenez, E.; Lagaron, J.M. Morphology and barrier properties of solvent cast composites of thermoplastic biopolymers and purified cellulose fibers. *Carbohydr. Polym.* **2008**, *71*, 235–244. [CrossRef]
73. Sanchez-Garcia, M.D.; Gimenez, E.; Lagaron, J.M. Novel pet nanocomposites of interest in food packaging applications and comparative barrier performance with biopolyester nanocomposites. *J. Plast. Film Sheeting* **2007**, *23*, 133–148. [CrossRef]
74. Hashemi Tabatabaei, R.; Jafari, S.M.; Mirzaei, H.; Mohammadi Nafchi, A.; Dehnad, D. Preparation and characterization of nano-SiO_2 reinforced gelatin-k-carrageenan biocomposites. *Int. J. Biol. Macromol.* **2018**, *111*, 1091–1099. [CrossRef] [PubMed]
75. Hassannia-Kolaee, M.; Khodaiyan, F.; Pourahmad, R.; Shahabi-Ghahfarrokhi, I. Development of ecofriendly bionanocomposite: Whey protein isolate/pullulan films with nano-SiO_2. *Int. J. Biol. Macromol.* **2016**, *86*, 139–144. [CrossRef] [PubMed]
76. Tongnuanchan, P.; Benjakul, S.; Prodpran, T. Properties and antioxidant activity of fish skin gelatin film incorporated with citrus essential oils. *Food Chem.* **2012**, *134*, 1571–1579. [CrossRef] [PubMed]
77. Aguirre, A.; Borneo, R.; León, A.E. Antimicrobial, mechanical and barrier properties of triticale protein films incorporated with oregano essential oil. *Food Biosci.* **2013**, *1*, 2–9. [CrossRef]
78. Atarés, L.; De Jesús, C.; Talens, P.; Chiralt, A. Characterization of SPI-based edible films incorporated with cinnamon or ginger essential oils. *J. Food Eng.* **2010**, *99*, 384–391. [CrossRef]
79. Atarés, L.; Chiralt, A. Essential oils as additives in biodegradable films and coatings for active food packaging. *Trends Food Sci. Technol.* **2016**, *48*, 51–62. [CrossRef]
80. Torres-Giner, S.; Gil, L.; Pascual-Ramírez, L.; Garde-Belza, J. Packaging: Food waste reduction. In *Encyclopedia of Polymer Applications*; Mishra, M., Ed.; CRC Press: Boca Raton, FL, USA, 2018; Volume 3, pp. 1990–2009.
81. Exner, M.; Bhattacharya, S.; Christiansen, B.; Gebel, J.; Goroncy-Bermes, P.; Hartemann, P.; Heeg, P.; Ilschner, C.; Kramer, A.; Larson, E.; et al. Antibiotic resistance: What is so special about multidrug-resistant gram-negative bacteria? *GMS Hyg. Infect. Control* **2017**, *12*, Doc05. [PubMed]
82. Castro-Mayorga, J.L.; Fabra, M.J.; Pourrahimi, A.M.; Olsson, R.T.; Lagaron, J.M. The impact of zinc oxide particle morphology as an antimicrobial and when incorporated in poly(3-hydroxybutyrate-co-3-hydroxyvalerate) films for food packaging and food contact surfaces applications. *Food Bioprod. Process.* **2017**, *101*, 32–44. [CrossRef]
83. Figueroa-Lopez, K.J.; Vicente, A.A.; Reis, M.A.M.; Torres-Giner, S.; Lagaron, J.M. Antimicrobial and antioxidant performance of various essential oils and natural extracts and their incorporation into biowaste derived poly(3-hydroxybutyrate-*co*-3-hydroxyvalerate) layers made from electrospun ultrathin fibers. *Nanomaterials* **2019**, *9*, 144. [CrossRef]
84. Li, Z.; Zhou, P.; Zhou, F.; Zhao, Y.; Ren, L.; Yuan, X. Antimicrobial eugenol-loaded electrospun membranes of poly(ε-caprolactone)/gelatin incorporated with REDV for vascular graft applications. *Colloids Surf. B Biointerfaces* **2018**, *162*, 335–344. [CrossRef] [PubMed]
85. Park, S.-Y.; Barton, M.; Pendleton, P. Mesoporous silica as a natural antimicrobial carrier. *Colloids Surf. A Physicochem. Eng. Asp.* **2011**, *385*, 256–261. [CrossRef]

nanomaterials

Article

Antimicrobial and Antioxidant Performance of Various Essential Oils and Natural Extracts and Their Incorporation into Biowaste Derived Poly(3-hydroxybutyrate-*co*-3-hydroxyvalerate) Layers Made from Electrospun Ultrathin Fibers

Kelly J. Figueroa-Lopez [1], António A. Vicente [2], Maria A. M. Reis [3], Sergio Torres-Giner [1] and Jose M. Lagaron [1,*]

1 Novel Materials and Nanotechnology Group, Institute of Agrochemistry and Food Technology (IATA), Spanish National Research Council (CSIC), Calle Catedrático Agustín Escardino Benlloch 7, Paterna, 46980 Valencia, Spain; kjfigueroal@iata.csic.es (K.J.F.-L.); storresginer@iata.csic.es (S.T.-G.)
2 Centre of Biological Engineering, University of Minho, Campus Gualtar, 4710-057 Braga, Portugal; avicente@deb.uminho.pt
3 UCIBIO-REQUIMTE, Departamento de Química, Faculdade de Ciências e Tecnologia, Universidade Nova de Lisboa, Campus de Caparica, 2829-516 Caparica, Portugal; amr@fct.unl.pt
* Correspondence: lagaron@iata.csic.es

Received: 4 January 2019; Accepted: 19 January 2019; Published: 23 January 2019

Abstract: In this research, the antibacterial and antioxidant properties of oregano essential oil (OEO), rosemary extract (RE), and green tea extract (GTE) were evaluated. These active substances were encapsulated into ultrathin fibers of poly(3-hydroxybutyrate-*co*-3-hydroxyvalerate) (PHBV) derived from fruit waste using solution electrospinning, and the resultant electrospun mats were annealed to produce continuous films. The incorporation of the active substances resulted in PHBV films with a relatively high contact transparency, but it also induced a slightly yellow appearance and increased the films opacity. Whereas OEO significantly reduced the onset of thermal degradation of PHBV, both the RE and GTE-containing PHBV films showed a thermal stability profile that was similar to the neat PHBV film. In any case, all the active PHBV films were stable up to approximately 200 °C. The incorporation of the active substances also resulted in a significant decrease in hydrophobicity. The antimicrobial and antioxidant activity of the films were finally evaluated in both open and closed systems for up to 15 days in order to anticipate the real packaging conditions. The results showed that the electrospun OEO-containing PHBV films presented the highest antimicrobial activity against two strains of food-borne bacteria, as well as the most significant antioxidant performance, ascribed to the films high content in carvacrol and thymol. Therefore, the PHBV films developed in this study presented high antimicrobial and antioxidant properties, and they can be applied as active layers to prolong the shelf life of the foods in biopackaging applications.

Keywords: PHBV; oregano; rosemary; green tea; electrospun nanofibers; antibacterial; antioxidant

1. Introduction

The packaging industry requires the development of new plastic materials with active properties, based on the demand by consumers for safer and more nutritive food [1]. Moreover, the growing concern over the environmental problems caused by petroleum-derived materials has led to the search for new renewable raw materials for the development of compostable packaging [2,3]. Polyhydroxyalkanoates (PHAs) are amongst the most promising biopolymers, being a group of totally renewable, biodegradable, and biocompatible aliphatic polyesters. PHAs are synthesized in the cytoplasm of a wide range of bacteria from glucose-rich substrates [4,5]. Some PHAs, such as poly(3-hydroxybutyrate) (PHB), poly(3-hydroxybutyrate-*co*-3-hydroxyvalerate) (PHBV), poly(3-hydroxybutyrate-*co*-4-hydroxybutyrate) (P(3HB-*co*-4HB)), and poly(3-hydroxybutyrate-*co*-3-hydroxyhexanoate) (PHBH) are currently being employed to develop bioplastic packaging articles, such as injection-molded pieces, compression-molded sheets, and films [6–9].

Active packaging technology is mostly related to materials and articles that are intended to extend food shelf life, and also to improve packaged food conditions by interacting with the food product and/or with its internal packaging environment. Active packaging materials are usually designed to deliberately incorporate components, which would then release and/or absorb substances into or from the packaged food or the environment surrounding the food [10]. Active packaging systems can therefore extend the shelf life of food products and reduce food waste by maintaining the quality of food products for longer, increasing product safety by securing the foods against pathogens, and enhancing the convenience of food processing, distribution, retailing, and consumption [11]. Concerning the active packaging materials, these are classified as either active scavenging types (e.g., oxygen scavengers) [12] or active releasing types (e.g., antioxidants) [13]. Active releasing-type packaging can provide novel "extra" functions, such as aromatic, antioxidant, and long-term antimicrobial properties [14]. In particular, active-releasing antimicrobial packaging applications are directly related to food microbial safety, as well as to shelf life extension, by preventing the growth of spoilage and/or pathogenic microorganisms [15,16]. The growth of spoilage microorganisms can not only reduce the food shelf life, but it can also endanger public health (particularly in the case of pathogenic microorganisms).

Active properties can be conferred by the incorporation into the packaging materials of substances with inherent antioxidant and antimicrobial properties, such as essential oils (EOs) [17], natural extracts (NEs) [18], and/or inorganic and metal nanoparticles [19]. EOs are volatile compounds obtained from aromatic plants that produced them naturally as secondary metabolites [20]. EOs and NEs are mainly composed of terpenoids, phenolic, and aromatic compounds, and their composition can widely vary depending on the edaphoclimatic characteristics of the plant, the part of the plant (i.e., flower, seed, leaves, fruits, stems, and others), and the extraction procedure [21]. There is great interest in the use of these natural products because they are classified as generally recognized as safe (GRAS) food additives by the Food and Drug Administration (FDA) [22].

In line with this, over the last few years, different EOs and NEs have been proposed as alternative sources of antimicrobials in packaging materials. Within the great variety of EOs, oregano essential oil (OEO) from *Origanum vulgare* is well known for its antioxidative and antimicrobial activities [23]. The EO content in the oregano plant fluctuates from 0.5–2% [24] up to 7% [25]. Its main constituents are the isomer phenols, carvacrol and thymol, which represent up to 80% and 64%, respectively [26]. In addition, up to 52% of each of their precursor monoterpenes, p-cymene and γ-terpinene, as well as terpinen-4-ol, linalool, β-myrcene, trans-sabinene hydrate, and β-caryophyllene, are also present [27]. Rosemary extract (RE), which is obtained from *Rosmarinus officinalis*, is an aromatic plant belonging to the Lamiaceae family [28], and it also presents strong antimicrobial and antifungal properties [29]. The active properties of RE are primarily conferred by its phenolic, and the volatile constituents carnosol, carnosic acid, and rosmarinic acid [30]. Its minor components may have a potential influence on biological activity due to the possibility of synergistic effects amongst their components [31]. Finally, green tea tree extract (GTE) obtained from *Camellia sinensis* has gained significant attention in recent years. GTE is mainly composed of gallic acid, theobromine, chlorogenic acid, and caffeic acid [32]. In view of the potential uses of these natural products as effective antimicrobial and antioxidants for food preservation, they can be great candidates for incorporation into PHA films to generate active packaging articles.

Since most EOs and NEs are volatile compounds, they require the use of manufacturing methods that are carried out at room temperature to preserve their original properties. In this sense, the electrospinning technique is an emerging technology in the food packaging field [33,34], which is based on the application of electrostatic forces to polymer solutions to generate polymer fibers with diameters ranging from below 100 nm to several micrometers. Owing to the high surface-to-volume ratio of the electrospun fibers and the controllable pore size of the electrospun mats, several active and bioactive applications have been proposed in recent years [35], including the development of novel antimicrobial systems [36]. Since the electrospinning technique is frequently performed at room temperature, it facilitates the processing of thermolabile substances [37]. In addition, in a packaging application context, the ultrathin electrospun PHA fiber mats can be further converted into continuous films through the application of a thermal post-treatment below the polymer's melting temperature (T_m), i.e., the so-called annealing [38,39].

The objective of this research was to develop, for the first time, electrospun PHBV films containing OEO, RE, and GTE, in order to obtain active packaging layers with antioxidant and antimicrobial properties. Likewise, the morphological, optical, and thermal properties of the electrospun biopolymer films were also evaluated.

2. Materials and Methods

2.1. Materials

PHBV copolyester was produced at a pilot-plant scale at the Universidade NOVA de Lisboa (Lisboa, Portugal). This biopolymer was obtained using mixed microbial cultures fed with fermented fruit waste derived from the manufacturing of fruit juice, supplied by SumolCompal S.A. (Lisbon, Portugal). The molar fraction of the 3-hydroxyvalerate (HV) in the copolyester used was 20%. The synthesis, purification, and characterization details of this biopolymer was thoroughly described in Reference [39].

Chloroform, reagent grade with 99.8% purity, and methanol, HPLC grade with 99.9% purity, were purchased from Panreac S.A. (Barcelona, Spain). Additionally, 1-Butanol, reagent grade with 99.5% purity, and 2,2-diphenyl-1-picrylhydrazyl radical (DPPH) were purchased from Sigma Aldrich S.A. (Madrid, Spain).

OEO had a purity >99% and a relative density of 0.925–0.955 g/mL. RE presented a relative density of 0.915–0.926 g/mL, an acidity index of $\leq$1 mg KOH/g, an iodine index of 80.0–145.0%, a saponification index of 180–200 mg KOH/g, and a peroxide index of $\leq$5.0 meqO_2/kg. GTE showed a relative density of 0.915–0.925 g/mL, an acidity index of $\leq$1 mg KOH/g, an iodine index of 80–145%, a saponification index of 188–195 mg KOH/g, and a peroxide index of $\leq$5.0 meq O_2/Kg. All natural products were obtained from Gran Velada S.L. (Zaragoza, Spain) and were processed as received.

2.2. Preparation of the Solutions

The PHBV solution was prepared by dissolving 10% (wt./vol.) of biopolymer in a chloroform/1-butanol 75:25 (vol./vol.) mixture, both reagent grades, at room temperature. The OEO, RE, and GTE, were all added to the solution at 10 wt.% in relation to the PHBV and stirred for 24 h until a single-phase solution was obtained.

2.3. Characterization of the Solution Properties

All the PHBV solutions were analyzed in terms of their viscosity, surface tension, and conductivity. The apparent viscosity (ηa) was determined at 100 s^{-1} using a rotational viscosity meter Visco BasicPlus L from Fungilab S.A. (San Feliu de Llobregat, Spain). The surface tension was measured following the Wilhemy plate method using an EasyDyne K20 tensiometer from Krüss GmbH (Hamburg, Germany). The conductivity was evaluated using a conductivity meter HI9819X from Hanna Instruments

(Woonsocket, Rhode Island, USA). All the measurements were carried out at room temperature and in triplicate.

2.4. Electrospinning

The PHBV solutions containing OEO, RE, and GTE were each electrospun for 3 h onto an aluminum foil using a high-throughput electrospinning/electrospraying pilot line Fluidnatek® LE-500 with temperature and relative humidity (RH) control, which was manufactured and commercialized by Bioinicia S.L. (Valencia, Spain). The solutions were then processed at 25 °C and 40% RH under a constant flow using a 24 emitter multi-nozzle injector, scanning vertically onto the metallic plate. A dual polarization added voltage of 38 kV, and a flow-rate of 4 mL/h per single emitter and a tip-to-collector distance of 20 cm were used. A neat PHBV solution was electrospun in identical conditions as the control.

Thereafter, an annealing treatment was applied to the resultant electrospun mats. This process was performed in a 4122-model press from Carver, Inc. (Wabash, IN, USA) at 125 °C, for 5 s, without pressure. The resultant film samples had an average thickness in the 60–80 μm range.

2.5. Characterization of the Electrospun Materials

2.5.1. Film Thickness

Before testing, the thickness of the PHBV films containing the natural products was measured using a digital micrometer (S00014, Mitutoyo, Corp., Kawasaki, Japan) with ± 0.001 mm accuracy. Measurements were performed and averaged at five different points, one in each corner and one in the middle.

2.5.2. Morphology

The morphology of the electrospun PHBV fibers and their films containing the OEO, RE, and GTE were examined by scanning electron microscopy (SEM). The micrographs were taken using a Hitachi S-4800 electron microscope (Tokyo, Japan), at an accelerating voltage of 10 kV and a working distance of 8–10 mm. The samples were previously sputtered with a gold-palladium mixture for 3 min under vacuum. The average fiber diameter was determined via the ImageJ software v 1.41 using at least 20 SEM images.

2.5.3. Transparency

The light transmission of the PHBV films was determined in specimens of 50 × 30 mm^2 by quantifying the absorption of light at wavelengths between 200 and 700 nm, using an ultraviolet–visible (UV–Vis) spectrophotometer VIS3000 from Dinko Instruments (Barcelona, Spain). The transparency value (T) of the films was calculated using Equation (1) [1], whereas their opacity value (O) was determined using Equation (2) [40]:

$$T = \frac{A_{600}}{L} \quad (1)$$

$$O = A_{500}L \quad (2)$$

where A_{500} and A_{600} are the absorbance values at 500 and 600 nm, respectively, and L is the film thickness (mm).

2.5.4. Color

The PHBV films color was determined using a chroma meter CR-400 (Konica Minolta, Tokyo, Japan). The color difference (ΔE^*) was calculated using the following Equation (3) [1], as defined by the Commission Internationale de l'Eclairage (CIE):

$$\Delta E^* = \left[(\Delta L^*)^2 + (\Delta a^*)^2 + (\Delta b^*)^2\right]^{0.5} \tag{3}$$

where ΔE^*, ΔL^*, Δa^*, and Δb^* correspond to the differences between the color parameters of the sample films and the values of the control film ($a^* = 0.87$, $b^* = -0.38$, $L^* = 89.82$).

2.5.5. Thermal Analysis

Thermogravimetric analysis (TGA) of the neat OEO, RE, and GTE in their liquid form, and the PHBV films, was performed under a nitrogen atmosphere in a Thermobalance TG-STDA Mettler Toledo model TGA/STDA851e/LF/1600 analyzer (Greifensee, Switzerland). The TGA curves were obtained after conditioning the samples in the sensor for 5 min at 30 °C. The samples were then heated from 25 °C to 700 °C, at a heating rate of 10 °C/min. All tests were carried out in triplicate.

2.5.6. Water Contact Angle Measurements

The PHBV films surface wettability was evaluated using dynamic water contact angle (WCA) measurements in an optical tensiometer (Theta Lite, Staffordshire, UK). Five droplets were seeded at 5 μL/s on the film surfaces of each studied material sizing of 2×5 cm^2, in triplicate, and the resulting average contact angle was calculated.

2.5.7. Antimicrobial Activity

Staphylococcus aureus (*S. aureus*) CECT240 (ATCC 6538p) and *Escherichia coli* (*E. coli*) CECT434 (ATCC 25922) strains were obtained from the Spanish Type Culture Collection (CECT, Valencia, Spain) and stored in phosphate buffered saline (PBS), with 10 wt.% tryptic soy broth (TSB, Conda Laboratories, Madrid, Spain) and 10 wt.% glycerol (99.5% purity, Sigma Aldrich S.A. Madrid, Spain) at −80 °C. Prior to each study, a loopful of bacteria was transferred to 10 mL of TSB and incubated at 37 °C for 24 h. A 100 μL aliquot from the culture was again transferred to the TSB and grown at 37 °C to the mid-exponential phase of growth. The approximate count of 5×10^5 CFU/mL of culture had an absorbance value of 0.20, as determined by the optical density at 600 nm (UV 4000 spectrophotometer, Dinko Instruments, Barcelona, Spain).

The minimum inhibitory concentration (MIC) and minimum bactericide concentration (MBC) of the OEO, RE, and GTE against the selected food-borne bacteria was tested following the plate micro-dilution protocol, as described in the Methods for Dilution Antimicrobial. Susceptibility Tests for Bacteria That Grow Aerobically; Approved Standard Tenth. Edition (M07-A10) by the Clinical and Laboratory Standards Institute (CLSI). For this, a 96-well plate with an alpha numeric coordination system (columns 12 and rows A-H) was used, where 10 μL of the tested samples were introduced into the wells with 90 μL of the bacteria medium. In the wells corresponding to A, B, C, E, F, and G columns, different concentrations of the natural products, that is, 0.312, 0.625, 1.25, 2.5, 5, 10, 20, 40, 80, 160 μL/mL, were tested, in triplicate, from rows 1 to 10. Columns D and H were used as control of the natural extracts in the TSB without bacteria. Row 11 was taken as a positive control, that is, only the TSB, and row 12 was used as a negative control, that is, *S. aureus* and *E. coli* in the TSB. The plates were incubated at 37 °C for 24 h. Thereafter, 10 μL of resazurin sodium salt (MP biologicals, Illkirch, France), a metabolic indicator, was added to each well and incubated again at 37 °C for 2 h. Upon obtaining the resazurin change, the wells were read through the color difference. The MIC and MBC values were determined as the lowest concentration of the natural products presenting bacteriostatic and bactericide effects, respectively [41].

The antimicrobial performance of the electrospun PHBV films was evaluated using a modification of the Japanese Industrial Standard (JIS) Z2801 (ISO 22196:2007) [42]. To this end, a microorganism suspension of *S. aureus* and *E. coli* was applied onto the test films, that is, containing the natural extracts, and the negative control film, that is, without the natural extracts, all sizing 1.5 × 1.5 cm^2. Tests were performed in either hermetically closed or open vials with a volume of 20 mL. After incubation at 24 °C and at a relative humidity (RH) of, at least, 95% for 24 h, the bacteria were recovered with PBS, 10-fold serially diluted, and incubated at 37 °C for 24 h in order to quantify the number of viable bacteria by a conventional plate count. The antimicrobial activity was evaluated after 1 (initial day), 8, and 15 days. The value of the antimicrobial reduction (R) was calculated following Equation (4):

$$R = \left[log\left(\frac{B}{A}\right) - log\left(\frac{C}{A}\right)\right] = log\left(\frac{B}{C}\right) \quad (4)$$

where A is the average of the number of viable bacteria on the control film sample immediately after inoculation, B is the average of the number of viable bacteria on the control film sample after 24 h, and C is the average of the number of viable bacteria on the test film sample after 24 h. Three replicate experiments were performed for each sample and the following assessment was conducted: Nonsignificant (R < 0.5), slight (R ≥ 0.5 and <1), significant (R ≥ 1 and <3), and strong (R ≥ 3) as in Reference [43].

2.5.8. Antioxidant Activity

The DPPH inhibition assay was used to evaluate the free radical scavenging activity of the neat OEO, RE, GTE in their oil forms, in the electrospun PHBV fibers (at day 1) and their corresponding films (at 1, 8, and 15 days). Samples were weighed in triplicate in cap vials and then an aliquot of the DPPH solution (0.05 g/L in methanol) was added to each one. Vials without samples were also prepared as controls. All the samples were prepared and immediately stored at room temperature for 2 h in darkness. After this, the absorbance of the solution was measured at 517 nm in the UV 4000 spectrophotometer from Dinko Instruments. Results were expressed as the percentage of inhibition to DPPH following Equation (5) [44] and µg equivalent of trolox per gram of sample, employing a previously prepared calibration curve of trolox.

$$Inhibition\ DPPH\ (\%) = \frac{A_{Control} - \left(A_{sample} - A_{blank}\right)}{A_{control}} * 100 \quad (5)$$

where $A_{control}$, A_{blank}, and A_{sample} are the absorbance values of the DPPH solution, methanol with the test sample, and the test sample, respectively.

2.5.9. Statistical Analysis

The solution properties, color, transparency, and opacity values, and contact angle values were evaluated through analysis of variance (ANOVA) using STATGRAPHICS Centurion XVI v 16.1.03 from StatPoint Technologies, Inc. (Warrenton, VA, USA). Fisher's least significant difference (LSD) was used at the 95% confidence level ($p < 0.05$). Mean values and standard deviations were also calculated.

3. Results

3.1. Solution Properties

The use of polymer solution with the adequate properties is a key parameter to obtain uniform fibers during electrospinning [45]. In Table 1, the viscosity, surface tension, and conductivity of the PHBV solutions containing OEO, RE, and GTE at 10 wt.% are shown. The neat PHBV solution, without OEO and NEs, showed the highest viscosity value, that is, 212.4 cP. This value was relatively similar to that reported by Melendez-Rodriguez et al. [39], who obtained a value of viscosity of 296.8 cP for a

PHBV solution in 2,2,2-trifluoroethanol (TFE) at 2 wt%. This difference could be mainly ascribed to the solvent type and, more likely, to the use of a biopolymer with a higher molecular weight (M_W). When OEO, RE, and GTE were added, the viscosity of the PHBV solution slightly decreased. This effect could be ascribed to a reduction of the molecular cohesion forces in the biopolymer due to the presence of the active substances. This result was in agreement with, for instance, previous research works reported by Arfa et al. [46] and Jouki et al. [47], showing that the addition of either OEO or its active components decreased the apparent viscosity of polymer solutions of mucilage and soy protein (SP). In any case, for all the here-prepared PHBV solutions, the viscosity values were within the range of values reported by other authors, that is, from 1 to 20 poise (P), for the formation of homogeneous fibers during electrospinning [48].

Table 1. Solution properties of poly(3-hydroxybutyrate-*co*-3-hydroxyvalerate) (PHBV) containing oregano essential oil (OEO), rosemary extract (RE), and green tea tree extract (GTE).

Samples	Apparent Viscosity (cP)	Surface Tension (mN/m)	Conductivity (μs)
PHBV	212.4 ± 0.04 [a]	25.3 ± 0.05 [a]	0.40 ± 0.01 [a]
PHBV + OEO	205.5 ± 0.01 [b]	25.5 ± 0.07 [a]	0.42 ± 0.02 [a]
PHBV + RE	208.1 ± 0.03 [c]	25.4 ± 0.09 [a]	0.41 ± 0.01 [a]
PHBV + GTE	206.9 ± 0.05 [d]	25.5 ± 0.06 [a]	0.39 ± 0.01 [a]

[a–d] Different letters in the same column indicate a significant difference ($p < 0.05$).

The surface tension and conductivity of the solutions showed no significant differences ($p > 0.05$). However, it is worthy to mention that other authors have observed changes in the latter parameters when homogenization treatments (e.g., ultrasound, sonication, etc.) were applied to polymer solutions containing different EOs and NEs [49,50]. Moreover, in the case of polymer emulsions, the incorporation of these natural extracts into the oil or water phases resulted in an increasing drop size that destabilized the emulsion [51]. Therefore, the similar values observed for the neat PHBV solution and the PHBV solutions containing the OEO, RE, and GTE suggested that a high homogenization was achieved in all cases. Therefore, it was considered that the resulting solutions presented the adequate values for being processed by electrospinning.

3.2. Morphology

The morphology of the electrospun ultrathin neat PHBV fibers and the PHBV fibers containing the OEO, RE, and GTE was analyzed by SEM and the images are shown in Figure 1. The neat PHBV fibers, without EOs and NEs, were relatively uniform and presented a mean diameter of approximately 1 μm, as seen in Figure 1A. The morphology of the here-obtained electrospun ultrathin PHBV fibers was similar to those fibers reported by Melendez-Rodriguez et al. [39], showing diameters of ~1.32 μm. The PHBV fibers containing 10 wt% OEO, RE, and GTE are presented in Figure 1B–D, respectively. The diameters of the fibers were relatively similar, with a mean size of approximately 0.8 μm. The reduction achieved in the fiber diameter could be related to the slightly lower viscosities observed for the PHBV solutions containing the active substances. It was also evident that all the electrospun fibers were uniform and smooth, without any superficial and structural defects, which indicated that the addition of both EOs and NEs did not alter the fiber formation during electrospinning.

Figure 2 shows the SEM images of the electrospun materials, after annealing at 125 °C, in their cross-section and top views. In all cases, one can observe that the thermal post-treatment on the electrospun mats resulted in the formation of a continuous film. Figure 2A corresponds to the cross-section of the neat PHBV film, that is, without OEO and NEs, which presented an average thickness of ~80 μm. In Figure 2B, one can observe that the film sample also exhibited a homogeneous surface without cracks and/or pores. Similar morphologies were reported, for instance, by Cherpinski et al. [38] for electrospun PHB fibers thermally post-treated at 160 °C. The particular change from fiber-based to film-like morphology was ascribed to a process of fibers coalescence during annealing. Figure 2C,E,G show the cross-sections of the electrospun PHBV films containing OEO, RE, and GTE, respectively. The thicknesses of all the film samples were kept at ~80 µm. The presence of a certain number of pores can be related to the partial evaporation of the oily materials enclosed in the PHBV film during the thermal post-treatment. Similar voids were observed in the electrospun PHBV films derived from biowaste by Melendez-Rodriguez et al. [39], when temperatures close to T_m were applied, which was ascribed to the partial material melting and/or degradation. However, in Figure 2D,F,H, showing the top view of the film samples containing the active substances, it can be seen that the PHBV films still showed a smooth and homogeneous surface without pores and cracks. Therefore, despite the fact that the active substances were partially released during the film-forming process, a good compatibility and then a high solubility of OEO and the NEs with the PHBV matrix was attained.

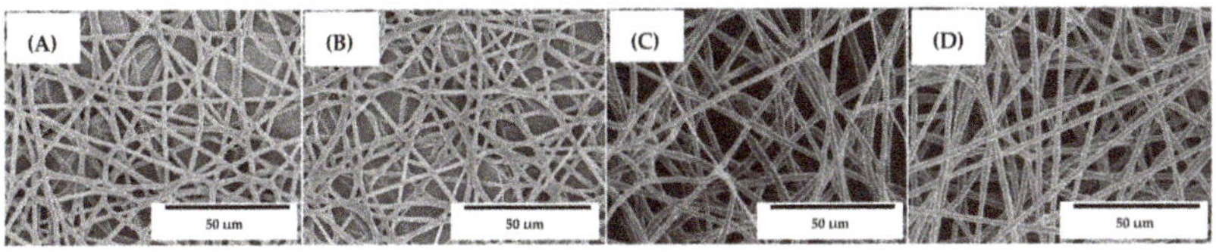

Figure 1. Scanning electron microscopy (SEM) micrographs of the electrospun fibers of: (**A**) Neat poly(3-hydroxybutyrate-*co*-3-hydroxyvalerate) (PHBV); (**B**) Oregano essential oil (OEO)-containing PHBV; (**C**) Rosemary extract (RE)-containing PHBV; (**D**) Green tea tree extract (GTE)-containing PHBV. Scale markers of 50 µm.

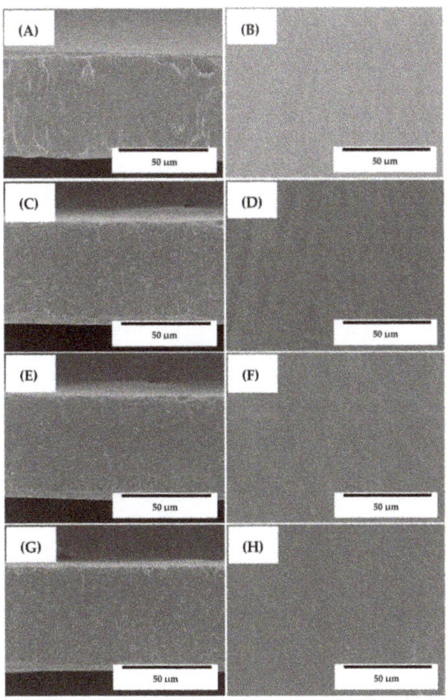

Figure 2. Scanning electron microscopy (SEM) micrographs of the electrospun films in their cross-section (left column) and top view (right column) of: (**A**,**B**) Neat poly(3-hydroxybutyrate-*co*-3-hydroxyvalerate) (PHBV); (**C**,**D**) Oregano essential oil (OEO)-containing PHBV; (**E**,**F**) Rosemary extract (RE)-containing PHBV; (**G**,**H**) Green tea tree extract (GTE)-containing PHBV. Scale markers of 50 µm.

3.3. Optical Properties

Figure 3 shows the visual aspect of the electrospun PHBV films to evaluate their contact transparency. The effects of the addition of OEO and the NEs on the color coordinates (L^*, a^*, b^*) and the values of ΔE, T, and O of the electrospun PHBV films are shown in Table 2. One can observe that all the here-prepared PHBV films presented a high contact transparency, but they also developed a slightly yellow appearance when the active substances were incorporated. The ΔE values of the active PHBV films with respect to the neat PHBV film were 8.36, 7.52, and 15.82 for the films with OEO, RE, and GTE, respectively. Therefore, the highest color change was observed for the GTE-containing PHBV film. The main changes observed were based on a decrease in brightness (L^*) and an increase in the b^* coordinate, that is, a yellower material, which was related to the intrinsic color of the added active substances.

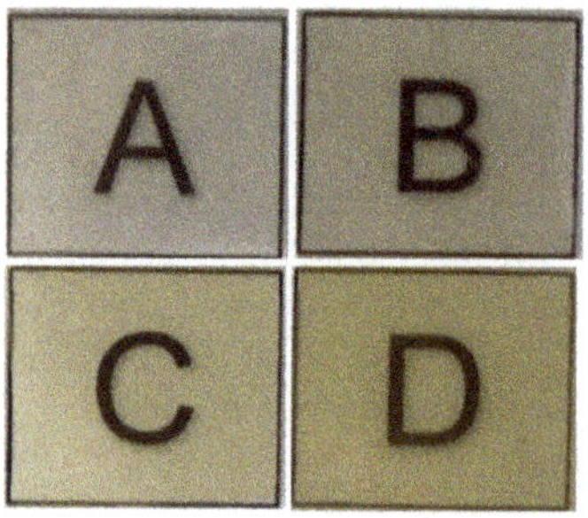

Figure 3. Visual aspect of the electrospun films of: (**A**) Neat poly(3-hydroxybutyrate-*co*-3-hydroxyvalerate) (PHBV); (**B**) Oregano essential oil (OEO)-containing PHBV; (**C**) Rosemary extract (RE)-containing PHBV; (**D**) Green tea tree extract (GTE)-containing PHBV. Films are 1.5 × 1.5 cm^2.

Table 2. Color parameters (ΔE^*, a^*, b^*, and L^*) and transparency characteristics of the electrospun films of poly(3-hydroxybutyrate-*co*-3-hydroxyvalerate) (PHBV) containing oregano essential oil (OEO), rosemary extract (RE), and green tea tree extract (GTE).

Samples	a^*	b^*	L^*	ΔE^*	T	O
PHBV	0.87 ± 0.07^a	-0.38 ± 0.02^a	89.82 ± 0.06^a	-	3.13 ± 0.02^a	0.016 ± 0.06^a
PHBV + OEO	1.13 ± 0.05^b	6.67 ± 0.03^b	85.35 ± 0.07^b	8.36 ± 0.08^a	3.55 ± 0.03^b	0.019 ± 0.08^b
PHBV + RE	0.04 ± 0.01^c	6.67 ± 0.08^b	87.33 ± 0.01^c	7.52 ± 0.06^b	6.44 ± 0.02^c	0.026 ± 0.05^c
PHBV + GTE	0.07 ± 0.09^c	14.45 ± 0.05^c	84.38 ± 0.03^d	15.82 ± 0.05^c	16.42 ± 0.06^d	0.067 ± 0.04^d

$\boldsymbol{a^*}$: red/green coordinates (+a red, −a green); $\boldsymbol{b^*}$: yellow/blue coordinates (+b yellow, −b blue); $\boldsymbol{L^*}$: Luminosity (+L luminous, −L dark); $\boldsymbol{\Delta E^*}$: color differences; T: transparency; O: opacity. $^{a–d}$ Different letters in the same column indicate a significant difference ($p < 0.05$).

One can also observe that the OEO-containing PHBV film presented a transparency similar to that of the neat PHBV film, both having T values in the range of 3–4, which indicated a greater passage of visible light through the material. However, the incorporation of RE and, particularly, of GTE resulted in an increase of T up to values of 6.4 and 16.4, respectively. Therefore, the capacity of transmission of visible and UV light of the films was significantly reduced by the addition of RE and GTE ($p < 0.05$), causing a phenomenon of light scattering due to the characteristic tones of the active substances. Similarly, whereas opacity was kept relatively low for the neat PHBV film and the OEO-containing PHBV films, which both had O values in the 0.015–0.02 range, these values increased up to 0.026 and 0.067 for the RE- and GTE-containing PHBV films, respectively. Then, the presence of the latter active substances, particularly GTE, reduced the transparency properties by blocking the passage of UV-Vis light and it increased the opacity of the films, caused by the scattering of light. However, as other authors have previously stated, this property can be also a desired characteristic in some packaging materials for the protection of foodstuff from light, especially UV radiation, which can cause lipid oxidation in the food products [1,40]. In this sense, the work reported by Gómez-Estaca et al. [52] also concluded that the addition of certain NEs to fish gelatin films decreased the transparency of the films and increased the opacity of the final material.

3.4. Thermal Stability

Figure 4 includes the weight loss curves of the free active substances and of the electrospun PHBV films obtained by TGA. The curves for the neat OEO, RE, and GTE are shown in Figure 4A, while the values of the onset degradation temperature, that is, the temperature at 5% weight loss ($T_{5\%}$), degradation temperature (T_{deg}), and residual mass at 700 °C are gathered in Table 3. One can observe that OEO presented the lowest thermal stability, showing values of $T_{5\%}$ and T_{deg} of 101.5 °C and 178.4 °C, respectively, with a respective weight loss of 74.16% at T_{deg}, corresponding to the volatilization and/or degradation of low-M_W volatile compounds present in the OEO (e.g., carvacrol, thymol, and pinene). In this sense, other authors have also reported that the EOs and NEs of oregano are among the most thermally unstable active substances. For instance, Barbieri et al. [53] reported that 96–97% of the OEO's weight was lost between 200 °C and 216 °C, attributed to its volatilization. In another work, Yang et al. [54] determined a significant degradation of all terpenes extracted oregano leaves in the 200–250 °C range. Similarly, Gibara Guimarães et al. [55] reported the fully thermal decomposition of carvacrol, the most representative active compound of oregano, at 168 °C. Opposite to OEO, both RE and GTE showed a high thermal stability with a similar mass loss profile. In particular, both active substances presented $T_{5\%}$ values over 350 °C, with T_{deg} values of 412.7 °C (52.45%) and 411.5 °C (49.89%) for RE and and GTE, respectively. Similar results, though slighlty lower, were reported for RE by Piñeros-Hernandez et al. [56], showing a significant mass loss at 300 °C, corresponding to the decomposition of phenolic diterpenes, that is, carnosic acid, carnosol, and rosmarinic acid. Likewise, Cordeiro et al. [57] obtained a mass loss as low as 6% up to 190 °C. In the case of GTE, López de Dicastillo et al. [58] determined that it remained stable up to the range of 200–400 °C, where the thermal degradation of partially glycosylated catechins occurs. Furthermore, all active substances produced a residual mass below 1%.

In Figure 4B, the weight loss curves of the electrospun PHBV films containing OEO, RE, and GTE are gathered. The neat PHBV film was thermally stable up to 251.5 °C, showing a T_{deg} value of 278.7 °C (47.74%) and a residual mass of 2.10%. While the incorporation of RE and GTE slightly reduced the thermal stability by 5–10 °C, the presence of OEO considerably reduced the onset of degradation, showing a $T_{5\%}$ value of 197.5 °C. It is also worthy to mention, however, that all active substances increased the thermally decomposed mass at T_{deg}, that is, the weight values decreased to the 60–70 % range. Therefore, the here-produced active PHBV films were stable up to 200 °C, which certainly opened up their application as an active food packaging interlayer and/or coating.

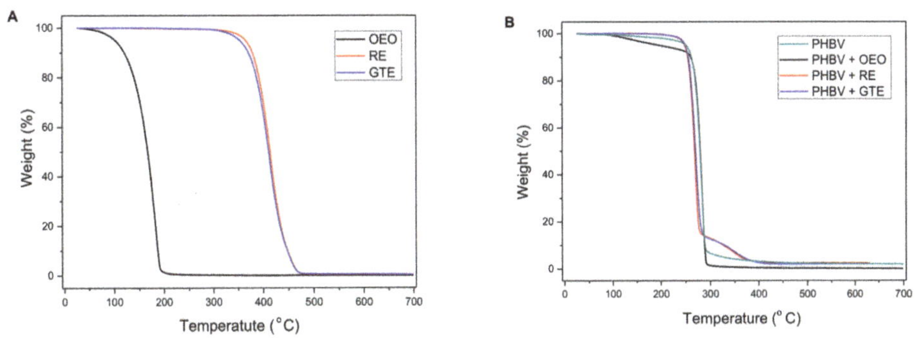

Figure 4. Weight loss as a function of temperature for: (**A**) Oregano essential oil (OEO), rosemary extract (RE), and green tea tree extract (GTE); (**B**) Electrospun films of neat poly(3-hydroxybutyrate-co-3-hydroxyvalerate) (PHBV) and PHBV containing OEO, RE, and GTE.

Table 3. Thermal properties of oregano essential oil (OEO), rosemary extract (RE), and green tea tree extract (GTE) and of the electrospun films of poly(3-hydroxybutyrate-*co*-3-hydroxyvalerate) (PHBV) containing OEO, RE, and GTE in terms of temperature at 5 % weight loss ($T_{5\%}$), degradation temperature (T_{deg}), and residual mass at 700 °C.

Sample	$T_{5\%}$ (°C)	T_{deg} (°C)	Mass Loss (%)	Residual Mass (%)
OEO	101.5	178.4	74.16	0.14
RE	364.0	412.7	52.45	0.48
GTE	352.7	411.5	49.89	0.56
PHBV	251.5	278.7	47.74	2.10
PHBV + OEO	197.5	283.6	69.58	0.16
PHBV + RE	248.5	270.8	60.94	2.49
PHBV + GTE	249.3	273.8	61.65	2.21

3.5. *Water Contact Angle*

The water contact angle refers to the degree of affinity of water with a surface, which defines the degree of hydrophilicity/hydrophobicity of a given polymer material [59]. In Figure 5, the water drop images on the films, as well as the values of their contact angles, are shown for the electrospun PHBV films. In Figure 5A, one can observe that the neat PHBV film presented an angle of 103.61°, which is characteristic of hydrophobic materials [60]. In all cases, the incorporation of the active substances resulted in a significant decrease in hydrophobicity ($p < 0.05$). Figure 5B shows that the OEO-containing PHBV film presented a water contact angle of 82.23°, whilst these values were even lower for both the films containing RE (Figure 5C), that is, 73.86°, and GTE (Figure 5D), that is, 71.26°. The reduction achieved could be related to the presence of the oily molecules on the surfaces of the PHBV films, which decreased the surface tension. A similar decrease in the water contact angles was observed by Galus and Kadzińska [61] in whey protein isolate (WPI) edible films when almond and walnut oils were added. In any case, following the terms "hydrophobic" and "hydrophilic", defined for $\theta > 65$ ° and ≤ 65 °, respectively [62], the angles for each of the films studied were still within the hydrophobic range.

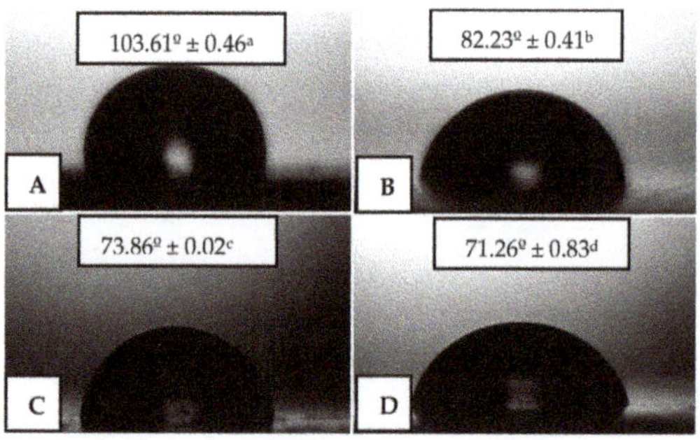

Figure 5. Water contact angle of the electrospun films of: (**A**) Neat poly(3-hydroxybutyrate-*co*-3-hydroxyvalerate) (PHBV); (**B**) Oregano essential oil (OEO)-containing PHBV; (**C**) Rosemary extract (RE)-containing PHBV; (**D**) Green tea tree extract (GTE)-containing PHBV.

3.6. *Active Properties*

3.6.1. Antimicrobial Activity

The EOs and NEs obtained from aromatic plants are constituted by a wide range of active compounds that are responsible for antimicrobial and antioxidant activity, which has promoted their application in active food packaging [1,41,42]. Table 4 shows the MIC and MBC values of the neat OEO, RE, and GTE against strains of *S. aureus* (Gram positive, G+) and *E. coli* (Gram negative, G-). The results showed that the OEO presented the highest antibacterial effect against both bacterial strains, having achieved identical MIC and MBC values, that is, 0.625 μL/mL, against *E. coli*, and 0.312 μL/mL, against *S. aureus*. The fact that the MIC and MBC values were identical can be related to the high effectivity of the natural compounds that were achieved, at the same time, the inhibition of microbial growth and the elimination of 99.9% of the microorganisms [63]. The antimicrobial activity of the OEO has been mainly ascribed to its high content in carvacrol and thymol [64,65]. RE presented MIC values of 10 μL/mL and 5 μL/mL against *E. coli* and *S. aureus*, respectively, whereas its MBC values were 20 μL/mL, against *E. coli*, and 10 μL/mL, against *S. aureus*. The main compounds responsible for the antimicrobial activity of RE were ∝-pinene, myrcene, camphor, 1,8-cineole, and camphene [29,66,67]. Likewise, GTE showed the lowest antimicrobial performance. This NE presented MIC values of 160 μL/mL, against *E. coli*, and 80 μL/mL, against *S. aureus*. Its MBC values were 160 μL/mL, against *E. coli*, and 40 μL/mL, against *S. aureus*. Gallic acid (GA), theobromine, chlorogenic acid, and caffeic acid are known to be responsible for its antimicrobial activity [32,68]. Most research related to MIC and MBC determination has been conducted with these EOs and NEs, finding that these compounds have a broad inhibition spectrum against G+ bacteria, but they are not as efficient against some G- bacteria [69,70]. The values of the MBC and MIC for OEO were the same, while for RE and GTE, the MBC values were higher than the MIC values. This fact is related to the effectiveness of the active compounds, the susceptibility of the microorganisms, and the variation in the penetration rate of the extracts through the cell wall and the structures of the cell membrane [71,72].

Table 4. Minimum inhibitory concentration (MIC) and minimum bactericidal concentration (MBC) of oregano essential oil (OEO), rosemary extract (RE), and green tea tree extract (GTE) against *S. aureus* and *E. coli*.

Sample	Bacteria	MIC	MBC
OEO	*E. coli*	0.625 μL/mL	0.625 μL/mL
	S. aureus	0.312 μL/mL	0.312 μL/mL
RE	*E. coli*	10 μL/mL	20 μL/mL
	S. aureus	5 μL/mL	10 μL/mL
GTE	*E. coli*	160 μL/mL	160 μL/mL
	S. aureus	40 μL/mL	80 μL/mL

The antimicrobial properties of the electrospun PHBV films containing OEO, RE, and GTE were also evaluated using the JIS Z2801 against *S. aureus* and *E. coli* bacteria, in both an open and closed system, for 1, 8, and 15 days. In relation to the open system, as shown in Table 5, one can observe that the films containing OEO showed the strongest inhibition. These films provided a strong reduction ($R \geq 3$), that is, with a reduction of 3 log units, against *S. aureus*, and also a high antimicrobial effect, though slightly lower, presenting *R* values of 2.7–2.9 against *E. coli*. In the case of the films containing RE and GTE, the films yielded a bacteriostatic effect ($1 \leq R < 3$) against both bacteria. The antimicrobial effect of RE was also approximately 1 log units higher than that observed for GTE. These results agreed with the MIC and MBC described above, where OEO inhibited the growth of *E. coli* and *S. aureus* at lower contents, whilst RE and GTE showed higher MIC and MBC values. It is also worthy to mention that in all cases, the bacterial reduction slightly increased over the days, which can be related to the slow release of the active compounds to the surface of the films. In comparison to the previous results of electrospun antimicrobial films reported by Jeong-Ann Parka and Song-Bae Kim [73] in open systems, it was observed that the inhibition of *S. aureus* increased from a 0.6 log of reduction, at the initial time, to a 1.2 log of reduction, after 120 min. In another study, Figueroa et al. [42] also reported an increase in the bacterial inhibition with the passage of storage days, showing a 3.9 log of reduction after 10 days against *S. aureus*.

Table 5. Antibacterial activity against *S. aureus* and *E. coli* of the electrospun poly(3-hydroxybutyrate-*co*-3-hydroxyvalerate) (PHBV) films containing oregano essential oil (OEO), rosemary extract (RE), and green tea tree extract (GTE) in the open system for up to 15 days.

Microorganism		Day	Control Sample log (CFU/mL)	Test Sample log (CFU/mL)	*R*
S. aureus	PHBV + OEO	1	6.91 ± 0.06	3.78 ± 0.08	3.13 ± 0.06
		8	6.88 ± 0.50	3.68 ± 0.03	3.20 ± 0.04
		15	6.89 ± 0.20	3.65 ± 0.10	3.24 ± 0.15
	PHBV + RE	1	6.87 ± 0.03	4.07 ± 0.07	2.80 ± 0.06
		8	6.88 ± 0.09	4.01 ± 0.03	2.87 ± 0.03
		15	6.87 ± 0.02	3.95 ± 0.01	2.92 ± 0.02
	PHBV + GTE	1	6.91 ± 0.10	5.00 ± 0.32	1.91 ± 0.20
		8	6.89 ± 0.23	4.94 ± 0.18	1.95 ± 0.13
		15	6.92 ± 0.11	4.93 ± 0.22	1.99 ± 0.19
E. coli	PHBV + OEO	1	6.95 ± 0.30	4.24 ± 0.09	2.71 ± 0.10
		8	6.90 ± 0.08	4.09 ± 0.10	2.81 ± 0.20
		15	6.87 ± 0.07	4.01 ± 0.03	2.86 ± 0.05
	PHBV + RE	1	6.89 ± 0.03	5.00 ± 0.06	1.89 ± 0.05
		8	6.90 ± 0.09	4.96 ± 0.07	1.94 ± 0.07
		15	6.88 ± 0.08	4.91 ± 0.09	1.97 ± 0.07
	PHBV + GTE	1	6.89 ± 0.15	5.70 ± 0.19	1.19 ± 0.17
		8	6.87 ± 0.33	5.63 ± 0.21	1.24 ± 0.23
		15	6.90 ± 0.46	5.62 ± 0.27	1.28 ± 0.31

As shown in Table 6, for all the samples, the reduction was slightly higher in the closed system than in the open one. While the films containing OEO presented the strongest inhibition ($R \geq 3$) after 15 days of storage in the closed system against the two bacterial strains, the films that contained RE and GTE showed a significant inhibition ($1 \leq R < 3$). This result could be attributed to the accumulation of volatile active compounds in the headspace of the closed chamber. There are a limited number of studies reporting the antimicrobial performance of the active films in closed systems, which indeed are more practical from the point of view of packaging and the design of containers to avoid deterioration of food products during storage. For instance, Torres-Giner et al. [74] developed a multilayer system, based on an electrospun coating of zein composite nanofibers containing thymol on a polylactide (PLA) film, that was evaluated against *Listeria monocytogenes* in a closed atmosphere in desiccators. It was reported that a concentration as low as 1.6 ppm was able to produce a decrease in the CFU of about 3 log units, whereas above 6.1 ppm, no CFU were detected. The high antimicrobial performance achieved was ascribed to the capacity of the electrospun material to release the bioactive in a sustained manner. The results are of potential interest in packaging applications, since the antimicrobial effect was not only successfully achieved in open packaging systems, but it also prolonged and improved over time in closed packaging systems, thereby extending the shelf life of perishable foods [75,76].

Table 6. Antibacterial activity against *S. aureus* and *E. coli* of the electrospun poly(3-hydroxybutyrate-*co*-3-hydroxyvalerate) (PHBV) films containing oregano essential oil (OEO), rosemary extract (RE), and green tea tree extract (GTE) in the closed system for up to 15 days.

Microorganism		Day	Control Sample log (CFU/mL)	Test Sample log (CFU/mL)	*R*
S. aureus	PHBV + OEO	1	6.91 ± 0.06	3.78 ± 0.08	3.13 ± 0.06
		8	6.93 ± 0.30	3.52 ± 0.90	3.41 ± 0.30
		15	6.92 ± 0.20	3.33 ± 0.08	3.59 ± 0.07
	PHBV + RE	1	6.87 ± 0.03	4.07 ± 0.07	2.80 ± 0.06
		8	6.89 ± 0.07	3.92 ± 0.05	2.91 ± 0.04
		15	6.88 ± 0.12	3.86 ± 0.15	3.02 ± 0.11
	PHBV + GTE	1	6.91 ± 0.10	5.00 ± 0.32	1.91 ± 0.20
		8	6.89 ± 0.15	4.89 ± 0.17	2.00 ± 0.11
		15	6.86 ± 0.20	4.78 ± 0.19	2.08 ± 0.21
E. coli	PHBV + OEO	1	6.95 ± 0.30	4.24 ± 0.09	2.71 ± 0.10
		8	6.90 ± 0.08	3.96 ± 0.10	2.94 ± 0.30
		15	6.92 ± 0.09	3.91 ± 0.07	3.01 ± 0.06
	PHBV + RE	1	6.89 ± 0.03	5.00 ± 0.06	1.89 ± 0.05
		8	6.87 ± 0.05	4.88 ± 0.08	1.99 ± 0.09
		15	6.88 ± 0.07	4.79 ± 0.03	2.09 ± 0.05
	PHBV + GTE	1	6.89 ± 0.15	5.70 ± 0.19	1.19 ± 0.17
		8	6.91 ± 0.11	5.62 ± 0.13	1.29 ± 0.15
		15	6.90 ± 0.28	5.53 ± 0.21	1.37 ± 0.19

3.6.2. Antioxidant Activity

The antioxidant activity of the EOs and NEs, obtained from aromatic plants, is conferred by their phenolic compounds. The DPPH free radical method is an antioxidant assay based on an electron-transfer that produces a violet solution in methanol. This free radical, stable at room temperature, is reduced in the presence of an antioxidant molecule (active compound), giving rise to a colorless solution [77]. The use of the DPPH assay provides an easy and rapid way to evaluate antioxidants using spectrophotometry. In the conservation of foods, substances with character antioxidants are of great interest because the main cause of food deterioration results from enzymatic reactions that trigger the oxidation of lipids and carbohydrates [78,79].

The percent inhibition and the equivalent concentration in micrograms of trolox per gram of sample of the neat OEO, RE, GTE, and of the electrospun fibers and films of the PHBV containing the active substances are shown in Table 7. One can observe that all the EOs and NEs showed DPPH radical scavenging activity. OEO presented the highest percentage of inhibition (91.96%) attributed to its main active compounds (e.g., carvacrol, thymol, p-cymene, γ-terpinene) [64]. Similarly, Chun et al. [80] reported an inhibition percentage of DPPH of 82% for oregano extracts. RE presented a percentage of inhibition of 75.24%, which was in agreement with, for instance, Bajalan et al. [81] who reported an inhibition percentage of DPPH of 73.69%. GTE showed an inhibition of 71.77%, conferred by the relative amount of catechins and GA [82]. Afroz Bakht et al. [83] studied the antioxidant activity from DPPH of five commercial teas, finding percentages of inhibition in the 24–71% range. In addition, Lu and Chen [84] reported percentages of inhibition between 33% and 62%. In this sense, it is important to highlight that antioxidant activity is dependent on the quantity of secondary metabolites that the plant manages to synthesize in its development stage, which is influenced by the variety of the plant, the environmental conditions [85–87], and the extraction method used [83].

One can observe that the antioxidant activity decreased in the electrospun PHBV fibers containing OEO, RE, and GTE. After the electrospinning process, the fibers with OEO showed a percentage of inhibition of 43.14%, whereas the fibers containing RE and GTE showed values of 25.82% and 22.12%, respectively. In all cases, there was a decrease in the antioxidant activity of between 20% and 30% compared to the pure OEO, RE, and GTE. The lower antioxidant inhibition of the active compounds inside the fibers could be related to polarity differences between the solvent and the polymer, the stirring process applied to the active solution prior to electrospinning, and the loss of volatiles compounds during the electrospinning process [44].

Table 7. Inhibition percentage (%) of 2,2-diphenyl-1-picrylhydrazyl radical (DPPH) and concentration (eq. trolox/g sample) of DPPH for oregano essential oil (OEO), rosemary extract (RE), and green tea tree extract (GTE) and the electrospun fibers of poly(3-hydroxybutyrate-*co*-3-hydroxyvalerate) (PHBV) containing OEO, RE, and GTE.

Sample	Inhibition Percentage (%)	Concentration (μg eq trolox/g Sample)
OEO	91.96 ± 0.03	84.34 ± 0.03
RE	75.24 ± 0.04	62.34 ± 0.03
GTE	71.77 ± 0.08	61.95 ± 0.07
PHBV + OEO	43.14 ± 0.07	28.56 ± 0.05
PHBV + RE	25.82 ± 0.07	18.31 ± 0.05
PHBV + GTE	22.12 ± 0.06	13.14 ± 0.04

Table 8 shows the percentages of inhibition and the equivalent concentration in micrograms of trolox per gram of sample of the electrospun PHBV films containing OEO, RE, and GTE, evaluated in the so-called open and closed systems on days 1, 8, and 15. Films containing OEO exhibited the highest inhibition of DPPH at day 1 (24.54%), followed by films with the RE (15.59%) and films with the GTE (11.14%). Over time, all the films decreased their antioxidant activity, obtaining no significant differences in the results between each storage system. Thus, after 15 days, an inhibition percentage in the range of 8.83–10.55% was obtained for the films containing OEO. At this time, the inhibition percentage ranges for the films containing RE and GTE were 6.91–7.31% and 5.45–6.68%, respectively. As can be observed, the antioxidant activity decreased when the EOs and NEs were included in the PHBV fibers and, more intensively, when the films were formed. Likewise, during the days of storage, a continuous release of the characteristic volatile compounds of each EOs and NEs was produced, which was evidenced by the low percentage of DPPH inhibition at day 15 for all the samples. Previous reports have indicated that the degree of antioxidant power of biodegradable films is generally proportional to the amount of added antioxidant additives, while the thermal process for obtaining the films also affects the bioactivity, since most of the bioactive compounds are sensitive to temperatures above 80 °C [47,88]. Regardless of this, all the films presented antioxidant performance and they can, therefore, be applied in antioxidant active packaging systems to extend the shelf life of packaged food products, thus minimizing the development of off-flavors, color and flavor changes, and nutritional losses [89,90].

Table 8. Inhibition percentage (%) of 2,2-diphenyl-1-picrylhydrazyl radical (DPPH) and concentration (eq. trolox/g sample) of DPPH for the electrospun poly(3-hydroxybutyrate-*co*-3-hydroxyvalerate) (PHBV) films containing oregano essential oil (OEO), rosemary extract (RE), and green tea tree extract (GTE).

Sample	Day	Open System		Closed System	
		Inhibition Percentage (%)	Concentration (μg eq trolox/g Sample)	Inhibition Percentage (%)	Concentration (μg eq trolox/g Sample)
PHBV + OEO	1	24.54 ± 0.04	26.48 ± 0.04	24.54 ± 0.04	26.48 ± 0.04
	8	16.08 ± 0.08	16.82 ± 0.09	17.43 ± 0.04	17.57 ± 0.04
	15	14.90 ± 0.06	15.75 ± 0.06	15.24 ± 0.01	16.47 ± 0.01
PHBV + RE	1	15.59 ± 0.02	16.31 ± 0.02	15.59 ± 0.02	16.31 ± 0.02
	8	10.42 ± 0.08	10.27 ± 0.08	13.50 ± 0.01	13.97 ± 0.01
	15	7.310 ± 0.04	7.710 ± 0.04	8.200 ± 0.15	8.120 ± 0.02
PHBV + GTE	1	11.14 ± 0.04	11.79 ± 0.04	11.14 ± 0.04	11.79 ± 0.04
	8	8.910 ± 0.10	8.760 ± 0.10	9.960 ± 0.02	9.820 ± 0.02
	15	6.680 ± 0.11	6.540 ± 0.11	7.800 ± 0.02	8.250 ± 0.02

4. Conclusions

The evaluation of the active properties, that is, antimicrobial and antioxidant properties, of OEO, RE, and GTE showed that OEO was the active substance that presented the highest antimicrobial activity against *S. aureus* and *E. coli*. This effect was mainly attributed to the effectiveness of its most representative active compounds, that is, carvacrol and thymol, showing identical MIC and MBC values of 0.312 and 0.625 μL/mL, respectively. The antioxidant activity of OEO was also higher than that for the RE and GTE. In particular, the percentage of inhibition of the DPPH was 91.96%. Thereafter, these active substances were incorporated at 10 wt.% into fruit waste derived PHBV by electrospinning. To this end, the solution properties of the PHBV containing OEO, RE, and GTE were first evaluated to determine the optimal conditions to obtain homogenous fibers. The diameters of the fibers were relatively similar, with a mean size of approximately ~0.8 μm, being uniform and smooth, without any superficial and structural defects. It was observed that the addition of the OEO and the NEs did not alter the fiber formation during electrospinning or the morphology of the electrospun ultrathin PHBV fibers. A good compatibility and, then, high solubility of the OEO and NEs with the PHBV matrix was considered.

In order to obtain an interesting active continuous layer to be applied in the design of biopackaging, the electrospun mats of the PHBV fibers containing the active substances were subjected to a thermal post-treatment at 125 °C. Continuous PHBV films of ~80 μm, with a smooth surface were obtained, though the presence of the active substances induced a slight porosity in their cross-section. The optical properties of the PHBV films were slightly impaired by the addition of the active substances, particularly GTE, reducing their transparency from 3.13 up to 16.42 through blocking the passage of UV-Vis light and increasing their opacity, which was caused by the scattering of light. In any case, all the PHBV films were contact transparent. All active substances also decreased the values of the T_{onset} by 54 °C for OEO, 3 °C for RE, and 2.2 °C for GTE, and the thermally decomposed mass at T_{deg} decreased to the 60–70% range. However, all the active PHBV films were stable up to 200 °C. Referring to the hydrophobicity of the films, the addition of active substances decreased the superficial tension of the PHBV films with respect to the control, but the angles for each of the films studied were still within the hydrophobic range. The PHBV films containing OEO, RE, and GTE showed antimicrobial activity against strains of *S. aureus* and *E. coli* in both the here-studied open and closed systems, where the bacterial reduction improved over time due to the release and accumulation of the active compounds on the film surface. The antimicrobial activity was higher in the case of the closed system due to the presence of volatiles stored in the headspace. The films containing OEO presented the highest reduction values against the two bacterial strains ($R \geq 3$), while the films containing RE and GTE showed lower reduction values ($1 \leq R < 3$), which agreed with the MIC and MBC values of the pure active substances. The antioxidant activity of the fibers and films was much lower than that of the neat active substances, which was related to entrapment and loss during electrospinning and film processing, and which was also reduced with the passage of days due to the continuous release of the active compounds.

The here-developed electrospun PHBV layers with OEO, RE, and GTE are potential candidates for use in the design of sustainable active multilayer biopackaging. The antimicrobial and antioxidant performance of these materials is advantageous to prolonging the shelf life of foods, delaying the proliferation of microorganisms, and the enzymatic oxidation of foodstuffs.

Author Contributions: K.J.F.-L. performed all the experiments, measurements, data analysis, and wrote the manuscript draft. S.T.-G. proposed, planned, guided the execution of the research work, and wrote and edited the manuscript. A.A.V. provided the active substances and co-wrote the manuscript. M.A.M. Reis synthesized and characterized the biopolymer. J.M.L. supervised the research and performed the project administration.

Funding: This research was supported by the Ministry of Science, Innovation, and Universities (MICIU) program number AGL2015-63855-C2-1-R and by the EU H2020 project YPACK (reference number 773872).

Acknowledgments: K.J.F.-L. is a recipient of a Santiago Grisolía (GRISOLIAP/2017/101) grant of the Generalitat Valenciana (GVA) and S.T.-G. is on a Juan de la Cierva - Incorporación contract (IJCI-2016-29675) from MICIU.

Conflicts of Interest: The authors declare no conflict of interest.

References

1. Figueroa-Lopez, K.J.; Andrade-Mahecha, M.M.; Torres-Varga, O.L. Development of antimicrobial biocomposite films to preserve the quality of bread. *Molecules* **2018**, *23*, 212. [CrossRef] [PubMed]
2. Atarés, L.; Chiralt, A. Essential oils as additives in biodegradable films and coatings for active food packaging. *Trends Food Sci. Technol.* **2016**, *48*, 51–62. [CrossRef]
3. Requena, R.; Vargas, M.; Chiralt, A. Obtaining antimicrobial bilayer starch and polyester-blend films with carvacrol. *Food Hydrocoll.* **2018**, *83*, 118–133. [CrossRef]
4. Lenz, R.W.; Marchessault, R.H. Bacterial polyesters: Biosynthesis, biodegradable plastics and biotechnology. *Biomacromolecules* **2005**, *6*, 1–8. [CrossRef] [PubMed]
5. Sudesh, K.; Abe, H.; Do, Y. Synthesis, structure and properties of polyhydroxyalkanoates: Biological polyesters. *Prog. Polym. Sci.* **2000**, *25*, 1503–1555. [CrossRef]
6. Torres-Giner, S.; Montanes, N.; Boronat, T.; Quiles-Carrillo, L.; Balart, R. Melt grafting of sepiolite nanoclay onto poly(3-hydroxybutyrate-co-4-hydroxybutyrate) by reactive extrusion with multi-functional epoxy-based styrene-acrylic oligomer. *Eur. Polym. J.* **2016**, *84*, 693–707. [CrossRef]
7. Torres-Giner, S.; Montanes, N.; Fombuena, V.; Boronat, T.; Sanchez-Nacher, L. Preparation and characterization of compression-molded green composite sheets made of poly(3-hydroxybutyrate) reinforced with long pita fibers. *Adv. Polym. Technol.* **2018**, *37*, 1305–1315. [CrossRef]
8. Arifin, W.; Kuboki, T. Effects of thermoplastic elastomers on mechanical and thermal properties of glass fiber reinforced poly(3-hydroxybutyrate-co-3-hydroxyhexanoate) composites. *Polym. Compos.* **2018**, *39*, E1331–E1345. [CrossRef]
9. Torres-Giner, S.; Hilliou, L.; Melendez-Rodriguez, B.; Figueroa-Lopez, K.J.; Madalena, D.; Cabedo, L.; Covas, J.A.; Vicente, A.A.; Lagaron, J.M. Melt processability, characterization, and antibacterial activity of compression-molded green composite sheets made of poly(3-hydroxybutyrate-co-3-hydroxyvalerate) reinforced with coconut fibers impregnated with oregano essential oil. *Food Packag. Shelf Life* **2018**, *17*, 39–49. [CrossRef]
10. López-Rubio, A.; Almenar, E.; Hernandez-Muñoz, P.; Lagarón, J.M.; Catalá, R.; Gavara, R. Overview of active polymer-based packaging technologies for food applications. *Food Rev. Int.* **2004**, *20*, 357–387. [CrossRef]
11. Torres-Giner, S.; Gil, L.; Pascual-Ramírez, L.; Garde-Belza, J.A. Packaging: Food waste reduction. In *Encyclopedia of Polymer Applications, 3 Volume Set*; Mishra, M., Ed.; CRC Press, Taylor and Francis Group: Boca Raton, FL, USA, 2018; pp. 1990–2009.
12. Cherpinski, A.; Gozutok, M.; Sasmazel, H.T.; Torres-Giner, S.; Lagaron, J.M. Electrospun oxygen scavenging films of poly(3-hydroxybutyrate) containing palladium nanoparticles for active packaging applications. *Nanomaterials* **2018**, *8*, 469. [CrossRef] [PubMed]

13. Domínguez, R.; Barba, F.J.; Gómez, B.; Putnik, P.; Bursać Kovačevićc, D.; Pateiro, M.; Santos, E.M.; Lorenzo, J.M. Active packaging films with natural antioxidants to be used in meat industry: A review. *Food Res. Int.* **2018**, *113*, 93–101. [CrossRef] [PubMed]
14. Ge, L.; Zhu, M.; Li, X.; Xu, Y.; Ma, X.; Shi, R.; Li, D.; Mu, C. Development of active rosmarinic acid-gelatin biodegradable films with antioxidant and long-term antibacterial activities. *Food Hydrocoll.* **2018**, *83*, 308–316. [CrossRef]
15. Irkin, R.; Esmer, O.K. Novel food packaging systems with natural antimicrobial agents. *J. Food Sci. Technol.* **2015**, *52*, 6095–6111. [CrossRef] [PubMed]
16. Appendini, P.; Hotchkiss, J.H. Review of antimicrobial food packaging. *Innov. Food Sci. Emerg. Technol.* **2002**, *3*, 113–126. [CrossRef]
17. Ribeiro-Santos, R.; Andrade, M.; Sanches-Silva, A. Application of encapsulated essential oils as antimicrobial agents in food packaging. *Curr. Opin. Food Sci.* **2017**, *14*, 78–84. [CrossRef]
18. Nguimjeu, C.; Moundanga, S.; Sadaka, F.; Dony, P.; Plasseraud, L.; Boni, G.; Brachais, L.; Vroman, I.; Brachais, C.H.; Tighzert, L.; et al. Suitability and effectiveness of active packaging for food contact: A study of sorption/desorption and antimicrobial performances of some natural extracts in packaging. In *18th IAPRI World Packaging Conference*; DEStech Publications, Inc.: Lancaster, PA, USA, 2012; pp. 15–22.
19. Hoseinnejad, M.; Jafari, S.M.; Katouzian, I. Inorganic and metal nanoparticles and their antimicrobial activity in food packaging applications. *Crit. Rev. Microbiol.* **2018**, *44*, 161–181. [CrossRef]
20. Teixeira, B.; Marques, A.; Ramos, C.; Neng, N.R.; Nogueira, J.M.F.; Saraiva, J.A.; Nunes, M.L. Chemical composition and antibacterial and antioxidant properties of commercial essential oils. *Ind. Crop. Prod.* **2013**, *43*, 587–595. [CrossRef]
21. Raut, J.S.; Karuppayil, S.M. A status review on the medicinal properties of essential oils. *Ind. Crop. Prod.* **2014**, *62*, 250–264. [CrossRef]
22. FDA. *Code of Federal Regulations—Title 21—Food Drugs. Chapter I—Food Drug Adm. Dep. Heal. Hum. Serv. Subchapter b—Food Hum. Consum. (Continued), Part 182—Subst. Gen. Recognized as Safe (Gras)*; FDA: Silver Spring, MD, USA, 2016.
23. Leyva-López, N.; Gutiérrez-Grijalva, E.; Vazquez-Olivo, G.; Heredia, J. Essential oils of oregano: Biological activity beyond their antimicrobial properties. *Molecules* **2017**, *22*, 989. [CrossRef]
24. Kokkini, S.; Karousou, R.; Dardioti, A.; Krigas, N.; Lanaras, T. Autumn essential oils of greek oregano. *Phytochemistry* **1997**, *44*, 883–886. [CrossRef]
25. Gounaris, Y.; Skoula, M.; Fournaraki, C.; Drakakaki, G.; Makris, A. Comparison of essential oils and genetic relationship of origanum x intercedens to its parental taxa in the island of crete. *Biochem. Syst. Ecol.* **2002**, *30*, 249–258. [CrossRef]
26. Shang, X.; Wang, Y.; Zhou, X.; Guo, X.; Donga, S.; Wang, D.; Zhang, J.; Pan, H.; Zhang, Y.; Miao, X. Acaricidal activity of oregano oil and its major component, carvacrol, thymol and p-cymene against psoroptes cuniculi in vitro and in vivo. *Vet. Parasitol.* **2016**, *226*, 93–96. [CrossRef] [PubMed]
27. Burt, S.A.; Vlielander, R.; Haagsman, H.P.; Veldhuizen, E.J.A. Increase in activity of essential oil components carvacrol and thymol against *Escherichia coli* o157:H7 by addition of food stabilizers. *J. Food Prot.* **2005**, *68*, 919–926. [CrossRef] [PubMed]
28. Begum, A.; Sandhya, S.; Ali, S.S.; Vinod, K.R.; Reddy, S.; Banji, D. An in-depth review on the medicinal flora rosmarinus officinalis (lamiaceae). *Acta Sci. Pol. Technol. Aliment.* **2013**, *12*, 61–73.
29. Jiang, Y.; Wu, N.; Fu, Y.-J.; Wang, W.; Luo, M.; Zhao, C.-J.; Zu, Y.-G.; Liu, X.-L. Chemical composition and antimicrobial activity of the essential oil of rosemary. *Environ. Toxicol. Pharmacol.* **2011**, *32*, 63–68. [CrossRef]
30. Arranz, E.; Mes, J.; Wichers, H.J.; Jaime, L.; Mendiola, J.A.; Reglero, G.; Santoyo, S. Anti-inflammatory activity of the basolateral fraction of caco-2 cells exposed to a rosemary supercritical extract. *J. Funct. Foods* **2015**, *13*, 384–390. [CrossRef]
31. Hussain, A.I.; Anwar, F.; Chatha, S.A.S.; Jabbar, A.; Mahboob, S.; Nigam, P.S. Rosmarinus officinalis essential oil: Antiproliferative, antioxidant and antibacterial activities. *Braz. J. Microbiol.* **2010**, *41*, 1070–1078. [CrossRef]

32. Fournier-Larente, J.; Morin, M.-P.; Grenier, D. Green tea catechins potentiate the effect of antibiotics and modulate adherence and gene expression in porphyromonas gingivalis. *Arch. Oral Biol.* **2016**, *65*, 35–43. [CrossRef]
33. Torres-Giner, S. Electrospun nanofibers for food packaging applications. In *Multifunctional and Nanoreinforced Polymers for Food Packaging*; Lagaron, J.M., Ed.; Woodhead Publishing Ltd.: Cambridge, UK, 2011; pp. 108–125.
34. Torres-Giner, S.; Busolo, M.; Cherpinski, A.; Lagaron, J.M. Electrospinning in the packaging industry. In *Electrospinning: From Basic Research to Commercialization*; Kny, E., Ghosal, K., Thomas, S., Eds.; The Royal Society of Chemistry: Cambridge, UK, 2018; pp. 238–260.
35. Torres-Giner, S.; Lagaron, J.M. Zein-based ultrathin fibers containing ceramic nanofillers obtained by electrospinning. I. Morphology and thermal properties. *J. Appl. Polym. Sci.* **2010**, *118*, 778–789. [CrossRef]
36. Torres-Giner, S. Novel antimicrobials obtained by electrospinning methods. In *Antimicrobial Polymers*; Lagaron, J.M., Ocio, M.J., López-Rubio, A., Eds.; John Wiley & Sons, Inc.: Hoboken, NJ, USA; pp. 261–285.
37. Altan, A.; Aytac, Z.; Uyar, T. Carvacrol loaded electrospun fibrous films from zein and poly(lactic acid) for active food packaging. *Food Hydrocoll.* **2018**, *81*, 48–59. [CrossRef]
38. Cherpinski, A.; Torres-Giner, S.; Cabedo, L.; Lagaron, J.M. Post-processing optimization of electrospun submicron poly(3-hydroxybutyrate) fibers to obtain continuous films of interest in food packaging applications. *Food Addit. Contam. Part A* **2017**, *34*, 1817–1830. [CrossRef] [PubMed]
39. Melendez-Rodriguez, B.; Castro-Mayorga, J.L.; Reis, M.A.M.; Sammon, C.; Cabedo, L.; Torres-Giner, S.; Lagaron, J.M. Preparation and characterization of electrospun food biopackaging films of poly(3-hydroxybutyrate-co-3-hydroxyvalerate) derived from fruit pulp biowaste. *Front. Sustain. Food Syst.* **2018**, *2*, 1–16. [CrossRef]
40. Kanatt, S.R.; Rao, M.S.; Chawla, S.P.; Sharma, A. Active chitosan–polyvinyl alcohol films with natural extracts. *Food Hydrocoll.* **2012**, *29*, 290–297. [CrossRef]
41. Figueroa-Lopez, K.J.; Andrade-Mahecha, M.M.; Torres-Vargas, O.L. Spice oleoresins containing antimicrobial agents improve the potential use of bio-composite films based on gelatin. *Food Packag. Shelf Life* **2018**, *17*, 50–56. [CrossRef]
42. Figueroa-Lopez, K.J.; Castro-Mayorga, J.L.; Andrade-Mahecha, M.M.; Cabedo, L.; Lagaron, J.M. Antibacterial and barrier properties of gelatin coated by electrospun polycaprolactone ultrathin fibers containing black pepper oleoresin of interest in active food biopackaging applications. *Nanomaterials* **2018**, *8*, 199. [CrossRef] [PubMed]
43. Torres-Giner, S.; Torres, A.; Ferrándiz, M.; Fombuena, V.; Balart, R. Antimicrobial activity of metal cation-exchanged zeolites and their evaluation on injection-molded pieces of bio-based high-density polyethylene. *J. Food Saf.* **2017**, *37*, 1–12. [CrossRef]
44. Busolo, M.A.; Lagaron, J.M. Antioxidant polyethylene films based on a resveratrol containing clay of interest in food packaging applications. *Food Packag. Shelf Life* **2015**, *6*, 30–41. [CrossRef]
45. Rošic, R.; Pelipenko, J.; Kocbek, P.; Baumgartner, S.; Bešter-Rogač, M.; Kristl, J. The role of rheology of polymer solutions in predicting nanofiber formation by electrospinning. *Eur. Polym. J.* **2012**, *48*, 1374–1384. [CrossRef]
46. Arfa, A.B.; Chrakabandhu, Y.; Preziosi-Belloy, L.; Chalie, P.; Gontard, N. Coating papers with soy protein isolates as inclusion matrix of carvacrol. *Food Res. Int.* **2007**, *40*, 22–32. [CrossRef]
47. Jouki, M.; Yazdi, F.T.; Mortazavi, S.A.; Koocheki, A. Quince seed mucilage films incorporated with oregano essential oil: Physical, thermal, barrier, antioxidant and antibacterial properties. *Food Hydrocoll.* **2014**, *36*, 9–19. [CrossRef]
48. Bhardwaj, N.; Kundu, S.C. Electrospinning: A fascinating fiber fabrication technique. *Biotechnol. Adv.* **2010**, *28*, 325–347. [CrossRef] [PubMed]
49. Kaltsa, O.; Yanniotis, S.; Mandala, I. Stability properties of different fenugreek galactomannans in emulsions prepared by high-shear and ultrasonic method. *Food Hydrocoll.* **2016**, *52*, 487–496. [CrossRef]
50. Mahdi Jafari, S.; He, Y.; Bhandari, B. Production of sub-micron emulsions by ultrasound and microfluidization techniques. *J. Food Eng.* **2007**, *82*, 478–488. [CrossRef]

51. Paximada, P.; Echegoyen, Y.; Koutinas, A.A.; Mandala, I.G.; Lagaron, J.M. Encapsulation of hydrophilic and lipophilized catechin into nanoparticles through emulsion electrospraying. *Food Hydrocoll.* **2017**, *64*, 123–132. [CrossRef]
52. Gómez-Estaca, J.; Montero, P.; Fernández-Martín, F.; Alemán, A.; Gómez-Guillén, M.C. Physical and chemical properties of tuna-skin and bovine-hide gelatin films with added aqueous oregano and rosemary extracts. *Food Hydrocoll.* **2009**, *23*, 1334–1341. [CrossRef]
53. Barbieri, N.; Sanchez-Contreras, A.; Canto, A.; Cauich-Rodriguez, J.V.; Vargas-Coronado, R.; Calvo-Irabien, L.M. Effect of cyclodextrins and mexican oregano (lippia graveolens kunth) chemotypes on the microencapsulation of essential oil. *Ind. Crop. Prod.* **2018**, *121*, 114–123. [CrossRef]
54. Yang, Y.; Kayan, B.; Bozer, N.; Pate, B.; Baker, C.; Gizir, A.M. Terpene degradation and extraction from basil and oregano leaves using subcritical water. *J. Chromatogr. A* **2007**, *1152*, 262–267. [CrossRef]
55. Gibara Guimarães, A.; Almeida Oliveira, M.; dos Santos Alves, R.; dos Passos Menezes, P.; Russo Serafini, M.; Antunes de Souza Araújo, A.; Pereira Bezerra, D.; Quintans Júnior, L.J. Encapsulation of carvacrol, a monoterpene present in the essential oil of oregano, with β-cyclodextrin, improves the pharmacological response on cancer pain experimental protocols. *Chem. Biol. Interact.* **2015**, *227*, 69–76. [CrossRef]
56. Pineros-Hernandez, D.; Medina-Jaramillo, C.; López-Córdoba, A.; Goyanes, S. Edible cassava starch films carrying rosemary antioxidant extracts for potential use as active food packaging. *Food Hydrocoll.* **2017**, *63*, 488–495. [CrossRef]
57. Cordeiro, A.M.T.M.; Medeiros, M.L.; Santos, N.A.; Soledade, L.E.B.; Pontes, L.F.B.L.; Souz, A.L.; Queiroz, N.; Souza, A.G. Rosemary (*Rosmarinus officinalis* L.) extract. *J. Therm. Anal. Calorim.* **2013**, *113*, 889–895. [CrossRef]
58. López de Dicastillo, C.; Castro-López, M.d.M.; López-Vilariño, J.M.; González-Rodríguez, M.V. Immobilization of green tea extract on polypropylene films to control the antioxidant activity in food packaging. *Food Res. Int.* **2013**, *53*, 522–528. [CrossRef]
59. Cherpinski, A.; Torres-Giner, S.; Vartiainen, J.; Peresin, M.S.; Lahtinen, P.; Lagaron, J.M. Improving the water resistance of nanocellulose-based films with polyhydroxyalkanoates processed by the electrospinning coating technique. *Cellulose* **2018**, *25*, 1291–1307. [CrossRef]
60. Zou, P.; Liu, H.; Li, Y.; Huang, J.; Yao, D. Surface dextran modified electrospun poly (3-hydroxybutyrate-co-3-hydroxyvalerate) (phbv) fibrous scaffold promotes the proliferation of bone marrow-derived mesenchymal stem cells. *Mater. Lett.* **2016**, *179*, 109–113. [CrossRef]
61. Galus, S.; Kadzińska, J. Whey protein edible films modified with almond and walnut oils. *Food Hydrocoll.* **2016**, *52*, 78–86. [CrossRef]
62. Medina-Jaramillo, C.; Ochoa-Yepes, O.; Bernal, C.; Famá, L. Active and smart biodegradable packaging based on starch and natural extracts. *Carbohydr. Polym.* **2017**, *176*, 187–194. [CrossRef] [PubMed]
63. Bravo Cadena, M.; Preston, G.M.; Van der Hoorn, R.A.L.; Townley, H.E.; Thompson, I.P. Species-specific antimicrobial activity of essential oils and enhancement by encapsulation in mesoporous silica nanoparticles. *Ind. Crop. Prod.* **2018**, *122*, 582–590. [CrossRef]
64. Castilho, P.C.; Savluchinske-Feio, S.; Weinhold, T.S.; Gouveia, S.C. Evaluation of the antimicrobial and antioxidant activities of essential oils, extracts and their main components from oregano from madeira island, portugal. *Food Control* **2012**, *23*, 552–558. [CrossRef]
65. Stefanakis, M.K.; Touloupakis, E.; Anastasopoulos, E.; Ghanotakis, D.; Katerinopoulos, H.E.; Makridis, P. Antibacterial activity of essential oils from plants of the genus origanum. *Food Control* **2013**, *34*, 539–546. [CrossRef]
66. Mohsenabadi, N.; Rajaei, A.; Tabatabaei, M.; Mohsenifar, A. Physical and antimicrobial properties of starch-carboxy methyl cellulose film containing rosemary essential oils encapsulated in chitosan nanogel. *Int. J. Biol. Macromol.* **2018**, *112*, 148–155. [CrossRef]
67. Okoh, O.O.; Sadimenko, A.P.; Afolayan, A.J. Comparative evaluation of the antibacterial activities of the essential oils of rosmarinus officinalis l. Obtained by hydrodistillation and solvent free microwave extraction methods. *Food Chem.* **2010**, *120*, 308–312. [CrossRef]
68. Saeed, M.; Naveed, M.; Arif, M.; Kakar, M.U.; Manzoor, R.; El-Hack, M.E.A.; Alagawany, M.; Tiwari, R.; Khandia, R.; Munjal, A.; et al. Green tea (camellia sinensis) and l-theanine: Medicinal values and beneficial applications in humans—A comprehensive review. *Biomed. Pharmacother.* **2017**, *95*, 1260–1275. [CrossRef] [PubMed]

69. Clarke, D.; Molinaro, S.; Tyuftin, A.; Bolton, D.; Fanning, S.; Kerry, J.P. Incorporation of commercially-derived antimicrobials into gelatin based films and assessment of their antimicrobial activity and impact. *Food Control* **2016**, *64*, 202–211. [CrossRef]
70. Rakmai, J.; Cheirsilp, B.; Mejuto, J.C.; Torrado-Agrasar, A.; Simal-Gandara, J. Physico-chemical characterization and evaluation of bio-efficacies of black pepper essential oil encapsulated in hydroxypropyl-beta-cyclodextrin. *Food Hydrocoll.* **2017**, *65*, 157–164. [CrossRef]
71. Cox, S.D.; Mann, C.M.; Markham, J.L.; Bell, H.C.; Gustafson, J.E.; Warmington, J.R.; Wyllie, S.G. The mode of antimicrobial action of the essential oil of melaleuca alternifolia (tea tree oil). *J. Appl. Microbiol.* **2000**, *88*, 170–175. [CrossRef]
72. Yong, A.-L.; Ooh, K.-F.; Ong, H.-C.; Chai, T.-T.; Wong, F.-C. Investigation of antibacterial mechanism and identification of bacterial protein targets mediated by antibacterial medicinal plant extracts. *Food Chem.* **2015**, *186*, 32–36. [CrossRef] [PubMed]
73. Park, J.-A.; Kim, S.-B. Preparation and characterization of antimicrobial electrospun poly(vinyl alcohol) nanofibers containing benzyl triethylammonium chloride. *React. Funct. Polym.* **2015**, *93*, 30–37. [CrossRef]
74. Torres-Giner, S.; Martinez-Abad, A.; Lagaron, J.M. Zein-based ultrathin fibers containing ceramic nanofillers obtained by electrospinning. Ii. Mechanical properties, gas barrier, and sustained release capacity of biocide thymol in multilayer polylactide films. *J. Appl. Polym. Sci.* **2014**, *131*, 9270–9276. [CrossRef]
75. Cruz-Gálvez, A.M.; Castro-Rosas, J.; Rodríguez-Marín, M.L.; Cadena-Ramírez, A.; Tellez-Jurado, A.; Tovar-Jiménez, X.; Chavez-Urbiola, E.A.; Abreu-Corona, A.; Gómez-Aldapa, C.A. Antimicrobial activity and physicochemical characterization of a potato starch-based film containing acetonic and methanolic extracts of hibiscus sabdariffa for use in sausage. *LWT* **2018**, *93*, 300–305. [CrossRef]
76. Khaneghah, A.M.; Bagher Hashemi, S.M.; Limbo, S. Antimicrobial agents and packaging systems in antimicrobial active food packaging: An overview of approaches and interactions. *Food Bioprod. Process.* **2018**, *111*, 1–19. [CrossRef]
77. Sentkowska, A.; Pyrzynska, K. Investigation of antioxidant interaction between green tea polyphenols and acetaminophen using isobolographic analysis. *J. Pharm. Biomed. Anal.* **2018**, *159*, 393–397. [CrossRef] [PubMed]
78. Lorenzo, J.M.; Sichetti Munekata, P.E. Phenolic compounds of green tea: Health benefits and technological application in food. *Asian Pac. J. Trop. Biomed.* **2016**, *6*, 709–719. [CrossRef]
79. Shahidi, F.; Zhong, Y. Novel antioxidants in food quality preservation and health promotion. *Eur. J. Lipid Sci. Technol.* **2010**, *112*, 930–940. [CrossRef]
80. Chun, S.-S.; Vattem, D.A.; Lin, Y.-T.; Shetty, K. Phenolic antioxidants from clonal oregano (origanum vulgare) with antimicrobial activity against helicobacter pylori. *Process Biochem.* **2005**, *40*, 809–816. [CrossRef]
81. Bajalan, I.; Rouzbahani, R.; Pirbalouti, A.G.; Maggi, F. Antioxidant and antibacterial activities of the essential oils obtained from seven iranian populations of rosmarinus officinalis. *Ind. Crop. Prod.* **2017**, *107*, 305–311. [CrossRef]
82. Hongjun Shao, S.Y.; Qi Wang, Z.Z.; Xingbin Yang, L.Z. Non-extractable polyphenols of green tea and their antioxidant, anti-α-glucosidase capacity, and release during in vitro digestion. *J. Funct. Foods* **2018**, *42*, 129–136.
83. Afroz Bakht, M.; Geesi, M.H.; Riadi, Y.; Imran, M.; Imtiyaz Ali, M.; Jawed Ahsan, M.; Ajmal, N. Ultrasound-assisted extraction of some branded tea: Optimization based on polyphenol content, antioxidant potential and thermodynamic study. *Saudi J. Biol. Sci.* **2018**, in press. [CrossRef]
84. Lu, M.-J.; Chen, C. Enzymatic modification by tannase increases the antioxidant activity of green tea. *Food Res. Int.* **2008**, *41*, 130–137. [CrossRef]
85. Asensio, C.M.; Grosso, N.R.; Juliani, H.R. Quality characters, chemical composition and biological activities of oregano (*Origanum* spp.) essential oils from central and southern argentina. *Ind. Crop. Prod.* **2015**, *63*, 203–213. [CrossRef]
86. Mechergui, K.; Jaouadi, W.; Coelho, J.P.; Larbi Khouja, M. Effect of harvest year on production, chemical composition and antioxidant activities of essential oil of oregano (*Origanum vulgare* subsp *glandulosum* (desf.) ietswaart) growing in north africa. *Ind. Crop. Prod.* **2016**, *90*, 32–37. [CrossRef]
87. Yan, F.; Azizi, A.; Janke, S.; Schwarz, M.; Zeller, S.; Honermeier, B. Antioxidant capacity variation in the oregano (*Origanum vulgare* L.) collection of the german national genebank. *Ind. Crop. Prod.* **2016**, *92*, 19–25. [CrossRef]

88. Shojaee-Aliabadi, S.; Hosseini, H.; Mohammadifar, M.A.; Mohammadi, A.; Ghasemlou, M.; Ojagh, S.M.; Hosseini, S.M.; Khaksar, R. Characterization of antioxidant-antimicrobial κ-carrageenan films containing satureja hortensis essential oil. *Int. J. Biol. Macromol.* **2013**, *52*, 116–124. [CrossRef] [PubMed]
89. Gómez-Estaca, J.; López-de-Dicastillo, C.; Hernández-Muñoz, P.; Catalá, R.; Gavara, R. Advances in antioxidant active food packaging. *Trends Food Sci. Technol.* **2014**, *35*, 42–51. [CrossRef]
90. Robertson, G.L. *Food Packaging: Principles and Practice*; CRC Press, Taylor and Francis Group: Boca Raton, FL, USA, 1998; Volume 6.

nanomaterials

MDPI

Article

Electrospun Poly(ethylene-*co*-vinyl alcohol)/Graphene Nanoplatelets Composites of Interest in Intelligent Food Packaging Applications

Sergio Torres-Giner [1], Yolanda Echegoyen [1], Roberto Teruel-Juanes [2], Jose D. Badia [2,3], Amparo Ribes-Greus [2] and Jose M. Lagaron [1,*]

1 Novel Materials and Nanotechnology Group, Institute of Agrochemistry and Food Technology (IATA), Spanish Council for Scientific Research (CSIC), Calle Catedrático Agustín Escardino Benlloch 7, 46980 Paterna, Spain; storresginer@iata.csic.es (S.T.-G.); yolanda.echegoyen@iata.csic.es (Y.E.)

2 Technological Institute of Materials (ITM), Universitat Politècnica de València (UPV), Camí de Vera s/n, 46022 Valencia, Spain; rotejua@upvnet.upv.es (R.T.-J.); aribes@ter.upv.es (A.R.-G.)

3 Department of Chemical Engineering, School of Engineering, Universitat de València, Camí de les Universitats s/n, 46800 Burjassot, Spain; jose.badia@uv.es

* Correspondence: lagaron@iata.csic.es

Received: 1 August 2018; Accepted: 19 September 2018; Published: 20 September 2018

Abstract: Graphene nanoplatelets (GNPs) were synthetized from graphite powder and, thereafter, embedded in poly(ethylene-*co*-vinyl alcohol) (EVOH) fibers by electrospinning in the 0.1–2 wt.-% range. The morphological, chemical, and thermal characterization performed on the electrospun nanocomposite fibers mats revealed that the GNPs were efficiently dispersed and rolled along the EVOH fibrilar matrix up to contents of 0.5 wt.-%. Additionally, the dielectric behavior of the nanocomposite fibers was evaluated as a function of the frequency range and GNPs content. The obtained results indicated that their dielectric constant rapidly decreased with the frequency increase and only increased at low GNPs loadings while the nanocomposite fiber mats became electrically conductive, with the maximum at 0.5 wt.-% GNPs content. Finally, the electrospun mats were subjected to a thermal post-treatment and dark films with a high contact transparency were obtained, suggesting that the nanocomposites can be used either in a nonwoven fibers form or in a continuous film form. This study demonstrates the potential of electrospinning as a promising technology to produce GNPs-containing materials with high electrical conductivity that can be of potential interest in intelligent packaging applications as "smart" labels or tags.

Keywords: EVOH; graphene; electrospinning; smart labels; intelligent packaging

1. Introduction

In the field of intelligent packaging, the use of electrically conductive polymer-based materials opens up new opportunities to create "smart" labels or tags. Intelligent or smart packaging is the umbrella term for a range of intelligent technologies that allow packaging to contain, evaluate, and transmit relevant information [1]. For instance, smart packages can enable monitoring of the conditions and quality of the packaged products (e.g., food freshness) from the production line to the end user. This includes relevant information and spoilage indicators such as time, temperature, and pH or the presence of different gases, chemical contaminants, pathogens, etc. Apart from this, smart labels can also include components that range from bar codes to radio frequency transmitters, i.e., radio frequency identification (RFID) devices and printed electronics [2]. The smart tags can be used to electronically transfer information from the packaging to the consumer about the packaged material through, for instance, refrigerator displays. Therefore, intelligent packaging systems can make packaging more informative and interactive whereas their global demand is expected to grow strongly to reach US $1.5 billion by 2025 [2].

Currently, however, the intelligent packaging technology habitually requires the use of a silicon chip as the substrate for the high-frequency electronics, limiting its application for packaging uses. In addition, there is an increasing necessity for the development of new thin-film transmitters that can be efficiently embedded in the packaging structure. In this sense, the use of poly(ethylene-*co*-vinyl alcohol) (EVOH) copolymers could represent an advantageous strategy since this family of copolymers is frequently employed in high-barrier packaging films in the form of inner layers with very low thickness (typically well below 10 μm). In addition, EVOH films are highly transparent and hydrophilic, yet water-insoluble. Therefore, EVOH is a suitable candidate to be efficiently employed for the inclusion of smart tags in the packaging structure and/or for the creation of patterns acting like a bar code. This would provide a unique response to electrical stimuli that give relevant information about the physicochemical properties of the foodstuff packaged and/or for traceability and improved supply chain management (SCM) purposes.

Graphene was the last carbon allotrope to be discovered, after fullerenes and carbon nanotubes (CNTs) [3]. This carbonaceous material is presented in the form of a unique two-dimensional (2D) macromolecular sheet of carbon atoms with a honeycomb-like structure, the so-called graphene nanoplatelets (GNPs), which has become one of the most promising materials available today. Compared to other nano-sized carbonaceous systems (e.g., CNTs), GNPs have attracted considerable attention because of their combination of outstanding mechanical flexibility, excellent electrical and thermal conductivity, optical transparency, and low density. The peculiar properties of single layers of graphene include a Young's modulus of ~1 TPa, an electrical conductivity (σ) of approximately 6000 S/cm, a thermal conductivity (λ) of up to 5300 W/m·k, a surface area of over 2600 m^2/g, a high chemical tolerance, and a broad electrochemical window [4,5]. As a result, GNPs have been widely considered as a perfect filler to develop novel carbon-based reinforced polymer nanocomposites with enhanced thermal, electrical, and mechanical properties [6,7]. In this sense, the use of GNPs-containing plastics is very advantageous for several applications in energy and electronics, but their use can also be originally focused on intelligent packaging strategies in combination to EVOH since it is a food contact polymer with excellent optical properties and polarity.

Electrohydrodynamic processing (EHDP) is a straightforward, versatile, and low-cost technique based on the application of high electrical fields to a viscoelastic polymer solution or melt via a metallic capillary orifice that allows to fabricate polymer nanostructures with different functionalities [8]. EHDP is habitually referred to as electrospinning when fiber-based morphologies are produced in which a wide range of polymers and biopolymers can be processed. Electrospun nanofibers can find several applications in the packaging industry, including the development of active and intelligent systems [9,10]. Although the electrospun materials are predominantly polymer based, certain amount of non-polymer contents (e.g., nano-sized fillers) can also be incorporated into the primary electrospinning solution to form hybrid ultrathin or nanocomposite fibers [11–13]. At present, the incorporation of GNPs has been achieved into electrospun fibers of poly(vinylpyrrolidone) (PVP) as a conductive additive to enhance the high-rate capabilities for lithium-ion batteries [4], polystyrene (PS) and polyvinyl chloride (PVC) to generate superhydrophobic surfaces [14], poly(vinyl acetate) (PVAC) to improve the optical absorption for ultrafast photonics [15], polyacrylonitrile (PAN) to produce carbon nanofibers (CNFs) [16], polyaniline/poly(methyl methacrylate) (PANi/PMMA) blends for conductive devices [17], and PAN/PVP blends as high capacitance materials [18].

In this work, for the first time, EVOH/GNPs nanofibers were prepared by electrospinning. The morphology, thermal properties, and electrical conductivity of the resultant hybrid nanofibers mats were characterized. Finally, the electrospun mats were subjected to a thermal post-treatment in order to generate films that could be applied for creating smart labels or tags with high electrical rate capabilities in the field of intelligent food packaging.

2. Experiment

2.1. Materials

EVOH containing 32 mol% ethylene content, i.e., EVOH32, was supplied by Nippon Gohsei (Osaka, Japan) as Soarnol™ DC3212B. The copolymer has a density of 1.19 g/m^3 (23 °C) and a melt flow rate (MFR) of 12 g/10 min (210 °C, 2160 g). Graphite powder, G282863 grade, was purchased from Sigma-Aldrich S.A. (Madrid, Spain). Sodium nitrate ($NaNO_3$), hydrogen peroxide solution (H_2O_2) at 30 wt.-%, hydrazine hydrate (N_2H_4) at 50–60 wt.-%, ammonia solution (NH_4OH) at 25 wt.-%, potassium permanganate ($KMnO_4$) of 97% purity, and isopropyl alcohol (IPA) with purity ≥99% were also purchased from by Sigma-Aldrich S.A. Sulfuric acid (H_2SO_4) of 96% purity was provided by Panreac S.A. (Barcelona, Spain).

2.2. Oxidation of Graphite

In a first stage, graphite oxide (GO) was prepared by oxidizing graphite powder based on the modification of the so-called Hummer's method described by Hirata et al. [19]. Briefly, 10 g of graphite powder and 7.5 g of $NaNO_3$ were placed into a 2000 mL round-bottom glass flask. Then, 621 g of H_2SO_4 was added and the mixture was stirred while being cooled in an ice water bath. Thus, 45 g of $KMnO_4$ was gradually added for about 1 h. Cooling was completed after 2 h and the mixture was allowed to stand for five days at about 20 °C with gentle stirring to obtain a highly viscous liquid. After this, 1000 cm^3 of 5 wt.-% H_2SO_4 aqueous solution was gradually added to the resultant solution for 1 h under continuous and gentle stirring. The resultant mixture was further stirred for 2 h. Then, 30 g of the H_2O_2 aqueous solution was added to the above liquid and the mixture was stirred for another 2 h. In order to remove the ions of oxidant origin, especially manganese ions, the resultant liquid was purified by repeating the following procedure cycle 15 times: Centrifugation, removal of the supernatant liquid, addition of a mixed aqueous solution of 3 wt.-% H_2SO_4/0.5 wt.-% H_2O_2, and shaking to re-disperse. The mixed solution amounted, in total, to about 13 g. The purification procedure was similarly repeated a further three times, except that the liquid to be added was replaced with water. The resultant mixture was allowed to stand for at least 24 h to precipitate thick particles, which were filtered and removed. The remaining dispersion was purified several times with water. Finally, a brown-black viscous flurry containing GO particles was obtained. The GO content was ca. 1 wt.-%, as determined from weight difference of the dispersion before and after drying at 150 °C for 1 h in an oven.

2.3. Reduction of Graphite Oxide

In a second stage, the resultant GO particles were reduced to graphene using NH_4OH based on previous methodology [20]. For this, 0.1 g of the above-obtained GO particles was mixed with 100 mL of deionized water in a 250 mL round-bottom glass flask, yielding an inhomogeneous yellow-brown dispersion. This dispersion was bath ultrasonicated using a Fisher Scientific FS60 Ultrasonic Cleaner (150 W) from Thermo Fisher Scientific (Waltham, MA, USA) until it became clear, i.e., with no visible particulate matter. Then, 10 mL of N_2H_4 and 10 mL of NH_4OH were added and the resultant solution was heated overnight in an oil bath at 90 °C under a water-cooled condenser, in which the reduced GO gradually precipitated out as a black solid. The final suspension was isolated by filtration and then vacuum-dried at 60 °C until the solids reached a concentration of ca. 0.1 wt.-%.

2.4. Preparation of Electrospun Fiber Mats

The polymer solution for electrospinning was prepared by fully dissolving 7 wt.-% EVOH in 4/1 (vol./vol.) IPA/water. The obtained EVOH solution was then added to the above-described water-based solution containing graphene, pre-heated at 50 °C, to reach the following weight contents of GNPs in EVOH: 0.1, 0.5, 1, and 2 wt.-%. The resultant GNPs dispersions in EVOH were then ultrasonicated for 15 min and transferred immediately to a 5 mL plastic syringe. A control solution of EVOH without GNPs was prepared in identical conditions.

The electrospinning process was performed using a Fluidnatek® LE-50 benchtop line from Bioinicia S.L. (Valencia, Spain) with a dual polarizer yielding a variable high-voltage ranging from 0–60 kV and with temperature and humidity control. The EVOH solutions containing the GNPs were pumped through a stainless-steel needle injector and collected on a grounded metallic flat plate. The applied voltage, flow-rate, and tip-to-collector distance were set at 15 kV, 0.5 mL/h, and 15 cm, respectively. All samples were electrospun in a controlled environmental chamber at 29 °C and 30% relative humidity (RH). The electrospun fiber mat thickness was ca. 200 microns. All the characterization work was carried out in the electrospun fibers mats.

Additionally, the obtained electrospun fibers mats were proven to be conformable into continuous films by annealing in a hydraulic press 4122-model from Carver, Inc. (Wabash, IN, USA). This step was optimally performed at 158 °C, without pressure, for 10 s. The resultant films were air cooled at room temperature.

2.5. Morphology

The morphology of the electrospun mats was examined by scanning electron microscopy (SEM) with a S-4800 from Hitachi (Tokyo, Japan). Prior to examination, samples were deposited on suitable beveled microscopy holders and sputtered using a gold-palladium mixture under vacuum. All SEM experiments were carried out at 8.0 kV.

Transmission electron microscopy (TEM) was performed using a JEOL 1010 from JEOL USA, Inc. (Peabody, MA, USA) equipped with a digital image acquisition system from Bioscan (Edmonds, WA, USA). TEM images were taken directly on mats electrospun onto the TEM observation grids. At least 25 SEM and TEM micrographs were analyzed for each sample using Adobe Photoshop 7.0 software from Adobe Systems Incorporated (San Jose, CA, USA) to determine the sizes from their original magnification.

2.6. Wide Angle X-ray Scattering

Wide angle X-ray scattering (WAXS) was performed using a D5000 X-ray Powder Diffractometer from Siemens AG (Munich, Germany). Radial scans of intensity vs. scattering angle (2θ) were recorded at room temperature in the range 2 to 50° (2θ), step size of 0.03° (2θ), scanning rate of 8 s/step, with identical settings of the instrument by using a filtered Cu Kα radiation (λ = 1.5406 Å), an operating voltage of 40 kV, and a filament current of 30 mA. Bragg's law ($n \cdot \lambda = 2 \cdot d \cdot \sin\theta$) was applied to calculate the basal spacing (*d*).

2.7. Raman Imaging

Raman images were taken with a Jasco NRS-3100 Confocal Micro-Raman spectrophotometer from Jasco Inc. (Easton, MD, USA) using a short working distance 100× objective, which according to the manufacturer provides a lateral and depth resolution of ca. 1 and 2 μm, respectively under high confocal conditions. The source was a red laser in the visible excitation tuned at 632.8 nm. Raman chemical images were carried out in the point by point mode by plotting the added area of the two graphene peaks present in the spectral 1780–918 cm^{-1} area and were constructed by taking 15 × 15 spectra equally spaced along a similar flat and continuous sample area in the case of the samples from 0.1 to 1 wt.-% GNPs loading and 20 × 20 spectra for the sample with 2 wt.-% GNPs content. Averaged spectra were taken in the samples with the 40× objective in a non-confocal mode.

2.8. Thermal Analysis

Thermal transitions of the electrospun EVOH nanofibers were evaluated by differential scanning calorimetry (DSC) using a DSC-7 analyzer from PerkinElmer, Inc. (Waltham, MA, USA) equipped with the refrigerating cooling accessory Intracooler 2. For this, ca. 2 mg samples were placed in 40-μL hermetic aluminum sealed pans, previously calibrated with an indium standard. The scanning rate was 10 °C/min. Samples were subjected to a first heating step from −25 to 200 °C, followed by a cooling step down to −25 °C, and a second heating step to 200 °C. An empty aluminum pan was used as a reference and all tests were carried out, at least, in triplicate.

2.9. Dielectrical Performance and Electrical Conductivity

The dielectric spectra of the samples were obtained using an alpha (α) mainframe frequency analyzer, in conjunction with an active cell Concept 40, from Novocontrol Technologies BmgH & Co. Kc (Hundsangen, Germany). The sample electrode assembly consisted of two stainless steel electrodes filled with the polymer. The diameters of the electrodes were 20 mm and the thickness values of each sample were determined for each measurement. A single-sweep experiment was performed. The spectra were measured in the frequency (f) range of 10^{-2}–10^{7} Hz, under isothermal conditions at a temperature of 25 °C in nitrogen atmosphere. The analysis was conducted through the complex dielectric permeability $\varepsilon^* = \varepsilon' - i \cdot \varepsilon''$, taking into account the real (ε') and imaginary (ε'') parts as well as the loss tangent ($\tan\delta = \varepsilon''/\varepsilon'$). In order to discriminate polarization and conductive effects [21,22], the complex electric modulus (M^*) was obtained as follows:

$$M^* = \frac{1}{\varepsilon^*} = \frac{1}{\varepsilon' - i \cdot \varepsilon''} = \frac{\varepsilon'}{\varepsilon'^2 + \varepsilon''^2} + i \cdot \frac{\varepsilon''}{\varepsilon'^2 + \varepsilon''^2} = M' + i \cdot M''$$

where

$$M' = \frac{\varepsilon'}{\varepsilon'^2 + \varepsilon''^2}$$

$$M'' = \frac{\varepsilon''}{\varepsilon'^2 + \varepsilon''^2}$$

The complex electrical conductivity $\sigma^* = \varepsilon^* \cdot \varepsilon_0 \cdot \omega$ was analyzed by means of the same experimental set. The values of the direct current electrical conductivity (σ_{dc}) were calculated by extrapolating the conductivity plateau to $f \rightarrow 0$ [23].

3. Results and Discussion

3.1. Characterization of GNPs

To confirm the detailed microstructure of the processed-graphite particles, the crystal phases of graphite, GO, and GNPs were analyzed by WAXS. From Figure 1 it can be observed that the XRD patterns of pristine graphite exhibits a strong peak centered at ~26.6° (2θ), which can be attributed to the typical graphite diffraction plane (002). By applying Bragg's law, this resulted in a d value of 3.35 Å, which is the characteristic interlayer distance between graphite layers [24]. Instead, a weak and broad basal diffraction peak was observed in the X-ray diffraction pattern of GO at ~11.4° (2θ), which corresponds to a d value between planes of 7.71 Å. This significant spacing increase reveals that the oxygen-containing functional groups were intercalated in the interlayer of graphite, then confirming the oxidation of graphite to GO. Furthermore, the lower intensity of this characteristic peak indicates that the degree of crystallinity in the GO structure was reduced. Indeed, GO is known to consist of heavily oxidized graphene sheets, which are loosely attached to each other [11]. Finally, the absence of peaks in the WAXS spectrum of the GNPs suggests that the graphene tactoids were randomly stacked.

Figure 2 shows a representative image, obtained by TEM, of a single layer of graphene. This provides direct evidence for the existence of a flat-like structure with a large surface and a diameter in the nanometric range, the so-called nanoplatelet. Previous studies have well described the typical morphology of a graphene particle, describing that it forms a 2D structure based on a cluster composed by several monolayers in which typical sizes of individual particles can reach several microns [25].

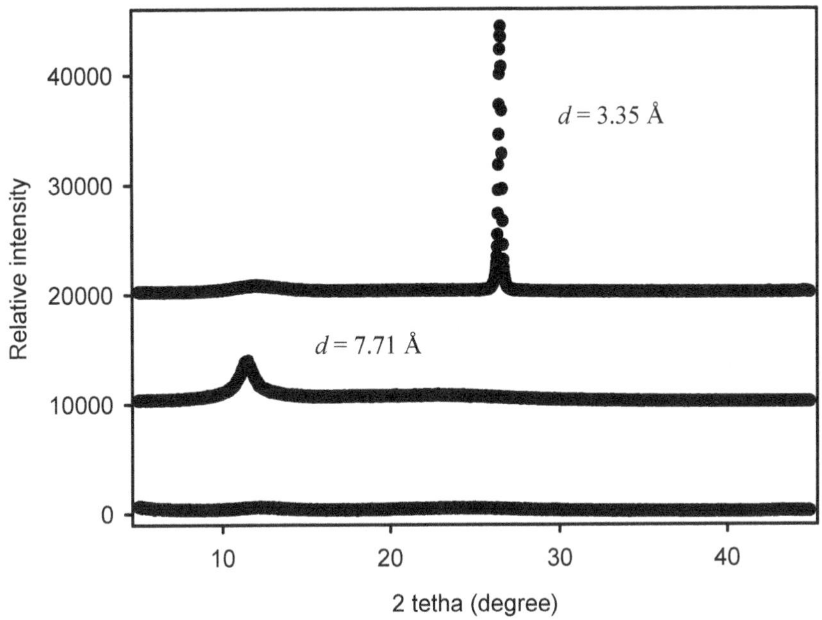

Figure 1. Wide angle X-ray scattering (WAXS) diffractograms of, from top to bottom, pristine graphite, graphite oxide (GO), and graphene nanoplatelets (GNPs). The interlayer distance is represented by d.

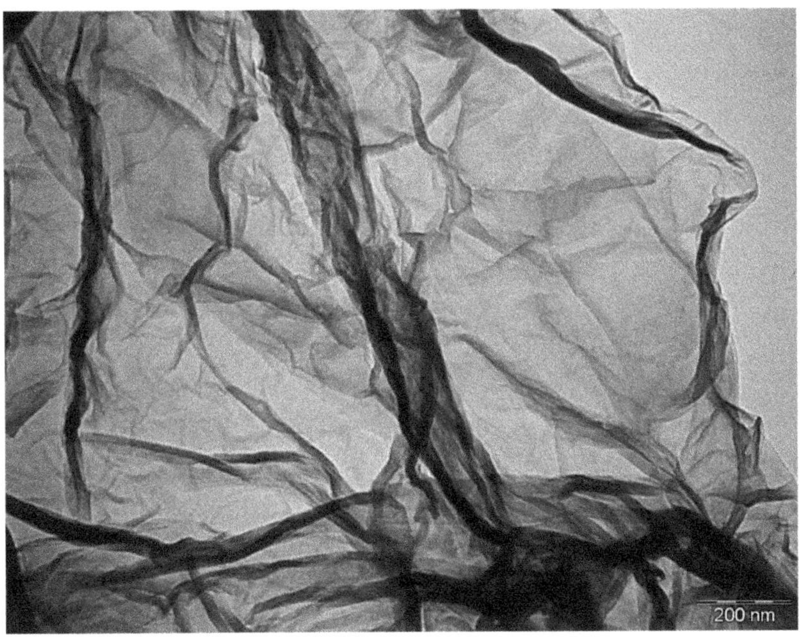

Figure 2. Transmission electron microscope (TEM) image of a single graphene nanoplatelet (GNP). Scale marker is 200 nm.

3.2. Morphology of EVOH/GNPs Fibers

Figure 3 shows the SEM images of the electrospun EVOH and EVOH/GNPs fibers. From Figure 3a it can be seen that the neat EVOH fibers presented a fibrilar morphology, completely free of beaded regions, with a mean diameter of approximately 675 nm. The resultant morphology differs from that previously reported in the study performed by Martínez-Sanz et al. [26], in which EVOH fibers presented beaded regions, this being mainly related to the lower polymer concentration used in the solution for electrospinning. As shown in Figure 3b–e, increasing the GNPs content led to a significant decrease in the electrospun fibers diameter. In particular, the average diameter was reduced from approximately 425 nm, for the fibers containing 0.1 wt.-% GNPs, to a value below 100 nm, for those fibers containing 2 wt.-% GNPs. This diameter decrease of the electrospun EVOH fibers can be related to an increase in the solution conductivity when the GNPs were added. In addition to a decrease in the fibers diameter, the GNPs incorporation gave rise to the formation of certain beaded fibers. This effect was particularly notable at the highest GNPs content, i.e., 2 wt.-%, where the bead morphology was the most prevalent. This observation preliminary suggests that GNPs agglomeration could occur at high contents inside the EVOH fibers during electrospinning.

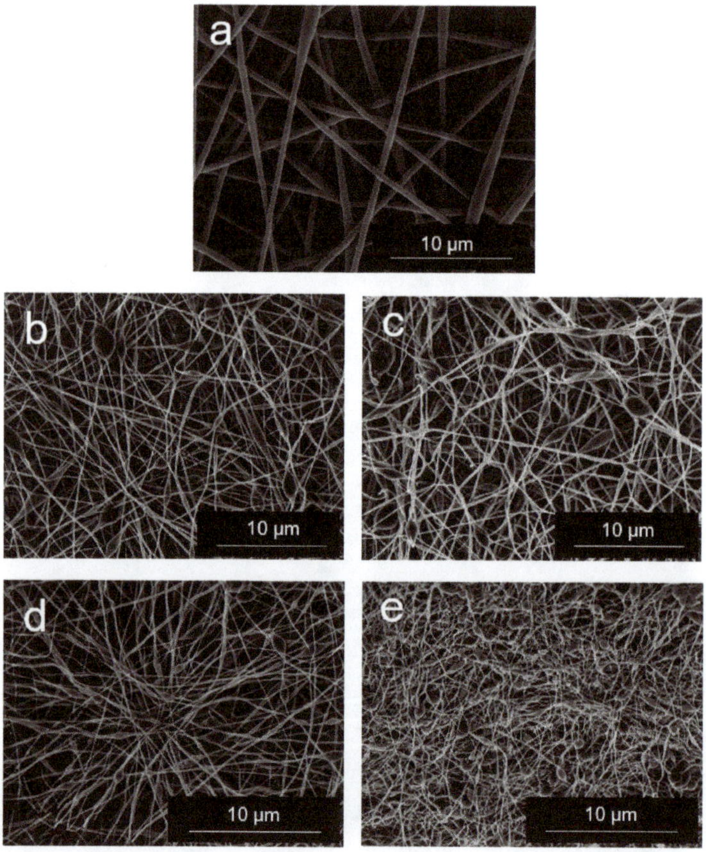

Figure 3. Scanning electron microscopy (SEM) images of electrospun fibers of: (**a**) Neat poly(ethylene-*co*-vinyl alcohol) (EVOH); (**b**) EVOH/graphene nanoplatelets (GNPs) at 0.1 wt.-%; (**c**) EVOH/GNPs at 0.5 wt.-%; (**d**) EVOH/GNPs at 1 wt.-%; (**e**) EVOH/GNPs at 2 w.-t%. Scale markers are 10 μm.

The presence and distribution of the GNPs in the electrospun EVOH fibers was also analyzed by TEM. Figure 4 displays a micrograph of the submicron EVOH fibers containing 0.5 wt.-% GNPs. In this image it can be observed that the GNPs were embedded and rolled up in the form of a continuous layer along the EVOH fiber axis. A good dispersion of the GNPs seems to be attained as a result of the favorable interfacial interaction between graphene and the EVOH matrix. Rolling and alignment of the GNPs can be ascribed to the inherent high resilience of the graphene layers combined with the extensional forces provided by the electrospinning process [4]. As a result, the GNPs were successfully incorporated by electrospinning into continuous submicron fibers of EVOH. The resultant roll-like morphology can be also advantageous to prevent restacking of the individual GNPs.

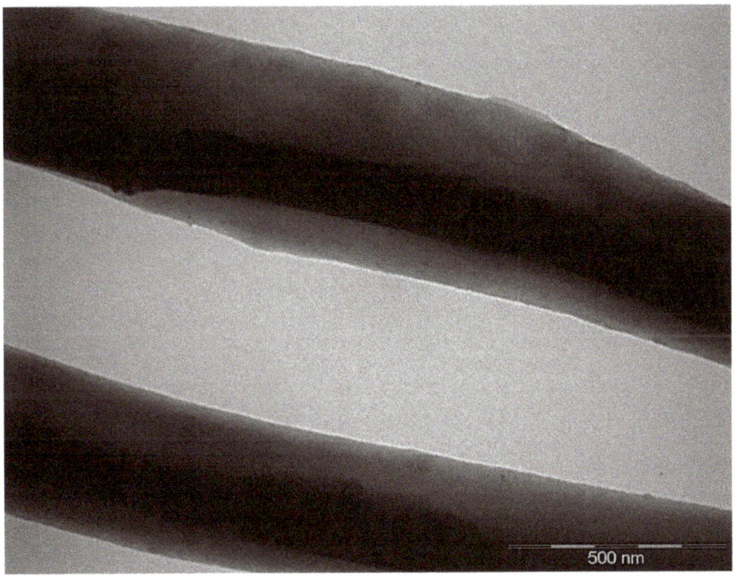

Figure 4. Transmission electron microscope (TEM) image of the electrospun poly(ethylene-*co*-vinyl alcohol) (EVOH) fibers containing graphene nanoplatelets (GNPs). Image corresponds to EVOH/GNP fibers at 0.5 wt.-%. Scale marker is 500 nm.

To further elucidate the GNPs dispersion in the submicron EVOH fibers, Raman analysis was carried out. Figure 5 shows typical Raman confocal images of the electrospun EVOH fiber mats reinforced with the different GNPs contents (Figure 5a–d) and also a typical non confocal spectrum of graphene taken in the electrospun mats (Figure 5e). The contour plots represent the added band areas of the two strong graphene bands across a defined sample area for the four nanocomposites. Thus, the more the yellow through to red color in the mapping, the stronger the presence of the GNPs. From the images, it seems that the dispersion was higher at lower nanofiller contents while the lowest dispersion, i.e., higher heterogeneity across the image, seemed to be for the sample with 2 wt.-% GNPs content, due to likely partial nanofiller agglomeration.

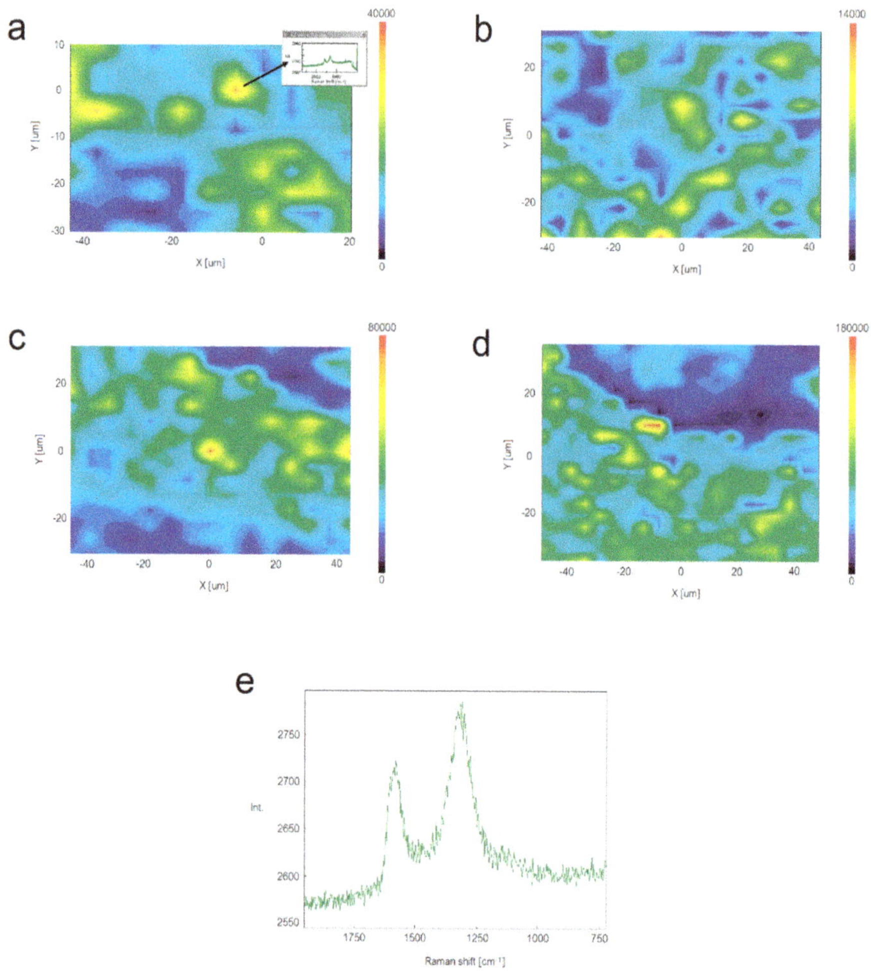

Figure 5. Raman spectroscopy contour plots of the electrospun poly(ethylene-*co*-vinyl alcohol) (EVOH) fibers containing graphene nanoplatelets (GNPs): (**a**) 0.1 wt.-%; (**b**) 0.5 wt.-%; (**c**) 1 wt.-%; (**d**) 2 wt.-%; (**e**) Typical non-confocal Raman spectrum of the sample EVOH/GNPs fibers at 2 wt.-%. As an example, Figure 5a shows as an inset the individual spectra that corresponds with the maximum signal of graphene in the image.

In relation to the individual spectra, the D-band at ~1316 cm^{-1} originates from the disordered carbon while the G-band, centered at ~1587 cm^{-1}, associates with the ordered graphitic carbon [27]. In particular, the D-peak presence indicates the breakdown of translational symmetry in the lattice, which is attributed to either bulk defects in the basal plane or edge defects [28]. Table 1 includes the relative intensity ratio of D-band to G-band, i.e., ID/IG, which is habitually referred to as the "R-value", for each electrospun fiber mat taken using a 40× objective in a non-confocal mode and, hence, averaging the graphene signal in the samples. This value can be used to quantitatively characterize the amount of structurally ordered graphite crystallites in carbonaceous materials. One can observe that the ID/IG ratio was kept constant at 0.98–0.99 for the electrospun EVOH fiber mats containing the GNPs in the 0.1–1 wt.-% range. This observation suggests that the GNPs were effectively dispersed as carbon nanofillers with ordered layers in the electrospun fibers. This is in good agreement with the recent findings by Li et al. [16], who additionally proposed that the embedded graphene particles played an important role in promoting graphitic crystallinity for PAN nanofibers. However, this value decreased to ~0.96 for the electrospun EVOH fibers filled with 2 wt.-% GNPs, which confirms that graphene was more agglomerated at the highest content. In addition, its higher separation in wavenumbers between the D-band and G-band ($d_{IG\text{-}ID}$), also shown in Table 1, may also indicate a reduction in the number of layers present in the graphene tactoids, i.e., a lower degree of exfoliation [27].

Table 1. Distance in wavenumbers ($d_{IG\text{-}ID}$) and relative intensity ratio of D-band to G-band (ID/IG) of the electrospun poly(ethylene-*co*-vinyl alcohol) (EVOH) fibers as a function of the graphene nanoplatelets (GNPs) content.

GNPs Content (wt.-%)	$d_{IG\text{-}ID}$ (cm^{-1})	ID/IG
0.1	261	0.986
0.5	257	0.991
1	267	0.981
2	270	0.962

3.3. Thermal Properties of EVOH/GNPs Fibers

Figure 6 shows the DSC curves of the electrospun EVOH mats with the different GNPs contents. In Figure 6a it can be seen that the neat electrospun EVOH mat presented a relatively low glass transition temperature (T_g), of around 40 °C, probably related to the presence of some remaining humidity in the fibers [29]. One can also observe that the T_g value increased with the GNPs content. Therefore, the presence of the carbonaceous nanoplatelets restricted the polymer chains motion. However, the highest T_g was observed at 54 °C, for the electrospun EVOH fibers containing 1 wt.-% GNPs, while the EVOH fibers with 2 wt.-% GNPs presented an intermediate T_g value of ~51 °C. This confirms that the GNPs agglomerated more strongly at their highest content, as supported by the above-described morphological and also chemical analyses. In addition, all samples presented a slight endothermic peak during glass transition, which can most likely be related to stress-relaxation mechanisms occurring in the vicinity of the T_g. Other researchers have ascribed this effect to a physical-aging process. With aging of the material after processing, the polymer chains exhibit slow thermodynamic changes in the amorphous region to attain a lower-free energy state. As a result of this segmental mobility, the aged sample has smaller free volume and lower potential energy than the just processed one [30]. Therefore, when the materials are reheated, more energy is required to surpass the glass transition, which results in a small endothermic peak [31]. This molecular rearrangement has been previously described for EVOH films after pressure-assisted thermal processing [32].

In addition to the thermal variations during glass transition, all the submicron EVOH fibers did not present any melting peak during the first heating scan. This observation confirms that the electrospinning process led to a fully amorphous structure. A similar effect of chain mobility restriction in the EVOH fibers due to the presence of the GNPs can be observed in the cooling scan during crystallization from the melt, shown in Figure 6b. For instance, the electrospun EVOH fibers with 0.5 wt.-% GNPs presented a crystallization temperature (T_c) of ~128 °C while the neat EVOH fibers showed a value of ~134 °C. This delay in crystallization supports the above-described statement that the embedded GNPs acted as an anti-nucleant agent, particularly at low contents due to their improved dispersion.

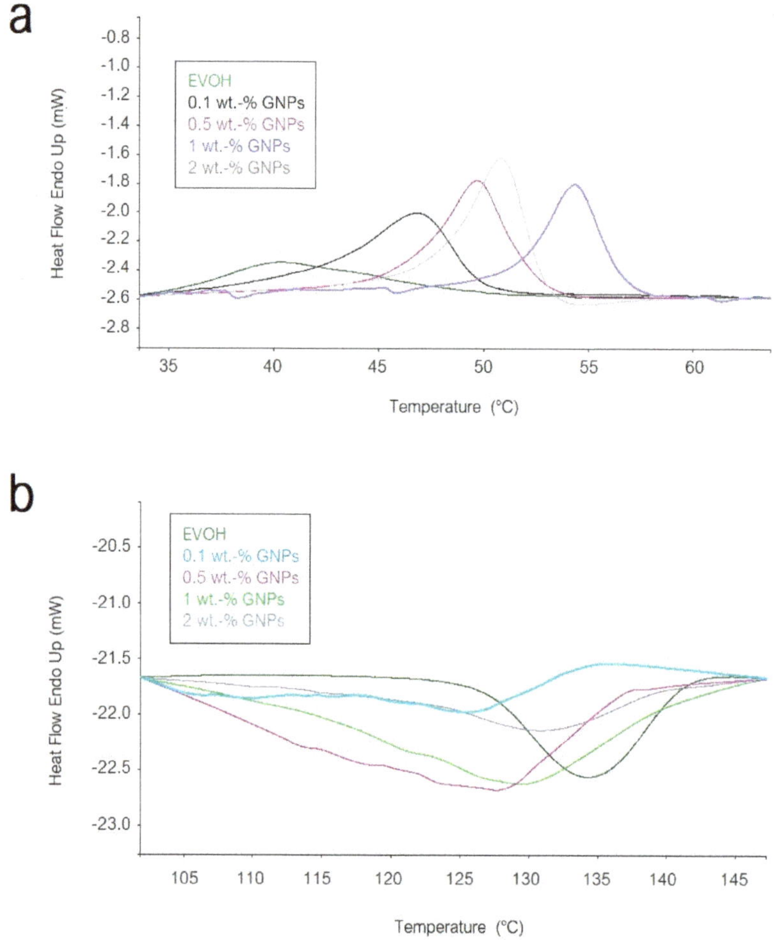

Figure 6. Differential scanning calorimetry (DSC) curves of the electrospun poly(ethylene-*co*-vinyl alcohol) (EVOH) fibers containing graphene nanoplatelets (GNPs): (**a**) First heating scan; (**b**) Cooling scan.

3.4. Dielectrical Performance and Electrical Conductivity of EVOH/GNPs Fibers

The dielectric analysis of polymer materials and composites permits understanding the macromolecular relaxations [33–36] and is also a valuable approach for evaluating their conductivity and electric response. In this sense, Figures 7–10 show the dielectric spectra of the EVOH and EVOH/GNP fibers in terms of ε', ε'', tan δ, and M″, respectively, in all the frequency range at 25 °C. In Figures 7 and 8 one can observe that, at low frequencies, tan δ and ε'' attained higher values though these values diminished rapidly with frequency. These results can be explained due to the fact that the alternation of the electric field was slow in the low frequency region, providing sufficient time for the permanent and induced dipoles to align themselves according to the applied field and, thus, leading to enhanced polarization. Broader dipolar polarization/relaxation processes were observed at higher frequencies, labelled as β-relaxation. This relaxation may be attributed to the local-mode relaxation in the crystalline regions of the copolymer and/or to the motion of their hydroxyl groups that could interact with each other. The neat EVOH fibers and EVOH/GNPs fibers exhibited almost identical values, with a slight increase when the GNPs were added. The maximum values were observed for a GNPs content of 0.5 wt.-%.

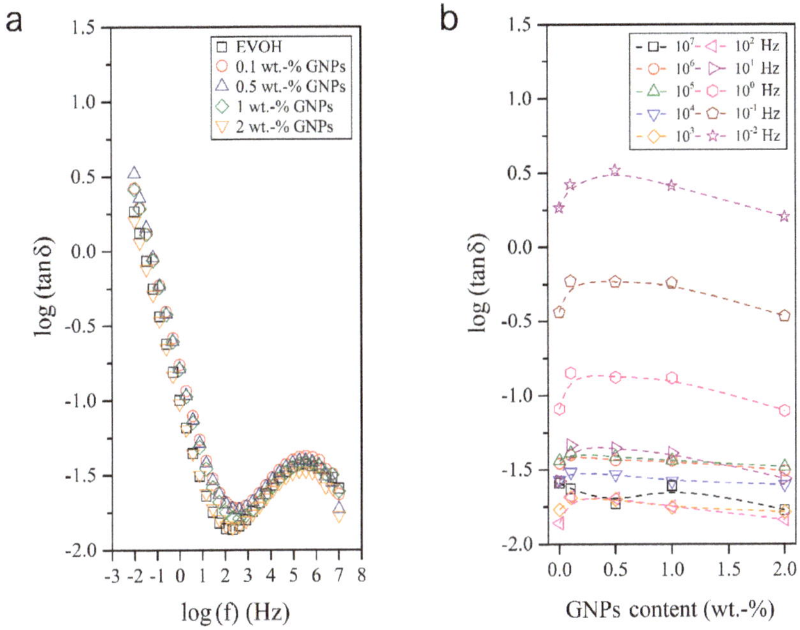

Figure 7. (**a**) Isothermal dielectric curves of the loss tangent (tan δ) of the electrospun poly(ethylene-*co*-vinyl alcohol) (EVOH) fibers containing graphene nanoplatelets (GNPs); (**b**) Influence of the GNPs content on tan δ for all frequency decades.

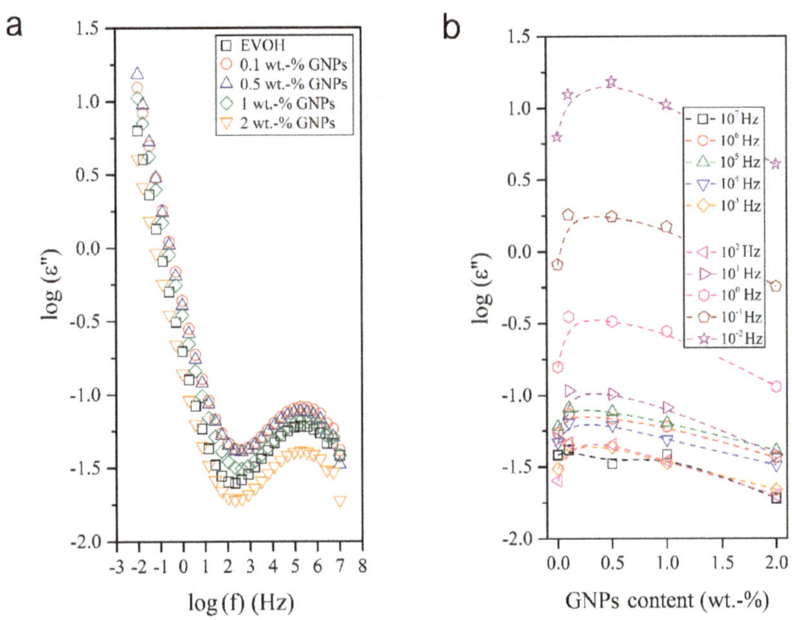

Figure 8. (**a**) Isothermal dielectric curves of the imaginary dielectric permeability (ε'') of the electrospun poly(ethylene-*co*-vinyl alcohol) (EVOH) fibers containing graphene nanoplatelets (GNPs); (**b**) Influence of the GNPs content on ε'' for all frequency decades.

Figure 9 shows the frequency response of the isotherms of the electrospun submicron neat EVOH fibers and EVOH/GNPs composite fibers in terms of ε'. Three stages can be observed with the frequency increase. In brief, these involve a first decrease at low frequencies, followed by a plateau related to the copolymer relaxation processes, and a second decrease at high frequencies. Although all samples presented the same behavior, a non-linear slight increase was observed when the GNPs content was increased. Regardless of the magnitude of the frequency, the influence of the GNPs content showed similar profiles, with maximum values of dielectric permittivity around 0.5 wt.-% GNPs, which may indicate the most relevant polarization enhancement.

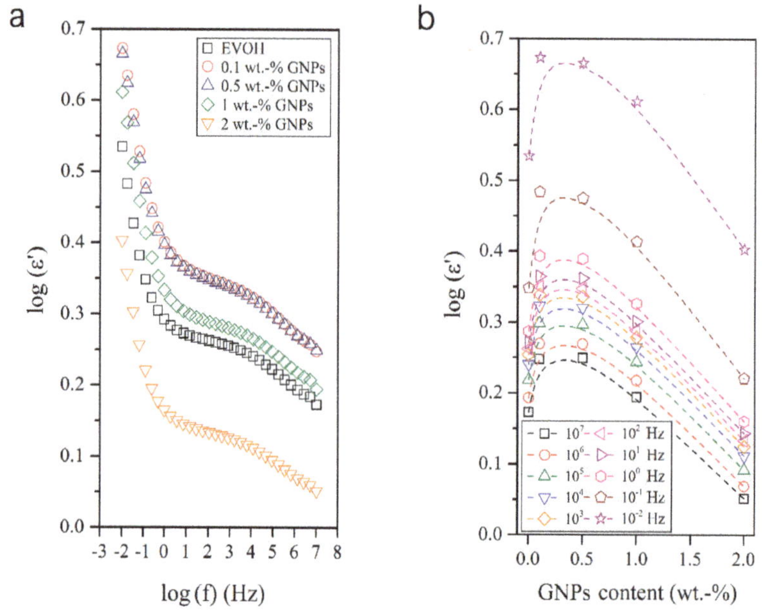

Figure 9. (**a**) Isothermal dielectric curves of the real dielectric permeability (ε') of the electrospun poly(ethylene-*co*-vinyl alcohol) (EVOH) fibers containing graphene nanoplatelets (GNPs); (**b**) Influence of the GNPs content on ε' for all frequency decades.

To ascertain the molecular origin of this effect, Figure 10 plots the variation of M" in the neat EVOH fibers and EVOH/GNPs nanocomposite fibers. When the conductive effects were minimized, two peaks could be recorded, that is, one prominent peak at low frequencies and a broader one at high frequencies. These peaks have been described by other authors as the α- and β-relaxations, respectively [37]. On the one hand, the α-relaxation is ascribed to the main chain segmental motion and reflects a transition from the glassy to the rubbery state, related to the so-called T_g. This is a large-scale cooperative process determined mainly by intermolecular interactions. One can observe that the presence of the GNPs shifted the α-relaxation peak to higher frequencies, showing a maximum of frequency for a content of 0.5 wt.-% GNPs in EVOH. It means that, at this particular composition, the segmental molecular mobility of the main chain was reduced and, therefore, the T_g value increased. However, when the GNPs composition was higher than 0.5 wt.-%, the frequency of the peak decreased. This further confirms that the nanoparticles may begin to agglomerate at higher contents, as supported by the above-described information during the morphological and thermal analyses. This fact highly influences the performance and especially the electrical properties of polymer nanocomposites [38]. On the other hand, the β-relaxation was attributed to the local-mode relaxation in the crystalline regions of the copolymer and/or to the motion of the hydroxyl groups. The β-relaxation peak also showed a maximum frequency value at 0.5 wt.-% GNPs.

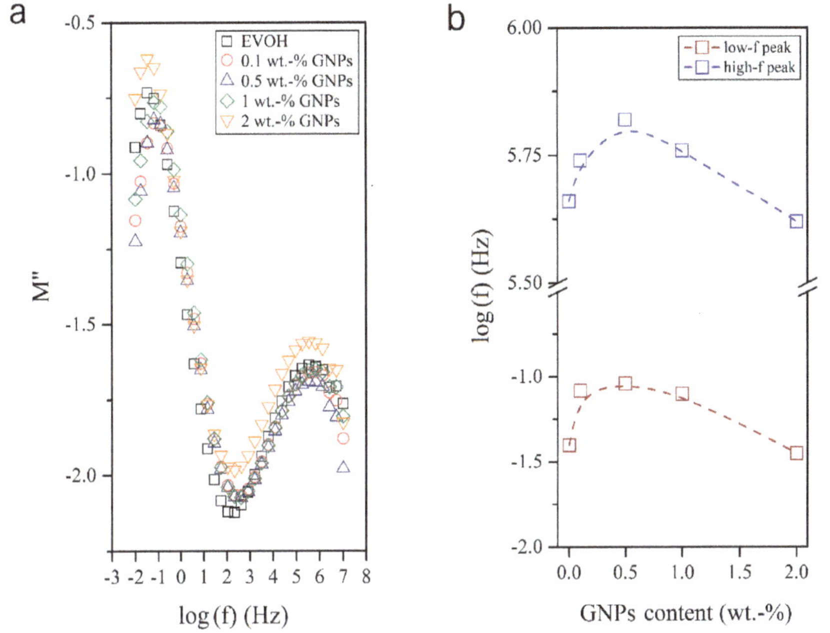

Figure 10. (**a**) Isothermal dielectric curves of the imaginary electric modulus (M") of the electrospun poly(ethylene-*co*-vinyl alcohol) (EVOH) fibers containing graphene nanoplatelets (GNPs); (**b**) Influence of the GNPs content on the frequency (f).

Figure 11a shows the evolution of σ in the 10^{-2}–10^7 Hz frequency range. This property is typical of electrode polarization, associated with the accumulation of charges at the interfaces between the electrodes and the polymer sample, which increases with increasing frequency. This motion of charged carriers is spatially limited within their potential wells [39–43]. At lower frequencies, plateau regions occur, which can be associated with the free-charge transfer in the rubbery state and they are related to σ_{dc}. However, at higher frequencies, the alternating current electrical conductivity (σ_{ac}) occurs. The transition from linear non-frequency dependent σ_{dc} to the frequency dependent range of the σ_{ac} regions corresponds to the change in the mechanism of electrical conduction, which can be described by the movement of charges at long distances [40,44]. In Figure 11b one can observe that the dependence of σ_{dc} of the nanocomposite fibers on the GNPs content was not linear. Table 2 shows the values of σ_{dc} as a function of the GNPs content. The incorporation of low nanofiller contents sharply increased the conductivity of the electrospun EVOH fibers, up to a content of 0.5 wt.-% GNPs, with a percolation threshold relatively close to 0.1 wt.-% GNPs. The conductivity behavior of the nanocomposite fibers then changed, at very low GNPs contents, from an electrical insulator to a semiconductor material. However, at the highest GNPs contents tested, the σ_{dc} values significantly decreased. As commented above, this phenomenon can be related to a low dispersion of the carbonaceous nanoparticles, for which no conducting clusters or bridges were formed inside the electrospun submicron EVOH fibers.

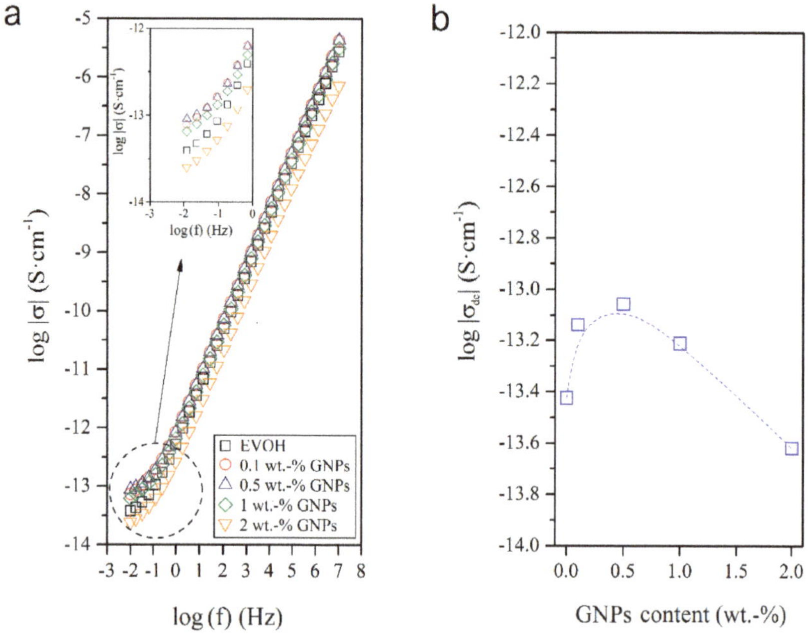

Figure 11. (**a**) Isothermal dielectric curves of the electrical conductivity (σ) of the electrospun poly(ethylene-*co*-vinyl alcohol) (EVOH) fibers containing graphene nanoplatelets (GNPs); (**b**) Influence of the GNPs content on the direct current electrical conductivity (σ_{dc}).

Table 2. Direct current electrical conductivity (σ_{dc}) of the electrospun poly(ethylene-*co*-vinyl alcohol) (EVOH) fibers as a function of the graphene nanoplatelets (GNPs) content.

GNPs Content (wt.-%)	σ_{dc} (S cm^{-1}) × 10^{14}
0.0	3.77
0.1	7.25
0.5	8.78
1.0	6.14
2.0	2.41

3.5. Electrospun EVOH/GNPs Films

Finally, Figure 12 shows the optical appearance of resultant EVOH/GNPs films obtained after thermal post-treatment at 158 °C, below the polymer's T_m, the so-called annealing, carried out on the electrospun fiber mats. Annealing applied on electrospun fibers mats results in continuous films that have significant potential for use in food packaging applications [45]. Simple naked eye examination of this image indicated that annealing produced continuous transparent films. This process has been recently ascribed to a compact packing rearrangement of the electrospun fibers by a phenomenon of fiber coalescence [46,47]. Another relevant observation is that, as the GNPs content increased, the resulting films became more opaque and developed a grey-like color, albeit contact transparency was preserved. This preliminary result indicates that the electrospun fibers mats can be turned into actual films, which will be the subject of further studies, since it may be advantageous for the application of developing labels or tags to have materials in a film format.

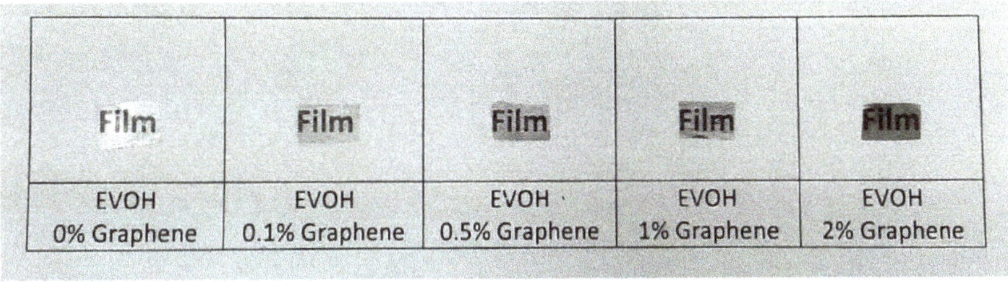

Figure 12. Electrospun films of poly(ethylene-*co*-vinyl alcohol) (EVOH) containing graphene nanoplatelets (GNPs) obtained by annealing.

4. Conclusions

EVOH fibers containing GNPs were successfully prepared by electrospinning. To this end, the GNPs were firstly synthetized from graphite powder by oxidation to GO according to the Hummer's method and, then, by reduction using NH_4OH. The resultant flat-like graphene nanoparticles, the so-called GNPs, were solution electrospun with EVOH to obtain nanocomposite fibers with sizes in the submicron range. The embedded GNPs were rolled up in the form of continuous layers along the EVOH fiber axis. Both morphological and Raman analyses revealed that the best GNPs dispersion was obtained for the electrospun EVOH fiber mats containing 0.5 wt.-% GNPs, while graphene is likely to increasingly agglomerate as its concentration was increased. Thermal characterization indicated that the incorporated GNPs acted as an anti-nucleant agent for the EVOH molecules, particularly at low contents due to its improved dispersion and high interaction with the polymer matrix. Finally, the dielectric behavior of the nanocomposite fibers was studied in the frequency 10^{-2}–10^{7} Hz range and as a function of the GNPs content. The dielectric constant was reduced with the frequency increase, in the whole range of frequencies, while it increased for GNPs contents up to 0.5 wt.-%, supporting the well dispersion of the nanoparticles at low loadings. In addition, the nanocomposite fibers presented high σ values in the 0.1–1 wt.-% GNPs range. The electrospun mat was, finally, thermally post-treated at 158 °C to produce continuous and contact transparent films. Applications of the resultant electrospun nanocomposite fiber mats and possibly also films in the intelligent packaging field as, for instance, smart labels or tags can be anticipated.

Author Contributions: Conceptualization was devised by J.M.L.; Methodology, Validation, and Formal Analysis was carried out by S.T.-G., Y.E., R.T.-J., J.D.B., A.R.-G., and J.M.L.; Investigation, Resources, Data Curation, and Writing-Original Draft Preparation was performed by S.T.-G. and J.D.B.; Writing-Review & Editing, S.T.-G.; Supervision, A.R.-G. and J.M.L.; Project Administration, J.M.L.

Funding: This research was funded by the European Regional Development Funds (ERDF) and Ministry of Science, Innovation, and Universities (MICIU) project numbers ENE2014-53734-C2-1-R, ENE2017-86711-C3-1-R, UPOV13-3E-1947, and AGL2015-63855-C2-1-R.

Acknowledgments: S.T.-G. acknowledges the MICIU for his Juan de la Cierva contract (IJCI-2016-29675). J.D.B. and R.T.-J. also thank Generalitat Valenciana (GV) for their projects APOSTD14/041 and GRISOLIA/2012/003, respectively.

Conflicts of Interest: The authors declare no conflict of interest.

References

1. Torres-Giner, S.; Gil, L.; Pascual-Ramírez, L.; Garde-Belza, J.A. Packaging: Food waste reduction. In *Encyclopedia of Polymer Applications*; Mishra, M., Ed.; CRC Press: Boca Raton, FL, USA, 2018.
2. Echegoyen, Y. Nano-developments for food packaging and labeling applications. In *Nanotechnologies in Food and Agriculture*; Rai, M., Ribeiro, C., Mattoso, L., Duran, N., Eds.; Springer International Publishing: Basel, Switzerland, 2015; pp. 141–166.
3. Novoselov, K.S.; Geim, A.K.; Morozov, S.V.; Jiang, D.; Zhang, Y.; Dubonos, S.V.; Grigorieva, I.V.; Firsov, A.A. Electric field effect in atomically thin carbon films. *Science* **2004**, *306*, 666–669. [CrossRef] [PubMed]

4. Zhu, N.; Liu, W.; Xue, M.; Xie, Z.; Zhao, D.; Zhang, M.; Chen, J.; Cao, T. Graphene as a conductive additive to enhance the high-rate capabilities of electrospun li4ti5o12 for lithium-ion batteries. *Electrochim. Acta* **2010**, *55*, 5813–5818. [CrossRef]
5. Balandin, A.A.; Ghosh, S.; Bao, W.; Calizo, I.; Teweldebrhan, D.; Miao, F.; Lau, C.N. Superior thermal conductivity of single-layer graphene. *Nano Lett.* **2008**, *8*, 902–907. [CrossRef] [PubMed]
6. Torres-Giner, S. Preparation of conductive carbon black-filled polymer nanocomposites *via* melt compounding. In *Conductive Materials and Composites*; Mitchell, V., Ed.; Nova Science Publishers, Inc.: New York, NY, USA, 2016; pp. 117–164.
7. Stankovich, S.; Dikin, D.A.; Dommett, G.H.B.; Kohlhaas, K.M.; Zimney, E.J.; Stach, E.A.; Piner, R.D.; Nguyen, S.T.; Ruoff, R.S. Graphene-based composite materials. *Nature* **2006**, *442*, 282. [CrossRef] [PubMed]
8. Torres-Giner, S.; Pérez-Masiá, R.; Lagaron, J.M. A review on electrospun polymer nanostructures as advanced bioactive platforms. *Polym. Eng. Sci.* **2016**, *56*, 500–527. [CrossRef]
9. Torres-Giner, S. Electrospun nanofibers for food packaging applications. In *Multifunctional and Nanoreinforced Polymers for Food Packaging*; Lagaron, J.M., Ed.; Woodhead Publishing Ltd.: Cambridge, UK, 2011; pp. 108–125.
10. Torres-Giner, S.; Busolo, M.; Cherpinski, A.; Lagaron, J.M. Electrospinning in the packaging industry. In *Electrospinning: From Basic Research to Commercialization*; Kny, E., Ghosal, K., Thomas, S., Eds.; The Royal Society of Chemistry: Cambridge, UK, 2018; pp. 238–260.
11. Torres-Giner, S.; Lagaron, J.M. Zein-based ultrathin fibers containing ceramic nanofillers obtained by electrospinning. I. Morphology and thermal properties. *J. Appl. Polym. Sci.* **2010**, *118*, 778–789. [CrossRef]
12. Torres-Giner, S.; Martinez-Abad, A.; Lagaron, J.M. Zein-based ultrathin fibers containing ceramic nanofillers obtained by electrospinning. II. Mechanical properties, gas barrier, and sustained release capacity of biocide thymol in multilayer polylactide films. *J. Appl. Polym. Sci.* **2014**, *131*, 9270–9276. [CrossRef]
13. Cherpinski, A.; Gozutok, M.; Sasmazel, H.T.; Torres-Giner, S.; Lagaron, J.M. Electrospun oxygen scavenging films of poly(3-hydroxybutyrate) containing palladium nanoparticles for active packaging applications. *Nanomaterials* **2018**, *8*, 469. [CrossRef] [PubMed]
14. Asmatulu, R.; Ceylan, M.; Nuraje, N. Study of superhydrophobic electrospun nanocomposite fibers for energy systems. *Langmuir* **2011**, *27*, 504–507. [CrossRef] [PubMed]
15. Bao, Q.; Zhang, H.; Yang, J.; Wang, S.; Tang, D.Y.; Jose, R.; Ramakrishna, S.; Lim, C.T.; Loh, K.P. Graphene–polymer nanofiber membrane for ultrafast photonics. *Adv. Funct. Mater.* **2010**, *20*, 782–791. [CrossRef]
16. Li, X.; Yang, Y.; Zhao, Y.; Lou, J.; Zhao, X.; Wang, R.; Liang, Q.; Huang, Z. Electrospinning fabrication and in situ mechanical investigation of individual graphene nanoribbon reinforced carbon nanofiber. *Carbon* **2017**, *114*, 717–723. [CrossRef]
17. Moayeri, A.; Ajji, A. Core-shell structured graphene filled polyaniline/poly(methyl methacrylate) nanofibers by coaxial electrospinning. *Nanosci. Nanotechnol. Lett.* **2016**, *8*, 129–134. [CrossRef]
18. Moayeri, A.; Ajji, A. High capacitance carbon nanofibers from poly(acrylonitrile) and poly(vinylpyrrolidone)-functionalized graphene by electrospinning. *J. Nanosci. Nanotechnol.* **2017**, *17*, 1820–1829. [CrossRef]
19. Hirata, M.; Gotou, T.; Horiuchi, S.; Fujiwara, M.; Ohba, M. Thin-film particles of graphite oxide 1: High-yield synthesis and flexibility of the particles. *Carbon* **2004**, *42*, 2929–2937. [CrossRef]
20. Stankovich, S.; Dikin, D.A.; Piner, R.D.; Kohlhaas, K.A.; Kleinhammes, A.; Jia, Y.; Wu, Y.; Nguyen, S.T.; Ruoff, R.S. Synthesis of graphene-based nanosheets via chemical reduction of exfoliated graphite oxide. *Carbon* **2007**, *45*, 1558–1565. [CrossRef]
21. van Turnhout, J.; Wübbenhorst, M. Analysis of complex dielectric spectra. Ii: Evaluation of the activation energy landscape by differential sampling. *J. Non-Cryst. Solids* **2002**, *305*, 50–58. [CrossRef]
22. Macedo, P.B.; Moynihan, C.T.; Bose, R. Role of ionic diffusion in polarization in vitreous ionic conductors. *Phys. Chem. Glasses* **1972**, *13*, 171–179.
23. Dyre, J.C. Some remarks on ac conduction in disordered solids. *J. Non-Cryst. Solids* **1991**, *135*, 219–226. [CrossRef]
24. Yasmin, A.; Daniel, I.M. Mechanical and thermal properties of graphite platelet/epoxy composites. *Polymer* **2004**, *45*, 8211–8219. [CrossRef]
25. Simon, D.A.; Bischoff, E.; Buonocore, G.G.; Cerruti, P.; Raucci, M.G.; Xia, H.; Schrekker, H.S.; Lavorgna, M.; Ambrosio, L.; Mauler, R.S. Graphene-based masterbatch obtained via modified polyvinyl alcohol liquid-shear exfoliation and its application in enhanced polymer composites. *Mater. Des.* **2017**, *134*, 103–110. [CrossRef]

26. Martínez-Sanz, M.; Olsson, R.T.; Lopez-Rubio, A.; Lagaron, J.M. Development of electrospun evoh fibres reinforced with bacterial cellulose nanowhiskers. Part i: Characterization and method optimization. *Cellulose* **2011**, *18*, 335–347. [CrossRef]
27. Sukumaran, S.S.; Rekha, C.R.; Resmi, A.N.; Jinesh, K.B.; Gopchandran, K.G. Raman and scanning tunneling spectroscopic investigations on graphene-silver nanocomposites. *J. Sci. Adv. Mater. Devices* **2018**. [CrossRef]
28. Lotya, M.; King, P.J.; Khan, U.; De, S.; Coleman, J.N. High-concentration, surfactant-stabilized graphene dispersions. *ACS Nano* **2010**, *4*, 3155–3162. [CrossRef] [PubMed]
29. Blanchard, A.; Gouanvé, F.; Espuche, E. Effect of humidity on mechanical, thermal and barrier properties of evoh films. *J. Membr. Sci.* **2017**, *540*, 1–9. [CrossRef]
30. Cowie, J.M.G.; Harris, S.; McEwen, I.J. Physical aging in poly(vinyl acetate). 2. Relative rates of volume and enthalpy relaxation. *Macromolecules* **1998**, *31*, 2611–2615. [CrossRef]
31. Hutchinson, J.M.; Smith, S.; Horne, B.; Gourlay, G.M. Physical aging of polycarbonate: Enthalpy relaxation, creep response, and yielding behavior. *Macromolecules* **1999**, *32*, 5046–5061. [CrossRef]
32. Dhawan, S.; Barbosa-Cànovas, G.V.; Tang, J.; Sablani, S.S. Oxygen barrier and enthalpy of melting of multilayer evoh films after pressure-assisted thermal processing and during storage. *J. Appl. Polym. Sci.* **2011**, *122*, 1538–1545. [CrossRef]
33. Martínez-Felipe, A.; Santonja-Blasco, L.; Badia, J.D.; Imrie, C.T.; Ribes-Greus, A. Characterization of functionalized side-chain liquid crystal methacrylates containing nonmesogenic units by dielectric spectroscopy. *Ind. Eng. Chem. Res.* **2013**, *52*, 8722–8731. [CrossRef]
34. Badia, J.D.; Reig-Rodrigo, P.; Teruel-Juanes, R.; Kittikorn, T.; Strömberg, E.; Ek, M.; Karlsson, S.; Ribes-Greus, A. Effect of sisal and hydrothermal ageing on the dielectric behaviour of polylactide/sisal biocomposites. *Compos. Sci. Technol.* **2017**, *149*, 1–10. [CrossRef]
35. Salaberría, A.M.; Teruel-Juanes, R.; Badia, J.D.; Fernandes, S.C.M.; de Juano-Arbona, V.S.; Labidi, J.; Ribes-Greus, A. Influence of chitin nanocrystals on the dielectric behaviour and conductivity of chitosan-based bionanocomposites. *Compos. Sci. Technol.* **2018**, *167*, 323–330. [CrossRef]
36. Badia, J.D.; Monreal, L.; de Juano-Arbona, V.S.; Ribes-Greus, A. Dielectric spectroscopy of recycled polylactide. *Polym. Degrad. Stab.* **2014**, *107*, 21–27. [CrossRef]
37. Bizet, A.; Nakamura, N.; Teramoto, Y.; Hatakeyama, T. The influence of moisture on the dielectric properties of poly(ethylene-*co*-vinyl alcohol). *Thermochim. Acta* **1994**, *241*, 191–198. [CrossRef]
38. Torres-Giner, S.; Chiva-Flor, A.; Feijoo, J.L. Injection-molded parts of polypropylene/multi-wall carbon nanotubes composites with an electrically conductive tridimensional network. *Polym. Compos.* **2016**, *37*, 488–496. [CrossRef]
39. Macdonald, J.R. Impedance spectroscopy. *Ann. Biomed. Eng.* **1992**, *20*, 289–305. [CrossRef] [PubMed]
40. Pissis, P.; Kyritsis, A. Electrical conductivity studies in hydrogels. *Solid State Ionics* **1997**, *97*, 105–113. [CrossRef]
41. Raistrick, I.D.; Franceschetti, D.R.; Macdonald, J.R. Theory. In *Impedance Spectroscopy*; John Wiley & Sons, Inc.: Hoboken, NJ, USA, 2005; pp. 27–128.
42. Miyairi, K. Low-frequency dielectric response of polyethylene terephthalate (pet) films. *J. Phys. D Appl. Phys.* **1986**, *19*, 1973–1980. [CrossRef]
43. Neagu, E.; Pissis, P.; Apekis, L. Electrical conductivity effects in polyethylene terephthalate films. *J. Appl. Phys.* **2000**, *87*, 2914–2922. [CrossRef]
44. Yamamoto, K.; Namikawa, H. Conduction current relaxation of inhomogeneous conductor I. *Jpn. J. Appl. Phys.* **1988**, *27*, 1845–1851. [CrossRef]
45. Lasprilla-Botero, J.; Torres-Giner, S.; Pardo-Figuerez, M.; Álvarez-Láinez, M.; Lagaron, J.M. Superhydrophobic bilayer coating based on annealed electrospun ultrathin poly(ε-caprolactone) fibers and electrosprayed nanostructured silica microparticles for easy emptying packaging applications. *Coatings* **2018**, *8*, 173. [CrossRef]

46. Cherpinski, A.; Torres-Giner, S.; Cabedo, L.; Lagaron, J.M. Post-processing optimization of electrospun submicron poly(3-hydroxybutyrate) fibers to obtain continuous films of interest in food packaging applications. *Food Addit. Contam.* **2017**, *34*, 1817–1830. [CrossRef] [PubMed]
47. Melendez-Rodriguez, B.; Castro-Mayorga, J.L.; Reis, M.A.M.; Sammon, C.; Cabedo, L.; Torres-Giner, S.; Lagaron, J.M. Preparation and characterization of electrospun food biopackaging films of poly(3-hydroxybutyrate-*co*-3-hydroxyvalerate) derived from fruit pulp biowaste. *Front. Sustain. Food Syst.* **2018**, *2*, 38. [CrossRef]

nanomaterials

MDPI

Article

Exploring Protein-Inorganic Hybrid Nanoflowers and Immune Magnetic Nanobeads to Detect *Salmonella* Typhimurium

Lei Wang [1], Xiaoting Huo [1], Ruya Guo [2], Qiang Zhang [3] and Jianhan Lin [1,*]

1 Key Laboratory of Agricultural Information Acquisition Technology, Ministry of Agriculture, China Agricultural University, Beijing 100083, China; wanglei123@cau.edu.cn (L.W.); huoxiaoting@cau.edu.cn (X.H.)

2 Key Laboratory of Modern Precision Agriculture System Integration Research, Ministry of Education, China Agricultural University, Beijing 100083, China; guoya@cau.edu.cn

3 Department of Biosystems Engineering, University of Manitoba, Winnipeg, MB R3T 2N2, Canada; Qiang.Zhang@umanitoba.ca

* Correspondence: jianhan@cau.edu.cn; Tel.: +86-10-6273-7599

Received: 30 October 2018; Accepted: 30 November 2018; Published: 4 December 2018

Abstract: Early screening of pathogenic bacteria is key to preventing and controlling outbreaks of foodborne diseases. In this study, protein-inorganic hybrid nanoflowers were synthesized for signal amplification and used with a calcium ion selective electrode (Ca-ISE) to establish a new enzyme-free assay for rapid and sensitive detection of *Salmonella*. Calcium hydrophosphate crystals were first conjugated with polyclonal antibodies against *Salmonella* to synthesize immune calcium nanoflowers (CaNFs), and streptavidin modified magnetic nanobeads (MNBs) were conjugated with biotinylated monoclonal antibodies against *Salmonella* to form immune MNBs. After target bacteria were separated using immune MNBs to form magnetic bacteria, immune CaNFs were conjugated with magnetic bacteria to form nanoflower conjugated bacteria. Then, hydrogen chloride was used to release calcium ions from nanoflower conjugated bacteria. After magnetic separation, the supernatant was finally injected as a continuous-flow to fluidic chip with Ca-ISE for specific detection of calcium ions. The supernatant's potential had a good linear relationship with bacteria concentration, and this assay was able to detect the *S.* Typhimurium cells as low as 28 colony forming units/mL within two hours. The mean recovery of target bacteria in spiked chicken samples was 95.0%. This proposed assay shows the potential for rapid, sensitive, and on-line detection of foodborne pathogens.

Keywords: protein-inorganic hybrid nanoflower; continuous-flow potentiometric detection; ion selective electrode; *Salmonella*

1. Introduction

Foodborne pathogens have become a major public health issue, increasingly attracting concern around the world. The WHO reported that almost 10% of the world population falls ill annually due to the consumption of foods and water contaminated by pathogenic microorganisms [1]. *Salmonella* is one of the main foodborne pathogens, which passes through the entire food supply chains from livestock farming to food processing, then to food logistics, and finally to retail. The clinical manifestations of *Salmonella* poisoning mainly include acute onset of fever, abdominal pain, diarrhea, nausea, and sometimes vomiting. Therefore, early screening of *Salmonella* and other common foodborne pathogens is vital to ensure food safety in food supply chains.

Ion selective electrode (ISE) is a type of electrochemical sensor that converts the activity of specific ions in a solution into the potential to determine the ion concentration [2,3]. Various ISEs have been developed and widely used in environmental monitoring [4–6], biochemical analysis [7–10], food safety [11,12], and other applications [13,14] due to their unique advantages, such as low cost, simple operation, small size, rapid response, and good selectivity. In addition, ISEs do not require external electrical excitement during electrochemical measurements; thus they are the most straightforward tool compared to other electrochemical methods [15,16]. ISEs are generally used to measure the specific ions in a static solution, but seldom under continuous-flow conditions. Recently, dynamic potentiometry, which was first introduced by Calvo et al. [17], has gained popularity. Some studies on the dynamic behaviors of ISEs were reported in developing new kinetic potentiometric methods [18]. Cuartero et al. reported a dynamical potentiometric ISE using 18-crown-6 as ionophore for continuous-flow detection of multiple ions in binary mixtures [19]. Thus, combining ISE-based potentiometric methods with the immune methods is showing promise for the development of novel assays using some specific ions as a signal.

In recent years, fast development of various nanomaterials has opened up new possibilities for biosensing signal amplification. Among them, protein-inorganic hybrid nanoflowers have been demonstrated to be able to maintain or even increase the activity of the proteins and effectively amplify the detection signals [20–22]. Ge et al. first proposed the coprecipitation method to synthesize hybrid organic-inorganic nanoflowers using copper ions as an inorganic component and proteins as an organic component. They successfully used laccase as the model protein to create the nanoflowers for detection of epinephrine [23]. Ye et al. developed Con A-invertase-$CaHPO_4$ hybrid nanoflowers to enhance the activity of the immobilized invertase and used them with a personal glucose meter for sensitive detection of *Escherichia coli* O157:H7 [24]. Ismail et al. reported the synthesis of hybrid nanoflowers using iron ions (Fe^{2+}) and horseradish peroxidase (HRP) and demonstrated that they increased the catalytic activity of HRP more than five-fold [25]. Therefore, the introduction of nanoflowers might further improve the sensitivity of the potentiometric assays.

In this study, we explore a new enzyme-free assay to detect *S.* Typhimurium using protein-inorganic hybrid nanoflowers for effective amplification of biological signals, magnetic nanobeads for immunomagnetic separation of target bacteria, and fluidic chip with the calcium ion selective electrode for continuous-flow detection of calcium ions. The objectives of this study were to: (1) develop a new enzyme-free, potentiometric assay for rapid and sensitive detection of *S.* Typhimurium, and (2) experimentally evaluate the sensitivity and applicability of the assay. As shown in Scheme 1, prior to testing, the streptavidin-modified magnetic nanobeads were conjugated with the biotinylated monoclonal antibodies against *Salmonella* to form the immune magnetic nanobeads through streptavidin-biotin binding, and calcium hydrophosphate crystals were conjugated with the polyclonal antibodies against *Salmonella* to synthesize the immune calcium nanoflowers through coprecipitation method. First, the immune magnetic nanobeads were used to separate the target bacteria from the sample to form the magnetic nanobeads-bacteria complexes (magnetic bacteria) that were concentrated in a smaller volume of phosphate buffered saline (PBS). Then, the immune calcium nanoflowers were conjugated with the magnetic bacteria to form the magnetic nanobeads-bacteria-nanoflower complexes (nanoflower conjugated bacteria). Successively, hydrogen chloride was used to release calcium ions from nanoflower conjugated bacteria. After magnetic separation, the supernatant was finally injected to the fluidic chip for continuous-flow detection of calcium ions to determine the concentration of the target bacteria.

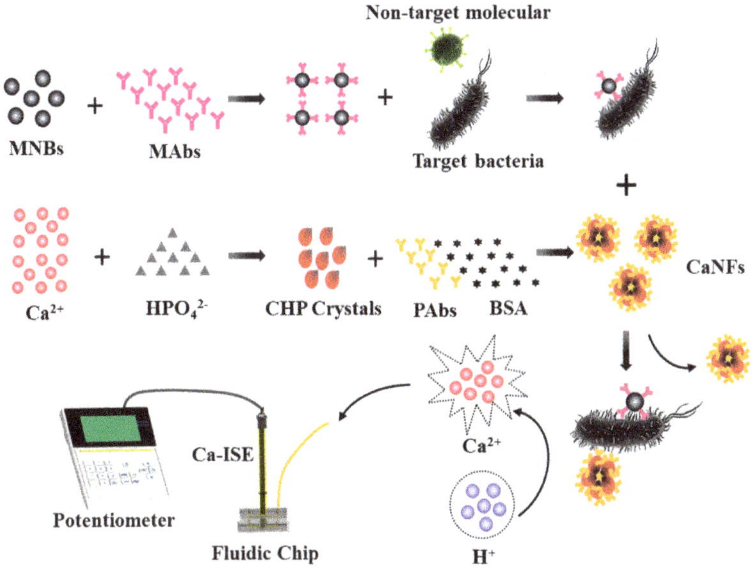

Scheme 1. Schematic of the proposed assay for sensitive detection of *Salmonella* Typhimurium.

2. Materials and Methods

2.1. Materials

The anti-*Salmonella* monoclonal antibodies (concentration: 1 mg/mL) from Abcam (Cambridge, MA, USA) and the anti-*Salmonella* polyclonal antibodies (concentration: 2.5 mg/mL) from Meridian (Memphis, TN, USA) were used for specific conjugation with the target *S.* Typhimurium cells. The long-arm biotin labeling kit from Elabscience Biotechnology (Wuhan, China) was used for the modification of biotin onto the monoclonal antibodies (MAbs). The streptavidin-modified magnetic nanobeads from Ocean Nanotech (SV0152, Fe content: 1 mg/mL, San Diego, CA, USA) were used with the biotinylated MAbs for immunomagnetic separation of the target bacteria. Calcium chloride from XiLong Scientific (Shantou, China) was used for the synthesis of the protein-inorganic hybrid nanoflowers. Phosphate buffered saline from Sigma Aldrich (10 times concentrated, St. Louis, MO, USA) was diluted with the deionized water to prepare the PBS solution. Bovine serum albumin (BSA) also from Sigma Aldrich (St. Louis, MO, USA) was used for blocking (1% and 10%, *w*/*v*). Tween 20 from Amresco (Solon, OH, USA) was used for washing. A silicone elastomer kit from Dow Corning (Sylgard 184, Auburn, MI, USA) was used for fabricating the poly(dimethoxy)silane (PDMS) channels. The printing material (Vero Whiteplus RGD835, Stratasys, Eden Prairie, MN, USA) was used with the Objet24 three-dimensional (3D) printer for fabricating the mold of the PDMS channels. The deionized water was produced by Millipore Advantage 10 (18.2 MΩ·cm, Billerica, MA, USA) and used to prepare all the solutions.

2.2. Synthesis of Protein-Inorganic Nanoflowers for Signal Amplification

The synthesis of protein-inorganic hybrid nanoflowers was based on the previously reported coprecipitation method with modifications [23,26]. First, 0.02 mg polyclonal antibodies (PAbs) against *Salmonella* and 0.18 mg BSA were simultaneously added into 1 mL PBS (3 mM). Then, 20 μL calcium chloride (200 mM) was added. After incubation at 15 rpm for 12 h, the immune calcium nanoflowers (CaNFs) were formed. To remove the surplus calcium ions and PAbs, the immune CaNFs were centrifuged at 15,000 rpm for 5 min, resuspended in the PBS (3 mM), and shaken evenly for 1 min to wash the immune CaNFs. The washing step was repeated 3 times to ensure that the free calcium ions and antibodies were washed away thoroughly. Finally, the immune CaNFs were resuspended in PBS and stored at 4 °C for further use.

2.3. Fabrication of Fluidic Chip for Potentiometric Detection

Fluidic chip with a calcium ion selective electrode (Ca-ISE) was developed for potentiometric detection of calcium ions in a continuous-flow condition. As shown in Figure S1, the fluidic chip consisted of Ca-ISE (9720BNWP, Thermo Fisher, Waltham, MA, USA) for potentiometric detection of calcium ions; a fluidic channel 30 mm long, 2 mm wide, and 2 mm tall for continuous-flow transportation of the supernatant; and a detection chamber with a 4 mm diameter for tight housing of the Ca-ISE. The 3D structural mold of the fluidic channel and the detection chamber were first designed using Solidworks (Dassault Systèmes Solidworks Corp., Waltham, MA, USA) and fabricated using the Objet24 3D printer. Then, the silicone elastomer base and curing agent were uniformly mixed at the ratio of 10:1 to cast the PDMS channel, which was bonded with the glass slide to fabricate the fluidic chip. Finally, the Ca-ISE was tightly inserted into the detection chamber, and the membrane of the Ca-ISE remained at the same level with the top of the channel for sensitive measurement and efficient washing.

2.4. Preparation of the Bacterial Cultures

Salmonella Typhimurium (ATCC14028) was used as the target bacteria, and *E. coli* O157:H7 (ATCC43888) and *Listeria monocytogenes* (ATCC13932) were used as the non-target bacteria. They were cultured in the Luria-Bertani (LB) medium (Aoboxing Biotech, Beijing, China) overnight at 37 °C with shaking at 180 rpm. The bacteria were 10-fold diluted with sterile PBS to obtain concentrations from 10^2 to 10^6 colony forming units (CFU)/mL. For bacteria enumeration, the bacteria were serially diluted with the sterile PBS and were grown on the LB plates. After incubation at 37 °C for 24 h, the visible colonies were counted to determine the concentration of the bacteria.

2.5. Bacteria Separation and Detection

Prior to testing, the anti-*Salmonella* monoclonal antibodies (MAbs) were modified with biotin through membrane dialysis using the long-arm biotin labeling kit according to the manufacturer's protocol to obtain the biotinylated MAbs. For magnetic separation and concentration of the target bacteria, 20 µL of the streptavidin modified magnetic nanobeads (MNBs) were first washed with 500 µL of PBST (PBS containing 0.05% Tween 20). Transmission electron microscopy (TEM) imaging was conducted to characterize the MNBs (Figure 1a). Then, 4 µL of the biotinylated MAbs were added to conjugate with the MNBs at 15 rpm for 45 min. After washing with PBST to remove the excessive antibodies, the immune MNBs were obtained. Finally, different concentrations of 500 µL of the target bacteria were added to resuspend the immune MNBs, and then incubated at 15 rpm for 45 min to form the magnetic bacteria samples of different concentrations. TEM imaging was conducted to characterize the magnetic bacteria (Figure 1b).

For detection of the magnetic bacteria, 100 µL of the immune CaNFs were added and incubated with the magnetic bacteria for 45 min, allowing the formation of the nanoflower conjugated bacteria. After the nanoflower conjugated bacteria were transferred to a new tube and washed with PBST 3 times, then 500 µL hydrogen chloride was added into the nanoflower conjugated bacteria, incubated at 15 rpm for 5 min and magnetically separated for 3 min to obtain the supernatant containing calcium ions through substitutional reaction. Finally, the supernatant was continuously injected using the syringe pump (Pump 11 Elite, Harvard Apparatus, Holliston, MA, USA) into the fluidic chip at a flow rate of 250 µL/min. The potential of the Ca-ISE was measured on-line using a potentiometer (M555P, Pinnacle, Corning, NY, USA) and recorded every 15 s.

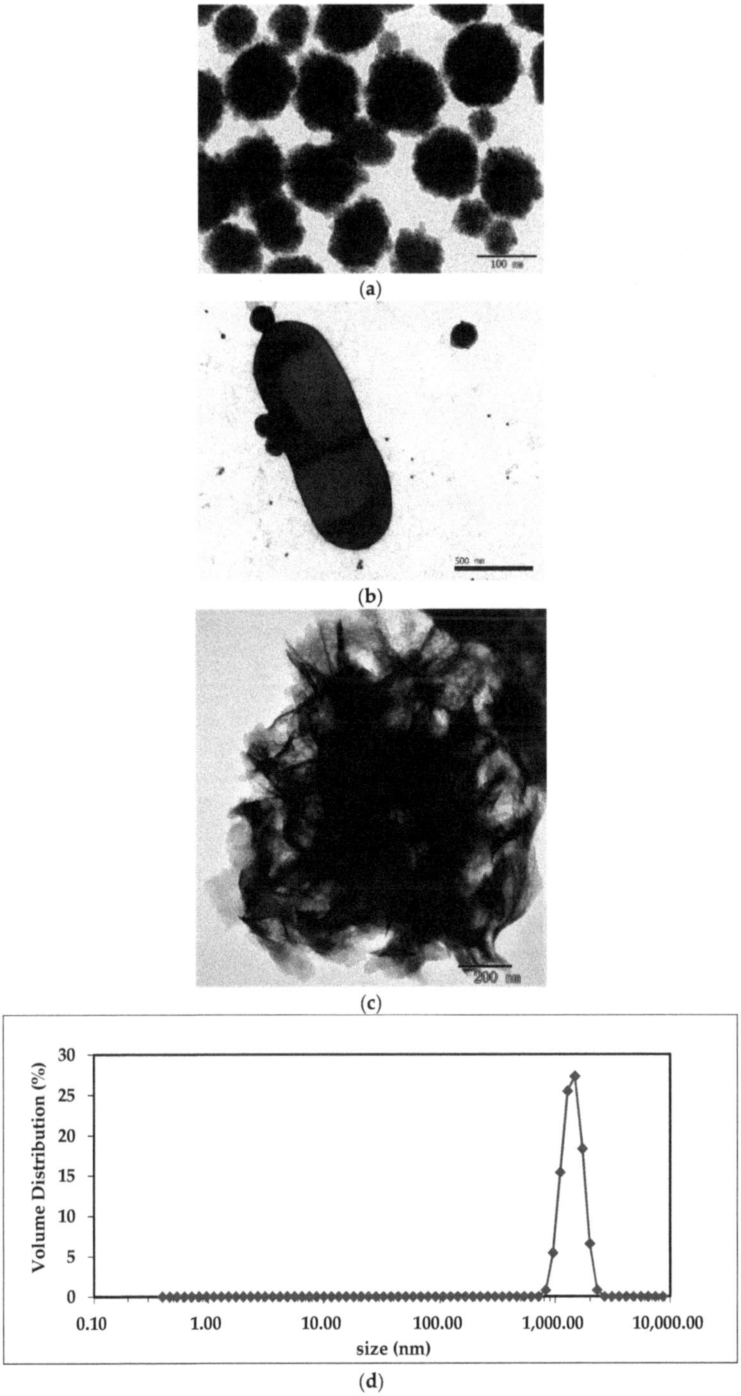

Figure 1. (**a**) TEM image of the immune magnetic nanobeads; (**b**) TEM image of the magnetic bacteria; (**c**) TEM image of the immune CaNFs; (**d**) Dynamic light scatting result of the immune CaNFs.

2.6. Bacteria Detection in Chicken Carcass

Prior to testing, each chicken carcass purchased from a local supermarket was placed in a sterile plastic bag containing 250 mL of PBS (pH 7.4, 10 mM), followed by vigorous shaking for 1 min to rinse the chicken carcass and standing for 10 min to obtain the supernatant. First, 1 mL of different concentrations of the target bacteria were added into 9 mL supernatant to prepare the spiked chicken samples with the bacterial concentrations from 10^2 to 10^6 CFU/mL. Then, the immune MNBs were used to magnetically separate the target bacteria from the spiked samples with different concentrations, respectively. After the magnetic bacteria were washed with PBST to avoid non-specific reaction, the immune CaNFs were added and incubated to form the nanoflower conjugated bacteria, and hydrogen chloride was used to resuspend the nanoflower conjugated bacteria to release calcium ions. Finally, the supernatant was obtained by magnetic separation and injected into the fluidic chip for potential measurement in a continuous-flow condition to determine the concentration of the target bacteria.

3. Results and Discussion

3.1. Characterization of Calcium Nanoflowers

The synthesis of the calcium nanoflowers based on the facile one-step coprecipitation method is the key to the development of the proposed assay. The primary crystals of calcium hydrophosphate were first formed through chemical reaction between calcium ions and hydrophosphate ions. After the proteins (BSA and PAbs) were added, the petals containing calcium ions were then formed through the coordination of amide groups on the protein backbone [23]. Finally, the immune calcium nanoflowers were completely formed by anisotropic growth, i.e., the proteins induced the nucleation of the calcium hydrophosphate (CHP) crystals to form the scaffold of the nanoflowers and acted as the cement to conjugate the petals. As observed from the TEM images (Figure 1c), the synthesized CaNFs had a diameter of ~1.50 μm. The dynamic light scatting technique was used to further characterize the diameter of the nanoflowers, and the result shown in Figure 1d indicates that the diameter of the nanoflowers was ~1.42 μm, which was consistent with the result of TEM imaging.

To further investigate the amount of calcium ions in the CaNFs, different concentrations of $CaCl_2$ ranging from 10 μM to 500 μM were measured using the flame photometer (Model 410, Sherwood Scientific, Cambridge, UK) and a linear calibration curve was developed (Figure S2). The calcium concentration of the synthesized CaNFs was measured using the flame photometer and calculated to be ~1.41 mM.

3.2. Evaluation on Ca-ISE for Continuous-Flow Detection of Calcium Ions

In our previous studies, interdigitated array microelectrodes [27–29], a screen-printed interdigitated electrode [30], and a printed circuit board electrode [31] were used as transducers for the development of electrochemical impedance biosensors. However, these electrodes could not distinguish the specific ions and often suffered from serious interference from both the sample background and the buffer solution, leading to low signal-to-noise ratios and sensitivity. The introduction of ISE with better selectivity to specific ions might improve the robustness and sensitivity of our previous electrochemical biosensors. Thus, the Ca-ISE was employed for specific detection of the signal (calcium ions) in this study.

Since this study aimed to develop a potentiometric assay for on-line detection of *S.* Typhimurium using calcium ions as a signal, it was essential to verify the feasibility of Ca-ISE for continuous-flow detection of calcium ions, which has traditionally been used in static conditions. Deionized water was first injected into the fluidic chip at the flow rate of 250 μL/min for cleaning and the measurements were used as the control. Then, different concentrations (10–500 μM) of $CaCl_2$ were injected into the chip from low concentration to high concentration at the same flow rate and the potentials were recorded every 15 s after continuous-flow injection for one minute. After each concentration of $CaCl_2$ was measured, the chip was thoroughly washed with deionized water for three minutes until the potential returned to the original level (the potential of the deionized water). As shown in Figure 2a, when the concentration of $CaCl_2$ changed from 10 μM to 500 μM, the potential measured in a continuous-flow condition increased from −73.7 mV to −28.2 mV. The potential remained fairly stable when the $CaCl_2$ solution was continuously injected, and returned to the original level after continuous-flow washing. As shown in Figure 2b, the average potential (E) of the Ca-ISE was found to have a good linear relationship with the concentration (C) of calcium ions, which could be expressed by $E = 11.89\ln(C) - 102.98$ ($R^2 = 0.98$). The successful implementation of continuous-flow detection of calcium ions enabled the integration of the Ca-ISE to develop a lab-on-a-chip assay for on-line monitoring of foodborne bacteria.

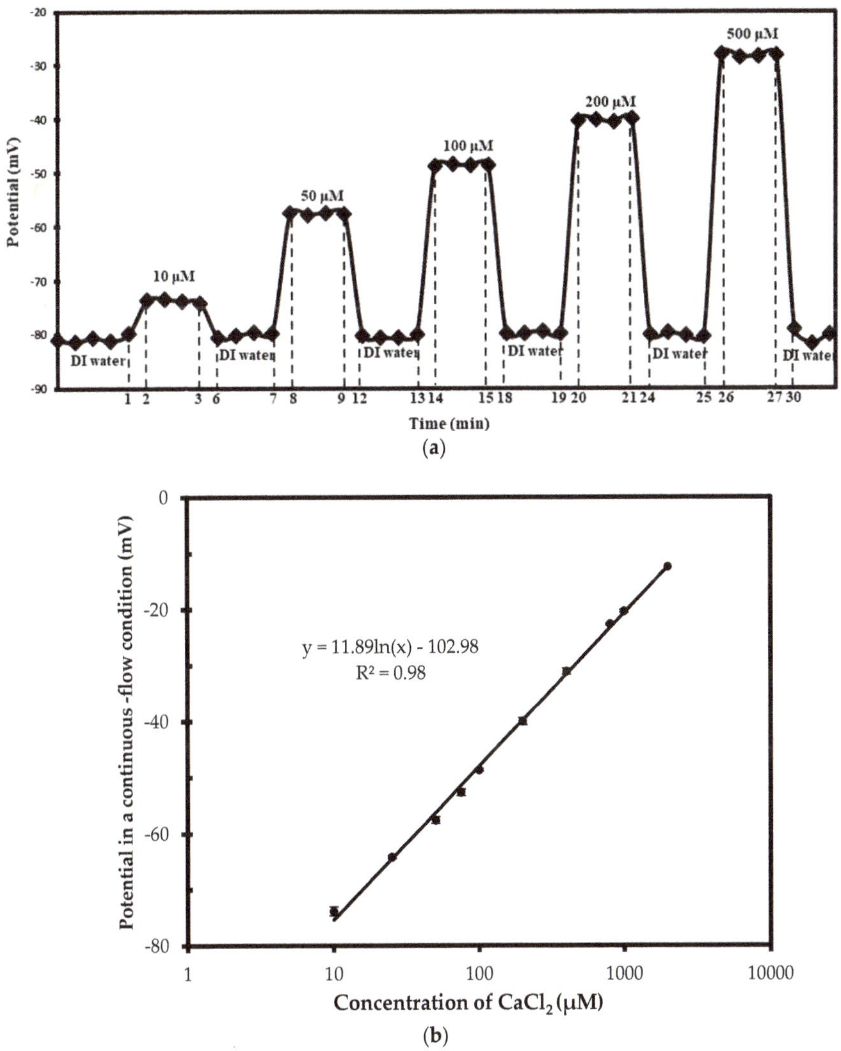

Figure 2. (**a**) The potential measured in a continuous-flow condition for different concentrations of calcium chloride (10 μm–500 μM); (**b**) The potential of the Ca-ISE for calcium chloride at the concentrations from 10 μM to 2 mM ($N = 3$).

3.3. Optimization of Proposed Assay

The release of calcium ions from different concentrations of the nanoflower conjugated bacteria was the basis of this proposed assay. Therefore, it was necessary to investigate the impact of the washing time of the CaNFs on the background noise control. The immune CaNFs were centrifuged at 15,000 rpm for 5 min to obtain the washing solution, and resuspended in 1 mL PBS (3 mM). As shown in Figure 3a, the potential of the washing solution decreased from −16.60 mV to −66.97 mV after three washes and changed little after the third wash, indicating that most non-specific absorbed calcium ions were washed away after washing three times. To minimize the impact of the noise from the background, the CaNFs were washed three times prior to use in this study.

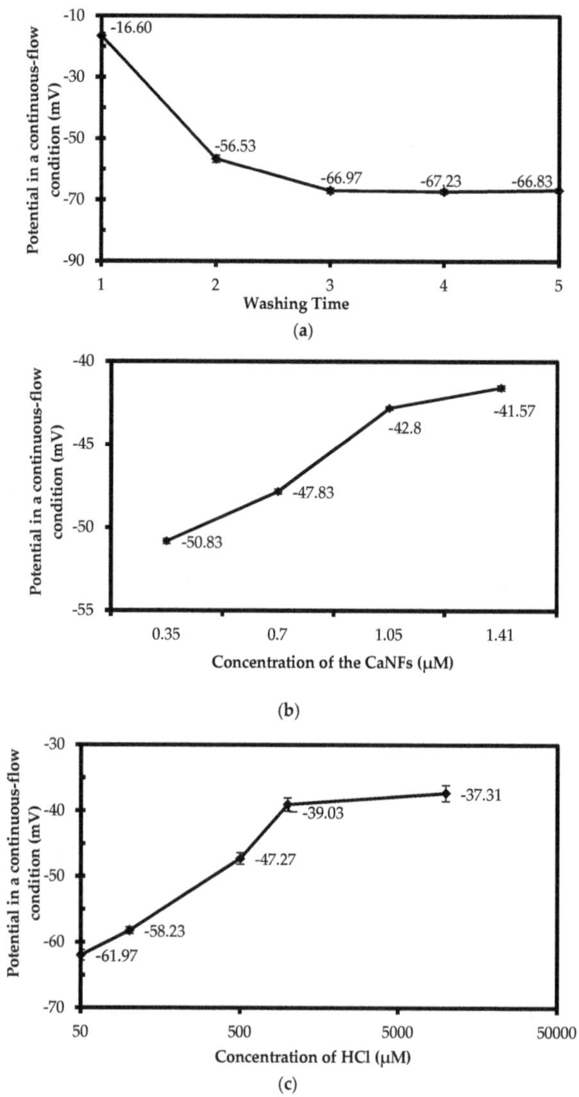

Figure 3. (**a**) Optimization of the washing time of the CaNFs ($N = 3$); (**b**) Optimization of the concentration of the CaNFs ($N = 3$); (**c**) Optimization of the concentration of hydrogen chloride ($N = 3$).

The amount of the CaNFs and the concentration of hydrogen chloride considerably impact the release of calcium ions from the nanoflower conjugated bacteria and are important to the sensitivity of the proposed assay. Different concentrations of the CaNFs (Ca^{2+} concentration: 0.35–1.41 mM, 100 μL) were used to conjugate with the *S.* Typhimurium cells at a concentration of 7.2×10^5 CFU/mL. The potential of the supernatant was measured in a continuous-flow condition using the fluidic chip. As shown in Figure 3b, when the concentration of CaNFs increased from 0.35 mM to 1.41 mM, the potential decreased from −50.83 mV to −41.57 mV, indicating that more CaNFs were conjugated with the target bacteria at higher concentrations of CaNFs. However, further increase in the concentration of the CaNFs resulted in little change (<3%) in the potential. Therefore, the optimal concentration of 1.41 mM for the CaNFs was used in this study.

Calcium ions on the nanoflower conjugated bacteria have to be released from the bacteria since the Ca-ISE only responds to free calcium ions. Thus, HCl was used in this study to replace calcium ions on the nanoflower conjugated bacteria with hydrogen ions. Different concentrations of HCl (50 μM–10 mM, 500 μL) were reacted with 100 μL CaNFs and the potential was dynamically measured by the fluidic chip. As shown in Figure 3c, the potential increased from −61.97 mV to −39.03 mV since more free calcium ions were replaced, while the concentration of HCl increased from 50 μM to 1 mM. However, further increase in the HCl concentration from 1 to 10 mM only resulted in a slight (4%) increase in the potential. Therefore, the optimal concentration of 1 mM for HCl was used in this study.

3.4. Detection of S. Typhimurium in Pure Cultures and Spiked Chicken Samples

To determine the unknown concentration of *S.* Typhimurium in a sample using this proposed assay, the calibration model between the potential and the concentration was established. Thus, three parallel tests on pure *S.* Typhimurium cells with different concentrations of 10^2–10^6 CFU/mL were conducted using this proposed assay. As shown in Figure 4a, a linear relationship between the potential (E) of the Ca-ISE in a continuous-flow condition and the concentration (C) of the *S.* Typhimurium cells was found, which could be expressed as $E = 2.33\ln(C) - 73.07$ ($R^2 = 0.98$). The low detection limit of this assay was determined to be 2.8×10^1 CFU/mL, according to three times of signal-to-noise ratio. The high sensitivity of this proposed assay could be attributed to three aspects: (1) effective amplification of the biological signals using the protein-inorganic hybrid nanoflowers with larger surface-to-volume ratio, resulting in more calcium ions on the nanoflower conjugated bacteria; (2) specific detection of calcium ions using the Ca-ISE, resulting in better control of the background noise; and (3) effective washing of the Ca-ISE due to continuous-flow and high-pressure flushing on the ion selective membrane, resulting in less interference from both the buffer solution and the sample background. TEM imaging confirmed the successful formation of the nanoflower conjugated bacteria, as shown in Figure 4b.

To further evaluate the practical applicability of the proposed assay, three parallel tests on different concentrations of the *S.* Typhimurium cells in the spiked chicken samples were conducted. As shown in Figure 4c, the potential of Ca-ISE for the spiked chicken samples containing the *S.* Typhimurium cells at different concentrations from 1.0×10^2 to 1.0×10^6 CFU/mL were slightly less than for the pure cultures at the same concentrations. This might be due to the interference from the background of the chicken samples, since a large number of proteins, fats, and other molecules in chicken might influence the separation of the target bacteria and the detection of calcium ions. The recovery of the target bacteria (R) was calculated as the ratio of the concentration of the target bacteria in the spiked chicken sample (Ns) to that of the bacteria in the pure culture (Nc), i.e., $R = Ns/Nc \times 100\%$. As shown in Table 1, the recoveries for different concentrations of the target bacteria ranged from 92.0% to 100.8%, with an average of 95.0%. This data clearly show that the proposed assay is adequate for detection of *S.* Typhimurium in the chicken samples.

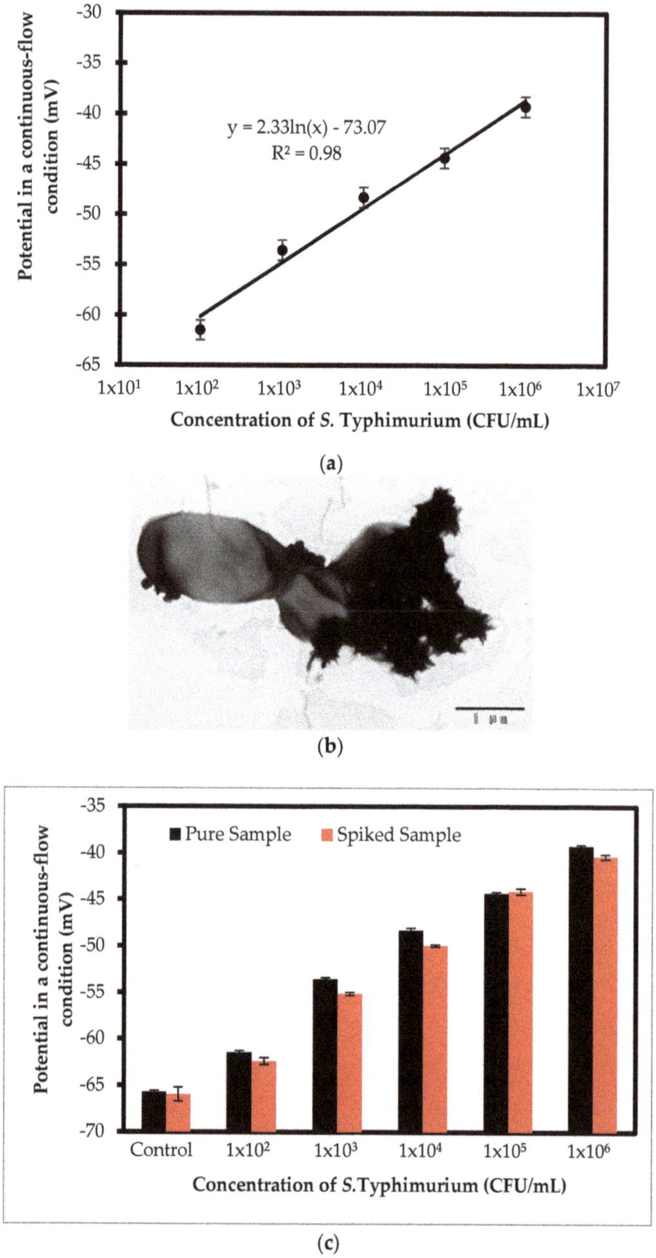

Figure 4. (**a**) Calibration curve of the proposed assay (N = 3); (**b**) TEM image of the nanoflower conjugated bacteria; (**c**) Detection of the spiked chicken samples containing the *S.* Typhimurium cells with the concentrations from 1.0×10^2 to 1.0×10^6 CFU/mL (N = 3).

3.5. Specificity of Proposed Assay

In this study, *L. monocytogenes* and *E. coli* O157:H7 were used as non-target bacteria to evaluate the specificity of the proposed assay. The target bacteria (*Salmonella* Typhimurium) and these two non-target bacteria at the same concentration of 1.0×10^4 CFU/mL and the negative controls (PBS) were detected using the proposed assay. As shown in Figure 5, the potentials of Ca-ISE for the negative controls, *L. monocytogenes*, and *E. coli* O157:H7 were −66.23 mV, −60.43 mV, and −58.58 mV, respectively, which were equivalent to the target bacteria with the respective concentrations of 1.9×10^1, 2.3×10^2, and 5.0×10^2 CFU/mL, respectively, based on the calibration model. The potential for *S.* Typhimurium was −49.20 mV, which was equivalent to the target bacteria with a concentration of 2.8×10^4 CFU/mL. The slight potential difference between *Listeria monocytogenes*, *E. coli* O157:H7, and the negative controls might be attributed to: (1) negligible cross reaction between the MAbs/PAbs and the *Listeria monocytogenes* and *E. coli* O157:H7 cells, resulting in a small amount of calcium ions; and (2) electrostatic adsorption of the CaNFs onto the non-target bacteria or impurities non-specifically captured by the MNBs, resulting in elevated background noises. The potential for the *S.* Typhimurium cells at the same concentration was much higher (more than two times compared to the control) than those for *Listeria monocytogenes* and *E. coli* O157:H7 due to the formation of the nanoflower conjugated bacteria resulting in the replacement of the nanoflowers to produce more calcium ions. This indicates that the proposed assay has good specificity.

Table 1. The recovery of the *S.* Typhimurium cells at the concentrations of 1.0×10^2–1.0×10^6 CFU/mL (N = 3).

Bacteria Concentration (CFU/mL)	Potential for the Spiked Chicken Sample (mV)	Potential for the Pure Culture Sample (mV)	Recovery
1.0×10^2	−62.40	−61.50	92.2%
1.0×10^3	−55.13	−53.57	92.0%
1.0×10^4	−49.96	−48.33	93.4%
1.0×10^5	−44.13	−44.37	100.8%
1.0×10^6	−40.37	−39.27	96.7%

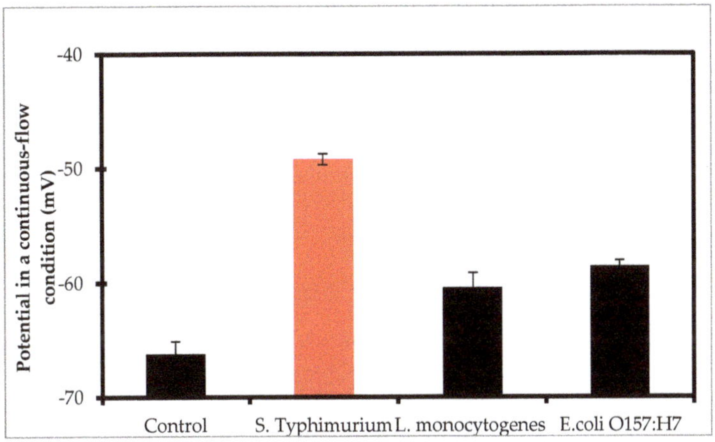

Figure 5. Specificity tests on the *S.* Typhimurium, *Listeria monocytogenes*, and *Escherichia coli* O157:H7 cells at the same concentration of 1.0×10^4 CFU/mL and the negative controls (N = 3).

4. Conclusions

A novel potentiometric assay based on effective amplification of biological signals using protein-inorganic hybrid CaNFs, efficient separation of target bacteria using immune MNBs, and continuous-flow detection of calcium ions using Ca-ISE was successfully developed for rapid and sensitive detection of *Salmonella*. The proposed assay was able to detect *S.* Typhimurium cells as low as 28 CFU/mL within two hours. The Ca-ISE embedded in the fluidic chip is capable of online monitoring calcium ions and has potential for the development of a lab-on-a-chip potentiometric biosensing device for sensitive detection of foodborne pathogens.

Supplementary Materials: The following are available online at http://www.mdpi.com/2079-4991/8/12/1006/s1, Figure S1: (a) The platform for proof of the proposed assay; (b) The fluidic chip with the Ca-ISE. Figure S2: Calibration curve of the flame photometer ($N = 3$).

Author Contributions: L.W. and J.L. designed the concept; L.W., Q.Z. and J.L. carried out Methodology, Validation, and Formal Analysis; Investigation and Resources were conducted by X.H. and R.G.; Writing—Original draft preparation was performed by L.W.; J.L. supervised all the work and revised the manuscript.

Funding: This research was funded by the National Science Foundation of China [31802219] and Walmart Foundation [SA1703161].

Conflicts of Interest: The authors declare no conflict of interest. The funders had no role in the design of the study; in the collection, analyses, or interpretation of data; in the writing of the manuscript, or in the decision to publish the results.

References

1. World Health Organization. Food Safety. Available online: http://www.who.int/news-room/fact-sheets/detail/food-safety (accessed on 31 October 2017).
2. López-López, M.; Bernal, E.; Moyá, M.L.; Sanchez, F.; López-Cornejo, P. Study of ionic surfactants interactions with carboxylated single-walled carbon nanotubes by using ion-selective electrodes. *Electrochem. Commun.* **2016**, *67*, 31–34. [CrossRef]
3. Eylem, C.C.; Tastekin, M.; Kenar, A. Simultaneous determination of copper and zinc in brass samples by PCR and PLS1 methods using a multiple ion-selective electrode array. *Talanta* **2018**, *183*, 184–191. [CrossRef] [PubMed]
4. Jasiński, A.; Guziński, M.; Lisak, G.; Bobacka, J.; Bocheńska, M. Solid-contact lead(II) ion-selective electrodes for potentiometric determination of lead(II) in presence of high concentrations of Na(I), Cu(II), Cd(II), Zn(II), Ca(II) and Mg(II). *Sens. Actuators B Chem.* **2015**, *218*, 25–30. [CrossRef]
5. Cuartero, M.; Bakker, E. Environmental water analysis with membrane electrodes. *Curr. Opin. Electrochem.* **2017**, *3*, 97–105. [CrossRef]
6. Crespo, G.A. Recent Advances in Ion-selective membrane electrodes for in situ environmental water analysis. *Electrochim. Acta* **2017**, *245*, 1023–1034. [CrossRef]
7. Ding, J.; Yu, N.; Wang, X.; Qin, W. Sequential and Selective Detection of Two Molecules with a Single Solid-Contact Chronopotentiometric Ion-Selective Electrode. *Anal. Chem.* **2018**, *90*, 1734–1739. [CrossRef]
8. Zhang, R.; Gu, Y.; Wang, Z.; Li, Y.; Fan, Q.; Jia, Y. Aptamer cell sensor based on porous graphene oxide decorated ion-selective-electrode: Double sensing platform for cell and ion. *Biosens. Bioelectron.* **2018**, *117*, 303–311. [CrossRef] [PubMed]
9. Kajisa, T.; Yanagimoto, Y.; Saito, A.; Sakata, T. Biocompatible Poly(catecholamine)-Film Electrode for Potentiometric Cell Sensing. *ACS Sens.* **2018**, *3*, 476–483. [CrossRef]
10. Guinovart, T.; Hernandez-Alonso, D.; Adriaenssens, L.; Blondeau, P.; Rius, F.X.; Ballester, P. Characterization of a new ionophore-based ion-selective electrode for the potentiometric determination of creatinine in urine. *Biosens. Bioelectron.* **2017**, *87*, 587–592. [CrossRef]
11. Fan, Y.; Xu, C.; Wang, R.; Hu, G.; Miao, J.; Hai, K. Determination of copper(II) ion in food using an ionic liquids-carbon nanotubes-based ion-selective electrode. *J. Food Compos. Anal.* **2017**, *62*, 63–68. [CrossRef]
12. Miao, J.; Wang, X.; Fan, Y.; Li, J.; Zhang, L.; Hu, G. Determination of total mercury in seafood by ion-selective electrodes based on a thiol functionalized ionic liquid. *J. Food Drug Anal.* **2018**, *26*, 670–677. [CrossRef] [PubMed]
13. Liu, S.; Ding, J.; Qin, W. Current pulse based ion-selective electrodes for chronopotentiometric determination of calcium in seawater. *Anal. Chim. Acta* **2018**, *1031*, 67–74. [CrossRef] [PubMed]
14. Cuartero, M.; Perez, S.; Garcia, M.S.; Garcia-Canovas, F.; Ortuno, J.A. Comparative enzymatic studies using ion-selective electrodes. The case of cholinesterases. *Talanta* **2018**, *180*, 316–322. [CrossRef] [PubMed]
15. Lu, M.; Ren, J. Polyaspartic acid-doped $CaCO_3$ nanospheres: A novel signal-generation tag for electrochemical immunosensing with calcium ion-selective electrode. *Electrochem. Commun.* **2017**, *84*, 10–13. [CrossRef]

16. Filotás, D.; Asserghine, A.; Nagy, L.; Nagy, G. Short-term influence of interfering ion activity change on ion-selective micropipette electrode potential; another factor that can affect the time needed for imaging in potentiometric SECM. *Electrochem. Commun.* **2017**, *77*, 62–64.
17. Calvo, D.; Durán, A.; del Valle, M. Use of pulse transient response as input information for an automated SIA electronic tongue. *Sens. Actuators B Chem.* **2008**, *131*, 77–84. [CrossRef]
18. Cuartero, M.; García, M.S.; Ortuño, J.A. Differential dynamic potentiometric responses obtained with anion-selective electrodes for perchlorate, thiocyanate, iodide, nitrate, sulfate, picrate and bis(trifluoromethylsulfonyl) imide. *Electrochim. Acta* **2013**, *93*, 272–278. [CrossRef]
19. Cuartero, M.; Ruiz, A.; Oliva, D.J.; Ortuño, J.A. Multianalyte detection using potentiometric ionophore-based ion-selective electrodes. *Sens. Actuators B Chem.* **2017**, *243*, 144–151. [CrossRef]
20. Altinkaynak, C.; Tavlasoglu, S.; Ozdemir, N.; Ocsoy, I. A new generation approach in enzyme immobilization: Organic-inorganic hybrid nanoflowers with enhanced catalytic activity and stability. *Enzyme Microb* **2016**, *93–94*, 105–112. [CrossRef]
21. Venkata, S.J.; Geetha, A.; Ramamurthi, K. Morphological and structural analysis of manganese oxide nanoflowers prepared under different reaction conditions. *Appl. Surf. Sci.* **2018**, *449*, 228–232. [CrossRef]
22. Cui, J.; Jia, S. Organic–inorganic hybrid nanoflowers: A novel host platform for immobilizing biomolecules. *Coord. Chem. Rev.* **2017**, *352*, 249–263. [CrossRef]
23. Ge, J.; Lei, J.; Zare, R.N. Protein-inorganic hybrid nanoflowers. *Nat. Nanotechnol.* **2012**, *7*, 428–432. [CrossRef] [PubMed]
24. Ye, R.; Zhu, C.; Song, Y. One-pot bioinspired synthesis of all-inclusive protein–protein nanoflowers for point-of-care bioassay: Detection of *E. coli* O157:H7 from milk. *Nanoscale* **2016**, *8*, 18980–18986. [CrossRef] [PubMed]
25. Ocsoy, I.; Dogru, E.; Usta, S. A new generation of flowerlike horseradish peroxides as a nanobiocatalyst for superior enzymatic activity. *Enzyme Microb* **2015**, *75–76*, 25–29. [CrossRef] [PubMed]
26. Wang, L.B.; Wang, Y.C.; He, R.; Zhuang, A.; Wang, X.; Zeng, J. A new nanobiocatalytic system based on allosteric effect with dramatically enhanced enzymatic performance. *J. Am. Chem. Soc.* **2013**, *135*, 1272–1275. [CrossRef] [PubMed]
27. Chen, Q.; Wang, D.; Cai, G.; Xiong, Y.; Li, Y.; Wang, M. Fast and sensitive detection of foodborne pathogen using electrochemical impedance analysis, urease catalysis and microfluidics. *Biosens. Bioelectron.* **2016**, *86*, 770–776. [CrossRef] [PubMed]
28. Yao, L.; Wang, L.; Huang, F.; Cai, G.; Xi, X.; Lin, J. A microfluidic impedance biosensor based on immunomagnetic separation and urease catalysis for continuous-flow detection of *E. coli* O157:H7. *Sens. Actuators B Chem.* **2018**, *259*, 1013–1021. [CrossRef]
29. Chen, Q.; Lin, J.; Gan, C.; Wang, Y.; Wang, D.; Xiong, Y. A sensitive impedance biosensor based on immunomagnetic separation and urease catalysis for rapid detection of *Listeria monocytogenes* using an immobilization-free interdigitated array microelectrode. *Biosens. Bioelectron.* **2015**, *74*, 504–511. [CrossRef] [PubMed]
30. Wang, D.; Chen, Q.; Huo, H.; Bai, S.; Cai, G.; Lai, W. Efficient separation and quantitative detection of *Listeria monocytogenes* based on screen-printed interdigitated electrode, urease and magnetic nanoparticles. *Food Control* **2017**, *73*, 555–561. [CrossRef]
31. Wang, L.; Huang, F.; Cai, G.; Yao, L.; Zhang, H.; Lin, J. An Electrochemical Aptasensor Using Coaxial Capillary with Magnetic Nanoparticle, Urease Catalysis and PCB Electrode for Rapid and Sensitive Detection of *Escherichia coli* O157:H7. *Nanotheranostics* **2017**, *1*, 403–414. [CrossRef]

nanomaterials

MDPI

Review

Optical Sensors Based on II-VI Quantum Dots

Anna Lesiak [1,2], Kamila Drzozga [1], Joanna Cabaj [1,*], Mateusz Bański [2], Karol Malecha [3] and Artur Podhorodecki [2,*]

1 Faculty of Chemistry, Wrocław University of Science and Technology, Wybrzeże Wyspiańskiego 27, 50-370 Wrocław, Poland; anna.lesiak@pwr.edu.pl (A.L.); kamila.drzozga@pwr.edu.pl (K.D.)
2 Faculty of Experimental Physics, Wrocław University of Science and Technology, Wybrzeże Wyspiańskiego 27, 50-370 Wrocław, Poland; mateusz.banski@pwr.edu.pl
3 Faculty of Microsystem Electronics and Photonics, Wrocław University of Science and Technology, Wybrzeże Wyspiańskiego 27, 50-370 Wrocław, Poland; karol.malecha@pwr.edu.pl
* Correspondence: joanna.cabaj@pwr.edu.pl (J.C.); artur.p.podhorodecki@pwr.edu.pl (A.P.); Tel.: +48-71-320-4641 (J.C.)

Received: 30 December 2018; Accepted: 28 January 2019; Published: 2 February 2019

Abstract: Fundamentals of quantum dots (QDs) sensing phenomena show the predominance of these fluorophores over standard organic dyes, mainly because of their unique optical properties such as sharp and tunable emission spectra, high emission quantum yield and broad absorption. Moreover, they also indicate no photo bleaching and can be also grown as no blinking emitters. Due to these properties, QDs may be used e.g., for multiplex testing of the analyte by simultaneously detecting multiple or very weak signals. Physico-chemical mechanisms used for analyte detection, like analyte stimulated QDs aggregation, nonradiative Förster resonance energy transfer (FRET) exhibit a number of QDs, which can be applied in sensors. Quantum dots-based sensors find use in the detection of ions, organic compounds (e.g., proteins, sugars, volatile substances) as well as bacteria and viruses.

Keywords: nanomaterials; colloidal quantum dots; sensors; detection mechanisms

1. Introduction

Among the most interesting and promising nanomaterials are colloidal semiconducting quantum dots (QDs). These nanostructures have found already several commercial applications in displays [1], light concentrators [2], photovoltaics [3] and as optical probes in various bio-applications [4,5]. The main reasons for their still growing success are: broad absorption band (several hundreds of nm), narrow emission band (below 40 nm), high quantum yield (up to 95%), possibility of emission band tuning over a wide range of wavelengths (350–2000 nm) and high resistivity of optical properties on external physico-chemical conditions, e.g., pH, temperature or power of the excitation beam [6]. QDs have a high surface to volume ratio, which can be controlled by QDs size but also by the nanostructures shape. This high surface area equips them with much more functional groups compared to organic compounds [7]. This makes QDs much more reactive and thus more effective in biological sensing [8].

The high quality QDs are typically grown as hydrophobic structures. In consequence, to make them useful for biological or medical application, additional post growth treatment is usually needed. This treatment includes QDs functionalization and in many cases bioconjugation (see Figure 1). One serious reason why QDs still do not dominate over organic markers (i.e., Green Fluorescent Protein, Rhodamine) lies in the absence of widely tested and already accepted protocols for QDs functionalization and bioconjugation [9]. There is also another reason to not use the QDs in biological sensing, especially in case of in vivo imaging, namely their toxicity. The QDs toxicity is a subject widely discussed recently in the literature [10]. The main reason affecting QDs toxicity is an aggregation of nanostructures in cells, organs, tissue etc. [11]. This is an even more serious problem than the chemical toxicity, due to dissociation of Cd atoms from CdSe/CdS QDs, and this is true for any types of nanostructures: semiconducting, dielectric or metallic nanostructures. Nevertheless, fortunately the QDs toxicity becomes a much less serious problem when the QDs are used for external sensing or for some in vitro applications when we can use their potential without barriers.

Concluding the above discussion, it can be seen that real benefits coming from extraordinary properties of QDs must be always compared to drawbacks of using inorganic probes in biological systems. In other words, for some specific applications QDs are an excellent choice, or very bad idea [12–14]. Among the applications where incontestably the advantages of QDs are utilized are sensing systems.

In this review, we present the main physical and chemical mechanisms used for detection of various species (bacteria, cells, nucleic acids, molecules, ions, etc.) with utilizing of QDs. We present the most successful examples of QDs applications in biology and medicine as optical and electrochemical sensors. Finally, we focus on the perspectives for further development in this field.

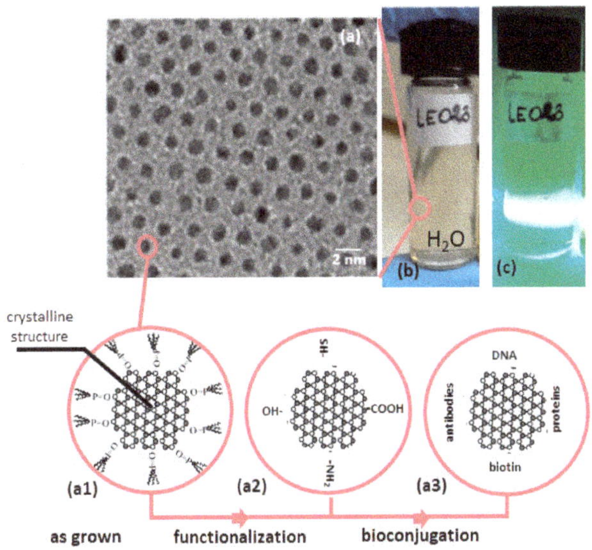

Figure 1. (**a**) TEM image of hydrophilic CdSe quantum dots. Schematic structure of selected quantum dot after synthesis (**a1**), after surface functionalization (with examples of most typical functional groups) (**a2**) and after bioconjugation (with examples of most common biomolecules used for detection/targeting) (**a3**). (**b**,**c**) Digital images of CdSe quantum dots dispersed in water with and without laser excitation.

2. Comparison between Optical Properties of Organic Dyes and QDs

In comparison to organic dyes, QDs have the spectral position of absorption and emission dependent on their size (so-called Quantum Size Effect) [15]. During synthesis, this effect enables continuous tuning of the emission peak position in a wide range of wavelengths (Figure 2a). Moreover, the broad absorption of QDs allows free selection of the excitation wavelength and thus straightforward separation of the excitation and emission signal (Figure 2b,c) [16]. The fluorescence lifetimes of organic dyes are commonly too short for efficient temporal discrimination of short-lived autofluorescence of biological objects. In the case of QDs, the emission decay time can be tuned or selected with a proper choice of QDs composition (giving times up to several microseconds). This enables straightforward temporal discrimination of the signal from cellular autofluorescence and scattered excitation light by time-gated measurements, thereby enhancing detection sensitivity [17]. In contrast to conventional dyes, QDs emitting different colors (and functionalized with different groups) can be simultaneously excited by a single excitation wavelength. This makes QDs suitable for multiplex testing by simultaneously detecting multiple signals [18]. Moreover, QDs characterize with extremely high chemical stability and photostability (stability against chemical reactions induced by the incoming radiation). In addition, QDs are free from photobleaching [19] (Figure 2d) what is one of their most important advantages.

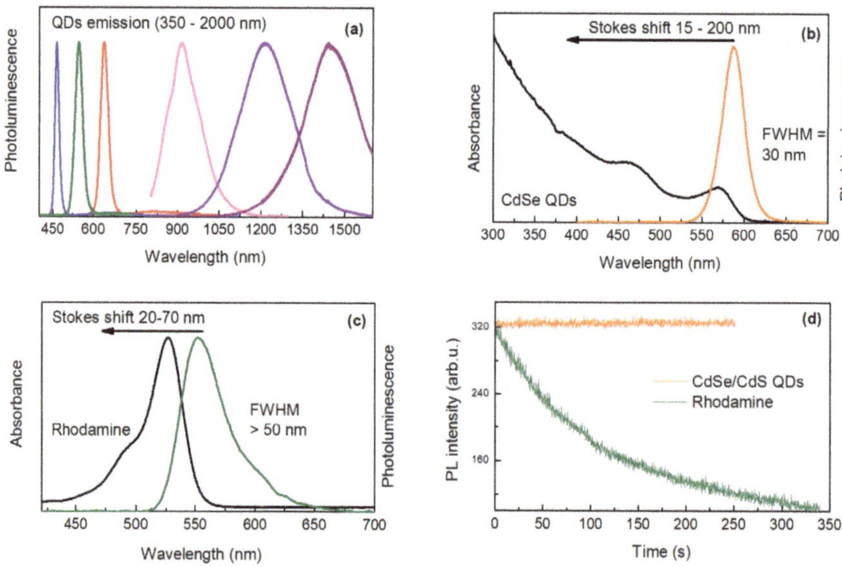

Figure 2. (**a**) Emission spectra from CdS QDs (left side) and PbS QDs (right side) with different size and chemical composition; (**b**) absorbance and emission spectra of CdSe/CdS quantum dots; (**c**) emission and absorption spectra of Rhodamine; (**d**) emission intensity vs illumination time for CdSe/CdS QDs and Rhodamine.

3. Fundamentals of QDs Sensing Phenomena

The unique optical properties of QDs make them attractive fluorophores that can be used both in vitro and in vivo in various biological studies, where traditional fluorescent labels based on organic molecules do not provide long-term stability, high enough intensity or where simultaneous detection of many signals is needed [7]. In sensors, the signal detection bases on a registration of the change in one of the physical properties (optical, thermal, mechanical, magnetic, electrical) of sensing material induced by the interaction with the analyte. Changing in optical properties of QDs like emission color, intensity, polarization or emission kinetics can be used as the principle in optical sensors system (Figure 3). In addition, obtained changes can be recorded directly by human senses or indirectly *via* the signal transformation, amplification, and visualization. All these factors determine sensors construction and their mechanism of action in the detection of various substances [20,21].

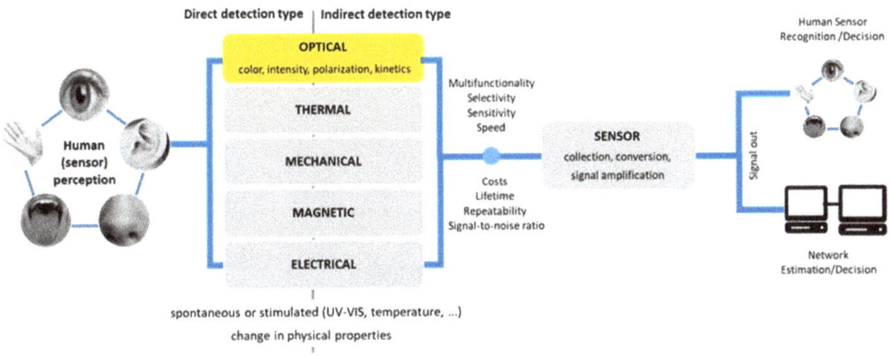

Figure 3. Signal processing characteristics for living organisms and sensor machines.

4. Basic Strategies for Analyte Detection

The use of QDs for sensors construction requires adjusting their optical properties adequately to the needs that arise their shape, size, the color of emission, position of the absorption band. Moreover, to get specificity of QDs in their sensing action, the surface modification—called functionalization—must be applied first [22,23]. Functionalization is the process of attaching, exchanging already attached chemical molecules present on the surface of quantum dots. Chemical and physical methods used for this purpose, include processes such as exchange of ligands, silanisation, the creation of additional coatings or dendrimeric structures [24]. The presence of ligands at the QDs surfaces affect their size, shape and physico-chemical properties, e.g., surface charge and chemical reactivity. Surface modifications allow the control of colloidal stability of QDs and their dispersion in non-polar environments (organic solvents in which they are most commonly synthesized) and polar (e.g., water, in which solubility is necessary for biological and medical applications). Moreover, the surface attached ligands determine the possibility of QDs conjugation to biological molecules (bioconjugation) or to determine their potential in applications where QDs must be embedded within the matrix [25,26].

In order to achieve high selectivity of QDs sensor, QDs are coupled to various vectors specific for an analyte. Wales et al. constructed a sensor for the selective detection of dicofol, a substance used to kill mites. For this purpose, they used CdS QDs with glutathione on their surface, whose both amino- and carboxyl- functional groups interact with chloride groups in the dicofol structure, thus leading to an increase in fluorescence intensity, which was directly proportional to the dicofol concentration in the studied sample [27,28].

The QDs-based sensors can be designed in several ways, depending on demands regarding their sensitivity, types of detected analytes, costs or complexity of their preparation. Figure 4 shows the examples of preparation protocols used in QDs-based optical sensors.

In all cases, the protocol starts with the appropriate modification of QDs surface selectivity. As a result, QDs are targeted to determine a particular analyte. An important aspect is also the preparation of substrates which can take an active part in the detection protocol. Strategies (a) and (b) differ in Stages III and IV, which occur in reverse order. While in Strategy (a) Step III is the deposition of QDs, in Strategy (b) it is Step IV. This stage can be made using methods such as layer-by-layer [29], sol-gel [30] or electrochemical method [31]. Stage IV in Strategy (a) and III in Strategy (b) are a conjugation of the analyte, which may be possible thanks to the previously prepared and targeted substrate. Jie et al. proposed the coupling of the analyte with the previously prepared substrate, based on CdSe nanocomposites, using antibodies selective for an antigen called human IgG [32].

The final step in all strategies is QDs stimulation, which is used to detect the analyte. As a result, both the qualitative and quantitative assessment of the presence of the designated substance is possible.

It is also possible to combine the first two strategies, resulting in Strategy (c), which uses the Förster resonance energy transfer between optical centers (QDs + QDs or dye).

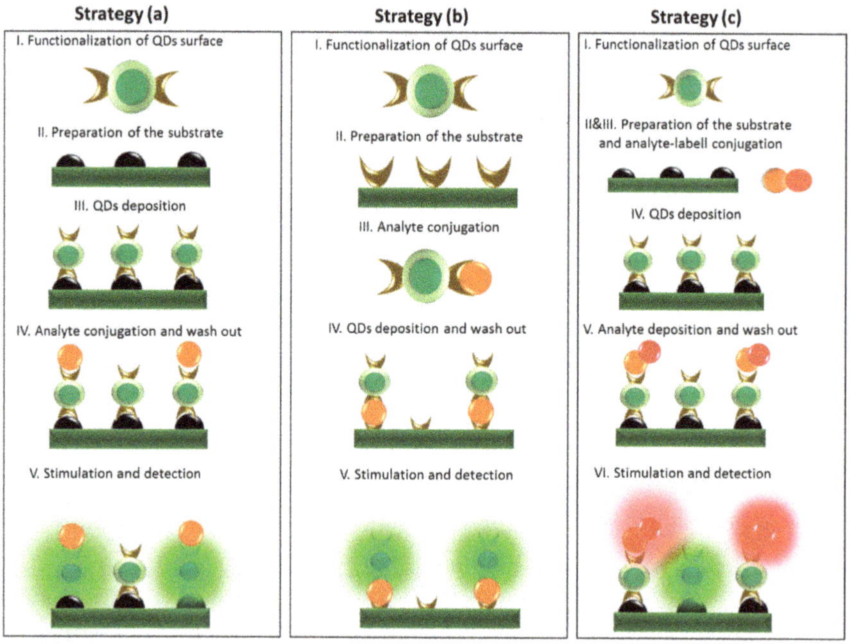

Figure 4. Three examples of the strategy of QDs-based optical sensors (strategy **a**—modification of substrate with QDs directed to detection of analyte, strategy **b**—modification of substrate for detection of analyte-QDs complex, strategy **c**—using the analyte labeled with appropriate fluorophore).

The presented strategies differ in the number of steps that complicate detection and require a lot of user experience.

5. Physico-Chemical Mechanisms Used for Analyte Detection

One of the most popular mechanisms used for detection of analyte relies on emission quenching from QDs. In this mechanism, due to the interaction of the QDs surface with the analyte, the QDs emission intensity decreases (Figure 5a) [33]. Another mechanism relies on an increase of QDs emission due to passivation of QDs surface by analyte (Figure 5b), e.g., addition of bovine serum albumin or nucleic acids resulted in increasing emission from CdS dots coated with mercaptoacetic acid [34].

The third mechanism, which can be used for analyte detection, is stimulated aggregation (Figure 5c). In this mechanism, due to an interaction of the analyte with the QDs surface, the surface ligands are detached and QDs aggregate. The aggregation can be also induced by analyte stimulated bonds formation between functionalized QDs [35].

There is also a very rarely used mechanism of analyte detection based on modification of the nanostructures' growth process by introduction of the analyte during the nanostructures' growth. Due to this perturbation, the nanostructures can have different emission or other properties which can be detected (Figure 5d). There is also the fifth mechanism commonly used for analyte detection based on changes in QDs optical properties. The changes come from excitation energy transfer from QDs to other optical center (QDs or dye). In consequences, the color of emission changes or emission decay time of donor is reduced (Figure 5e) [36,37].

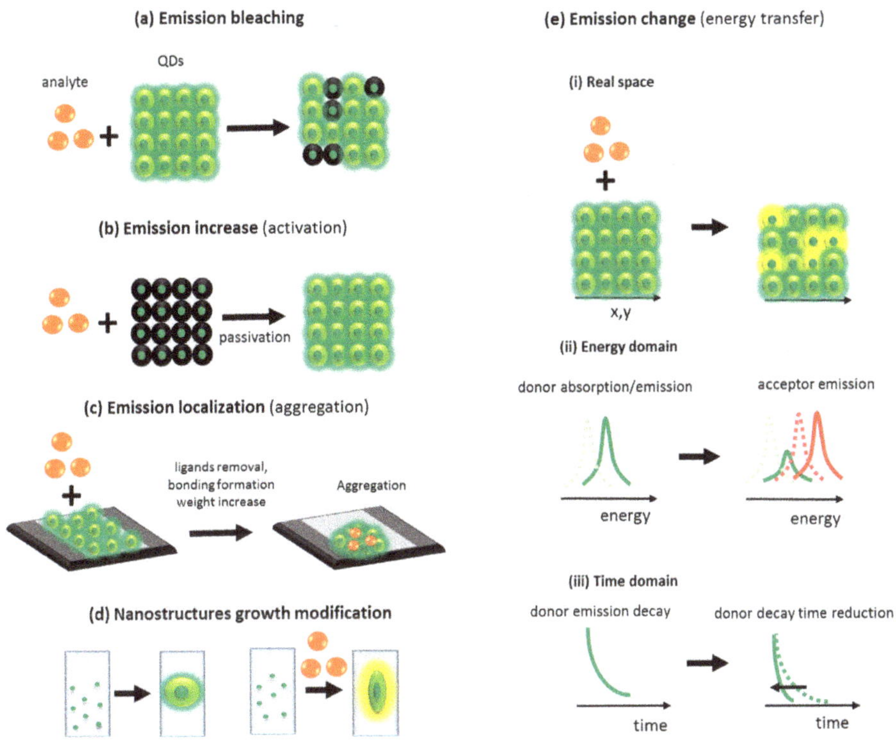

Figure 5. Examples of physico-chemical mechanisms used for analyte optical detection—emission bleaching (**a**), increase of emission (**b**), emission localization (**c**), nanostructures growth's modification (**d**), emission change (**e**).

5.1. Emission Bleaching

One of the basic mechanisms of an analyte detection is emission bleaching (Figure 5a). This mechanism can be induced by different physical phenomena schematically shown in Figure 6. In the first example, the absorption band of the analyte overlaps with the emission band of QDs. In such a case, the emission from QDs is absorbed by an analyte, which results in bleaching QDs emission. (Figure 6a). Another mechanism (Figure 6b) bases on a charge transfer from an analyte to QDs. When the additional electron or hole appears in an excited QDs the Auger processes lead to QDs ionization or charging. In both cases, the dot became non-emissive (dark). The third mechanism, which is responsible for QDs emission quenching under the interaction with the analyte is analyte-induced degradation of the QDs surface providing to emission decrease. As an example, thiol coated ZnS QDs in the presence of peptides showed a significant emission quenching [38].

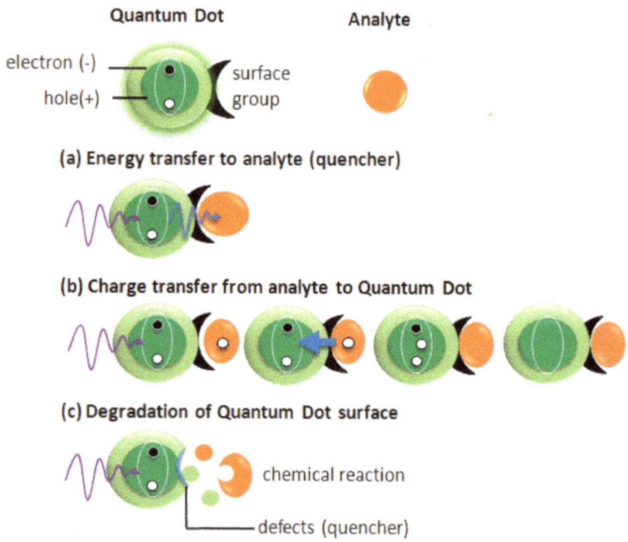

Figure 6. Examples of physico-chemical mechanisms responsible for quantum dots emission quenching—energy transfer from QD to analyte (**a**), charge transfer from analyte to QD (**b**), degradation of QD surface (**c**).

5.2. Emission Change—Nonradiative Förster Resonance Energy Transfer (FRET)

When two optically active centers (donor and acceptor) are in close proximity to each other, (typically 1–10 nm) and an absorption spectrum of the acceptor overlaps an emission spectrum of the donor, the non-radiative transfer of the excitation energy from the donor to acceptor appears. This phenomenon is called FRET (Förster Resonance Energy Transfer). For the experiments using the FRET approach, photostable emitters must be used, characterized by a quantum efficiency greater than 0.1 and a high brightness (one in which the absorbance coefficient ε is greater than 50.000 $M^{-1}\cdot cm^{-1}$) [39]. All these conditions are perfectly fulfilled by the quantum dots. The effectiveness of FRET is inversely proportional to the distance between the donor and acceptor and defined as:

$$K_{FRET} = \frac{1}{1+\left(\frac{R}{R_0}\right)^6}, \tag{1}$$

where: K_{FRET}—FRET efficiency; R—distance between donor and acceptor; R_0—distance at which the FRET efficiency is equal to 0.5 [40].

The highest sensitivity of FRET signal is for a distance between the donor and acceptor in the range from $0.5R_0$ to $1.5R_0$ [40]. For years researchers have been using the FRET mechanism to monitor intracellular interactions, due to its sensitivity to molecular rearrangements in the 1–10 nm range (this is the scale correlating with the size of biological macromolecules and the possibility of creating bonds between them) [41].

The universality of FRET method allows its use in nanosystems as well [42]. FRET yield is typically measured by observing one of the three parameters of the fluorescent donor: fluorescence intensity, spectral response or average fluorescence lifetime. Moreover, FRET has found application in many sensing systems giving the possibility of applying it to three analyte detection strategies. Figure 7 shows different processes which can be detected with use of FRET. The first mechanism uses analyte as optically active acceptor. In this case, the analyte attachment as well analyte removal can be observed as a change in the optical signal. The other strategy uses the analyte as the emission quencher and was also discussed in the previous paragraph [43]. The third strategy is more complex and uses a multistep energy-transfer phenomenon. Detection using FRET between the QDs, as donors, directed to a linker with an acceptor, associated, e.g., with a receptor protein, is widely used to study the receptor-ligand interactions and changes in protein conformation after binding to the target analyte [44]. Thanks to this, in analytics consisting of several acceptors, QDs can interact with only one of them, which significantly improves the efficiency and sensitivity of the FRET method [45].

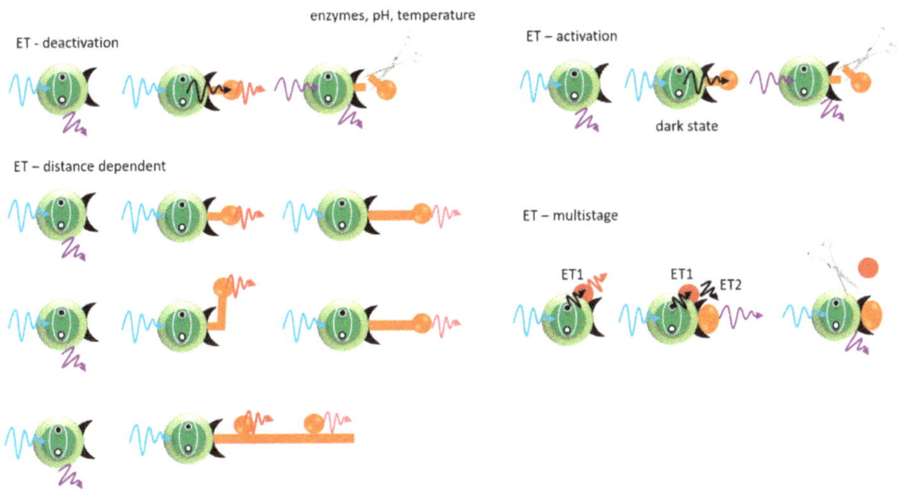

Figure 7. Different detection possibilities with use of nonradiative energy-transfer phenomena.

5.3. *Analyte Stimulated QDs Aggregation*

The colloidal QDs solution is sensitive to the presence of additional charges either on QDs surface or in the solvent, which may result in QDs aggregation. The charges may be introduced or induced by an analyte, which ultimately is manifested by QDs aggregation [46–48]. In the absence of analyte, a fluorescence comes from the whole volume of the QDs solution, while after aggregation caused by the analyte, the emission is localized [19]. This type of stimulation belongs to qualitative tests.

6. Photoelectrochemical and Electrochemical Methods of Analyte Detection

The chemical detection method is usually signaled by the following ways: competition binding assay of labelled and unlabeled analytes, using labelled molecules specific for immobilized analytes, sandwich formation or enzyme immunoassay, where enzymatically active substrate is added that changes color or fluorescence after interaction with enzyme-related analyzes [49].

Mo et al. [50] used a redox mechanism in the detection of hydroquinone in water samples. They have observed that ZnS QDs cannot react with hydroquinone. When hydroquinone and $K_2S_2O_8$ were added into ZnSe QDs solution, no new photoluminescence (PL) peak was observed. Comparing with the pure ZnSe QDs solution, the PL intensity of the mixture decreased. This result reveals that hydroquinone oxidation product can efficiently quench the fluorescence emission of ZnSe QDs by energy transfer in electrochemiluminescence mechanism.

The development of new, reliable, fast and efficient methods for detecting anthropogenic and natural substances, both organic and inorganic, is a huge challenge for modern analytical chemistry and diagnostics. An alternative to such methods are electrochemical strategies using semiconducting quantum dots. The growing interest in the construction of electrochemical devices using quantum dots results from their aforementioned properties. Due to these features, small changes in the external environment lead to great changes in particle properties and electron transfer. Based on these significant changes, quantum dots are prone to engaging in heterogeneous redox chemistry with the surrounding environment. QDs are also used as carriers of biomacromolecules in bioanalytics. For this purpose, the chemical functionalization of QDs is carried out by means of a functional cap layer that allows the molecules to be trapped. The immobilization of biomolecules (e.g., the enzyme catalyzing the redox reaction) on the surface of semiconductor QDs causes QDs to promote direct electron transfer between biomolecules and the surface of the electrode, which significantly affects the operation of the system by enhancing the sensitivity due to signal amplification. A tremendous increase of development of electrochemical sensors based on QDs has been observed over the past decades due to the simplicity of implementation, high selectivity, and specificity of the system, low cost and the possibility of miniaturization [51]. Moreover, research carried out by Bard et al. revealed that CdS QDs could also act as multi-electron donors or acceptors at a given potential due to trapping of holes and electrons within the particle [52]. On the other hand, the surface structures of QDs also play a key role in determining the properties of the particles [53].

Liu et al. [54] described an electrochemical assay strategy for specific recognition of tumor cells. For this purpose, gold nanoparticles (AuNPs) have been assembled onto the indium tin oxide (ITO) substrate to create a specific, biocompatible interface to effective capture of tumor cells. CdSe/ZnS QDs labelled on the cell surface have been used as an amplified signal during the square wave stripping voltammetry (SWSV). The developed biosensing platform shown good analytical performance with a broad linear range, good selectivity and low limit of detection (LOD).

Electrochemiluminescence (ECL) is a method which aims to convert electric energy into radiation energy, in which electrochemically generated intermediate products undergo a high energy electron transfer reaction to generate excited states, resulting in the emission of a measurable luminescence signal [55]. As a form of luminescence (light emission without heat), ECL is characterized by the fact that light emission occurs when an appropriate potential is applied to the electrode, as a result of which the oxidation or reduction reaction takes place. There are several features that distinguish ECL from other techniques, e.g., chemiluminescence (CL). It is clear that the electrochemical reaction that takes place allows for precise time control. This means that the emission of light can be delayed to the desired moment, e.g., an immune reaction or an enzymatic reaction. Another advantage of ECL is the ability to control the location of the reaction, which means that there is the possibility of limiting the emission of light to a specific area relative to the detector. Electrochemiluminescence may occur as a result of two independent processes: annihilation of ions and co-reactant ECL. The annihilation of ions consists in creating states of excited molecules due to the transfer of electrons between radical ions on the surface of the electrode. The ECL co-reactant is due to the use of anode or cathodic potential in a solution containing phosphor and co-agent molecules. Depending on the potential application, the phosphor or co-reactant molecules can be reduced or oxidized to form radical ions and medium compounds, followed by decomposition and formation of excited states that cause light emission [56].

7. Applications of QDs-Based Sensors

7.1. Detection of Ions

Fast and reliable detection and recognition of ions in the environment is extremely important in modern medicine and environmental protection. Among the ions, heavy metal ions such as mercury, cadmium and lead due to their high toxicity and negative health effects (cardiovascular diseases, cancer, liver, kidney and central nervous system disorders, reproductive and neurological disorders) require constant control concentration and rapid response in view of its possible reduction [57–61].

Hydrophilic QDs have been demonstrated to be a promising sensor probe for fluorescence-based sensing of heavy metal ions [57] such as Pb^{2+} [62], Cd^{2+} [63], Cu^{2+} [64], Hg^{2+} [65], Fe^{3+} [62], etc. Table 1 shows an exemplary strategies for heavy metal ions determination with using of QDs.

Table 1. QDs-based sensors for heavy metal ions determination.

Marker	Sensing Platform	Transduction Type	LOD *	References
Cu^{2+}	CdS QDs	Change in fluorescence intensity	0.8 µM	[13]
Fe^{3+}	CdTe QDs	Change in fluorescence intensity	14 nM	[57]
Hg^{2+}	CdSe/ZnS QDs	Change in fluorescence intensity	15 nM	[58]
Pb^{2+}	CdSe/CdS QDs	Change in fluorescence intensity	0.006 nM	[59]
Hg^{2+}	TGA-CdTe QDs	FRET	20 nM	[60]
Pb^{2+}	AuNPs & CA_CdTE QDs	FRET	30 ppb	[61]

* LOD—limit of detection

Detection of ions with the use of photoluminescent-induced changes in the QDs involves the use of a number of ligands—derivatives of thioalkyl, mercaptoacetic or dihydrolipoic acids. The affinity of the thiol group to QDs results in self-assembly of the ligands on the surface of the dots, as a result of which the hydrophilic carboxylic groups are exposed on the surface towards the surrounding aqueous solution [6]. Chen and Rosenzweig proposed a method for detection of Zn^{2+} and Cu^{2+} ions. They exploit the fact that surface-modified QDs with mercaptoacetic acid show high sensitivity and selectivity to Cu^{2+} copper ions present in the mixture. The result of adding Cu^{2+} to the ligand-QDs complex is a reduction of the intensity of PL QDs. Such constructed sensor exhibited high LOD 0.8 µM [13]. Selective quenching PL was also used by Li et al. [58] They constructed sensor sensitive to the presence of mercury ions Hg^{2+} In this measuring system, CdSe/ZnS QDs have been modified with sulfur calixarene (S-Calix). The linear range of this system was found as $0–3 \times 10^{-5}$ M with a LOD 15 nM.

Zhou et al. [57] presented a ratiometric fluorescence sensor for real-time and on-site detection of Fe^{3+} ions based on CdTe QDs-doped hydrogel optical fiber with a broad linear range from 0 to 3.5 µM and high LOD 14 nM. The ratiometric configuration of the proposed sensor provides a built-in calibration to eliminate the analyte-independent interferences. Two types of CdTe QDs, which possessed different emission bands, have been synthesized for ratiometric measurements. One of the QDs, coated with thioglycolic acid, exhibits green emission and is insensitive to metal ions, thus serving as a reference. The other QDs as the specific recognition element, coated with N-acetyl-l-cysteine, are red emissive and show high selectivity of fluorescence quenching towards Fe^{3+} ions. To avoid mutual interference, the green emissive QDs and red emissive QDs are doped in discrete sections of the hydrogel optical fiber. As a result, it has been observed a decrease in PL intensity.

7.2. pH Detection

Among the group of chemical sensors, the pH sensor is the object of greatest interest of scientists, because pH is one of the most important parameters in biochemical industrial processes [66–70]. Properly modified QDs, using organic ligands, may gain sensitivity to changes in pH. This property has a promising application in the design of a variety of luminescence sensors, examples of which are shown in Table 2.

Table 2. QDs-based sensors for pH determination.

Marker	Sensing Platform	Transduction Type	References
pH	CdSe/ZnS-methacrylic acid QDs	FRET	[12]
pH in range 3–11	CdSe/ZnS- [1,3] oxazine QDs	Change in fluorescence intensity	[66]
pH	CdSe/ZnS-phosphine oxide NC	FRET	[67]
pH in range 4–6	CdTe-thioglycolic acid NC	Change in luminescence intensity	[68]
pH	CdTe QDs	Change in fluorescence intensity	[69]
Urea	CdSe/ZnS-mercaptosuccinig acid QDs	Change in fluorescence intensity	[70]

In the work of Tomasulo et al. [66], the adsorption of pH-sensitive 1,3-oxazine on the surface of CdSe/ZnS QDs gives the possibility of changing the luminescence of inorganic nanoparticles by means of chemical stimulation: 1,3-oxazine rings open in an acidic or basic environment to form nitrobenzyl phenolate chromophores. This transformation activates the energy-transfer path from excited quantum dots to ligands and facilitates energy transfer in the opposite direction. As a result, the intensity of PL QDs decline. Such a system can be used in aqueous solutions for pH changes in the range of 3–11.

The Snee group proposed a sensor based on signal transduction by FRET between the QDs and a fluorescent pH-sensitive squaraine dye attached to the surface of the QDs. The detection system process consisted in modulating the FRET efficiency resulting from the overlap of the absorption spectrum of squaraine with the emission of QDs. The emission of QDs (donor) was inhibited by the presence of squaraine, acting as an acceptor. Lowering the pH value caused a rise in the photoluminescence intensity of QDs [67].

Many works present pH-sensitive sensors that use semiconductor QDs combined with thiol compounds [6]. The emission of QDs fluorescence with mercaptoacetic acid on the surface allowed in vitro detection (but only in an acidic environment) and in live cells. The increase in intracellular pH has given an increased intensity signal of PL QDs [12]. In contrast, QDs combined with mercaptosuccinic acid (MSA) proved to be a simple system for the detection of urea. Hydrolysis of urea generates hydroxide anions, gradually raising the pH of the solution. With increasing urea concentration, the intensity of PL QDs increased [70].

7.3. Detection of Organic Compounds

7.3.1. Proteins

Proteins are among the most important biomolecules found in the body. In addition to the basic building, transporting and regulating functions, proteins also act as biological catalysts–enzymes. The function of proteins is also invaluable in the immune system-acting as immunoglobulins. Due to the extremely important functions of proteins, it is necessary to monitor their concentration and the processes in which they take part [19,71–75]. QDs-based FRET nanosensors have been developed to monitor a variety of enzymes including alkaline phosphatase, ATPase renin, protein kinase, DNA methyltransferase, DNA glycosylase, and telomerase [76]. A different strategies for determination of proteins are presented in Table 3.

Table 3. QDs-based sensors for proteins determination.

Marker	Sensing Platform	Transduction Type	LOD	References
Protein kinases	CdTe QDs	Change in fluorescence intensity	5.0 fM	[19]
C-reactive protein	CdSe/ZnS QDs	Change in fluorescence intensity	0.47 mU·μL^{-1}	[71]
Exonuclease III	DNA-ZnS:Mn^{2+}	Change in fluorescence intensity	2.8 ng/mL	[72]
Micrococcal nuclease	CdSe/CdS	FRET	0.06 μ·mL^{-1}	[73]
Caspase	CdSe/ZnS	FRET	20 pM	[74]
Thrombin	PbS QDs	FRET	1 nM	[75]

Xu et al. have presented a novel label-free fluorescent assay for monitoring the activity and inhibition of protein kinases based on the aggregation behavior of unmodified CdTe QDs with very high LOD 5.0 fM. In this assay, cationic substrate peptides induce the selective aggregation of unmodified QDs with an anionic surface charge, whereas phosphorylated peptides do not. Phosphorylation by kinase alters the net charge of peptides and subsequently inhibits the aggregation of unmodified QDs, causing an enhanced QDs fluorescence [19].

Lv et al. [71] have proposed detection of C-reactive protein (CRP) based on fluorescence changes by CdSe/ZnS QDs, where QDs surfaces were modified with monoclonal antibodies. The fluorescence intensity has increased with the increasing of antigens concentration. The assay for the detection of CRP can provide a wide analytical range of 1.56–400 ng/mL with the LOD 0.46 ng/mL and the limit of quantification = 1.53 ng/mL.

Another example of a protein biosensor was developed by Zhang group. Using DNA-ZnS:Mn^{2+} QDs as the energy donor and WS_2 as the energy acceptor. DNA-ZnS:Mn^{2+} QDs were hybridized with biotin-DNA to obtain dsDNA. When Exonuclease III was added into the system, the biotin-DNA was hydrolyzed for the stepwise removal of mononucleotides from the 30-hydroxyl termini of dsDNA, releasing DNA-ZnS:Mn^{2+} QDs into solution. After incubation with WS_2 nanosheets, DNA-ZnS:Mn^{2+} QDs were absorbed on the surface of WS_2 due to their stronger affinity towards ssDNA than that of dsDNA. As the results, the fluorescence intensity was reduced with the increasing concentration of Exonuclease III. There is a good linear relationship between the fluorescence intensities and the concentration of SA (biotin-streptavidin) in the range of 5–150 ng/mL. The LOD was calculated as 2.8 ng/mL [72].

7.3.2. Sugars

Among the most popular groups of biomolecules that are the object of scientists' interest are sugars. Particularly noteworthy is glucose, the determination of which is important both in the pharmaceutical and food industries. Importantly, glucose monitoring is essential in the treatment of diabetes that is characterized by long-lasting hyperglycemia, making strict blood glucose control so important [77–81]. Table 4 shows an exemplary QDs-based sensors for sugars determination.

Table 4. QDs-based sensors for sugars determination.

Marker	Sensing Platform	Transduction Type	LOD	References
Glucose	CdSe/ZnS-TGA	Change in fluorescence intensity	-	[77]
Glucose	QDs-ConA-β-CDs-AuNPs	FRET	50 nM	[78]
Glucose and fructose	CdSe/ZnS QDs	Change in fluorescence intensity	1 μM	[79]
Glucose	Mn-ZnS QDs	Change in phosphorescence intensity	3 μM	[80]
Glucose	CdTe QDs	Change in fluorescence intensity	0.10 μM	[81]

Monitoring of glucose in human blood and urine is essential for the diagnosis and treatment of diabetes. The Sarana group has developed a biosensor for the detection of glucose based on cadmium quantum dots with a thiol ligand on the surface. This system has been coupled with glucose oxidase-a catalyst for glucose oxidation reaction, which releases hydrogen peroxide [77]. In constructing this biosensor, a capture mechanism was used, involving the charge transfer. The electron released in the process of reducing H_2O_2 to O_2 has been moved towards the exciton of QDs, acting as an acceptor. As a result of this process, a QDs-ion was formed and decreases of fluorescence intensity were observed [82].

Another sensing method of glucose approach is based on FRET between CdTe QDs as an energy donor and gold nanoparticles (AuNPs) as an energy acceptor. The specific combination of concanavalin A(ConA)-conjugated QDs and thiolated-cyclodextrins (b-SH-CDs)-modified AuNPs assembles a hyperefficient FRET nanobiosensor. In the presence of glucose, the AuNPs-b-CDs segment of the nanobiosensor is displaced by glucose which competes with b-CDs on the binding sites of ConA, resulting in the fluorescence recovery of the quenched QDs. Experimental results show that the increase in fluorescence intensity is proportional to the concentration of glucose in a linear range of 0.10–50 μM under the optimized experimental conditions. In addition, the sensor has high sensitivity with a LOD as low as 50 nM, and has excellent selectivity for glucose over other sugars and most biological species present in serum [78].

Riedel et al. [79] investigated the light-triggered reaction of the redox molecules, hexacyanoferrate, and ferrocenecarboxylic acid, at CdSe/ZnS quantum dot modified gold electrodes for light-driven applications. Here, electron transfer between QDs and redox mediators has been found to be feasible. Additionally, photoluminescence measurements in solution demonstrate the strong interaction between the QDs and the redox species by quenching of QD fluorescence. Subsequently, the established QD–mediator systems have been combined with the enzymes, pyrroloquinoline quinone-dependent glucose dehydrogenase and fructose dehydrogenase, to the feasibility of electrically contacted enzyme/QD biohybrids. This demonstrates the photoelectrochemical principle displays applicability for sensing and for driving QD electrodes by biocatalytic sugar consumption.

7.3.3. Nucleic Acids

One of the most important biomolecules is deoxyribonucleic acid (DNA), which is responsible for determining inherited traits and storing genetic information necessary for the replication of living organisms. The sensors that use DNA molecules are a great tool not only to detect individual DNA or RNA molecules but also molecules belonging to other classes of biomolecules [83–86]. Nucleic acids do not have properties that would be useful for their direct detection, so their detection requires the use of, e.g., fluorescent markers [14,87]. Table 5 shows an exemplary QDs-based sensors for nucleic acids determination.

Table 5. QDs-based sensors for nucleic acids determination.

Marker	Sensing Platform	Transduction Type	LOD	References
DNA in urine	CdTe QDs	Change in fluorescence intensity	3 ng·mL^{-1}	[83]
Mitoxantrone and ribonucleic acid	CdTe QDs	Change in fluorescence intensity	0.1 ng/µL	[84]
dsDNA	CdTe QDs	FRET	-	[85]
DNA, micro-RNA	CdTe/CDs QDs	Change in fluorescence intensity	1 fM	[86]

Nejdl et al. reported a systematic study of the self-assembly of CdTe QDs stabilized by mercaptosuccinic acid (MSA) These QDs were used for the preparation of a fluorescent (off–on) probe based on methylene blue as a quencher for the specific determination of nucleic acid from urine. Using this technique, it was possible to determine the DNA isolated from the urine and decide whether the amount of DNA was in an acceptable range. The LOD was calculated as 0.003 µg·mL^{-1} DNA. Such constructed sensing systems can be used for very sensitive detection of DNA [83].

Mohammadinejad group has prepared mercaptosuccinic acid-capped CdTe quantum dots, which were successfully fabricated as a simple synthesized and sensitive fluorescence sensor for tandem determination of mitoxantrone and ribonucleic acid and also monitoring their interaction. Due to the adsorption of positively-charged mitoxantrone on the surface of negatively-charged quantum dots through electrostatic interactions, the fluorescence intensity of mercaptosuccinic acid-capped CdTe QDs can be effectively quenched by mitoxantrone. After addition of ribonucleic acid to mitoxantrone–QDs solution, mitoxantrone mainly bound to the uracil (C=O) and adenine (C=N) sites of ribonucleic acid. A complex which was formed between mitoxantrone and ribonucleic acid, prevented more interactions between quantum dots and anticancer drug resulted in enhancing of fluorescence intensity. Quantitative results were obtained for all combinations with a linear range of 20–10,000 pM and a LOD of 3–52 pM [87].

7.3.4. Neurotransmitters

Neurotransmitters play a key role in acting as mediators of the autonomic system in the human body. Detection of biological abnormalities related to neurotransmitters (their concentration or metabolites) in the biological fluid is of fundamental importance in medical diagnostics. Differences in the level of neurotransmitters may be related to the occurrence of various diseases substrates such as schizophrenia, Parkinson's disease, Alzheimer's disease, Huntington's chorea, adrenocortical cancer and other cancers and depression. Monitoring the concentration and products of the neurotransmitters synthesis pathway is a promising strategy for early detection and thus preventing the development of these diseases. Table 6 shows an exemplary strategies based on QDs for neurotransmitters determination [88–92].

Table 6. QDs-based sensors for neurotransmitters determination.

Marker	Sensing Platform	Transduction Type	LOD	References
Dopamine	QD@SiO_2	FRET	12.5 nM	[89]
Dopamine	SnSe QDs	Photoelectrochemical assay	3 nM	[90]
Norepinephrine	CdTe@SiO_2@MIP	Change in fluorescence intensity	8 nM	[91]
Serotonin	ZnS QDs	Change in fluorescence intensity	0.69 ng·mL^{-1}	[30]
Acetylcholine	CdSe/ZnS	Change in fluorescence intensity	-	[92]

Wang et al. [89] described a fluorescence assay for the fluorometric determination of dopamine (DA). It is based on the use of silica-coated CdTe quantum dots (QD@SiO_2). When dopamine is added to a solution of the QD@SiO_2 and then oxidized by oxygen under the catalytic action of tyrosinase to form dopamine quinone, the fluorescence of QD@SiO_2 decreases, due to an electron transfer quenching process. Linear relationship over the range from 0.05 to 30 μM DA and high LOD of 12.5 nM. This suggested that the novel assay provided a promising possibility for further utilizing as an efficient platform for measuring DA in biological and environmental applications.

Another example of dopamine detection prepared by Hun et al. described photoelectrochemical sensor for dopamine which yields a signal upon irradiation with visible light. The electrons of SnSe QDs were excited under irradiation with visible light and transformed from valence band to conduction band. Dopamine, as an electron donor, provided the electrons to SnSe QDs. As a result, the enhanced photocurrent was obtained. This sensing system responds linearly to DA in the 0.01 μM to 10 μM concentration range and with a 3 nM LOD [90].

A novel molecular imprinted sensor based on CdTe@SiO2 QDs has been developed by Wei et al., for norepinephrine (NE) recognition. The synthesized nanosensor had a distinguished selectivity and high binding affinity to NE. Under optimal conditions, the relative fluorescence intensity of CdTe@SiO_2@Molecular-imprinted-polymer linearly decreased with an increase in the concentration of NE in the range of 0.04–10 μM. The LOD was calculated as 8 nM [91].

A very promising alternative for construction of new micro-devices is LTCC (Low Temperature Co-fired Ceramic [93]) technology, consisting in the creation of three-dimensional structures of electronic systems based on pressed and co-poured ceramic foils with printed functional layers [94,95]. This method was used by Baluta et al. They proposed a convenient fluorescence dopamine-sensing strategy based on polydopamine formed on the surface of graphene quantum dots (GQDs). This sensing system utilized the catalytic oxidation of DA to dopamine-o-quinone (DOQ), and then to poly(DA), which can selectively quench the strong luminescence of GQDs due to FRET. Such constructed biosensor exhibited a broad linear range from 1 μM up to 200 μM with LOD 80 nM [96].

7.3.5. Pesticides

Pesticides are chemical compounds increasingly used in agricultural production, which play an important role in ensuring optimal efficiency and maximizing income. Despite the positive impact on economic aspects, excessive use of pesticides can lead to the production of harmful chemical intermediates in vegetables, fruits and other agricultural products, which is a serious threat to food safety and human health. In order to ensure constant control of the concentration of these substances in agricultural products, it is necessary to find quick methods to confirm the presence and define their concentration [97–102]. Table 7 presents an exemplary fluorescent strategies for pesticides determination with using of QDs.

Table 7. QDs-based sensors for pesticides determination.

Marker	Sensing Platform	Transduction Type	LOD	References
Glyphosate	CdTe QDs	Change in fluorescence intensity	0.5 $\mu g \cdot mL^{-1}$	[97]
Trichlorfon	CdSe/ZnS	Change in fluorescence intensity	9.0 $\mu g \cdot L^{-1}$	[98]
Organophosphorus	CdTe-TGA QDs	Change in fluorescence intensity	0.68 nM	[99]
Thiram	ZnS:Mn^{2+}	Change in phosphorescence intensity	25 nM	[100]
Doxycycline	TGA/CdTe QDs	Change in fluorescence intensity	1.1×10^{-7} $mol \cdot L^{-1}$	[101]
Paraquat	CdSe/ZnS QDs	Change in fluorescence intensity	3.0 $ng \cdot L^{-1}$	[102]

Kanagasubbulakshmi et al. [99] show the thioglycolic acid (TGA) capped CdTe QDs were highly dispersed and uniform in nature. The TGA surface modification of CdTe did not lead to the agglomeration of QDs. But when an interaction occurred with malathion, the aggregation was formed due to the functional group detachment. The linearity was obtained in the range of 3–21 nM with the LOD 0.68 nM. While Jiménez-López et al. [97] has proposed a multi-commutated flow analysis method for the determination of glyphosate, based on the quenching effect produced by this herbicide on the fluorescence of CdTe quantum dots with the LOD 0.52 $\mu g \cdot mL^{-1}$.

Another example of a sensitive direct competitive biomimetic immunosorbent assay method for pesticide detection was proposed by Liu et al. [98], using the hydrophilic imprinted film as artificial antibody and CdSe/ZnS QD label as a marker. A decrease in the fluorescence of the CdSe/ZnS QD conjugate was observed when QDs has attached to the Trichlorfon. Under optimal conditions, the LOD and sensitivity of the biomimetic immunosorbent assay method were found as 9.0 $\mu g \cdot L^{-1}$ and 5.0 $mg \cdot L^{-1}$.

7.3.6. Toxins

Among the biggest threats to our community are toxins, whose quick and sensitive detection in aqueous solutions, body fluids, food, and drinking water enables the immediate application of appropriate remedies. Among the most common sources of toxins are bacteria whose toxins can be detected in trace amounts in urine or blood after poisoning. However, saliva and nasal swabs can be tested to confirm exposure to toxins even before the onset of symptoms. Another source of toxins in the environment is industrial, military or agricultural activity. Pollutants, such as pesticides and residues from explosives, pollute soil and groundwater and can easily enter the human body [103–107]. Table 8 shows an exemplary QDs-based sensors for toxins determination.

Table 8. QDs-based biosensors for toxins determination.

Marker	Sensing Platform	Transduction Type	LOD	References
Saxitoxin	CdS/CdSe/ZnS QDs	Change in fluorescence intensity	0.3 $\mu g \cdot kg^{-1}$	[103]
Botulinum neurotoxin	Carboxyl QDs	FRET	0.2 $ng \cdot mL^{-1}$	[104]
Cholera toxin, ricin, shiga-like toxin 1, staphylococcal enterotoxin B	CdSe/ZnS QDs	Change in photoluminescence intensity	-	[105]
Ochratoxin A	CdS QDs	Square-Wave Stripping Voltammetry	0.2 $pg \cdot mL^{-1}$	[106]

A new type of molecularly imprinted silica layers appended to CdS/CdSe/ZnS QDs (MIP-QDs) for saxitoxin were fabricated through the surface grafting technique. Sun et al. demonstrated that the synthesized MIP-QDs exhibited excellent selective fluorescence quenching to saxitoxin because of the complementary imprinted cavities on the surface of MIP-QDs. Such constructed MIP-QDs sensor exhibited excellent linearity in the range of 20.0–100.0 $\mu g/L$ with LOD 0.3 $\mu g/kg$ [103].

Wang et al. [104] reported a novel FRET-based nanobiosensor that uses luminescent QDs and dark quencher-labelled peptide probes to rapidly (on the order of hours) detect and quantify biologically active Botulinum neurotoxin (BoNT) and differentiate serotypes A and B, which is based on quantifiable differences in the photoluminescence (PL) intensity of QD reporters. The biorecognition elements for these probes are peptides that contain an amino acid sequence specific for BoNT/A or /B cleavage, a poly(histidine) sequence at the C-terminal for assembly on the QDs, and a dark quencher label (a dye with no native fluorescence) that quenches the QD PL only when the peptide chain is uncleaved (i.e., in the absence of the target BoNT). The sensor signal scaled linearly with the analyte concentration over a range of 8–200 nM, with 4 pM as the LOD.

7.3.7. Volatile Substances

Volatile organic compounds (VOCs) are organic chemical compounds with an evaporation temperature close to room temperature. These substances are commonly used in industry and as household products. Too high exposure to volatile organic compounds can have both short- and long-term adverse effects on health, such as respiratory failure [108–111]. An exemplary QDs-based sensors for volatile substances determination are presented in Table 9.

Table 9. QDs-based sensors for volatile substances determination.

Marker	Sensing Platform	Transduction Type	LOD	References
Ammonia gas	PbS QDs/TiO_2 NTARs	Change in fluorescence intensity	2 ppm	[109]
Methanol	CdS QDs	Photoelectrochemical assay	0.14 $\mu g \cdot L^{-1}$	[110]
Acetone	Mn^{2+}-ZnS QDs	Change in room temperature phosphorescence intensity	0.2 $mg \cdot L^{-1}$	[111]
Ethanol, 2-propanol, acetone	CdTe QDs	Change in photoluminescence intensity	-	[108]

Liu et al. [109] has presented work in which PbS-QDs/TiO_2-nanotubes arrays (PbS QDs/TiO_2 NTARs) are prepared by successive ionic layer adsorption and reaction, which are used to fabricate the gas sensor. The gas sensing performance shows that PbS QDs/TiO_2 NTARs possess a good response towards ammonia gas at room temperature. The enhanced sensing mechanism lies in the fact that PbS QDs in PbS QDs/TiO_2 NTARs may provide more sites to absorb the ammonia molecules and increase the depletion layer. The well-combined interface may provide effective transportation of the electrons as well as the direct transportation of the electrons along the TiO_2 NTARs axis. This sensing strategy exhibited linearity in the range from 2 to 100 ppm at room temperature, with a LOD 2 ppm.

Barroso et al. [110] presented a new strategy for the detection of methanol using fluorescence spectroscopy and photoelectrochemical (PEC) analysis. The analytical system is based on the oxidation of cysteine (CSH) with hydrogen peroxide (H_2O_2) enzymatically generated by alcohol oxidase (AOx). H_2O_2 oxidizes capping agent CSH, modulating the growth of CSH-stabilized CdS QDs. Disposable screen-printed carbon electrodes (SPCEs) modified with a conductive osmium polymer (Os-PVP) complex were employed to quantify resulting CdS QDs. This polymer facilitates the "wiring" of in situ enzymatically generated CdS QDs, which photocatalyzed oxidation of 1-thioglycerol (TG), generating photocurrent as the readout signal. As a result, an increase of the fluorescence intensity was observed.

Sotelo-Gonzalez et al. [111] has prepared colloidal Mn^{2+}-doped ZnS nanoparticles exhibiting room temperature phosphorescence (RTP) emission and water solubilized by capping the QDs surface with l-cysteine. Such coating of the nanoparticle with cysteine groups allows their analytical application for acetone determination (selected as model ketone species) in aqueous media (by measuring the quenching on the RTP emission of such QDs after direct interaction with the analyte). It was observed that the rise of acetone concentration efficiently quenches of the phosphorescence emission. The linear range of the developed methodology turned out to be at least up to 600 $mg \cdot L^{-1}$ with the LOD for acetone dissolved in an aqueous medium of 0.2 $mg \cdot L^{-1}$.

7.3.8. Vitamins

Vitamins, which are found in many animal and plant tissues, play an essential role in proper metabolism and maintenance of body cells. Disorders in their synthesis or metabolism may be the cause of many serious diseases, hence the invention of sensitive and fast sensor devices is significant [112–116]. An exemplary QDs-based sensors for determination of vitamins are presented in Table 10.

Table 10. QDs-based sensors for vitamins determination.

Marker	Sensing Platform	Transduction Type	LOD	References
Vitamin E	CdSe/ZnS QDs, GOx	Change in fluorescence intensity	3.5 nM	[112]
Folic acid	ZnS:Cu/ZnS-MPA QDs	Change in fluorescence intensity	11 μM	[113]
Vitamin B_{12}	CDs	FRET	0.1 μg·mL^{-1}	[114]
Vitamin B_6	CdTe-TGA QDs	Change in fluorescence intensity	-	[115]
Ascorbic Acid	CdS-diphenylcarbazide QDs	FRET	2 nM	[116]

Liu et al. [112] prepared a novel optosensing material based on quantum dots and graphene oxide for specific determination of Vitamin E. Ultra-high specific surface was obtained by synthesis of molecular imprinted polymer (MIP), which was stocked for specific Vitamin E reaction area. Under optimal condition, the fluorescence intensity of MIP was decreased linearly with the increasing concentration of Vitamin E. Such constructed sensor exhibited good linear range from 2.30×10^{-2}–9.20×10^{2} μM with a LOD 3.5 nM.

Another strategy was used by Geszke–Moritz et al. [113] They used high fluorescence sensitivity to folic acid due to the high affinity of nitrogen atoms and carboxyl groups to doped QDs. Due to the quenching of fluorescence intensity QDs, it is possible to detect folic acid concentrations from LOD 11 μM.

Ganiga and others presented an optical sensor that uses FRET for fast and sensitive detection of ascorbic acid (AA). For this purpose, CdS QDs and diphenylcarbadiazone (DPCD) were used. In the presence of AA, the DPCD was transformed into diphenylcarbazide (DPC), which resulted in the recovery of fluorescence. Changes in fluorescence intensity enabled the detection and determination of AA concentration in the linear range of 60–300 nM with LOD 2 nM [116].

7.4. Detection of Bacteria and Viruses

The identification of pathogenic bacteria and viruses in food, water, air, and body fluids is extremely important because of their drastic impact on our society. The result of the human body's contact with pathogenic bacteria or viruses is serious gastrointestinal infections that can lead to patient death without a doctor's control. Importantly, pathogenic bacteria also produce toxins that are responsible for the occurrence of serious diseases, such as hemorrhagic colitis, characterized by painful abdominal cramps and bloody diarrhea or hemolytic-uremic syndrome, the most severe effect of which is an acute renal failure. Fast and sensitive detection of pathogenic bacteria and viruses is necessary to prevent the occurrence of epidemics or severe forms of the disease [117–121]. Table 11 shows an exemplary strategies for bacteria and viruses determination.

Table 11. QDs-based sensors for bacteria and viruses determination.

Marker	Sensing Platform	Transduction Type	LOD	References
E. coli	CdSe/ZnS QDs	Change in fluorescence intensity	1.4×10^{1} CFU·mL^{-1}	[117]
E. coli	MPA-ZnSe/ZnS QDs	Change in fluorescence intensity	10^{1} CFU·mL^{-1}	[118]
HIV and HPV	CdTe QDs	Change in fluorescence intensity	100 nM	[119]
E. coli	CdSe/ZnS QDs	Change in fluorescence intensity	2.08×10^{7} CFU·mL^{-1}	[120]
E. coli	CdSe/ZnS QDs	Change in fluorescence intensity	2.3 CFU·mL^{-1}	[121]

Xue et al. [120] presented a novel fluorescent biosensor for ultra-sensitive and rapid detection of *E.coli* O157:H7 with LOD 14 CFU/mL. The proposed fluorescent biosensor used the double-layer channel with the immune magnetic nanoparticles (MNPs) for specific separation and efficient concentration of the target bacteria, and the immune CdSe/ZnS QDs with a portable optical system for quantitative detection of the bacteria. Initially, the bacteria were captured by the immune MNPs in the channel at the presence of the high gradient magnetic fields (HGMFs) to form the MNP-bacteria complexes. Then, the immune QDs were used to react with the target bacteria to form the MNP-bacteria-QDs complexes in the channel. Finally, the enriched complexes were collected and detected using the portable optical system to obtain increase the fluorescence intensity for final determination of the *E.coli* O157:H7 cells in the sample. Wu et al. [118] has prepared modified ZnSe/ZnS QDs by 3-mercaptopropionic acid and established a rapid fluorescence method to detect the *E. coli* cells count by using MPA-ZnSe/ZnS QDs as a fluorescence probe. The fluorescence peak intensity increases with increasing cells count of bacteria. Compared with the traditional fluorescent detection methods, this one is more convenient and useful in the bacterial count determination with LOD 10^1 CFU·mL^{-1}.

In addition to bacteria, QDs are also used for virus detection. Jimenez et al. reported work which was focused on the development of a nano-system for simultaneous identification of HIV and HPV viruses with 1 nM of LOD. Their construction and characterization were carried doubt using magnetic glass particles (MGPs) which joined with target DNA oligonucleotides and the second part of the construction formed by the conjugation of red and green CdTe QDs with oligodeoxyribonucleotides complementary probe, derived from these two viruses, that encode respectively their capsid and oncoproteins. As a result, after the conjugating, the fluorescent intensity was slightly reduced in both cases [119].

8. Conclusions and Perspectives

Quantum dots have remarkable optical properties, which make them among the most useful nanomaterials [6]. They may be utilized in a wide range of applications, e.g., in new types of fluorescent probes and as active components of nanostructure-biomolecule complexes [122]. Various schemes for the application of optical transduction QDs have been successfully tested, allowing a wide range of detection, high selectivity and sensitivity in the tested samples. The development of analytical methods for the detection of various chemical or biological compounds allows the use of QDs in sensors for determining the presence of ions, molecules and pH changes. The results of discussed studies lead to the improvement of existing detection devices and the design of new detection devices that allow more sensitive and faster analysis. Quantum dots-based detection technologies can be adapted to precision medical technologies by overturning point-of-care (POC) and personalized diagnostics. This engineering can supply high-throughput and mobile diagnostic platforms for screening pathogens and toxins immediately in field and POC clinical settings. Several of these technologies tender multiplexing capacities for simultaneous examination of multiple analytes with unexpectedly high sensitivity that can notably lower costs and detection time. Nevertheless, a universal sensor for different types of medicinal/or, i.e., food samples, is a challenge because of the inherent complexity of biological samples. Evaluation with numerous of biological (i.e., food, body fluids) samples and comparison with well-established techniques may assist to direct this challenge.

Among further fields that could exploit some of the advantages of QDs are fluorescent immunosensors designed as integrated devices. Due to the relevant improvement on the execution of fluorescent immunosensors and recent advances in miniaturization processes, it is believed that in the near future small and advanced fluorescent mobile analytical platforms, which combine steps of the immunoassay pathway, will be available. In conclusion, there is potential for further investigations of QDs in multiplex detection, particularly via continued miniaturization and integration into lab-on-chip platforms.

Author Contributions: Conceptualization, A.P. and J.C.; Formal Analysis, A.P., J.C. and K.M.; Investigation, A.L. and K.D.; Resources, A.L. and K.D.; Writing-Original Draft Preparation, A.L. and K.D.; Writing-Review & Editing, A.L. and K.D.; Visualization, A.L., A.P. and K.D.; Supervision, J.C., A.P., M.B. and K.M.; Project Administration, A.P., J.C., K.M.; Funding Acquisition, J.C. and K.M.

Funding: This research received no external funding.

Acknowledgments: The authors gratefully acknowledge the financial support from Wroclaw University of Science and Technology (statutory activity no. 0401/0145/18 and 0401/0137/18).

Conflicts of Interest: The authors declare no conflicts of interest.

References

1. Luo, Z.; Chen, Y.; Wu, S.-T. Wide color gamut LCD with a quantum dot backlight. *Opt. Express* **2013**, *21*, 26269–26284. [CrossRef] [PubMed]
2. Song, H.-J.; Jeong, B.G.; Lim, J.; Lee, D.C.; Bae, W.K.; Klimov, V.I. Performance Limits of Luminescent Solar Concentrators Tested with Seed/Quantum-Well Quantum Dots in a Selective-Reflector-Based Optical Cavity. *Nano Lett.* **2018**, *18*, 395–404. [CrossRef] [PubMed]
3. Liu, H.; Lia, S.; Chen, W.; Wang, D.; Lia, C.; Wu, D.; Hao, J.; Zhoua, Z.; Wang, X.; Wang, K. Scattering enhanced quantum dots based luminescent solar concentrators by silica microparticles. *Sol. Energy Mater. Sol. Cells* **2018**, *179*, 380–385. [CrossRef]
4. Zhang, Y.; Clapp, A. Overview of Stabilizing Ligands for Biocompatible Quantum Dot Nanocrystals. *Sensors* **2011**, *11*, 11037–11055. [CrossRef] [PubMed]
5. Larson, D.R.; Zipfel, W.R.; Williams, R.M.; Clark, S.W.; Bruchez, M.P.; Wise, F.W. Water-Soluble Quantum Dots for Multiphoton Fluorescence Imaging in Vivo. *Science* **2003**, *300*, 1434–1436. [CrossRef]
6. Frasco, M.F.; Chaniotakis, N. Semiconductor Quantum Dots in Chemical Sensors and Biosensors. *Sensors* **2009**, *9*, 7266–7286. [CrossRef]
7. Medintz, I.L.; Uyeda, H.T.; Goldman, E.R.; Mattoussi, H. Quantum dot bioconjugates for imaging, labelling and sensing. *Nat. Mater.* **2005**, *4*, 435–446. [CrossRef]
8. Sapsford, K.E.; Algar, R.W.; Berti, L.; Gemmill, K.B.; Casey, B.J.; Oh, E.; Stewart, M.H.; Medintz, I.L. Functionalizing Nanoparticles with Biological Molecules: Develpoing Chemistries that Facilitate Nanotechnology. *Chem. Rev.* **2013**, *113*, 1904–2074. [CrossRef]
9. Foubert, A.; Beloglazova Rajkovic, A.; Sas, B.; Madder, A.; Goryacheva, I.Y.; Saeger, S.D. Bioconjugation of quantum dots: Review & impact on future application. *Trends Anal. Chem.* **2016**, *83*, 31–48. [CrossRef]
10. Hardman, R. A Toxycologic Review of quantum Dots: Toxicity Depends on Physicochemical and Environmental Factors. *Environ. Health Perspect.* **2006**, *114*, 165–172. [CrossRef]
11. Lovrić, J.; Bazzi, H.S.; Cuie, Y.; Fortin, G.R.; Winnik, F.M.; Maysinger, D. Differences in subcellular distribution and toxicity of green and red emitting CdTe quantum dots. *J. Mol. Med.* **2005**, *83*, 377–385. [CrossRef] [PubMed]
12. Liu, Y.S.; Sun, Y.; Vernier, P.T.; Liang, C.H.; Chong, S.Y.C.; Gundersen, M.A. pH-sensitive photoluminescence of CdSe/ZnSe/ZnS quantum dots in human ovarian cancer cells. *J. Phys. Chem. C* **2007**, *111*, 2872–2878. [CrossRef] [PubMed]
13. Chen, Y.; Rosenzweig, Z. Luminescent CdS quantum dots as selective ion probes. *Anal. Chem.* **2002**, *74*, 5132–5138. [CrossRef] [PubMed]
14. Cissell, K.A.; Campbell, S.; Deo, S.K. Rapid, single-step nucleic acid detection. *Anal. Bioanal. Chem.* **2008**, *391*, 2577–2581. [CrossRef] [PubMed]
15. Dabbousi, B.O.; Rodriguez-Viejo, J.; Mikulec, F.V.; Heine, J.R.; Mattoussi, H.; Ober, R.; Jensen, K.F.; Bawendi, M.G. (CdSe)ZnS Core-Shell Quantum DotsL Synthesis and Characterization of a Size Series of Highly Luminescent Nanocrystallites. *J. Phys. Chem. B* **1997**, *101*, 9463–9475. [CrossRef]
16. Wang, X.; Qu, L.; Zhang, J.; Peng, X.; Xiao, M. Surface-related emission in highly luminescent CdSe QDs. *Nano Lett.* **2003**, *3*, 1103–1106. [CrossRef]
17. Resch-Genger, U.; Grabolle, M.; Cavaliere-Jaricot, S.; Nitschke, R.; Nann, T. Quantum dots versus organic dyes as fluorescent labels. *Nat. Methods* **2008**, *5*, 763–775. [CrossRef]
18. Goldman, E.R.; Medintz, I.L.; Mattoussi, H. Luminescent quantum dots in immunoassays. *Anal. Bioanal. Chem.* **2006**, *384*, 560–563. [CrossRef]
19. Xu, X.; Liu, X.; Nie, Z.; Pan, Y.; Guo, M.; Yao, S. Label-Free Fluorescent Detection of Protein Kinase Activity Based on the Aggregation Behavior of Unmodified Quantum Dots. *Anal. Chem.* **2011**, *83*, 52–59. [CrossRef]
20. Algar, R.W.; Tavares, A.J.; Krull, U.J. Beyond labels: A review of the application of quantum dots as integrated components of assays, bioprobes, and biosensors utilizing optical transduction. *Anal. Chim. Acta* **2010**, *673*, 1–25. [CrossRef]
21. Gründler, P. *Chemical Sensors: An Introduction for Scientists and Engineers*; Springer: Berlin/Heidelberg, Germany, 2007; ISBN 978-3-540-45742-8.

22. Nowak Wenger, W.; Bates, F.S.; Aydil, E.S. Functionalization of Cadmium Selenide Quantum Dots with Poly(ethylene glycol): Ligand Exchange, Surface Coverage, and Dispersion Stability. *Langmuir* **2017**, *33*, 8239–8245. [CrossRef] [PubMed]
23. Zhu, H.; Hu, M.Z.; Shao, L.; Yu, K.; Dabestani, R.; Zaman Md, B.; Liao, S. Synthesis and Optical Properties of Thiol Functionalized CdSe/ZnS (Core/Shell) Quantum Dots by Ligand Exchange. *J. Nanomater.* **2014**. [CrossRef]
24. Liu, X.; Luo, Y. Surface Modifications Technology of Quantum Dots Based Biosensors and Their Medical Applications. *Chin. J. Anal. Chem.* **2014**, *42*, 1061–1069. [CrossRef]
25. Blanco-Canosa, J.B.; Wu, M.; Susumu, K.; Petryayeva, E.; Jennings, T.L.; Dawson, P.E.; Algar, W.R.; Medintz, I.L. Recent progress in the bioconjugation of quantum dots. *Coord. Chem. Rev.* **2014**, *263–264*, 101–137. [CrossRef]
26. Zhou, J.; Liu, Y.; Tang, J.; Tang, W. Surface ligands engineering of semiconductor quantum dots for chemosensory and biological applications. *Mater. Today* **2017**, *20*, 360–376. [CrossRef]
27. Nsibande, S.A.; Forbes, P.B.C. Fluorescence detection of pesticides using quantum dot materials—A review. *Anal. Chim. Acta* **2016**, *945*, 9–22. [CrossRef]
28. Walia, S.; Acharya, A. Fluorescent cadmium sulfide nanoparticles for selective and sensitive detection of toxic pesticides in aqueous medium. *J. Nanopart. Res.* **2014**, *14*, 1–10. [CrossRef]
29. De Bastida, G.; Arregui, F.J.; Javier Goicoechea, J.; Matias, I.R. Quantum Dots-Based Optical Fiber Temperature Sensors Fabricated by Layer-by-Layer. *IEEE Sens. J.* **2006**, *6*, 1378–1379. [CrossRef]
30. Wang, Z.; Zhang, Y.; Zhang, B.; Lu, X. Mn^{2+} doped ZnS QDs modified fluorescence sensor based on molecularly imprinted polymer/sol-gel chemistry for detection of Serotonin. *Talanta* **2018**, *190*, 1–8. [CrossRef]
31. Jiang, H.; Ju, H. Electrochemiluminescence Sensors for Scavengers of Hydroxyl Radical Based on Its Annihilation in CdSe Quantum Dots Film/Peroxide System. *Anal. Chem.* **2007**, *79*, 6690–6696. [CrossRef]
32. Jie, G.; Zhang, J.; Wang, D.; Cheng, C.; Chen, H.-Y.; Zhu, J.-J. Electrochemiluminescence Immunosensor Basedon CdSe Nanocomposites. *Anal. Chem.* **2008**, *80*, 4033–4039. [CrossRef] [PubMed]
33. Shang, L.; Zhang, L.; Dong, S. Turn-on fluorescent cyanide sensor based on copper ion-modified CdTe quantum dots. *Analyst* **2009**, *134*, 107–113. [CrossRef] [PubMed]
34. Wang, L.Y.; Wang, L.; Gao, F.; Yu, Z.Y.; Wu, Z.M. Application of functionalized CdS nanoparticles as fluorescence probe in the determination of nucleic acid. *Analyst* **2002**, *127*, 977–980. [CrossRef] [PubMed]
35. Kavosia, B.; Navaee, A.; Salimi, A. Amplified fluorescence resonance energy transfer sensing of prostate specific antigen based on aggregation of CdTe QDs/antibody and aptamer decoratedof AuNPs-PAMAM dendrimer. *Luminescence* **2018**, *204*, 368–374. [CrossRef]
36. Anikeeva, P.O.; Madigan, C.F.; Halpert, J.E.; Bawendi, M.G.; Bulović, V. Electronic and excitonic processes in light-emitting devices based on organic materials and colloidal quantum dots. *Phys. Rev. B* **2008**, *78*, 085434. [CrossRef]
37. Van Sark WGJ, H.M.; Frederix PL, T.M.; Bol, A.A.; Gerritsen, H.C.; Meijerink, A. Blueing, Bleaching, and Blinking of Single CdSe/ZnS Quantum Dots. *Chemphyschem* **2002**, *3*, 871–879. [CrossRef]
38. Wang, L.Y.; Kan, X.W.; Zhang, M.C.; Zhu, C.Q.; Wang, L. Fluorescence for the determination of protein with functionalized nano-ZnS. *Analyst* **2002**, *127*, 1531–1534. [CrossRef]
39. Loura, L.M.S.; Prieto, M. FRET in membrane biophysics: An overview. *Front. Physiol.* **2011**, *2*, 82. [CrossRef] [PubMed]
40. Clapp, A.R.; Medintz, I.L.; Mattoussi, H. Förster Resonance Energy Transfer Investigations Using Quantum-Dot Fluorophores. *ChemPhysChem* **2006**, *7*, 47–57. [CrossRef]
41. Miyawaki, A. Visualization of the spatial and temporal dynamics of intracellular signaling. *Dev. Cell* **2003**, *4*, 295–305. [CrossRef]
42. Miyawaki, A. Green Fluorescent Protein Glows Gold. *Cell* **2008**, *135*, 987–990. [CrossRef] [PubMed]
43. Lasota, S.; Baster, Z.; Witko, T.; Zimoląg, E.; Sroka, J.; Rajfur, Z.; Madeja, Z. Zastosowanie biosensorów typu FRET w badaniach mikroskopowych procesu migracji komórkowej. *Postępy Biochemii* **2017**, *63*, 16–33. Available online: http://www.postepybiochemii.pl/pdf/1_2017/16.pdf (accessed on 13 November 2018). [PubMed]

44. Clapp, A.R.; Medintz, I.L.; Mauro, J.M.; Fisher, B.R.; Bawendi, M.G.; Mattoussi, H. Fluorescence resonance energy transfer between quantum dot donors and dye-labeled protein acceptors. *J. Am. Chem. Soc.* **2004**, *126*, 301–310. [CrossRef] [PubMed]
45. Patolsky, F.; Gill, R.; Weizmann, Y.; Mokari, T.; Banin, U.; Willner, I. Lighting-Up the Dynamics of Telomerization and DNA Replication by CdSe-ZnS Quantum Dots. *J. Am. Chem. Soc.* **2003**, *125*, 13918–13919. [CrossRef] [PubMed]
46. Coto-García, A.M.; Sotelo-González, E.; Fernández-Argüelles, M.T.; Pereiro, R.; Costa-Fernández, J.M.; Sanz-Medel, A. Nanoparticles as fluorescent labels for optical imaging and sensing in genomics and proteomics. *Anal. Bioanal. Chem.* **2011**, *399*, 29–42. [CrossRef] [PubMed]
47. Freeman, R.; Girsh, J.; Willner, I. Nucleic Acid/Quantum Dots (QDs) Hybrid Systems for Optical and Photoelectrochemical Sensing. *ACS Appl. Mater. Interfaces* **2013**, *5*, 2815–2834. [CrossRef] [PubMed]
48. Zhang, H.; Zhang, L.; Liang, R.-P.; Huang, J.; Qiu, J.-D. Simultaneous Determination of Concanavalin A and Peanut Agglutinin by Dual-Color Quantum Dots. *Anal. Chem.* **2013**, *85*, 10969–10976. [CrossRef]
49. Waggonera, P.S.; Craighead, H.G. Micro- and nanomechanical sensors for environmental, chemical, and biological detection. *Lab Chip* **2007**, *7*, 1238–1255. [CrossRef]
50. Mo, G.; He, X.; Zhou, C.; Ya, D.; Feng, J.; Yu, C.; Deng, B. Sensitive detection of hydroquinone based on electrochemiluminescence energy transfer between the exited ZnSe quantum dots and benzoquinone. *Sens. Actuator B Chem.* **2018**, *266*, 784–792. [CrossRef]
51. Yao, J.; Li, L.; Li, P.; Yang, M. Quantum dots: From fluorescence to chemiluminescence, bioluminescence, electrochemiluminescence, and electrochemistry. *Nanoscale* **2017**, *9*, 13364–13383. [CrossRef]
52. Haram, S.K.; Quinn, B.M.; Bard, A.J. Electrochemistry of CdS Nanoparticles: A Correlation between Optical and Electrochemical Band Gaps. *J. Am. Chem. Soc.* **2001**, *123*, 8860–8861. [CrossRef] [PubMed]
53. Huang, H.; Zhu, J.J. The electrochemical applications of quantum dots. *Analyst* **2013**, *138*, 5855–5865. [CrossRef]
54. Liu, Y.; Zhu, L.; Kong, J.; Yang, P.; Liu, B. A quantum dots-based electrochemical assay towards the sensitive detection of tumor cells. *Electrochem. Commun.* **2013**, *33*, 59–62. [CrossRef]
55. Xu, Y.; Liu, J.; Gao, C.; Wang, E. Applications of carbon quantum dots in electrochemiluminescence: A mini review. *Electrochem. Commun.* **2014**, *48*, 151–154. [CrossRef]
56. Bertoncello, P.; Ugo, P. Recent Advances in Electrochemiluminescence with Quantum Dots and Arrays of Nanoelectrodes. *ChemElectroChem* **2017**, *4*, 1663–1676. [CrossRef]
57. Zhou, M.; Guo, J.; Yang, C. Ratiometric fluorescence sensor for Fe3+ ions detection based on quantum dot-doped hydrogel optical fiber. *Sens. Actuator B Chem.* **2018**, *264*, 52–58. [CrossRef]
58. Li, H.; Zhang, Y.; Wang, X.; Xiong, D.; Bai, Y. Calixarene capped quantum dots as luminescent probes for Hg^{2+} ions. *Mater. Lett.* **2007**, *61*, 1474–1477. [CrossRef]
59. Zhao, Q.; Rong, X.; Ma, H.; Tao, G. Dithizone functionalized CdSe/CdS quantum dots as turn-on fluorescent probe for ultrasensitive detection of lead ion. *J. Hazard. Mater.* **2013**, *250*, 45–52. [CrossRef]
60. Li, J.; Mei, F.; Li, W.Y.; He, X.W.; Zhang, Y.K. Study on the fluorescence resonance energy transfer between CdTe QDs and butylrhodamine B in the presence of CTMAB and its application on the detection of Hg(II). *Spectrochim. Acta Part A* **2008**, *70*, 811–817. [CrossRef]
61. Wang, X.; Guo, X. Ultrasensitive Pb^{2+} detection based on fluorescence resonance energy transfer (FRET) between quantum dots and gold nanoparticles. *Analyst* **2009**, *134*, 1348–1354. [CrossRef]
62. Wang, C.; Huang, Y.; Jiang, K.; Humphrey, M.G.; Zhang, C. Dual-emitting quantum dot/carbon nanodot-based nanoprobe for selective and sensitive detection of Fe3+ in cells. *Analyst* **2016**, *141*, 4488–4494. [CrossRef] [PubMed]
63. Brahim, N.B.; Mohamed, N.B.H.; Echabaane, M.; Haouari, M.; Chaâbane, R.B.; Negrerie, M.; Ouadaa, H.B. Thioglycerol-functionalized CdSe quantum dots detecting cadmium ions. *Sens. Actuators B Chem.* **2015**, *220*, 1346–1353. [CrossRef]
64. Yao, J.; Zhang, K.; Zhu, H.; Ma, F.; Sun, M.; Yu, H.; Sun, J.; Wang, S. Efficient ratiometric fluorescence probe based on dual-emission quantum dots hybrid for on-site determination of copper ions. *Anal. Chem.* **2013**, *85*, 6461–6468. [CrossRef] [PubMed]
65. Xu, H.; Zhang, K.; Liu, Y.; Xie, M. Visual and fluorescent detection of mercury ions by using a dually emissive ratiometric nanohybrid containing carbon dots and CdTe quantum dots. *Microchim. Acta* **2017**, *184*, 1199–1206. [CrossRef]

66. Tomasulo, M.; Yildiz, I.; Kaanumalle, S.L.; Raymo, F.M. pH-Sensitive Ligand for Luminescent Quantum Dots. *Langmuir* **2006**, *22*, 10284–10290. [CrossRef] [PubMed]
67. Snee, P.T.; Somers, R.C.; Nair, G.; Zimmer, J.P.; Bawendi, M.G.; Nocera, D.G. A Ratiometric CdSe/ZnS Nanocrystals pH Sensor. *J. Am. Chem. Soc.* **2006**, *128*, 13320–13321. [CrossRef] [PubMed]
68. Susha, A.S.; Javier, A.M.; Parak, W.J.; Rogach, A.L. Luminescent CdTe nanocrystals as ion probes and pH sensors in aqueous solutions. *Colloids Surf. A* **2006**, *281*, 40–43. [CrossRef]
69. Wang, Y.; Ye, C.; Zhu, Z.H.; Hu, Y.Z. Cadmium telluride quantum dots as pH-sensitive probes for tiopronin determination. *Anal. Chim. Acta* **2008**, *610*, 50–56. [CrossRef]
70. Huang, C.P.; Li, Y.L.; Chen, T.M. A highly sensitive system for urea detection by using CdSe/ZnS core-shell quantum dots. *Biosens. Bioelectron.* **2007**, *22*, 1835–1838. [CrossRef]
71. Lv, Y.; Wu, R.; Feng, K.; Li, J.; Mao, Q.; Yuan, H.; Shen, H.; Chai, X.; Li, L.S. Highly sensitive and accurate detection of C-reactive protein by CdSe/ZnS quantum dot-based fluorescence-linked immunosorbent assay. *J. Nanobiotechnol.* **2017**, *15*, 35. [CrossRef]
72. Zhang, C.; Ding, C.; Zhou, G.; Xue, Q.; Xian, Y. One-step synthesis of DNA functionalized cadmium-free quantum dots and its application in FRET-based protein sensing. *Anal. Chim. Acta* **2017**, *957*, 63–69. [CrossRef] [PubMed]
73. Huang, S.; Xiao, Q.; He, Z.K.; Liu, Y.; Tinnefeld, P.; Su, X.R.; Peng, X.N. A high sensitive and specific QDs FRET bioprobe for MNase. *Chem. Commun.* **2008**, *45*, 5990–5992. [CrossRef] [PubMed]
74. Boeneman, K.; Mei, B.C.; Dennis, A.M.; Bao, G.; Deschamps, J.R.; Mattoussi, H.; Medintz, I.L. Sensing caspase 3 activity with quantum dot-fluorescent protein assemblies. *J. Am. Chem. Soc.* **2009**, *131*, 3828–3829. [CrossRef] [PubMed]
75. Choi, J.H.; Chen, K.H.; Strano, M.S. Aptamer-capped nanocrystal quantum dots: A new method for label-free protein detection. *J. Am. Chem. Soc.* **2006**, *128*, 15584–15585. [CrossRef] [PubMed]
76. Hu, J.; Wang, Z.Y.; Li, C.; Zhang, C. Advances in single quantum dot-based nanosensors. *Chem. Commun.* **2017**, *53*, 13284–13295. [CrossRef] [PubMed]
77. Saran, A.D.; Sadawana, M.M.; Srivastava, R.; Bellare, J.R. An optimized quantum dot-ligand system for biosensing applications: Evaluation as a glucose biosensor. *Colloids Surf. A* **2011**, *384*, 393–400. [CrossRef]
78. Tang, B.; Cao, L.; Xu, K.; Zhuo, L.; Ge, J.; Li, Q.; Yu, L. A New Nanobiosensor for Glucose with High Sensitivity and Selectivity in Serum Based on Fluorescence Resonance Energy Transfer (FRET) between CdTe Quantum Dots and Au Nanoparticles. *Chem. Eur. J.* **2008**, *14*, 3637–3644. [CrossRef] [PubMed]
79. Riedel, M.; Sabir, N.; Scheller, F.W.; Parak, W.J.; Lisdat, F. Connecting quantum dots with enzymes: Mediator-based approaches for the light-directed read-out of glucose and fructose oxidation. *Nanoscale* **2017**, *9*, 2814–2823. [CrossRef]
80. Wu, P.; He, Y.; Wang, H.F.; Yan, X.P. Conjugation of Glucose Oxidase onto Mn-Doped ZnS Quantum Dots for Phosphorescent Sensing of Glucose in Biological Fluids. *Anal. Chem.* **2010**, *82*, 1427–1433. [CrossRef]
81. Cao, L.; Ye, J.; Tong, L.; Tang, B. A new route to the considerable enhancement of glucose oxidase (GOx) activity: The simple assembly of a complex from CdTe quantum dots and GOx and its glucose sensing. *Chem. Eur. J.* **2008**, *14*, 9633–9640. [CrossRef]
82. Hua, M.; Tiana, J.; Lua, H.T.; Weng, L.X.; Wang, L.H. H_2O_2-sensitive quantum dotsfor the label-free detection of glucose. *Talanta* **2010**, *82*, 997–1002. [CrossRef] [PubMed]
83. Nejdl, L.; Zelnickova, J.; Vaneckova, T.; Hynek, D.; Adam, V.; Vaculovicova, M. Rapid preparation of self-assembled CdTe quantum dots used for sensing of DNA in urine. *New J. Chem.* **2018**, *42*, 6005–6012. [CrossRef]
84. Mohammadinejad, A.; Es'haghi, Z.; Abnous, K.; Mohajeri, S.A. Tandem determination of mitoxantrone and ribonucleic acid using mercaptosuccinic acid-capped CdTe quantum dots. *J. Lumin.* **2017**, *190*, 254–260. [CrossRef]
85. Dai, Z.; Zhang, J.; Dong, Q.; Guo, N.; Xu, S.; Sun, B.; Bu, Y. Adaptation of Au Nanoparticles and CdTe Quantum Dots in DNA Detection. *Chin. J. Chem. Eng.* **2007**, *15*, 791–794. [CrossRef]
86. Su, S.; Fan, J.; Xue, B.; Yuwen, L.; Liu, X.; Pan, D.; Fan, C.; Wang, L. DNA-Conjugated Quantum Dot Nanoprobe for High-Sensitivity Fluorescent Detetction of DNA and micro-RNA. *ACS Appl. Mater. Interfaces* **2014**, *6*, 1152–1157. [CrossRef]
87. Pei, X.; Yin, H.; Lai, T.; Zhang, J.; Liu, F.; Xu, X.; Li, N. Multiplexed Detection of Attomoles of Nucleic Acids Using Fluorescent Nanoparticle Counting Platform. *Anal. Chem.* **2018**, *90*, 1376–1383. [CrossRef]

88. Ghasemi, F.; Hormozi-Nezhad, M.R.; Mahmoudi, M. Identification of catecholamine neurotransmitters using fluorescence sensor array. *Anal. Chim. Acta* **2016**, *917*, 85–92. [CrossRef]
89. Wang, B.; Chen, M.; Zhang, H.; Wen, W.; Zhang, X.; Wang, S. A simple and sensitive fluorometric dopamine assay based on silica-coated CdTe quantum dots. *Microchim. Acta* **2017**, *184*, 3189–3196. [CrossRef]
90. Hun, X.; Wang, S.; Mei, S.; Qin, H.; Zhang, H.; Luo, X. Photoelectrochemical dopamine sensor based on a gold electrode modified with SnSe nanosheets. *Microchim. Acta* **2017**, *184*, 3333–3338. [CrossRef]
91. Wei, F.; Wu, Y.; Xu, G.; Gao, Y.; Yang, J.; Liu, L.; Zhou, P.; Hu, Q. Molecularly imprinted polymer based on CdTe@SiO2 quantum dots as a fluorescent sensor for the recognition of norepinephrine. *Analyst* **2014**, *139*, 5785–5792. [CrossRef]
92. Jin, T.; Fujii, F.; Sakata, H.; Tamura, M.; Kinjo, M. Amphiphilic p-sulfonatocalix[4]arene-coated CdSe/ZnS quantum dots for the optical detection of the neurotransmitter acetylcholine. *Chem. Commun.* **2005**, *0*, 4300–4302. [CrossRef] [PubMed]
93. Malecha, K.; Macioszczyk, J.; Słobodzian, P. Application of microwave heating in ceramic-based microfluidic module. *Microelectron. Int.* **2018**, *35*, 126–132. [CrossRef]
94. Malecha, K. The Implementation of Fluorescence-Based Detection in LTCC (Low-Temperature-Co-Fired -Ceramics) Microfluidic Modules. *Int. J. Appl. Ceram. Technol.* **2016**, *13*, 69–77. [CrossRef]
95. Golonka, L.J. Technology and applications of Low Temperature Cofired Ceramic (LTCC) based sensors and microsystems. *Bull. Pol. Acad. Sci. Tech. Sci.* **2006**, *54*, 221–231.
96. Baluta, S.; Malecha, K.; Zając, D.; Sołoducho, J.; Cabaj, J. Dopamine sensing with fluorescence strategy based on low temperature co-fired ceramic technology modified with conducting polymers. *Sens. Actuator B Chem.* **2017**, *252*, 803–812. [CrossRef]
97. López, J.; Llorent-Martinez, J.; Ortega-Barrales, P.; Ruiz-Medina, A. Multicommutated Flow System for the Determination of Glyphosate Based on Its Quenching Effect on CdTe-Quantum Dots Fluorescence. *Food Anal. Methods* **2018**, *11*, 1840–1848. [CrossRef]
98. Liu, Q.; Jiang, M.; Ju, Z.; Qiao, X.; Xu, Z. Development of direct competitive biomimetic immunosorbent assay basedon quantum dot label for determination of trichlorfon residues in vegetables. *Food Chem.* **2018**, *250*, 134–139. [CrossRef] [PubMed]
99. Kanagasubbulakshmi, S.; Kathiresan, R.; Kadirvelu, K. Structure and physiochemical properties based interaction patterns of organophosphorous pesticides with quantum dots: Experimental and theoretical studies. *Colloids Surf. A* **2018**, *549*, 155–163. [CrossRef]
100. Zhang, C.; Zhang, K.; Zhao, T.; Liu, B.; Wang, Z.; Zhang, Z. Selective phosphorescence sensing of pesticide based on the inhibition of silver(I) quenched ZnS:Mn^{2+} quantum dots. *Sens. Actuator B Chem.* **2017**, *252*, 1083–1088. [CrossRef]
101. Tashkhourian, J.; Absalan, G.; Jafari, M.; Zare, S. A rapid and sensitive assay for determination of doxycycline using thioglycolic acid-capped cadmium telluride quantum dots. *Spectrochim. Acta Part A* **2016**, *152*, 119–125. [CrossRef]
102. Durán, G.M.; Contento, A.M.; Ríos, Á. Use of Cdse/ZnS quantum dots for sensitive detection and quantification of paraquat in water samples. *Anal. Chim. Acta* **2013**, *801*, 84–90. [CrossRef] [PubMed]
103. Sun, A.; Chai, J.; Xiao, T.; Shi, X.; Li, X.; Zhao, Q.; Li, D.; Chen, J. Development of a selective fluorescence nanosensor based on molecularly imprinted-quantum dot optosensing materials for saxitoxin detection in shellfish samples. *Sens. Actuator B Chem.* **2018**, *258*, 408–414. [CrossRef]
104. Wang, Y.; Fry, H.C.; Skinner, G.E.; Schill, K.M.; Duncan, T.V. Detection and Quantification of Biologically Active Botulinum Neurotoxin Serotypes A and B Using a Förster Resonance Energy Transfer-Based Quantum Dot Nanobiosensor. *ACS Appl. Mater. Interfaces* **2017**, *9*, 31446–31457. [CrossRef] [PubMed]
105. Goldman, E.R.; Clapp, A.R.; Anderson, G.P.; Uyeda, H.T.; Mauro, J.M.; Medintz, I.L.; Mattoussi, H. Multiplexed Toxin Analysis Using Four Colors of Quantum Dot Fluororeagents. *Anal. Chem.* **2004**, *76*, 684–688. [CrossRef] [PubMed]
106. Tong, P.; Zhao, W.-W.; Zhang, L.; Xua, J.-J.; Chen, H.-Y. Double-probe signal enhancing strategy for toxin aptasensing based on rolling circle amplification. *Biosens. Bioelectron.* **2012**, *33*, 146–151. [CrossRef] [PubMed]
107. Ligler, F.S.; Taitt, C.R.; Shriver-Lake, L.C.; Sapsford, K.E.; Shubin, Y.; Golden, J.P. Array biosensor for detection of toxins. *Anal. Bioanal. Chem.* **2003**, *377*, 469–477. [CrossRef] [PubMed]

108. Bakar, N.A.; Rahmi, A.; Umar, A.A.; Salleh, M.M.; Muhammad Yahaya, M. Fluorescence Gas Sensor Using CdTe Quantum Dots Film to Detect Volatile Organic Compounds. *Mater. Sci. Forum* **2011**, *663–665*, 276–279. [CrossRef]
109. Liua, Y.; Wanga, L.; Wanga, H.; Xionga, M.; Yanga, T.; Zakharova, G.S. Highly sensitive and selective ammonia gas sensors based on PbS quantum dots/TiO2 nanotube arrays at room temperature. *Sens. Actuator B Chem.* **2016**, *236*, 529–536. [CrossRef]
110. Barroso, J.; Díez-Buitrago, B.; Saa, L.; Möller, M.; Briz, N.; Pavlov, V. Specific bioanalytical optical and photoelectrochemical assays for detectionof methanol in alcoholic beverages. *Biosens. Bioelectron.* **2018**, *101*, 116–122. [CrossRef]
111. Sotelo-Gonzalez, E.; Fernandez-Argüelles, M.T.; Costa-Fernandez, J.M.; Sanz-Medel, A. Mn-doped ZnS quantum dots for the determination of acetone by phosphorescence attenuation. *Anal. Chim. Acta* **2012**, *712*, 120–126. [CrossRef]
112. Liu, H.; Fang, G.; Zhu, H.; Li, C.; Liu, C.; Wang, S. A novel ionic liquid stabilized molecularly imprinted optosensing material based on quantum dots and graphene oxide for specific recognition of vitamin E. *Biosens. Bioelectron.* **2013**, *47*, 127–132. [CrossRef] [PubMed]
113. Geszke-Moritz, M.; Clavier, G.; Lulek, J.; Schneider, R. Copper- or manganese-doped ZnS quantum dots as fluorescent probes for detecting folic acid in aqueous media. *J. Lumin.* **2012**, *132*, 987–991. [CrossRef]
114. Wang, J.; Wei, J.; Su, S.; Qiu, J. Novel fluorescence resonance energy transfer optical sensors for vitamin B_{12} detection using thermally reduced carbon dots. *New J. Chem.* **2015**, *39*, 501. [CrossRef]
115. Sun, J.F.; Ren, C.L.; Liu, L.H.; Chen, X.G. CdTe quantum dots as fluorescence sensor for the determination of vitamin B6 in aqueous solution. *Chin. Chem. Lett.* **2008**, *19*, 855–859. [CrossRef]
116. Ganiga, M.; Cyriac, J. An ascorbic acid sensor based on cadmium sulphide quantum dots. *Anal. Bioanal. Chem.* **2016**, *408*, 3699–3706. [CrossRef] [PubMed]
117. Xue, L.; Zheng, L.; Zhan, H.; Jinc, X.; Lin, J. An ultra sensitive fluorescent biosensor using high gradient magnetic separation and quantum dots for fast detection of foodborne pathogenic bacteria. *Sens. Actuator B Chem.* **2018**, *265*, 318–325. [CrossRef]
118. Wu, P.; Huang, R.; Li, G.; He, Y.; Chen, C.; Xiao, W.; Ding, P. Optimization of Synthesis and Modification of ZnSe/ZnS Quantum Dots for Fluorescence Detection of *Escherichia coli*. *J. Nanosci. Nanotechnol.* **2018**, *18*, 3654–3659. [CrossRef]
119. Jimenez, A.M.J.; Moulick, A.; Richtera, L.; Krejcova, L.; Kalina, L.; Datta, R.; Svobodova, M.; Hynek, D.; Masarik, M.; Heger, Z.; et al. Dual-color quantum dots-based simultaneous detection of HPV-HIV co-infection. *Sens. Actuator B Chem.* **2018**, *258*, 295–303. [CrossRef]
120. Hahn, M.A.; Tabb, J.S.; Krauss, T.D. Detection of Single Bacterial Pathogens with Semiconductor Quantum Dots. *Anal. Chem.* **2005**, *77*, 4861–4869. [CrossRef]
121. Liu, Y.; Brandon, R.; Cate, M.; Peng, X.; Stony, R.; Johnson, M. Detection of pathogens using luminescent CdSe/ZnS dendron nanocrystals and a porous membrane immunofilter. *Anal. Chem.* **2007**, *79*, 8796–8802. [CrossRef]
122. Hu, P.; Chen, L.; Kang, X.; Chen, X. Surface Functionalization of Metal Nanoparticles by Conjugated Metal–Ligand Interfacial Bonds: Impacts on Intraparticle Charge Transfer. *Acc. Chem. Res.* **2016**, *49*, 2251–2260. [CrossRef] [PubMed]

nanomaterials

MDPI

Review

Potential of Nanomaterial Applications in Dietary Supplements and Foods for Special Medical Purposes

Josef Jampilek [1,2,*], Jiri Kos [3] and Katarina Kralova [4]

1 Division of Biologically Active Complexes and Molecular Magnets, Regional Centre of Advanced Technologies and Materials, Faculty of Science, Palacky University, Slechtitelu 27, 783 71 Olomouc, Czech Republic
2 Institute of Neuroimmunology, Slovak Academy of Sciences, Dubravska cesta 9, 845 10 Bratislava, Slovakia
3 Department of Pharmaceutical Chemistry, Faculty of Pharmacy, Comenius University, Odbojarov 10, 832 32 Bratislava, Slovakia; jirikos85@gmail.com
4 Institute of Chemistry, Faculty of Natural Sciences, Comenius University, Ilkovicova 6, 842 15 Bratislava, Slovakia; kata.kralova@gmail.com
* Correspondence: josef.jampilek@gmail.com

Received: 8 January 2019; Accepted: 15 February 2019; Published: 19 February 2019

Abstract: Dietary supplements and foods for special medical purposes are special medical products classified according to the legal basis. They are regulated, for example, by the European Food Safety Authority and the U.S. Food and Drug Administration, as well as by various national regulations issued most frequently by the Ministry of Health and/or the Ministry of Agriculture of particular countries around the world. They constitute a concentrated source of vitamins, minerals, polyunsaturated fatty acids and antioxidants or other compounds with a nutritional or physiological effect contained in the food/feed, alone or in combination, intended for direct consumption in small measured amounts. As nanotechnology provides "a new dimension" accompanied with new or modified properties conferred to many current materials, it is widely used for the production of a new generation of drug formulations, and it is also used in the food industry and even in various types of nutritional supplements. These nanoformulations of supplements are being prepared especially with the purpose to improve bioavailability, protect active ingredients against degradation, or reduce side effects. This contribution comprehensively summarizes the current state of the research focused on nanoformulated human and veterinary dietary supplements, nutraceuticals, and functional foods for special medical purposes, their particular applications in various food products and drinks as well as the most important related guidelines, regulations and directives.

Keywords: bioactive agents; dietary supplements; foodstuffs; feed; nanoparticles; nanoformulations; nanoemulsions; nutraceuticals; encapsulation

---

1. Introduction

Fortification of edible products (e.g., food, food constituents, or supplements) with nutrients or non-nutrient bioactive components can help to balance the total nutrient profile of a diet and supplement nutrients lost in processing and thus to correct or prevent insufficient nutrient intake and associated deficiencies [1]. Compounds of natural origin, such as curcumin (CUR) occurring in turmeric, ω-3-fatty acid in fish oil, vitamins from fruits, when encapsulated in an appropriate nanocarrier, will be released after consumption of the food in the target organ and utilized according to its nutritional property [2].

Basic types of preparations/materials influencing human health or condition can be classified as follows: (i) drug products, (ii) homeopathics, (iii) dietary supplements (DISs), (iv) medical devices, (v) cosmetics, and (vi) biocidal products. Dietary (food) supplements are products that look similar to medicines (can be sold in pharmacies) but are a special category of foods. They contain vitamins, minerals, amino acids, essential fatty acids, natural products, probiotics, etc., as active ingredients. The purpose of a DIS is to keep the human body functioning properly by delivering compounds that are needed by the human body but could not be received sufficiently from a regular diet. According to the manufacturers, DISs have beneficial effects on health conditions. They are manufactured in the form of pills, capsules, tablets, or liquids [3,4]. DISs are regulated by many guidelines, regulations and directives, for example, by the European Commission directives 2002/46/EC and 2006/37/EC, European regulations 1924/2006, 1137/2008, 1170/2009, 1161/2011, 119/2014, 2015/414, 2017/1203 [4], and by a number of documents published by the U.S. Food and Drug Administration (FDA) [5], to ensure the quality and safety of these products, to protect consumers against potential health risks from such products, and to ensure that they are not provided with misleading information. In addition, DISs are regulated by the European Food Safety Authority (EFSA), national legislation (e.g., on food) and regulations (e.g., requirements for food supplements and food enrichment), etc. For example, in the EU market, approx. 30 approved nutrition claims (meaning that specific requirements are to be met) can be found, and a product can be marketed as a DIS only if it meets the so-called health claims, which is any statement about a relationship between food and health. The European Commission approves various health claims, which have to be easily understood by consumers, based on scientific evidence. The EFSA is responsible for evaluating the scientific evidence supporting health claims, the types of which are as follows: (i) 'Function Health Claims' (relating to the growth, development and functions of the body, or referring to psychological and behavioral functions, or on slimming or weight-control), (ii) 'Risk Reduction Claims' (on reducing a risk factor in the development of a disease), and (iii) Health 'Claims referring to children's development' [6].

It is important to note that DISs are not food additives, which are special excipients added to foods for modifications of their flavor, color, or longevity [7].

Foods for special medical purposes (FSMPs), based on the definition of the EFSA, "are designed to feed patients who, because of a particular disease, disorder, or medical condition, have nutritional needs that cannot be met by consuming standard foodstuffs. Specifically, according to EU legislation, they are intended for patients with a limited, impaired, or disturbed capacity to take, digest, absorb, metabolize, or excrete ordinary foods, or certain nutrients or metabolites; or with other medically nutrient requirements whose dietary management cannot be achieved by modification of the normal diet alone" [8]. It means that FSMPs are foods that are intended for nourishment at: (i) certain groups of people whose digestive process or metabolism is impaired, (ii) certain groups of persons in a particular physiological state, which therefore, may have specific benefits from controlled consumption of certain substances in food, or (iii) healthy infants and young children. Therefore, the following categories of FSMPs can be distinguished: (i) food for infant and follow-on nourishment and nutrition of small children, (ii) food for cereal and other non-cereal food for infant and young children, (iii) low-energy foods designed to reduce body weight, (iv) food without phenylalanine, (v) gluten-free foods, (vi) foods for people with disorders of carbohydrate metabolism (diabetics), (vii) low lactose or lactose-free foods, (viii) foods with low protein content, (ix) foods intended for athletes and for persons with increased physical performance. FSMPs are advised to be used only under medical supervision and have to be provided with labels with information about their intended use. The European Commission also issued several documents for the regulation of FSMPs, e.g., Commission Directive 1999/21/EC, Commission Regulations No. 953/2009, 609/2013, 2016/128 [9], and so-called "medical foods" are regulated also by FDA [10].

A functional food or functional ingredient is any food or food component providing health benefits beyond basic nutrition, and natural bioactive compounds as functional ingredients showing beneficial effects for health become increasingly popular in the diet [11]. Therefore, functional foods are similar to traditional conventional foods but have more advantageous properties in relation to healthy physical condition. Nutraceuticals are based on both food and herbal or other natural products and are used in the form of pharmaceutical formulations, i.e., tablets, capsules, drops, or liquids, and have physiological benefits. The main focus of all these products is to improve health and reduce the risk of disease. In contrast to drugs, in all these cases, the active substance or a mixture of active compounds is present in low concentration [12].

Nanotechnology is a rapidly growing field that ensures the development of materials with new dimensions, novel properties, and a wider range of applications. U.S. National Nanotechnology Initiative defines nanoparticles (NPs) in the range of 1–100 nm [13]. According to the Recommendation on the definition of a nanomaterial adopted by the European Commission, the term "nanomaterial" means "a natural, incidental or manufactured material containing particles, in an unbound state or as an aggregate or as an agglomerate and where, for 50% or more of the particles in the number size distribution, one or more external dimensions is in the size range 1–100 nm. In specific cases and where warranted by concerns for the environment, health, safety or competitiveness the number size distribution threshold of 50% may be replaced by a threshold between 1 and 50%" [14]. However, in pharmacy, particles of 10–500 nm have been used, rarely up to 700 nm. From the aspect of passage through vessels, the inside diameter of which is in the range from 25 mm (aorta) to 5 μm (capillaries), the ideal size of NPs should be <300 nm to ensure efficient transport for targeted distribution of drugs [15–20].

NPs can be prepared from both inorganic and organic materials [21–23], and currently, especially encapsulation to various biodegradable nature-based biopolymers is more and more frequently used [19,20,24,25]. NPs can be generated by either top-down methods (dispergation, fluidization, homogenization processes, or emulsifying technologies) or bottom-up methods (precipitation/condensation processes, evaporation techniques, various controlled sol-gel syntheses) [21,26]. NPs produced using mechanical approaches are usually, in the range 100–1000 nm; to produce NPs of size 10–100 nm, chemical and bottom-up methods are used [21,27]. "Green" synthesis of NPs or innovative biotechnological approaches related to the synthesis of NPs are summarized by Singh and Shukla et al. [22,23].

The physical, chemical, and biological properties of nanoscale materials are significantly, different from those of bulk materials and single atoms or molecules; therefore, different properties of active pharmaceutical ingredients have been modified in such a way [14,28–33]. In biomedical branches, NPs can be used for nanodiagnostics, as nanomaterials for tissue engineering, as drug carriers for specific delivery/targeted biodistribution or controlled release, and as agents/drugs for prevention/treatment of diseases. Therefore, application of nanotechnology can be considered as an excellent tool for modification of parameters of bioactive agents. Modification of properties using nanosystems/nanoformulations helps to enhance the bioavailability of active substances and change the route of administration when needed. Therefore, smaller amounts of substances can be used, which allows decreasing dose-dependent toxicity and various side effects. In addition, many formulations also protect bioactive molecules from degradation [19,20,24,25,28,29,31,34–42]. The enhancement of bioavailability could be achieved by the improved solubility of bioactive compounds under gastrointestinal (GI) conditions, their protection from the chemical conditions in the GI tract, and controlled release within the GI tract, or by an improved transfer through the intestinal wall, and the particle size, surface properties, and physical state of the nanomaterials used in food supplements are crucial characteristics affecting their final nutritional value [43]. Recent findings and advancements related to lipid nanoscale cargos for the protection and delivery of food bioactive ingredients and nutraceuticals were overviewed by Akhavan et al. [44]. Nanoemulsion (NE) compositions, types of active ingredients, applications in different types of food systems, toxicological and safety aspects, and future directions were summarized by Kumar and Sarkar [45]. For encapsulating drugs/nutraceuticals and fortification of food products, especially beverages with water insoluble nutraceuticals, nanostructured lipid carriers (NLCs) could be successfully applied [46]. Micro- and nano bio-based delivery systems (DESs) for food applications were discussed also by Simoes et al. [47].

The above-mentioned nanoformulations can be found in many drug classes, and so it is not surprising that supplements and FSMPs have also started being formulated in the nanoscale, especially with the aim to improve bioavailability, protect active ingredients against degradation, or reduce side effects. Therefore, this contribution summarizes the current state of the research focused on nanoformulated human and veterinary DISs and FSMPs.

2. Types of Formulations and Used Materials

Nutraceuticals' functionality in food products can be stabilized and enhanced using bio-based nanoscaled DESs that help to improve their bioavailability and protect valuable nutraceuticals at food processing or digestion, see Figure 1, where individual most frequently applied nanoformulations are mentioned. Selected nanoformulations are discussed below in the following subchapters. Advances in nutraceutical DESs with focus on the formulation design for the enhancement of nutraceuticals' bioavailability with the purpose to ensure effective preservation or maximization of their bioactivity and safety inside the human body were summarized by Goncalves et al. [48]. A review of recent research developments related to nanocarrier-based delivery of nutraceuticals for cancer prevention and treatment was presented by Arora and Jaglan [49]. Recent findings related to advances made in the nanoencapsulation of lipophilic and hydrophilic vitamins, safety issues, and health risks regarding the consumption of these products, which would result in widespread utilization of nanoencapsulated vitamins in the food and beverage products in the future, were summarized by Katouzian and Jafari [50]. The intelligent DESs for bioactive compounds in foods designed to improve their low solubility, poor stability, and low permeability in the GI tract and improving their oral bioavailability were discussed by Chai et al. [51] from the aspect of physicochemical and physiological conditions, absorption mechanisms, obstacles, and responsive strategies.

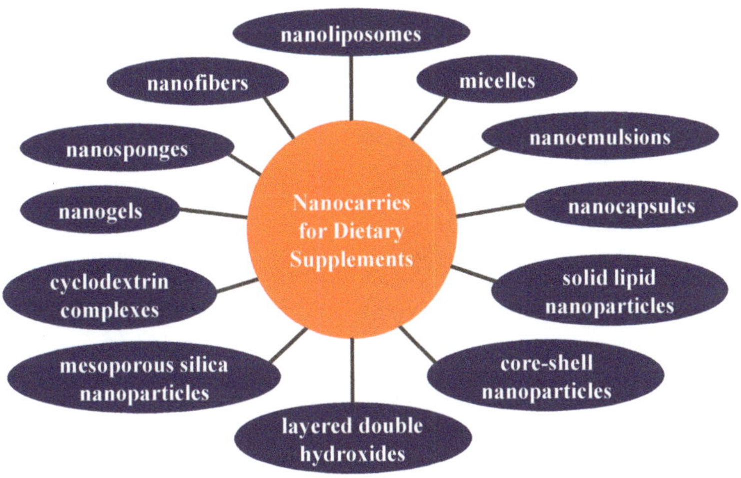

Figure 1. Most frequently used nanoformulation types of dietary supplements and foods for special medical purposes.

Gleeson et al. [52] looked into the potential of certain delivery strategies for the improvement of the oral bioavailability of different types of nutraceuticals, such as fatty acids, bioactive peptides, micronutrients, and phytochemicals, and emphasized that nutraceutical and pharmaceutical industries could leverage approaches to oral delivery formulations, which would result in synergies for nutraceutical and pharmaceutical molecules. For example, microfluidization could be considered as an efficient emulsification technique resulting in fish oil encapsulated powder producing emulsions at the nanoscale range (d_{43} of 210–280 nm) with the lowest unencapsulated oil at the surface of particles [53]. At investigating the effect of excipient emulsions with different surface-weighted mean droplet diameters d_{32} = 0.15 μm (small), 0.40 μm (medium), and 22.3 μm (large) on the bioaccessibility of carotenoids from tomatoes using a simulated GI tract, it was found that the bioaccessibility of carotenoids decreased with an increase of initial droplet size, which could be attributed to more efficient extraction of carotenoids from tomato tissues by smaller droplets that were digested faster. This caused faster mixed micelle formation and, consequently, enhanced solubilization of carotenoids in intestinal fluids. Moreover, when tomatoes were boiled with emulsions, the bioaccessibility of carotenoids was higher than when they were boiled alone and subsequently added to emulsions [54]. Electrospinning and electrospraying technologies constitute useful and modern techniques used for the encapsulation and controlled release of bioactive compounds, including drugs and health-promoting agents. Both electrospinning, mostly used for fibres, and electrospraying, mostly used for particles, are voltage-driven fabrication technologies enabling tight control of fibres and particles in the micro-, submicro- and nanoscale dimensions suitable for a wide range (polymers, proteins, inorganic) of materials. Both processes are able to replace traditional techniques, e.g., spray-drying or lyophilisation, as they propose several benefits such as (i) production of dry products in a single step, (ii) room temperature operation (suitable for labile components, e.g., antioxidants, omega-3 oils, living cells, etc.), (iii) enabling to produce single-phase or multi-component fibres and particles, and (iv) high effectivity of encapsulation [53,55–60], as mentioned below.

2.1. Liposomes and Nanoscale Emulsions

Nanoliposomes, or nanometric bilayer phospholipid vesicles, have a very promising potential for the nutraceutical industry, because they can encapsulate simultaneously lipophilic and hydrophilic materials, ensuring a synergistic effect, and can protect sensitive bioactive compounds, enhance their bioavailability, ensure sustained-release, and improve storage stability. The unique properties of nanoliposomes predestine them to be used in DISs for effective disease prevention and health promotion [61].

Nanophytosome is one of the newest lipid-based nanocarriers enabling the delivery of botanical based nutraceuticals, which could be potentially used in food products for designing novel functional foods and beverages [62]. Phytosomes-phosphatidylcholine (PC)—rutin complexes prepared by the encapsulation of rutin with PC using rutin:PC molar ratio 1:3 were found to provide the highest physical and chemical stability (during 30 days of storage) with fine particle sizes (<100 nm) and the encapsulation efficiency (EE) of 99%, and due to the ability of masking undesirable features of rutin, they may be applied in fortification of food products with water insoluble nutraceuticals [63].

At the preparation of liposomes starting from multilamellar large vesicles with a diameter range 2.9–5.7 μm using an ultrasound-assisted approach based on the thin-film hydration method, unilamellar vesicles with diameter sizes ranging from 40 nm to 51 nm were achieved, showing the EE of 56% for cobalamin, 76% for α-tocopherol, and 57% for ergocalciferol. The nanovesicles and their content were kept intact for >10 days when incubated at simulated conditions of extracellular environment thanks to the used lipid composition [64]. The investigation of the encapsulation and preservation of quercetin (Q) with cyclodextrins (CDs), conventional liposomes composed of three different types of phospholipids (unsaturated egg Lipoid E80, unsaturated soybean Lipoid S100, and saturated soybean Phospholipon 90H), and drug-in-CD-in-liposomes showed that the application of Lipoid E80-liposomes resulted in a better protection of Q against UV irradiation, and its photostability was additionally improved when encapsulated in drug-in-CD-in-liposomes (sulfobutylether β-CD/Q inclusion complex in Lipoid E80 liposomes) [65].

Compared to CUR liposomes, the Pluronic® modified CUR liposomes showed a slower release rate and lower cumulative release percentage for CUR, enhanced pH stability and thermal stability, and pronouncedly improved absorption in simulated GI tract in vitro, suggesting that both types of liposomes could be used as carriers of CUR in nutraceuticals and functional foods. The best bioaccessibility was observed for CUR liposomes modified with Pluronic® F-127 [66].

Stimuli-sensitive (smart) nano DESs for nutraceuticals of both a nutritional and pharmaceutical value are of great importance for the formulation of novel functional foods. Because the best effect on the human health was observed when the weight ratio of ω-6/ω-3 polyunsaturated fatty acids (PUFAs) is in the range between 1:1 and 5:1, Semenova et al. [67] focused their attention on the molecular design of DESs on the basis of nanoscale complexes formed between a covalent conjugate (sodium caseinate (SCas) + maltodextrin; dextrose equivalent = 2) and combinations of polyunsaturated lipids that are mutually complementary in the content of ω-6 and ω-3 PUFAs: α-linolenic acid (α-LNA) + α-linoleic acid (α-LLA); liposomes of soy PC + α-LNA, and micelles of soy lyso-PC + α-LNA. The researchers concluded that thanks to the EE of all these lipid combinations by the conjugate, lipids were highly protected against oxidation, and their high solubility in an aqueous medium was reached. Dey et al. [68] designed ω-3 PUFA enriched biocompatible NE with sesame protein isolate (SPI) as a natural surfactant. NE with 0.5% (*w*/*v*) SPI and Tween 20 and Span 80 used in 1:1 ratio having the hydrodynamic droplet size of 89.68 ± 2.38 nm effectively enhanced the shelf-life stability of NEs, and the fatty acid release from NE droplets was ≥90% during 120 min of simulated two-step in vitro digestion.

The short-chain triglyceride-based NE encapsulating vitamin E did not physically withstand temperatures exceeding 25 °C, while with long-chain triglyceride-based NEs, good vitamin E retention even at 40 °C was observed, and the retention was increased when the NEs were stored in the dark [69]. Vitamin D NEs with small droplet diameters (d < 200 nm) fabricated by spontaneous emulsification using medium chain triglycerides (MCT) and Tween 80 at surfactant-to-oil ratio ≥1 at high stirring speeds (800 rpm) were found to be relatively stable at ambient temperatures and unstable at heating (T > 80 °C), but the application of a cosurfactant (sodium dodecyl sulfate) could improve their thermal stability [70]. The investigation of the effect of excipient NEs formulated from long or medium chain triglycerides (LCT or MCT) on β-carotene (β-Car) bioaccessibility from commercial DISs (tablets or soft gels) studied using an in vitro GI tract model showed that the application of LCT NEs enhanced β-Car bioaccessibility from tablets and soft gels by 20% and 5%, respectively, while the effect of MCTs was minor. This could be connected with the fact that large carotenoid molecules could be incorporated only into large mixed micelles formed by LCT digestion, and thus, excipient NEs could be applied to improve nutraceutical bioavailability from DISs [71]. NEs prepared using three LCT oils (flaxseed, olive and corn oil) increased the bioaccessibility of astaxanthin (AST) compared to the control due to the formation of mixed micelles that solubilized the hydrophobic carotenoids. The final amount of free fatty acids released affected lipid digestion and AST bioaccessibility, which decreased in the following order: olive oil > flaxseed oil > corn oil, and free fatty acids unsaturation and chain length affected lipid digestion and micelle formation [72]. Saxena et al. [73] increased the bioavailability of the model bioactive compound α-tocopherol as a food supplement using edible (coconut) oil NEs. The prepared NEs were found stable and biocompatible, and the contribution of kinetic-controlled release was found to be approx. 70%, while that of diffusion-controlled release was approx. 30%, suggesting the potential of the use of edible oil NEs in food and beverages.

A saponin coated NE with mean droplet diameter 277 nm encapsulating vitamin E was found to be more stable to droplet coalescence at thermal processing (30–90 °C), long-term storage, and mechanical stress than a conventional emulsion with mean droplet diameter 1.285 μm. At application of both emulsion formulations to male Wistar rats, droplet flocculation and coalescence during in vivo digestion was observed, however, the higher in vivo oral bioavailability of vitamin E encapsulated in the NE was reflected in a 3-fold increase in the area under the curve (AUC) compared to the conventional emulsion [74]. The lowest particle diameters (d_{32}) of vitamin E NEs fabricated using natural surfactants, quillaja saponin, and lecithin and high-pressure homogenization were 0.13 μm for lecithin and 0.12 μm for quillaja saponin at vitamin E to orange oil ratio 50:50%. At pH 7, both systems were stable in the temperature range 3–90 °C but unstable at pH 2 or in the presence of NaCl

(>100 mM NaCl for lecithin and ≥400 mM NaCl for quillaja saponin) [75]. The encapsulation of CUR in saponin-coated CUR NEs fabricated using a simple pH-driven loading method improved CUR solubility and bioavailability, and its in vitro bioaccessibility was approx. 3.3-fold higher compared to free CUR. In an in vivo study, oral administration of these NPs to Sprague Dawley rats resulted in approx. 8.9-fold higher in vivo bioavailability than that estimated with free CUR [76].

Zheng et al. [77] subjected CUR loaded oil-in-water (O/W) NEs prepared using the conventional oil-loading method, the heat-driven method, and the pH-driven method and three commercial CUR supplements (Nature Made, Full Spectrum, and CurcuWin) to a simulated GI tract model consisting of mouth, stomach, and small intestine phases and found that the three tested NEs showed similar CUR bioaccessibility (74–79%) with the highest absolute amount of CUR in the mixed micelle phase of the NE fabricated by the pH-driven method. The concentration of CUR in mixed micelles decreased as follows: CurcuWin ≈ pH-driven method > heat-driven method > conventional method >> full spectrum > nature made, and CUR encapsulated in small lipid particles had an improved absorption in GI tract.

Cholecalciferol (vitamin D_3) minitablets and an optimized bile salt/lipase alginate-glycerin film provided unique oral components for inclusion in a bioactive association platform (BAP) capsule designed to deliver the active nutraceutical ingredient from the formulation framework resulting in the enhanced in vitro and in vivo performance of cholecalciferol. The in vivo experiment showed that cholecalciferol bioavailability from the BAP was 3.2-fold greater than that of the conventional product, and improved and maintained serum levels of 25-hydroxyvitamin D_3 were observed as well, suggesting that BAP could be considered as an ideal oral vehicle for enhanced delivery of cholecalciferol [78].

β-Car enriched O/W emulsions, in which chlorogenic acid-lactoferrin-polydextrose conjugate was used as an emulsifier to stabilize lipid droplets, showed improved stability to droplet aggregation under simulated GI tract conditions, resulting in increased β-Car bioaccessibility, suggesting that the ternary conjugate-stabilized emulsions could be used as protectors and carriers of hydrophobic drugs, supplements, and nutraceuticals [79]. On the other hand, excipient NEs had much less effect on the bioaccessibility of phenolic compounds, probably due to their smaller and more polar molecules, which could be more easily solubilized in aqueous intestinal fluids [80].

Among NEs prepared using soy protein isolate (132 nm), whey protein concentrate (190 nm), maltodextrin (266 nm), and gum arabic (468 nm), the soy protein isolate NE showing the smallest droplet size provided the highest protection of vitamin D (85%) at 4 wt % concentration, pH 7, and 25 °C [81].

2.2. *Lipid-Based Carriers*

Nanocapsules based on lipid formulations having larger surface area than microsized carriers can more effectively enhance solubility, bioavailability, and controlled release of nanoencapsulated phenolic compounds and could be successfully applied in functional foods [82]. For example, the physical stability of β-Car nanocapsules (>300 nm) showed only minor changes during storage, suggesting that they could be used in functional beverages and foods as well as nutraceutical products [83]. At the investigation of the impact of solid domain properties on the rate of compound release from NLCs using Monte Carlo simulations, it was found that the release of encapsulated bioactive compounds by solid impenetrable domains at the particle/solution interface is hindered only when the domain size is much smaller than the size of NPs, even if a considerable proportion of the interface is covered by these domains, with the rate of release depending also on the geometry of the solid domains [84]. The preparation and characterization of vitamin A palmitate-loaded NLCs as DESs for food products was reported by Kong et al. [85].

Solid lipid microparticles (MPs) loaded with 0.1% of vitamin D_3 by spray chilling with mean diameter 83.0–98.6 μm that were fabricated using vegetable fat as a carrier and beeswax (1% of the formulation) showed increased vitamin stability at 25 °C, and 86.3% of vitamin were detected after

65 days of storage compared to 60.8% estimated with non-immobilized vitamin at same conditions, suggesting the potential of using such formulation in foods [86].

Prolonged physical stability at room and refrigerated temperature conditions as well as an increase in the bioavailability of encapsulated CUR compared to that of CUR suspensions was shown by chitosan (CS) coated solid lipid NPs incorporating this nutraceutical after oral administration [87].

In a systematic review, Nunes et al. [88] focused on the use of solid lipid NPs as oral DESs of phenolic compounds that allow overcoming the pharmacokinetic limitations of these compounds and ameliorate their nutraceutical potential.

2.3. Polysaccharide Matrices

Polysaccharides that have various enzymatic susceptibilities to ensure specific degradation in the small or large intestine when used as a NP coating can efficiently retard the nonspecific release of encapsulated bioactive compounds until the coating is exposed to its intended environment of release, and such coated NPs can be potentially targeted to different GI tract organs and taken up by the enterocytes, providing improved oral bioavailability [89].

High amylose corn and potato starches nanocarriers with granular structure and particle sizes ranging from 32.04 to 99.2 nm were used to encapsulate vitamin D_3, and their EE ranged from 22.34 to 94.8%. By using ultrasonic treatment, an increase of the hydrocarbon chain length was observed resulting in van der Waals and *H*-bonds of vitamin D_3 with the potato starch and greater thermal stability [90]. Low-molecular-weight octenyl succinic anhydride modified starches were reported to be suitable to form stable vitamin E nanocapsules for potential application in beverages [91].

Recent findings concerning the use of cellulosic nanomaterials for food and nutraceutical needs were summarized by Khan et al. [92]. The addition of cellulose nanocrystals and lecithin into alginate microbeads improved the viability of encapsulated probiotic (*Lactobacillus rhamnosus* ATCC 9595) during gastric passage and storage, and at 25 and 4 °C storage conditions, a decrease in the viability of *L. rhamnosus* by 1.23 and 1.08 log, respectively, was estimated, while at encapsulation of the probiotic with alginate microbeads, a 3.17 and 1.93 log reduction, respectively, was observed [93]. The oligo-hyalurosomes nanoscale DES based on oligo-hyaluronic acid-CUR polymer co-loaded with both CUR and resveratrol (RES) showing the average particle size of 134.5 ± 5.1 nm, spherical shape, and zeta potential of −29.4 ± 1.2 mV at pH 7.4 phosphate buffer conditions exhibited excellent stability and sustained release character and higher radical scavenging activity compared to the single formulations and liposomes suggesting that this system could be considered as a promising nanofood DES applicable in juice, yoghourt and nutritional supplements [94].

CS/tripolyphosphate-nanoliposomes core-shell nanocomplexes as vitamin E carriers showed vitamin E retention rate >80% during the 30-day storage and 92% and 97% after heating at 65 °C for 30 min and at 80 °C for 16 s, respectively, and based on the enhanced stability of liposomes against temperature stress reflected in reduced particle aggregation, zeta potential inversion, and membrane fluidity, this formulation could be considered as appropriate for commercial use in the food industry [95]. CS hydrochloride/carboxymethyl CS nanocomplexes loaded with anthocyanins with particle size 178.1 nm, zeta potential +25.6 mV, and polydispersity index 0.315 showed a higher stability when placed at different conventional storage temperatures, various L-ascorbic acid concentrations, varying pH, or white fluorescent light, suggesting that such nanocomplexes could be applied in food ingredients associated with stable anthocyanins in functional foods and nutraceutical applications [96].

In food-grade alginate/CS nanolaminates obtained by the layer-by-layer technique, in which folic acid (FA) was incorporated by post-diffusion, a higher stability of FA under ultraviolet light exposure compared to free FA was estimated, and the higher rate and concentration of FA released from nanolaminates at pH 7 in comparison with that at pH 3 suggested that nanolaminates containing hydrophilic active compounds can be used for food applications [97].

Insulin encapsulated in antacid-loaded calcium alginate microgels (diameter 280 μm) had higher biological activity in simulated gastric conditions than free insulin, and considerably increased Akt phosphorylation at Thr308 and Ser473 in L6 myotubes was observed [98].

Papagiannopoulos and Vlassi [99] reported preparation of multi-functional stimuli-responsive NPs for food and biomedical applications by combining electrostatic complexation between proteins and polysaccharides with following thermal protein denaturation for the production of chondroitin sulfate/bovine serum albumin NPs. The irreversible protein–protein contacts upon temperature treatment provide the complexes with properties of nanogels, and the surface charge of the prepared NPs reversed at pH 5.3, while their size depended on the solution ionic strength and pH. Protein-polysaccharide-surfactant ternary complex particles prepared by anti-solvent co-precipitation using zein, propylene glycol alginate, and either rhamnolipid or lecithin pronouncedly improved the photostability and bioaccessibility of CUR suggesting that they could be used to deliver hydrophobic nutraceuticals for applications in foods, supplements, and pharmaceuticals [100].

2.4. Protein-Based Carriers

The state of the art of protein-based nanoencapsulation approaches as well as protein modification approaches in order to extend their functionality in nanocarrier systems to achieve an improvement in encapsulation, retention, protection, and release of bioactive agents was summarized by Fathi et al. [101]. A review paper discussing the latest findings concerning the nanoscale phenomena of whey protein denaturation and aggregation, which could contribute to the design of protein nanostructures with new or improved properties for the incorporation of nutraceuticals in food matrices and their release was presented by Ramos et al. [102]. Using whey protein isolate as an encapsulating agent, Parthasarathi and Anandharamakrishnan [103] presented a spray freeze-drying based microencapsulation technique as a promising strategy to enhance the oral bioavailability of poorly water-soluble bioactive compounds like vitamin E.

Significant aggregation and sedimentation of zein NPs encapsulating lutein (ZLNPs) with hydrodynamic radius approx. 75 nm were observed at gastric digestion conditions, and the ZLNPs that were not fully digested by gastric enzymes adhered to lipid droplets; however, the aggregation was reduced and digestion was stimulated when salt (i.e., high ion concentration) was left out. On the other hand, thanks to the encapsulation of lutein into NPs, its digestive stability was increased [104].

The size of egg albumin (Alb)-FA nanocomplexes prepared by mixing egg Alb NPs with FA did not change after adjusting the pH from 3 to 4, but showed considerable increase after adjusting pH to 5, 6, or 7; however, the bioavailability of FA in the form of digested nanocomplexes for *Lactobacillus rhamnosus* was improved [105].

The degree of FA binding to β-lactoglobulin (β-Lglb) and type A gelatin carriers was affected by their pH-dependent zeta-potential, which indicated the occurrence of ionic bonds, and the binding of FA reached 100% at pH 3. At pH 3, particle size considerably increased at increasing the molar FA/protein ratio; however, shifting back the pH to 7 totally reversed it, which means that these formulations could protect FA at pH 3 prevailing in the stomach, but they are strongly favorable for its delivery to the duodenum (pH 7) [106]. β-Lglb nanostructures were reported to be suitable carriers for riboflavin and its controlled release in an in vitro GI system: approx. 11% was released during their passage through the stomach, while 35%, 38%, and 5% of the total riboflavin were released during their passage through duodenum, jejunum, and ileum, respectively. At food simulant conditions (yoghurt simulant, 3% acetic acid), β-Lglb nanostructures were stable for more than 14 days and had protective impact on riboflavin activity, releasing it in a 7-day period [107].

Isolated 7S and 11S globulins (Glbs) obtained from defeated soy flour, which were complexed with FA and included in culture media, showed higher bacterial growth of *Lactobacillus casei* BL23. Therefore, Glbs-FA based nanocomplexes have potential to be used in nutraceutical, pharmaceutical, and food industries [108]. *Lactobacillus casei* BL23 produces microvesicles carrying proteins that have been connected with its probiotic effect, and, using a proteomic approach, Rubio et al. [109] identified proteins described as mediators of *Lactobacillus*' probiotic effects, namely p40, p75, and the product of LCABL_31160, which was annotated as an adhesion protein. The expression and subsequent encapsulation of proteins into microvesicles of bacteria generally considered as safe could be also used in applications of foods and nutraceuticals.

Negatively charged (−41 mV) sophorolipid-coated CUR NPs with the particle size of 61 nm showing relatively high EE and loading capacity for CUR that was present in an amorphous state exhibited 2.7–3.6-fold higher bioavailability than free CUR crystals, which was connected primarily with their higher bioaccessibility [110].

Protein–lipid composite NPs having a three-layered structure (barley protein layer, α-tocopherol layer, and phospholipid layer) and an inner aqueous compartment to load the hydrophilic nutraceutical vitamin B_{12} exhibited controlled release behavior in simulated GI media, and in an in vivo experiment, the NPs loaded with vitamin B_{12} increased serum vitamin B_{12} levels in rats upon their oral administration and reduced the level of methylmalonic acid more efficiently than the free vitamin B_{12} form without any toxicity of the formulation observed during 14 days. These NPs could be used for increasing vitamin B_{12} absorption upon oral administration [111].

The enhanced physicochemical stability and in vitro bioaccessibility of vitamin D_3 in corn protein hydrolysate-based vitamin D_3 nanocomplexes showing spherical structure with sizes 102–121 nm was reported by Lin et al. [112]. In vitamin D–potato protein co-assemblies, the nanocomplexation provided pronounced protection and reduced vitamin D losses during pasteurization and also under several different sets of storage conditions, suggesting that potato protein could be used as a protective carrier for hydrophobic nutraceuticals suitable for enrichment of clear beverages and other food or drink products with beneficial impact on human health [113].

After drying and reconstitution, vitamin D-loaded re-assembled casein micelles (r-CMs) were found to improve the in vitro bioavailability of vitamin D in a Caco-2 cell model and showed strong protective effect against its gastric degradation, providing 4-fold higher bioavailability compared to free vitamin D [114]. Ghayour et al. [115] encapsulated CUR and Q using a hierarchical approach (binding of ligand to SCas with subsequent re-assembling of micellar nanostructures or formation of casein NPs). r-CMs had smaller mean particle size than casein NPs, and the entrapment efficiency of both ligands was >90%. An incorporated phenolic compound showed notably improved chemical stability during an accelerated shelf-life test. The aqueous solubility of CUR and Q after loading in r-CMs was higher than that of free polyphenol molecules, and the viability of treated MCF-7 human breast cancer cells decreased as follows: free polyphenol molecules >> non-digested polyphenol-loaded carriers > digested polyphenol-loaded r-CMs. Based on the investigation of the stability and bioavailability of CUR in mixed SCas and pea protein isolate NEs, Yerramilli et al. [116] reported that pea proteins could be used to partially replace SCas as an emulsion stabilizer for the protection and delivery of oil-soluble bioactive compounds. Based on in vitro proteolysis, it was found that in low-fat yogurt supplemented with the spray- and freeze-dried casein micelles loaded with vitamin D_2, 90% of the vitamin remained active compared to 67% estimated with free vitamin [117].

FA–loaded casein NPs of 150 nm fabricated with the use of a coacervation process, stabilized with lysine or arginine, and finally dried by spray-drying were administered to laboratory animals p.o. at dose 1 mg FA/kg and ensured considerably higher serum levels of the vitamin than an aqueous solution of FA administered to animals, and the release profile and oral bioavailability of FA were not affected by the treatment of casein NPs by high hydrostatic pressure [118].

2.5. Inorganic Matrices

Inorganic porous materials, such as various silica- or aluminosilicate-based materials/composites, clays, calcium carbonate, calcium phosphate, layered double hydroxides (LDHs), etc., have become good candidates for the delivery of a range of drugs, providing some advantages in formulation and engineering. They have suitable architecture, large surface area, and stability in biological fluids; thus, they are used for high loading capacity, controllable release, and improved targeting [119–121].

Comparison of four different capped SiO_2 mesoporous particles (MSPs) (i.e., hollow silica shells, MCM-41, SBA-15 and UVM-7) showed that they were able to hinder the delivery of FA at low pH (to stomach) and deliver large amounts of the vitamin at neutral pH (to intestine); nevertheless, the usage of supports with large pore entrance ensured an initial fast release, while the mesoporous material MCM-41 demonstrated a sustained release over the time [122]. The amine-capped MSPs also hindered the release of FA in gastric fluids (pH 2) and progressively delivered it in the presence of a simulated intestinal juice (pH 7.5) [123]. Similarly, the in vitro digestion procedure showed that mesoporous silica support loaded with FA and functionalized with amines inhibited the release of FA in acidic solution at pH 2 (stomach) and enabled its controlled release in neutral pH (intestine), thereby modulating the bioaccessibility [124]. Ruiz-Rico et al. [125] investigated controlled FA delivery and stability in fruit juices to reduce potential for over-fortification risks by using dated MSPs and observed that the encapsulation of FA into MSPs resulted in considerably improved vitamin stability and contributed to controlled release after consumption by modifying FA bioaccessibility.

RES encapsulated in mesoporous silica (MCM-48) NPs with the particle size of 90 nm did not alter its bioactivity and, at lower concentration, i.e., 5 μg/mL, exhibited higher anti-inflammatory activity compared to RES suspension or its solution [126]. Similar findings concerning the enhancement of the biological activity of RES by colloidal mesoporous silica NPs were reported also by Summerlin et al. [127]. Singh et al. [128] reviewed causes and consequences of micronutrient deficiencies and the bioavailability of nutrients, vitamins, minerals, and silica for food and outlined that the release of nutrients from silica in simulated intestinal fluid is better than in simulated gastric fluid.

Simple powders and tablets of inorganic–organic nanostructured hybrids prepared by intercalating FA in the MgAl-LDH and ZnAl-LDH exhibited enhanced FA release compared to crystalline FA, suggesting that such hybrids could be used to enhance the active ingredient dissolution at low pH values in effective nutraceutical products [129].

The analysis and speciation of selenium in nutritional supplements based on next-generation Se ingredients, i.e., Se forms with lower toxicity, higher bioavailability, and controlled release, such as selenium NPs (SeNPs) and selenized polysaccharides, was presented by Constantinescu-Aruxandei et al. [130].

3. Antioxidants

Reactive oxygen species (ROS) is a term used for oxygen containing free radicals (such as $O_2^\bullet$, $HO^\bullet$, $HO_2^\bullet$, $RO^\bullet$, $ROO^\bullet$) or reactive oxygen-containing compounds (such as H_2O_2, O_3, $^1O_2$), depending on their reactivity and oxidizing ability. ROS participate in diverse chemical reactions (oxidative stress) resulting in the decomposition of biologically active compounds or biomolecules. Antioxidants that are able to protect other molecules from the damaging effects of such ROS can be used as excipients in formulations or as biologically active compounds preventing oxidative stress [131]. Bioactive compounds like polyphenols, flavonoids, and vitamins showing antioxidant properties are suitable to be used for the fortification of food products to enhance their functionality, and therefore, encapsulation systems for the delivery of such nutraceuticals are necessary to overcome their low stability and bioavailability [132]. The choice of the appropriate encapsulation method is essential, because the modification of bioactivity (increase, preservation, or decrease) is affected by interactions established between the functional groups of the encapsulated compound and the encapsulating nanomaterial [133].

3.1. Nanoformulations with Antioxidant Capacity

A significant property of nanoformulations is the possibility to co-encapsulate antioxidants together with an active ingredient and thus increase the stability and extend expiration. As antioxidant excipients, PUFAs, carotenoids, antioxidant plant extracts, CUR, and catechins can be used, and as formulations, nano/micro emulsions, NLCs, NPs/MPs, and liposomes can be applied.

The antioxidant capacity of α-LNA loaded microemulsion was strongly enhanced after the introduction of carbon dots, which were distributed mainly at the oil-water interface, suggesting a "turn off" effect of the interface [134]. Benzylisothiocyanate nutraceutical encapsulated in a stable α-tocopherol-based O/W NE stabilized with a nontoxic, biodegradable surfactant, sodium stearoyl lactate, showed better antioxidant activity than pure and CUR encapsulated NEs, however CUR entrapped in the NE was effectively protected from UV light-induced degradation [135].

AST-loaded NLCs with the Z-average size of 94 nm containing α-tocopherol and EDTA as antioxidants that were stabilized using Tween 80 and lecithin and mixed with non-pasteurized CO_2-free beer at the volume ratio of 3:97 showed improved stability at low storage temperature of 6 °C [136].

Anionic sphere-shaped core-shell NPs with zein-epigallocatechin gallate (EGCG) conjugates as the hydrophobic core and a biosurfactant (rhamnolipid) as a shell with average diameters <200 nm co-loaded with CUR and RES protected these nutraceuticals from degradation, simultaneously preserving their antioxidant activity, and by mixing these NPs with lipid droplets, the bioaccessibility of both encapsulated compounds pronouncedly increased [137]. The incorporation of polysaccharides as a second polymer matrix can provide stability in zein NPs used as DESs for antioxidants in the prevention of chronic degenerative diseases [138].

Ethyl cellulose MPs with encapsulated hydroxytyrosol, a constituent of olive oil showing antioxidant properties, produced by double emulsion solvent evaporation (average particle size ranging from 156.6 ± 6.9 μm to 304.0 ± 16.0 μm) demonstrated the effectiveness of their gastro-resistance and the antioxidant capacity preservation of >50%, indicating possible applications of this formulation in foods, drugs, and nutraceuticals [139]. *Citrus reticulata* Blanco cv. unshiu peel extract (CPE) flavonoids encapsulated by pectin NPs with particle size 271.5 ± 5.3 nm released only 28.78% of flavonoids in simulated gastric fluid within 2 h compared to naked CPE and showed higher antioxidant activity than blank pectin NPs and free CPE [140]. The replacement of 30% of pectin (low charge density) with alginate (high charge density) forming shell around zein NPs significantly improved the aggregation stability at pH 5–7 and high ionic strengths (2.0 μM NaCl), and CUR encapsulated in these core-shell NPs was characterized by higher antioxidant and radical scavenging activities than CUR dissolved in ethanol solutions [141].

The comparison of liposomes and CS coated liposomes co-loaded with vitamin C and FA with the mean particle size of 138 nm and 249 nm, respectively, showed the higher EEs of both drugs as well as the higher antioxidant activity of CS coated liposome nanoformulation, suggesting that it could be applied as a promising DES in the food industry [142]. The deposition of CS and alginate layers on CUR NEs improved CUR antioxidant capacity during in vitro digestion and showed a better control of the rate and extent of lipid digestibility by decreasing free fatty acids release compared to uncoated NEs [143].

3.2. *Supplements with Antioxidant Effect*

Oxidative stress is able to generate an imbalance between the production and accumulation of ROS in cells and tissues and thus modify the ability of a biological system to detoxify these reactive products. ROS have several physiological roles (i.e., cell signaling), and they are normally generated as by-products of oxygen metabolism; despite this, environmental stressors (i.e., UV, ionizing radiations, pollutants, and heavy metals) and xenobiotics (i.e., antiblastic drugs) contribute significantly to ROS production, thus causing an imbalance that leads to cell and tissue damage (oxidative stress) [144,145]. At present, various oral (flavonoids, carotenoids, vitamin C, and synthetic) antioxidants, see Figure 2, are available on the market and are generally recommended to be used. However, these supplements should be used in accordance with recommendations of a conscious physician or health care professional to avoid their adverse effect—pro-oxidant activity depending on the specific set of conditions (their dosage, redox conditions, the presence of free transition metals in cellular milieu) [144–148]

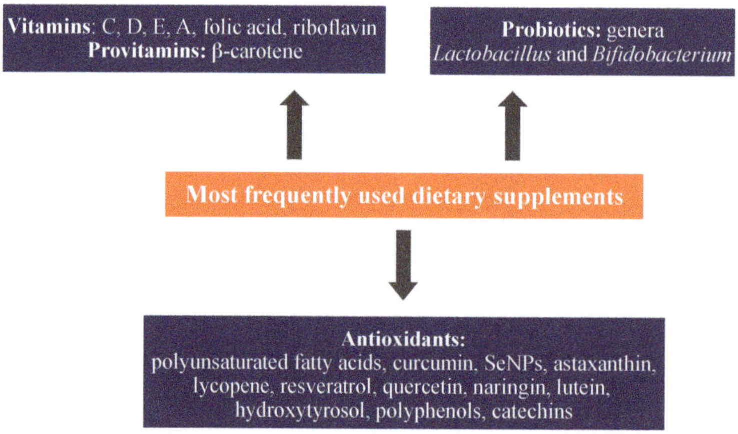

Figure 2. Most frequently used human dietary nanosupplements.

An improvement in aqueous solubility, antioxidant and other health-promoting properties, in vitro GI release profile, and protection against process and environment harsh conditions (e.g., light, oxygen, high temperatures, and humidity) of hydrophobic food bioactive compounds suitable as DISs could be achieved by nanoencapsulation using different nanoencapsulation DESs, including inclusion complexes of CDs, amylose, yeast cells, nanogels, NEs, nanofibers, nanosponges, nanoliposomes, and NPs made with lipids [149].

A CUR-β-CD inclusion complex and iron oxide NPs were co-encapsulated in liposomes, and these CUR-in-β-CD-in-nanomagnetoliposomes with mean particle size 67 nm and 71% CUR EE showed a radical scavenging property exceeding that of conventional CUR liposome and iron oxide NPs [150].

β-Car, the most important dietary source of provitamin A, is necessary for optimum human health. To increase its solubility and bioaccessibility, 0.1% β-Car was dispersed in corn oil (5 or 10%) and homogenized with 2% SCas solution at 100 MPa, and the prepared NEs had particle sizes <200 nm. β-Car stability towards oxidation decreased with the decreasing droplet diameter, and the extent of lipolysis in an in vitro system was higher and linearly related to the inverse of the droplet diameter [151]. The updated understanding of emulsion-based DESs for β-Car was overviewed by Mao et al. [152] who focused their attention also on emulsion design enabling the delivery of β-Car in complex food systems and fulfilling its benefits in functional foods. Lipid droplets in β-Car enriched O/W emulsions stabilized with surface-active chlorogenic acid-lactoferrin-polydextrose conjugate used as an emulsifier with the mean particle diameter of <400 nm across the pH range 2–9 (except pH value around 6.0) exhibited better stability against droplet aggregation under simulated GI tract conditions (mouth, stomach, and small intestine) than other systems, which resulted in improved β-Car bioaccessibility, and such formulations could be potentially applied as protectors and carriers of hydrophobic drugs, supplements, and nutraceuticals [79]. Encapsulation of β-Car into solid lipid MPs of palm stearin stabilized with hydrolyzed soy protein isolate and containing α-tocopherol with the mean diameter of about 1.2 μm preserved approx. 75% of the encapsulated β-Car after 45 d of storage, and the formulation withstand treatments with higher temperatures (>60 °C), while showing low stability after different ionic strength stresses [153].

DESs of Q, one of the most well-known flavonoids that was included in human diet long ago due to health benefits associated with its antioxidant, anti-inflammatory, antiviral, and anticancer activities as well as Q biological activities themselves, its chemical stability, metabolism, and positive impact on some cardiovascular diseases (CVD) (i.e., heart disease, hypertension, and high blood cholesterol) were overviewed by Wang et al. [154]. An extract of tartary buckwheat rich in flavonoids (TBFs) incorporated in spherical biocompatible lipid–polymer hybrid NPs of 61.25 ± 1.83 nm showed higher antioxidant activity and significant suppression of the pro-inflammatory cytokine secretion in RAW 264.7 macrophage compared to free TBFs and exhibited immune-enhancing efficacy in immunosuppressed mice, suggesting that such nanosystem loaded with TBFs is suitable for nutraceutical applications [155]. Nanoliposomes incorporating olive leaf extract containing high levels of phenolic compounds and oleuropein, showing antioxidant and antimicrobial activities, with average particle size 25–158 nm, negative charge, and EE 70.7–88.2%, which were supplemented to yogurt, improved its antioxidant activity, and no significant changes in color and sensorial attributes were observed, suggesting that olive leaf phenolics can be entrapped in nanoliposomes and could increase the nutritional value of products like yogurt [156].

RES encapsulated in zein/pectin core-shell NPs with mean diameter approx. 235 nm and polydispersity index 0.24 showed improved in vitro antioxidant activity as well as lower IC_{50} values (by 32%) related to antiproliferative activity tested using human hepatocarcinoma Bel-7402 cells compared to free RES, suggesting that such nanoformulation of RES could be used in functional foods and beverages as well as in DISs and pharmaceutical products [157]. The aqueous solubility of RES from α-lactalbumin (α-Lalb)-RES nanocomplexes was 32-fold higher than that of free RES, and the nanocomplexes considerably improved the antioxidant chemical stability under storage, especially at pH 8.0 and high temperature, and showed superb in vitro antioxidant activity compared to free RES, suggesting that α-Lalb as a nanoscale carrier could effectively deliver lipophilic nutraceuticals in the functional food, biomedical, and pharmaceutical products [158]. The protection of light-sensitive AST, a carotenoid with the most potent antioxidant activity, from photodegradation achieved by its inclusion in different hierarchically assembled nano- and microstructures in order to produce model foods for humans and fishes decreased as follows: NEs > carrageenan-coated NEs > CS coated NEs, and CS beads provided higher protection to AST than alginate beads. These hierarchically assembled materials represent ideal platforms to create foods for humans and animal species, because their flexibility enables also the incorporation of other active molecules such as proteins, PUFAs, antibiotics, antiparasitics, etc. [159]. The average size (94 nm) of AST-loaded NLCs containing α-tocopherol and ethylenediaminetetraacetic acid as antioxidants and stabilized with Tween 80 and lecithin increased at pH $\leq$ 5, high NaCl concentrations ($\geq$50 mM), and slightly at simulated gastric juice, which was connected mainly with decreasing zeta-potential, while increasing at treatment at 80/90 °C. On the other hand, cryoprotectant glycerol prevented the aggregation of AST-NLCs during freeze-thawing. Therefore, this nanoformulation could be used as a DIS [160].

Nanoscale thymoquinone [161], which was found to improve the anticancer roles of doxorubicin by upregulation of P53 and downregulation of Bcl2 and potentiate paclitaxel's apoptosis in MCF-7 breast cancer cells, could protect also against diabetes, inflammation, central nervous system, and hepatotoxicity primarily by enhancing the antioxidant status of organs and could be considered as a promising nutraceutical for human health [162].

It should be noted that microencapsulation of riboflavin, a water-soluble vitamin acting as cofactor in various processes of oxidation-reduction in a cellular system, with galactomannan biopolymer and Pluronic® F127 resulted in its slower release in both acidic and basic media compared to free vitamin [163].

Digested kenaf (*Hibiscus cannabinus* L.) seed O/W NEs stabilized by a SCas, Tween 20, and β-CD complex demonstrated good lipid digestion, significant bioaccessibility of antioxidants (tocopherols and total phenolic contents), and lower phytosterol degradation rate compared to digested bulk oil, which indicates the possibility of their future application in food and nutraceutical industries [164].

4. Other Functional Applications of Human Supplements

This chapter is focused on all other applications of DISs and FSMPs except antioxidants, i.e., supplements affecting intestine, various nutraceuticals with beneficial effect against tumor cells or constituting nutritional support therapy in the treatment of cancer, supplements supporting mental and psychomotor development, supplements for prophylaxis of metabolic syndrome, supplements for osteoporosis management, and supplements against iron deficiency.

4.1. Supplements Affecting Intestine and Absorption

It is indisputable that for proper functioning of the intestines, nutrient absorption and prevention of malnutrition, it is necessary to have a suitable composition of the "good" intestinal microflora. Different diseases and subsequent treatments may change the composition of the intestinal microflora, which may result in various, initially intestinal problems. These problems can be avoided by using different products. Prebiotics are compounds in food that induce the growth or the activity of beneficial microorganisms. Probiotics are live microorganisms (in general, bacteria of the genera Lactobacillus and Bifidobacterium) intended to provide health benefits when consumed, generally by improving or restoring the gut flora. Synbiotics refer to food ingredients or DISs combining probiotics and prebiotics in the form of synergism [165,166].

Probiotics, prebiotics, and synbiotics could suppress enteric pathogens, because they can compete with pathogenic microbiota for adhesion sites, inhibit the growth of pathogens, or stimulate, modulate, and regulate the immune response of the host by initiating the activation of specific genes in and outside the intestinal tract. Moreover, it was also shown that probiotics regulate fat storage and stimulate intestinal angiogenesis [167]. Probiotic encapsulation technology was developed rapidly in the past decade. Based on this technology, a wide range of microorganisms have been immobilized within semipermeable and biocompatible materials that modulate the delivery of cells [168].

Co-encapsulation of probiotic strains *Staphylococcus succinus* (MAbB4) and *Enterococcus fecium* (FIdM3) in alginate (2 g/100 mL) resulted in a significant improvement ($P < 0.05$) in the survival of co-encapsulated cells when exposed to acidic (pH 2.0–3.0) and bile (0.3, 0.6 and 0.8 g/100 mL) conditions. Viability was maintained throughout the storage period and ranged from 8.1 log cfu/mL (Colony Forming Unit) to 7.9 log cfu/mL for about a period of 30 days at 4 °C [169].

Dietary factors such as prebiotics (e.g., inulin) play important roles in the growth of intestinal microbiota and may impact the intestinal health [170]. The encapsulation of the probiotic *Pediococcus pentosaceus* Li05 in alginate-gelatin microgels loaded with MgO NPs enhanced its viability by filling pores inside the microgels and thus, the ability of O_2 and H^+ ions to access the probiotic could be inhibited due to the neutralization of H^+ ions in the gastric fluids by MgO NPs, thereby suppressing its acid-induced degradation. Such formulation could be considered as an appropriate DES for improving the efficacy of orally administered probiotics [171]. The effect of a prebiotic matrix consisting of inulin in concentrations 0%, 5%, 10%, 15% and 20% (w/v) in alginate beads on the viability of encapsulated probiotic strains *Pediocucus acidilactici*, *Lactobacillus reuteri*, and *Lactobacillus salivarius* was investigated by Atia et al., and the researchers found that the beads with 5% w/v inulin were the most effective in bacterial protection against bile-salts and acidity [172].

Probiotics are also affected by prebiotics apart from other things. Therefore, Peredo et al. [173] investigated the influence of natural prebiotics potato starch, *Plantago psyllium*, and inulin co-encapsulated with alginate on the viability of *Lactobacillus casei* Shirota and two strains of *Lactobacillus plantarum* Lp33 and Lp17. The results showed a higher encapsulation yield when *P. psyllium* (94% for Lp17) and inulin (78% in Lp33) were used; *P. psyllium* ensured a higher viability of the bacteria during storage at 4 °C and the best protection in GI conditions.

Recent findings related to the production of probiotics, prebiotics, and nutraceuticals using a nanotechnology approach with respect to the functional foods was presented by Mishra et al. [174].

4.2. Anticancer Nutraceuticals

Nutraceuticals, such as soya bean, garlic, ginger, green tea, propolis, honey, RES, Q, EGCG, etc., may have chemopreventive effects. They are able to induce the apoptosis of cancer cells. These special foods can be used for chemoprevention or as a supportive therapy at treatment of tumor by standard anticancer chemotherapeutics [175–180].

In a review paper, McClements and Xiao [181] focused their attention on some most important anticancer nutraceuticals found in foods, the main factors affecting their bioaccessibility, absorption, and transformation, and different types of DESs and excipient systems improving the overall bioavailability of anticancer nutraceuticals.

(-)-Epigallocatechin-3-gallate is the most abundant catechin and also the most effective cancer chemopreventive polyphenol in green tea. This EGCG pronouncedly inhibited β-Car degradation in both MCT and corn O/W emulsions in a dose dependent manner and did not adversely affect lipid oxidation, while α-Lalb was not able to protect β-Car in MCT emulsions; their combination had a similar effect as EGCG alone [182]. A comprehensive review related to findings concerning the encapsulation of EGCG by means of nanocarriers was presented by Granja et al. [183].

Biopolymer core-shell NPs consisting of hydrophobic protein (zein) as the core and a hydrophilic polysaccharide (pectin) as the shell fortified with CUR showing strong anticancer activity with the diameter of 250 nm, which were converted into a powdered form resulting in good water dispersibility, were reported to be suitable for incorporating CUR into functional foods and beverages as DISs, and pharmaceuticals [184]. Bioavailable NEs loaded with nutraceuticals (CUR and fresh and dry tomato extracts rich in lycopene) with the hydrodynamic size of NEs approx. 100 nm applied in combination with doxorubicin enhanced cell viability in cardiomyoblasts (H9C2 cells) by 35–40% compared to that observed in cardiomyoblasts treated with doxorubicin alone, provided protection against oxidative stress, inhibited the release of IL-6, IL-8, IL-1, TNF-α, and nitric oxide by approx. 35–40%, and increased IL-10 production by 25–27% compared to cells without NE treatment. The best cardioprotective profile was showed by a lycopene-rich NE capable to effectively protect against doxorubicin-induced cardiotoxicity by reducing inflammation and lipid oxidative stress [185].

Both a cinnamon oil NE and a vitamin D encapsulated cinnamon oil NE with particle sizes 40.52 and 48.96 nm, respectively, arrested the cell cycle progression in the G_0/G_1 phase, showed an increased expression of Bax, capase3, and caspase-9, and decreased the expression of BcL2 proteins along with a considerable increase of apoptotic cell population and loss of mitochondrial membrane potential. The NE with cinnamon oil as a carrier for a lipophilic nutraceutical like vitamin D showing potential anticancer activity in human alveolar carcinoma cells could be also used in the food industry [186].

A nanonutraceutical formulation of ω-3 PUFAs (fish oil) could effectively inhibit the release of ROS and reactive nitrogen species from human neutrophils and murine macrophages, the production of the proinflammatory cytokines TNF-α and MCP1, and tumor-cell proliferation in FaDu head and neck squamous carcinoma and 4T1 breast cancer cells in in vitro cultures. The ω-liposomes, in which docosahexaenoic acid (DHA) was formulated, could be used for intravenous delivery of fish oil fatty acids resulting in beneficial effects in the treatment of inflammatory disorders and cancer [187]. It is known that DHA (ω-3 PUFA), a component of fish oil, suppresses rat mammary carcinogenesis, reduces cell growth, and induces apoptosis in human breast cancer cell lines. An acid stable liposome formulation of DHA with the use of ether and phytanyl lipids similar in structure to those found in Archaea having the mean particle size of 137 ± 12 nm and a slightly negative charge was resistant to oxidation and stable over the pH range of 1.0–7.4 at 37 °C for two hours. Cell viability in MCF-7 cells and apoptosis in both MCF-7 and MDA-MB-231 cells were reduced more effectively by this liposomal formulation than by free DHA, suggesting that it could be potentially used in breast cancer prevention [188].

The investigation of nanosized complexes prepared using high amylose corn starch and flax seed oil processed to powder of MPs by spray-drying and subsequently incorporated into bread formulation showed a considerable reduction of lipid oxidation in breads during baking due to the encapsulation as well as a decreased formation of carcinogen acrylamide, suggesting a beneficial effect of this nanoformulation on the final product quality and safety [189].

4.3. Supplements Supporting Mental and Psychomotor Development

Fermented soybean nanonutraceuticals administered to rats intoxicated with colchicine and showing impairment in learning and memory and decreased activity of acetylcholinesterase (AChE) caused an increase of AChE activity (42%), a reduced activity of GSH (42%), SOD (43%), and catalase

(41%), and decreased lipid peroxidation (28%) and protein carbonyl contents (30%), which suggests a possible neuroprotective efficiency of the nanonutraceuticals, and in addition, a significant amyloid-β and BACE-1 inhibition activity was demonstrated in an in silico study. The beneficial effect of the discussed nanonutraceuticals is associated with their strong antioxidant activity, and it could be assumed that they could also positively influence cognitive defects associated with Alzheimer's disease [190].

Encapsulation in bovine-milk exosomes could protect cargos against enzymatic and nonenzymatic degradation. RNAs encapsulated in exosomes could be delivered to circulating immune cells in humans, and some microRNAs and mRNAs in bovine-milk exosomes could regulate human gene expression and be translated into protein. Gene expression can be altered by low concentrations of dietary microRNAs through noncanonical pathways, such as the accumulation of exosomes in the immune cell microenvironment and microRNA binding to Toll-like receptors. In mice, the proliferation of intestinal cells was promoted by porcine-milk exosomes, suggesting that milk exosomes and their cargos could be used in human nutrition. Therefore, it was suggested that milk modified in this way could contribute to better mental, psychomotor, and functional development of infants [191].

Natural compounds that are commonly present in foods and beverages are regarded as promising molecules in a nutraceutical approach associated with life-long healthy diets. An increased attention is devoted to food molecules that are candidates to enter clinical trials as such or after targeted molecular engineering and could have a beneficial effect on amyloid neurodegenerative diseases. Natural phenols abundant in healthy food products, such as green tea, red berries, extra virgin olive oil, red wine, and spices, could be considered particularly promising [192]. Biodegradable poly(lactic-co-glycolic acid) NPs encapsulating ginsenoside Rg3 (an important constituent of ginseng, playing a significant role in memory and improving cognition) and thioflavin T, which showed neuroprotective effects, were reported to be a theranostic material for the detection and treatment of Alzheimer's disease [193].

4.4. Supplements for Metabolic Syndrome Prophylaxis

Sodium alginates could be used for the management of GI tract disorders and the attenuation of components of the metabolic syndrome such as obesity, type 2 diabetes, hypertension, non-alcoholic fatty liver disease, and dyslipidemia. They could also protect cells during transplantation from immune responses of the host, and, in combination with antacid alginates, be applied in the treatment of gastric reflux disease. Moreover, alginates decrease food intake by inducing satiety, increase weight loss in patients on a calorie-restricted diet, and reduce both glucose and fatty acid uptake, and a decrease in blood pressure by alginates in rat models of hypertension was reported as well [194].

Using advanced proteomic and bioinformatic approaches, Kar et al. [195] characterized the protein components of six different protein sources (casein, partially delactosed whey powder, spray-dried porcine plasma, soybean meal, wheat gluten meal, and yellow meal worm) and predicted the bioactive properties of these protein sources after in silico digestion with monogastric proteolytic enzymes. The tested protein sources were potentially rich in bioactive peptides, in particular, angiotensin-converting enzyme inhibitors and peptides with antioxidant properties, and could be used as alternative sources of protein in animal feeds for monogastrics.

Temporal improvements in vitamin D status by vitamin D supplementation resulted in an increase in serum 25-hydroxyvitamin D concentrations and reduction of serum homocysteine concentrations suggesting that such treatment could reduce risk factors for CVD and may potentially contribute to the primary prevention of CVD [196]. Liposome-in-alginate beads were used to encapsulate the oyster hydrolysates showing antihypertensive effect to improve their bioavailability, protect them from degradation, and obtain sustained release; the release time of the oyster hydrolysate in the simulated GI fluid was up to 16 h [197]. Encapsulation of naringin, a flavonoid that occurs naturally in citrus fruits and possesses strong health benefits (recommended for the prevention of CVD and diabetes), in ternary NPs consisting of amylose, α-LLA, and β-Lglb resulted in a gradual release of naringin from the ternary NP–naringin inclusion complex in simulated gastric and intestinal fluids, and ternary NPs effectively improved the bioavailability of bioflavonoid [198].

Mahmoud et al. [199] studied the impact of dietary camel whey protein administered as a supplement to streptozotocin (STZ)-induced diabetic pregnant mice on the efficiency of the immune system of the offspring and verified its protective role in decreasing the tendency of the offspring to develop diabetes and related complications. A comparison of prophylactic effects of α-eleostearic acid rich nano and conventional bitter gourd seed oil emulsions in induced diabetic rats showed that the maximum efficiency in suppressing oxidative stress was achieved with a diet supplementation of 0.5% (w/v) NE with bioactive lipid-conjugated α-LNA, suggesting that such nanoformulation could be used as an appropriate nutraceutical against diabetes mellitus strongly attenuating an adverse impact of excessive ROS [200].

Although natural nanosized clinoptilolite and/or metformin did not affect pronouncedly the levels of serum glucose, minerals, and lipid profile in rats with high-fat-diet/STZ induced diabetes, the co-treatment of clinoptilolite with the drug notably increased high-density lipoprotein (HDL) cholesterol, while Cu and Ca levels increased only in the metformin group [201]. In STZ induced diabetic rats treated with nanosized clinoptilolite, blood glucose was found to decrease to near normal levels (12.4 vs. 27.5 mmol/L), but no significant impact on oxidative stress markers was estimated [202]. Nanosized clinoptilolite injected to STZ induced diabetic rats caused a partial improvement in their weight status and lack of undesirable effects, although beneficial changes in lipid profile were not detected, which could be connected with short study duration [203].

CS NPs loaded with *Stevia rebaudiana* leaf extract caused a considerable decrease of the mean fasting blood glucose level of treated diabetic rats in comparison with the diabetic control group, and serum levels of different enzymes and some antioxidants, e.g., catalase, reduced glutathione (GSH), and superoxide dismutase (SOD), were closer to normal levels in the group treated with NPs than in the control group [204].

By adding *Catathelasma ventricosum* polysaccharides (CVPs) to the redox system of selenite and ascorbic acid, spherical CVPs–selenite NPs with particle size approx. 50 nm were prepared, and based on serum profiles and antioxidant enzyme levels, it could be concluded that CVPs–selenite NPs showed a notably higher antidiabetic activity ($p < 0.05$) than other SeNPs, selenocysteine, and Na_2SO_3 [205].

To control the release of the anti-hyperglycemic agent fisetin for nutraceutical and/or therapeutic applications, an oral controlled release system consisting of polymeric NPs (140–200 nm) based on poly-(ε-caprolactone) and poly(lactic-*co*-glycolic acid)-polyethylene glycol-COOH encapsulating fisetin was designed, which protected and preserved the release of the active compound in gastric simulated conditions, controlled the release in the intestinal medium, and showed an improved α-glucosidase inhibiting activity of fisetin compared to that of the commercial formulation acarbose [206].

NLCs loaded with betasitosterol, a phytosterol showing beneficial effects on reducing total cholesterol and low-density lipoprotein (LDL), with particle size 165 nm, zeta potential −13.5 mV, and EE 99.96%, which were incorporated in butter, showed good stability during three months' storage period and increased the antioxidant property of enriched butter during the storage period, suggesting the suitability of such nanoformulation for functional dairy products [207]. Multilayer CS–alginate–CUR NEs could be important for functional food development for combating obesity, because they increase satiety by retarding lipid digestion [143].

4.5. Supplements for Osteoporosis Management

Many supplements in pharmacies serve to prevent or mitigate the effects of osteoporosis. Calcium (Ca) has clearly been shown to have some positive effect on osteoporosis, although the bioavailability of Ca from classical preparations is approximately 10–15% [208–210]. Recently, a number of scientific teams investigated Ca supplementation by nano-Ca either as solid peroral DISs or as nano-Ca from the fortified milk. Experiments performed in vivo on ovariectomized (OVX) rats demonstrated much greater absorption (up to 89%) and overall bioavailability (up to 42%) of preparations with nano-$CaCO_3$, citrate, or organically bound in shell oyster. Therefore, by in vivo studies it was confirmed that the application of nanosized Ca could improve Ca and even phosphorous content in bones [211–215].

Ca alginate NPs (200–500 nm in diameter) loaded with collagen peptide chelated Ca with the average diameter of approximately 150 nm and the Ca content of up to 130.4 g/kg notably enhanced Ca absorption and significantly increased femur bone mineral density and femur Ca content in rats, suggesting that they could prevent Ca deficiency and could be used as a new Ca supplement in the food industry [216]. A nanocomposite of whey protein hydrolysate chelated with Ca showed superb stability and absorbability under both acidic and basic conditions, which was beneficial for Ca absorption in the GI tract of the human body. Its pronouncedly higher Ca absorption on Caco-2 cells compared with Ca gluconate and $CaCl_2$ in vitro suggested a possible increase in Ca bioavailability and thus its potential to be used as DIS for improving bone health of humans [217].

It is also important to remember that oral administration of Ca hydroxyapatite microcrystals can accelerate fracture healing and repair and even prevent osteoporosis [218]. Moreover, Zhang et al. [219] described the benefit of nanohydroxyapatite/CS composite for bone regeneration when it was administered by injection. In addition, these nanocomposites showed antistaphylococcal activity [220].

Additionally, $CaCO_3$ from eggshell can be used as a Ca supplement [221]. Chicken eggshell powder became an attractive source of Ca for human nutrition. It can be added to food or drinks. For example, chocolate cakes were fortified by 3%, 6%, and 9% of them, and the results indicated that with respect to the Ca content, texture and sensory properties of the cakes, 6% eggshell supplementation (i.e., increased Ca content to 816.8 mg/100 g) was the best [222]. The preventive effects of nanopowdered eggshell (NPES) on postmenopausal osteoporosis in OVX rats was also studied, and the results were surprising. NPES fed rats showed an increase in bone mineral densities (BMD) by about 7% compared to OVX rats. Only powdered eggshell led to an increase of BMD by 2%. Serum analysis showed that NPES fed rats had a 22.4% higher osteocalcin level than OVX rats. Therefore, NPES attenuated the bone loss induced by ovariectomy in rats [223]. High-calcium yogurt as food for combat with osteoporosis was prepared using its fortification with 10-nm crystals of NPES. The addition of NPES up to 0.3% gave cow and buffalo's milk yogurts with acceptable composition, textural properties and sensory attributes, and this additive increased the Ca content of yogurt by about 15% [224].

4.6. Supplements against Iron Deficiency

Iron-deficiency anemia is the most common nutritional disorder worldwide with impact on health and economy. In spite of a number of commercially available supplements, this deficiency is a global public health problem due to the poor tolerability of the standard care soluble iron salts (such as ferrous sulfate), which results in non-compliance and ineffective correction of iron-deficiency anemia. On the other hand, poorly water-soluble compounds cause less sensory changes, but are not well absorbed [225]. Nanoformulations of iron were proposed to fortify food and feed to address these issues due to enhanced bioavailability, good product stability, limited side effects and the absence of changes of taste and color of the fortified foods [226]. In addition, in vitro and in vivo experiments have shown that iron NPs can be considered safe [227]. Ferritin, which is well absorbed [228] is itself composed of an iron oxide nanocore surrounded by a protein shell. Recently, Powell et al. [229] synthesized tartrate-modified, nano-disperse ferrihydrite having small primary particle size and enlarged or strained lattice structure (about 2.7 Å for the main Bragg peak versus 2.6 Å for synthetic ferrihydrite) that was able to efficiently provide GI delivery of soluble Fe(III) without the risk of free radical generation in murine models, where GI delivery did not depend on luminal Fe(III) reduction to Fe(II), and absorption was similar to that of $FeSO_4$. This nanoformulation could be considered as a potentially side effect-free form of Fe supplementation to human suffering from anemia.

The most promising preparation (iron hydroxide adipate tartrate: IHAT) showed ~80% relative bioavailability to $FeSO_4$ in humans and, in a rodent model, IHAT was equivalent to $FeSO_4$ at repleting hemoglobin. Moreover, IHAT did not accumulate in the intestinal mucosa and, unlike

$FeSO_4$, promoted a beneficial microbiota. In an in vitro study, IHAT was 14-fold less toxic than Fe(II) sulfate/ascorbate. The results of IHAT NPs observed from three-arm, double-blind, randomized, placebo-controlled trial conducted in Gambian children 6–35 months of age in relation to ferrous sulfate and non-inferiority in relation to placebo in terms of diarrhea incidence and prevalence confirmed the hypothesis that supplementation with IHAT eliminates iron deficiency and improves hemoglobin levels without inducing GI adverse effects [230,231].

Poorly water-soluble nanosized $FePO_4$ with specific surface area approx. 190 m^2/g made by scalable flame aerosol technology possesses in vivo iron bioavailability in rats comparable to $FeSO_4$ and causes less color change in reactive food matrices than conventional iron fortificants. The addition of Zn or Mg oxides to nano $FePO_4$ increases Fe absorption and also improves their color [232]. Additionally, Srinivasu et al. prepared nano ferric pyrophosphate (particle size 10–30 nm) as a potential food fortificant in iron-deficiency anemia and found that the peroral bioavailability of ferric pyrophosphate NPs in rats, calculated using hemoglobin regeneration efficiency, was 103.02% with respect to the reference salt, ferrous sulfate, while the NPs did not show any significant toxicity [233].

Salaheldin and Regheb biosynthesized biocompatible Fe_3O_4 NPs capped with vitamin C, and thus intestinal villi absorbed the NPs as vitamin C and not as an iron, because iron was coated with vitamin C. Clinical and histopathological studies on rats recommended the use of fortified biscuits with concentrations of 10 ppm and 30 ppm of nano iron; hemoglobin concentration increased from 9.9 ± 1.2 g/dL to 14.6 ± 1.1 and 16.7 ± 1.6 g/dL, respectively [234].

5. Veterinary Nanoscale Nutraceuticals and Dietary Supplements

As mentioned above for humans, nanoformulated DISs can also be applied for animals. In general, these veterinary DISs are regulated by the FDA's Center for Veterinary Medicine [235,236]. Nutraceuticals have become popular with the veterinary community; worldwide estimates of sales approach $100 billion [237]. Therefore, many different products can be found for veterinary applications. For example, the use of clinoptilolite (natural zeolite comprising a microporous arrangement of silica and alumina tetrahedra) showing unique antibacterial properties as a DIS in food and unifying properties of an immunomodulator and nutraceutical could represent an alternative to antibiotic growth promoters in animals of veterinary importance. Valpotic et al. [238] focused their attention mainly on clinoptilolite potentials and limitations in cattle related to metabolic and endocrine status, oxidative stress, and systemic local inflammatory responses involved in reproductive and metabolic disorders of dairy cows.

Zinc (Zn), copper (Cu) and selenium (Se) are essential nutrients for animals and humans, because these metals occur in various metaloenzymes as co-factors [239]. Zn is a nutritionally indispensable trace element that is required for normal growth, bone development, feathering, appetite regulation, metabolic functioning of nearly 300 biochemical enzymes, hormone production, cell division, protein and DNA synthesis for all avian species [239,240]; so, it can affect animals production and reproduction performance [241]. Zinc deficiency in animals caused a decrease in feed intake, growth, serum insulin-like growth factor-I, and growth hormone (GH) and lowered the hepatic production of insulin-like growth factor-I, GH receptor, and GH binding protein [241–244]. In addition, Zn is used to decrease fermentation of digestible nutrients in intestines and improve nutrients digestibility and appetite. Dietary Zn supplementation stimulates feed intake probably caused by increased ghrelin secretion [245]. It was observed that it caused an increase in insulin-like growth factor expression in the small intestine mucosa [246]. Increased Zn concentration in the intestines influences their structure and function. The growth-stimulating properties of dietary Cu have been attributed to its antimicrobial action, however, it was shown that also intravenous injection of Cu to weanling piglets stimulated their growth [247]. It seems that the growth-promoting properties of high dietary concentrations of Cu complement its antimicrobial action [248].

Se is very important in animal nutrition, because it functions as an anti-oxidant assisted by vitamin E; e.g., Se is a cofactor of glutathione peroxidase (GSHpx), deiodinases, thioredoxin reductases, selenophosphate synthatase, selenoprotein P, selenoprotein W, etc. Se deficiency can be a major problem that can be reduced or prevented by supplementation with inorganic or organic sources of Se. On the other hand, Se in high concentration is toxic to human and animal [249]. Recent knowledge related to beneficial biological effects of SeNPs in the organism, absorption mechanisms, and nanotechnological applications for peroral administration were summarized by Hosnedlova et al. [250]. The applications of the above-mentioned nutrients (Zn, Cu, Se) in nanoscale formulations allow increased efficacy, enhanced absorption, lower overall doses, etc. [251,252]. In the following subchapters, an attention is mainly focused on the beneficial effect of some inorganic NPs (Zn, ZnO, Cu, CuO, Se, Ag) and nanoscale formulations containing organic active compounds (e.g., essential oils, vitamins) on growth performance and some important biochemical parameters of aquatic animals, poultry, pigs and other domestic animals like cattle, sheep, and rabbits. Nanoformulations have also found their way into the fortification of animal feeds [253,254].

5.1. Aquatic Animals

The major challenge facing fish farming is the availability of relatively cheap but high-quality feed. Regular fish diet blended with nanosized mineral nutrients has beneficial impact on growth and overall health of fish, because they can pass across the gut tissue into cells more readily than bulk nutrients, and thus their assimilation processes in the fish are accelerated, resulting in improved growth [255].

Thyme essential oil at doses 400 and 800 mg/kg used as a DIS was found to reduce oxidative stress of gibel carp (*Carassius auratus gibelio*; average weight of 8.73 ± 2.1 g), and exposure to a sub-acute toxicity level of AgNPs for a period of 96 h after six weeks of a feeding trial confirmed the resistance of the carp to non-fatal effects of AgNPs [256]. A fish diet containing 1% *Aloe vera* NPs improved the growth factors (weight gain, initial body weight, condition factor, feed conversion ratio, specific growth rate) of Siberian sturgeon [257]. CS NPs showing spherical shape, particle size 185 nm, and positive zeta potential used to carry vitamin C through the GI tract of rainbow trout (*Oncorhynchus mykiss*) exhibited in vivo controlled release until 48 h and increased lysozyme and complement contents in the fish serum [258].

Shrimps (*Litopenaeus vannamei*) reared in clear water and in a biofloc system and receiving feed supplemented with nanocapsules containing lipoic acid showed increased final weight, higher GSH levels in the hepatopancreas and decreased percentage of hyaline hemocytes, while increased levels of granular hemocytes. Increased glutathione *S*-transferase activity in the gills and hepatopancreas was estimated only in shrimps reared in the biofloc system and fed with encapsulated antioxidant, while decreased levels of thiobarbituric acid reactive substances were estimated in the gills and muscles of the shrimps maintained in clear water [259].

A 60-day feeding of red sea bream (*Pagrus major*) with CuNPs (2 mg/kg) or/and vitamin C (800–1200 mg/kg) improved its growth and health, and higher final weight, weight gain, specific growth rate, protein gain, protein retention, feed intake, protease and bactericidal activities, and higher tolerance against stress than in controls was estimated as well. The feed and protein efficiency ratios and the body lipid content were considerably higher at treatment with 0/1200, 2/800, 2/1000 and 2/1200 mg CuNPs/vitamin C per kg, while the application of 2/800, 2/1000 and 2/1200 mg CuNPs/vitamin C per kg resulted in considerably enhanced body protein and higher tolerance against stress compared to the control was estimated as well [260]. Wang et al. [261] reported that for the dietary Cu requirements of Russian sturgeon (9.82 ± 0.08 g) fed with diets containing different forms of Cu for 8 weeks, Cu-methionine (Met) and CuO NPs were 1.5–2-fold more bioavailable than $CuSO_4$, optimal doses being approx. 5 mg/kg for Cu-Met or CuO NPs and 8 mg/kg for $CuSO_4$.

In *Pangasius hypophthalmus* fed with a diet incorporating 10 and 20 mg/kg ZnNPs and exposed to abiotic stress (sublethal dose of Pb 4ppm and temperature 34 °C), a considerably enhanced growth performance and improved immunological parameters (total protein, Alb, Glb, and Alb/Glb ratio) were observed, and reduced oxidative stress reflected in lower levels of blood glucose, cortisol, and HSP 70 suggested that the supplementation of dietary ZnNPs could alleviate abiotic stress in

P. hypophthalmus [262]. In Mozambique tilapia (*Oreochromis mossambicus*) receiving a diet supplemented with 0.004% of *Portunus pelagicus* β-1,3-glucan binding protein based ZnO NPs, considerable increases in growth performance and in cellular and humoral immune responses were estimated. Moreover, when after 30 days of a feeding trial, the fish was challenged with aquatic fish pathogen *Aeromonas hydrophila* (1×10^7 cells/mL) through intraperitoneal injection, a reduced mortality rate was observed in fish fed with the diet containing such ZnO NPs, suggesting a potential beneficial impact of the NPs on the immune system and survival of *O. mossambicus* [263]. Beneficial effects of Zn-proteinate, $ZnSO_4$, and ZnO NPs applied at dose 50 mg/g of Zn sources in an early diet of rainbow trout larvae with average weight of 82.3 ± 11.6 mg for 70 days enhanced the growth performance of the larvae [264].

Common carp (*Cyprinus carpio*) juveniles (9.7 ± 0.1 g), the diet of which was supplemented with SeNPs (0.7 mg Se/kg), showed the highest weight gain of $97.2 \pm 10.8\%$ and feed efficiency ratio $42.4 \pm 0.8\%$, the highest serum hemolytic activity, total immunoglobulin, and total protein and Alb contents as well as the lowest serum total cholesterol and LDL levels after 8 weeks of feeding compared to the carp fed with Na_2SeO_3, Se-Met, and the control. Carps fed with SeNPs or Se-Met showed also pronouncedly higher activities of serum glutathione peroxidase (GPx) and SOD and an increase in white blood cell counts, neutrophil percentage, and serum lysozyme activity compared to the control group and the Na_2SeO_3 group [265]. Dietary treatments of crucian carp, *Carassius auratus gibelio*, with SeNPs and Se-Met showed higher Se levels in muscle (16.42 ± 1.07 μg/g and 13.52 ± 1.31 μg/g, respectively) compared to carps fed with basal feed (6.10 ± 0.78 μg/g). Although the survival rate and the feed conversion ratio were not affected by the dietary treatments, GPx activities in Se-treated carp plasma and liver differed significantly from those of the control [266]. Dietary SeNPs supplementation at the dose of 0.68 mg/kg to juvenile mahseer (*Tor putitora*) considerably increased red blood cell count, hemoglobin level, hematocrit values, and lysozyme activity as well as serum GH levels, tissue total protein content, and GPx activity in liver and muscle tissues of *T. putitora* [267]. Chinese mitten crabs (*Eriocheir sinensis*) fed with a diet containing 0.2 mg/kg SeNPs in a 60 d feeding trial had a considerably higher weight gain rate and a reduced feed coefficient. When juvenile Chinese mitten crabs were kept under the condition of hypoxia, the up-regulative effects of SeNPs on antioxidant capacity, hemocyte counts, and hemocyanin expression were further amplified. Hypoxia exposure increasing mortality in crabs infected with *A. hydrophila* bacteria was also alleviated when crabs received a diet containing 0.2 mg/kg SeNPs, suggesting the importance of dietary SeNPs in regulating the immunity and disease resistance in crabs kept under hypoxia stress [268]. Naderi et al. [269] who investigated the impact of dietary SeNPs (1 mg/kg), vitamin E (500 mg/kg), and their combination on the humoral immune status and serum parameters of rainbow trout under high-density condition (80 kg/m^3) reported that the positive effects observed in the performance following the combine treatment may be due to vitamin E alone, because supplementation with SeNPs did not markedly affect the performance in rainbow trout under high-density conditions. In addition, the immuno-protective role of biologically synthesized dietary SeNPs applied at the dose of 1 mg/kg against multiple stressors (Pb level of 4 ppm, high temperature of 34 °C) in *Pangasinodon hypophthalmus* was reported by Kumar et al. [270].

5.2. Poultry

Typical poultry diets are commonly enriched by feed additives containing vitamins and minerals to support rapid growth and a favorable feed conversion ratio, and nanosized feed additives characterized by a high surface area to volume ratio and high absorption in the body could be incorporated in vaccines and nutrient supplements and directly transported to targeted organs or systems without degradation resulting in health benefits. Current state of NPs use as poultry feed supplements was reviewed by Gangadoo et al. [271].

AgNPs received in drinking water containing 1000 mg AgNPs/kg significantly reduced the body weights of the broilers after 42 days of administration, and this adverse effect could not be mitigated with a basal diet supplemented with Zn (60 and 120 mg/kg) and vitamin E (α-tocopherol acetate; 100 and 200 mg/kg). On the other hand, the increased activity of CuZn-SOD observed in AgNPs-treated broilers was not recorded in birds fed with the basal diet supplemented with 200 mg/kg vitamin E, suggesting its antioxidant effect. Moreover, Zn supplementing enhanced catalase and GPx activities in the jejunal mucosa resulting in increased malondialdehyde (MDA) levels in the animals. Therefore, it could be concluded that the dietary Zn and vitamin E supplementation was able to attenuate intestinal oxidative stress in AgNPs-treated broiler chickens, although it did not mitigate the growth reduction caused by AgNPs [272]. In ovo feeding was found to reduce post-hatch mortality and skeletal disorders and increase muscle growth and breast meat yield. Sawosz et al. [273] used AgNPs as a protective carrier for adenosine triphosphate (ATP) as well as an active agent, which may penetrate tissues and cells and localize inside cells. They injected AgNPs, ATP, or a complex of AgNPs + ATP (AgNPs/ATP) in broiler eggs, and on day 20 of incubation, the embryos were evaluated. An increased expression of fibroblast growth factor 2, vascular endothelial growth factor, and Na^+/K^+ transporting ATPase were estimated at the application of ATP or AgNPs to chicken embryos. Moreover, AgNPs also upregulated the expression of myogenic differentiation 1, affecting cell differentiation. Based on the above-mentioned findings, it could be concluded that an extra energy source in the form of ATP addition enhanced molecular mechanisms of muscle cell proliferation, and ATP and AgNPs could accelerate the growth and maturation of muscle cells [273].

The in ovo injection of CuNPs using the dose of 50 mg/kg CuNPs improved broiler performance more efficiently than the injection of 50 mg/kg $CuSO_4$ or the provision of CuNPs or $CuSO_4$ in drinking water containing 20 mg/kg CuNPs or $CuSO_4$ to growing chickens. In another experiment, which was carried out with 126 one-day-old broiler chickens from day 1 to 35 post-hatching, the in ovo application of Cu enhanced the final body weight, average daily gain, and feed conversion ratio compared to control animals and resulted in a considerable improvement in energy and nitrogen utilization, mainly for CuNPs application. The CuNPs treatment also reduced cholesterol, urea, and glucose levels in the blood [274]. The supplementation of a Cu deficient basal diet of chickens with CuNPs in drinking water to the level of Cu exceeding the National Research Council (NRC) recommendation by 54% resulted in the increased antioxidant potential of the organism and the inhibition of lipid peroxidation. Antioxidant and immune defenses of chickens were simultaneously increased in chickens receiving diet supplemented with CuNPs up to 12 mg per bird during 6 weeks of feeding, i.e., up to a level exceeding the NRC recommendation for growing broiler chickens at the most by 7%. It could be mentioned that at a higher CuNPs supplementation, a deterioration in red blood cell parameters and the stimulation of the immune system reflected in an increase in interleukin-6, immunoglobulin A (IgA), IgM, and IgY was observed [275].

The in ovo administration of NPs could be considered as a new method of nano-nutrition to supply an additional quantity of nutrients to embryos. ZnNPs, CuNPs and SeNPs supplemented in ovo at doses 20, 40, 60, and 80 μg ZnNPs/egg, 4, 8, 12, and 16 μg CuNPs/egg, and 0.075, 0.15, 0.225, and 0.3 μg SeNPs/egg (18th day incubation, amniotic route) did not show any adverse effect on the developing embryo and did not influence the hatchability, best feed efficiency being observed with 40 μg ZnNPs/egg, 4 μg CuNPs/egg and 0.225 μg SeNPs/egg. Moreover, the application of 12 μg CuNPs/egg resulted in considerably higher breast muscle percentage [276].

In white Leghorn laying hens (68-week old) receiving a diet supplemented with Zn-Met, bulk ZnO, and ZnO NPs reaching the level of 60 mg Zn/kg in the diet, pronouncedly higher Zn retention, serum GH concentration, and carbonic anhydrase activity were observed at the application of the ZnO NPs and Zn-Met compared to the control, and the ZnO NPs enhanced eggshell thickness as well [241]. Laying hens at 64 weeks of age fed with a basal diet supplemented with 80 mg/kg of bulk ZnO, ZnO NPs, and Zn-Met showed considerably higher egg production and egg mass as well as SOD activity in the liver, pancreas, and plasma when Zn-Met and the ZnO NPs were applied, while the greatest increase in eggshell thickness and shell strength was observed at the ZnO NPs application. The Zn supplementation resulted in reduced egg loss and lower MDA content and had a beneficial effect on serum total protein, Alb, glucose, alkaline phosphatase activity, carbonic anhydrase activity, and Zn level, which was reflected in an improved performance of laying hens. Due to the enhanced Zn absorption in the intestine of aged layers at the application of ZnO NPs, they could be considered as a more suitable source of Zn in diets than bulk ZnO [277]. At the dietary supplementation of Zn-Met, ZnO, ZnO NPs or polyglutamic acid (PGA)–ZnO NPs reaching the level of 80 mg Zn/kg in the diet, increased Zn content in eggshells, serum Zn concentration, ghrelin and IgG levels of 64-week old brown layers were observed at the application of the ZnO NPs and the PGA-ZnO NPs, exceeding that observed at the application of bulk ZnO, and serum carbonic anhydrase activity and ghrelin levels were also increased compared to Zn-Met, suggesting that the ZnO NPs alone or in combination with PGA show beneficial impact on the Zn status of aged layers [278].

Investigation of the effects of SeNPs on performance, meat quality, immune function, oxidation resistance, and tissue Se content in broilers performed with 1-day old male Arbor Acres broilers showed that the supplementation of corn-soybean meal-based diets with 0.3–0.5 mg SeNPs/kg was found to be the best, and the maximum supplementation of SeNPs could not exceed 1.0 mg SeNPs/kg [279]. The adverse effects of oxidative stress in broiler chickens induced by *tert*-butyl hydroperoxide were attenuated when the animals received a diet supplemented with 0.3 mg SeNPs/kg. In stressed chicks fed with SeNPs, the heterophil:lymphocyte ratio was lower than in the groups, the diet of which was supplemented with bulk inorganic or organic Se, suggesting a higher effectiveness of SeNPs in the mitigation of oxidative stress [280]. Supplementation of SeNPs (0.1–0.5 mg/kg) in broiler diets could improve growth performance, carcass components, and immune function of the animals, and no adverse effects on internal organs, other carcass parameters, and GI parts were observed. SeNPs dietary supplementation resulted in significantly improved weight gain and feed conversion ratio during the whole period of experiment (42 days) and more efficient energy and protein utilization compared to the control group [281].

Rahmatollah et al. [282] reported that 1.2 mg/kg cysteine-coated Fe_3O_4 NPs were found to be required and sufficient for quails' optimal maintenance and growth suggesting that cysteine-Fe_3O_4 NPs can be used as a Fe source in the quail diet.

Cr utilization in 32 three-week-old broilers fed with a diet supplemented with Cr at the 1200g/kg level using $CrCl_3$, chromium picolinate (CrPic), and CrPic NPs decreased as follows: CrPic NPs > CrPic > $CrCl_3$ > control groups, and significant differences between individual groups were estimated. When one-day-old broilers were fed with diet supplemented with the above-mentioned Cr compounds, the feed intake of 4–5 weeks showed better results in the $CrCl_3$ group compared to the CrPic group, while the LDL-cholesterol in the CrPic NPs groups was lower than in the CrPic group, and CrPic NPs and CrPic groups showed considerably enhanced serum Cr concentration compared to the control and $CrCl_3$ groups. Based on the above results, it could be concluded that the CrPic NPs supplementation has advantages compared to the bulk CrPic supplementation, because it not only increases Cr utilization but also results in a lower serum LDL-cholesterol level in broilers [283].

5.3. *Pigs*

ZnO and Cu salts, traditionally used in high doses as supplements to piglet's diet, stimulate piglet's daily gain and decrease feed conversion factor. However, the application of high concentrations of these metal additives could result in increased environmental pollution of soil and tap water; on the other hand, Zn applied at doses 2500–3000 mg/kg feed can contribute to the development of antimicrobial resistance and may regulate the expression of genes that modify piglets' immune response. Consequently, higher bioavailability which could be achieved by applying nanosized ZnO/Cu particles could notably reduce the dietary inclusion rate and environmental pollution with preserving beneficial impact on pig's health [244].

The degree of the reduction of piglet diarrhea incidence observed with a low dose of ZnO NPs (600 mg Zn/kg) supplemented to the basal diet of weaning piglets was comparable with that observed with the dose of 2000 mg Zn/kg when bulk ZnO was used, which could be connected with improved intestinal microbiota and inflammation response in piglets at the ZnO NPs application. Moreover, the application of ZnO NPs could contribute to reduced Zn environmental pollution [284]. ZnO NPs used as a DIS increased Zn digestibility, serum GH levels, and carbonic anhydrase activity and enhanced the immune response of weanling piglets [285].

Cr(III) belongs to essential elements in the nutrition of both animals and humans, and in many animal species its deficiency results in reduced feed intake, lower weight gains, reproductive disorders, and increased lipid levels, and a moderate Cr deficiency represents a risk factor of ischemic heart disease with myocardial infarction and coronary artery disease. Cr improves lean body mass in animals, increases growth rate and feed conversion, and improves feed intake and energy efficiency, and the dietary Cr requirement of an animal body is probably 300 μg Cr/kg d.w. of feed. A pronouncedly increased Cr content in the blood, longissimus muscle, heart, liver, kidneys, jejunum, and ileum was observed in pigs receiving dietary Cr nanocomposite supplementation [286]. Dietary CrPic NPs supplementation at 400 ppb increased feed intake in finisher gilts during mid-summer and was able to improve some of the adverse effects of heat stress in pigs, through decreasing circulating cortisol levels [287]. Pigs with initial body weight of 66.10 $\pm$ 1.01 kg receiving basal diet supplemented with 200 or 400 μg/kg of Cr from Cr-loaded CS NPs (Cr–CS NPs) for 35 d showed increased carcass lean ratio and longissimus muscle area, decreased carcass fat ratio and backfat thickness as well as increased serum free fatty acids, lipase activity, and serum insulin-like growth factor I, while a decreased level of serum insulin was estimated. Moreover, decreased activities of fatty acid synthase and malate dehydrogenase and increased activity of hormone-sensitive lipase in subcutaneous adipose tissue were observed in treated pigs. These results indicate a favorable impact of Cr in the form of the Cr–CS NPs on growth, carcass characteristics, pork quality, and lipid catabolism in finishing pigs [288]. The control diet supplemented with 200 μg Cr from Cr nanocomposite pronouncedly reduced serum levels of glucose, urea nitrogen, triglyceride, cholesterol, and nonesterified fatty acid, increased total protein, HDL, and lipase activity in finishing pigs with initial weight 64.8 $\pm$ 0.83 kg fed for 35 d, and considerably increased serum insulin-like growth factor I, while reducing serum insulin and cortisol levels. Moreover, it affected immune status in finishing pigs, which was reflected in notable increments of IgM and IgG contents in plasma [289].

5.4. Other Pets

Dried matrices of CS Cu chelate gels designed as a multimicronutrient feed additive for cattle were loaded with vitamin riboflavin. Following restricted rehydration in simulated rumen fluid, they exhibited sustained release of riboflavin without releasing Cu in these neutral conditions for up to 24 h, demonstrating Cu rumen bypass. A sustained release of the mineral was observed in abomasal conditions of pH 2 over a 3 h period suggesting that this formulation could supply nutritionally relevant levels of the free mineral in these conditions, as required for effective supplementation in cattle [290].

A decreasing trend in serum Fe concentration was observed in Lori-Bakhtiary sheep, which orally received SeNPs and Na_2SO_3 (1 mg/kg) for 10 consecutive days, particularly during the early and middle stages of supplementation (0–20 days) in contrast to the increasing levels of total iron binding capacity, suggesting that the expression of transferrin and its receptor genes was considerably increased. However, after this period, the expression of the transferrin and transferrin binding receptor genes showed a notable decrease, especially in $SeO_3{}^{2-}$ treated animals [291].

Improved rumen fermentation and feed utilization, stimulation of rumen microbial activity, digestive microorganisms, and enzyme activity by supplementation of SeNPs in basal diet of sheep with optimum dose approx. 3.0 g/kg dietary dry mater was reported by Shi et al. [292]. Using the rumen simulation technique, it was observed that nanoemulsified soyabean oil modulated the PUFAs proportions in ruminal cultures, which was reflected in markedly increased proportions of oleic acid, α-LLA, and α-LNA in the fermentation fluid without any negative effect on rumen fermentation parameters [293].

Nano-copper as a new growth promoter in the diet of growing New Zealand white rabbits was reported by Refaie et al. [294]. Similarly, the male five-week-old New Zealand white rabbits fed with the basal diet supplemented with 60 and 30 mg ZnO NPs /kg diet had higher body weight, daily weight gain, daily feed intake, serum total protein, globulin, IgG, and SOD compared to control animals and rabbits fed with a diet supplemented with bulk ZnO (60 mg/kg diet), suggesting that traditional zinc sources in rabbit diets could be replaced by 30 mg ZnO NP/kg diet [295]. On the other hand, Ismail and El-Araby [296] recommended the combined use of bulk and nanosized ZnO at ratio 1:1 in the dietary system of rabbit's farms ensuring dietary supplementation with Zn as an essential element but reducing adverse effects such as lipid peroxidation and oxidant stress induced by the whole dose of ZnO NPs. A complete overview of the used nanoscale veterinary dietary supplements is shown in Figure 3.

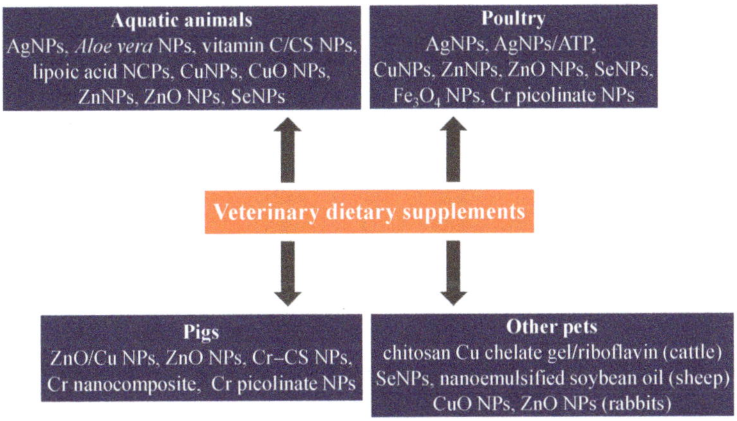

Figure 3. Summary of most frequently used veterinary dietary nanosupplements.

6. Conclusions

The importance of balanced nutrition containing important nutrients, for example, vitamins or antioxidants, in sufficient amount needed for health of humans and animals is indisputable. Moreover, recently, physicians attribute an increasing significance to the consumption of food products containing special effective ingredients to prevent and improve the health of people suffering from certain diseases (e.g., cancer, diabetes, hyperlipidemia, mental disorders, osteoporosis, various malabsorption, etc.). The fortification of food products with such dietary supplements/nutraceuticals can be easily used in practice when an appropriate stability of the active ingredient in the formulation could be guaranteed at least until the date of consumption (expiration date). For these purposes, nanoformulations of active compounds prepared using biodegradable nature- or semisynthetic-based nanocarriers, such as polymeric matrices, micelles, liposomes, nanoemulsions, solid lipid NPs, nanostructured lipid carriers, or appropriate inorganic matrices are especially favorable, securing not only enhanced stability but also frequently controlled release of nutrients. Definitely, based on previous thorough experiments, for supplementation of food products (e.g., bread, butter, yogurt, cake, biscuit) or beverages (e.g., milk, juice) with individual healthy ingredients, the most convenient nanoformulation could be selected and used. Unlike nutraceuticals that are available for sale in pharmacy and could be overused by some careless consumers resulting in possible harmful side effects, at fortification of food products with dietary supplements/nutraceuticals, the excessive consumption of these compounds is excluded. However, even though nanoformulations enhance the bioavailability and increase the stability of individual active ingredients, all nanoscale materials applied in food industry should be used advisedly and only after in-depth investigation of cytotoxicity due to possible increased nanosize-based toxicity effects (e.g., surface reactivity of NPs), which could result in unspecified toxic effects also in humans or animals. Therefore, an increased attention should be paid to the influence of risk factors associated with their applications and possible adverse/hazardous effects to humans and animals, observing the relevant guidelines, regulations, and directives issued by the European Commission and the EFSA.

Author Contributions: J.J., J.K. and K.K. composition and writing.

Funding: This study was supported by the Slovak Research and Development Agency (projects APVV-17-0373 and APVV-14-0547), and by the Ministry of Education of the Czech Republic (LO1305).

Conflicts of Interest: The authors declare no conflicts of interest.

Abbreviations

AChE (acetylcholinesterase); Alb (albumin); AST (astaxanthin); ATP (adenosine triphosphate); α-Lalb (α-lactalbumin); α-LLA (α-linoleic acid); α-LNA (α-linolenic acid); BAP (bioactive association platform); BMD (bone mineral densities); β-Car (β-carotene); β-Lglb (β-lactoglobulin); CDs (cyclodextrins); CPE (*Citrus reticulata* Blanco cv. unshiu peel extract); CrPic (chromium picolinate); CS (chitosan); CUR (curcumin); CVD (cardiovascular disease); CVPs (*Catathelasma ventricosum* polysaccharides); DESs (delivery systems); DHA (docosahexaenoic acid); DISs (dietary supplements); EE (encapsulation efficiency); EFSA (European Food Safety Authority); EGCG (epigallocatechin gallate); FA (folic acid); FDA (U.S. Food and Drug Administration); FSMPs (foods for special medical purposes); GH (growth hormone); GI (gastrointestinal); Glb (globulin); GPx (glutathione peroxidase); GSH (glutathione); HDL (high-density lipoprotein); LCT (long-chain triglycerides); LDHs (layered double hydroxides); LDL (low-density lipoprotein); MCT (medium chain triglycerides); MDA (malondialdehyde); Met (methionine); MPs (microparticles); MSPs (mesoporous particles); NE (nanoemulsion); NLCs (nanostructured lipid carriers); NPs (nanoparticles); NPES (nanopowdered eggshell); NRC (National Research Council); OVX (ovariectomized); PC (phosphatidylcholine); PGA (polyglutamic acid); PUFAs (polyunsaturated fatty acids); Q (quercetin); r-CMs (re-assembled casein micelles); RES (resveratrol); SCas (sodium caseinate); SOD (superoxide dismutase); SPI sesame protein isolate); STZ (streptozotocin); TBFs (tartary buckwheat rich in flavonoids); ZLNPs (zein NPs encapsulating lutein).

References

1. Dwyer, J.T.; Wiemer, K.L.; Dary, O.; Keen, C.L.; King, J.C.; Miller, K.B.; Philbert, M.A.; Tarasuk, V.; Taylor, C.L.; Gaine, P.C.; et al. Fortification and health: Challenges and opportunities. *Adv. Nutr.* **2015**, *6*, 124–131. [CrossRef] [PubMed]
2. Rao, P.J.; Naidu, M.M. Nanoencapsulation of Bioactive Compounds for Nutraceutical Food. In *Nanoscience in Food and Agriculture 2. Sustainable Agriculture Reviews*; Ranjan, S., Dasgupta, N., Lichtfouse, E., Eds.; Springer: Cham, Germany, 2016; Volume 21, pp. 129–156.
3. National Institutes of Health. Dietary Supplements: Background Information. Available online: https://ods.od.nih.gov/factsheets/DietarySupplements-HealthProfessional/ (accessed on 20 November 2018).
4. European Commission. Food Supplements. Available online: https://ec.europa.eu/food/safety/labelling_nutrition/supplements_en (accessed on 20 November 2018).
5. U.S. Food and Drug Administration. Dietary Supplements Guidance Documents & Regulatory Information. Available online: https://www.fda.gov/Food/GuidanceRegulation/GuidanceDocumentsRegulatoryInformation/DietarySupplements/default.htm (accessed on 20 November 2018).
6. European Commission—Health Claims. Available online: https://ec.europa.eu/food/safety/labelling_nutrition/claims/health_claims_en (accessed on 20 November 2018).
7. Kuhnert, P. Foods, 3. Food Additives. In *Ullmann's Encyclopedia of Industrial Chemistry*; Wiley-VCH: Weinheim, Germany, 2016; Available online: 10.1002/14356007.a11_561.pub2 (accessed on 20 November 2018).
8. EFSA. Foods for Special Medical Purposes. Available online: https://www.efsa.europa.eu/en/press/news/151126 (accessed on 20 November 2018).
9. European Commission—Foods for Specific Groups. Available online: https://ec.europa.eu/food/safety/labelling_nutrition/special_groups_food_en (accessed on 20 November 2018).
10. FDA. Medical Foods Guidance Documents & Regulatory Information. Available online: https://www.fda.gov/Food/GuidanceRegulation/GuidanceDocumentsRegulatoryInformation/MedicalFoods/default.htm (accessed on 20 November 2018).
11. Corradini, C.; Lantano, C.; Cavazza, A. Innovative analytical tools to characterize prebiotic carbohydrates of functional food interest. *Anal. Bioanal. Chem.* **2013**, *405*, 4591–4605. [CrossRef] [PubMed]

12. Shahidi, F. Nutraceuticals, functional foods and dietary supplements in health and disease. *J. Food Drug Anal.* **2012**, *20*, 226–230.
13. National Nanotechnology Initiative. *Big Things from a Tiny World*; National Nanotechnology Initiative: Arlington, VA, USA, 2008.
14. European Commission. Definition of a Nanomaterial. Available online: http://ec.europa.eu/environment/chemicals/nanotech/faq/definition_en.htm (accessed on 3 December 2018).
15. Mody, V.V.; Siwale, R.; Singh, A.; Mody, H.R. Introduction to metallic nanoparticles. *J. Pharm. Bioallied Sci.* **2010**, *2*, 282–289. [CrossRef]
16. Couvreur, P. Nanoparticles in drug delivery: Past, present and future. *Adv. Drug Deliv. Rev.* **2013**, *65*, 21–23. [CrossRef] [PubMed]
17. Kateb, B.; Heiss, J.D. *The Textbook of Nanoneuroscience and Nanoneurosurgery*; CRC Press, Taylor & Francis Group: Boca Raton, FL, USA, 2014.
18. Vaculikova, E.; Placha, D.; Jampilek, J. Toxicology of drug nanocarriers. *Chem. Listy* **2015**, *109*, 346–352.
19. Jampilek, J.; Kralova, K. Application of nanobioformulations for controlled release and targeted biodistribution of drugs. In *Nanobiomaterials: Applications in Drug Delivery*; Sharma, A.K., Keservani, R.K., Kesharwani, R.K., Eds.; CRC Press: Warentown, NJ, USA, 2018; pp. 131–208.
20. Jampilek, J.; Kralova, K. Nanotechnology based formulations for drug targeting to central nervous system. In *Nanoparticulate Drug Delivery Systems*; Keservani, R.K., Sharma, A.K., Eds.; Apple Academic Press & CRC Press: Warentown, NJ, USA, 2019; pp. 151–220.
21. Bhushan, B.; Luo, D.; Schricker, S.R.; Sigmund, W.; Zauscher, S. *Handbook of Nanomaterials Properties*; Springer: Berlin/Heidelberg, Germany, 2014.
22. Singh, O.V. *Bio-Nanoparticles: Biosynthesis and Sustainable Biotechnological Implications*; Wiley-Blackwell: Hoboken, NJ, USA, 2015.
23. Shukla, A.; Iravani, S. *Green Synthesis, Characterization and Applications of Nanoparticles*; Elsevier: Amsterdam, The Netherlands, 2018.
24. Jampilek, J.; Kralova, K. Nano-antimicrobials: Activity, benefits and weaknesses. In *Nanostructures for Antimicrobial Therapy*; Ficai, A., Grumezescu, A.M., Eds.; Elsevier: Amsterdam, The Netherlands, 2017; pp. 23–54.
25. Jampilek, J.; Kralova, K. Nanomaterials for delivery of nutrients and growth-promoting compounds to Plants. In *Nanotechnology: An Agricultural Paradigm*; Prasad, R., Kumar, M., Kumar, V., Eds.; Springer: Singapore, 2017; pp. 177–226.
26. Brayner, R.; Fievet, F.; Coradin, T. *Nanomaterials: A Danger or a Promise? A Chemical and Biological Perspective*; Springer: London, UK, 2013.
27. Acosta, E. Bioavailability of nanoparticles in nutrient and nutraceutical delivery. *Curr. Opin. Colloid Interface Sci.* **2009**, *14*, 3–15. [CrossRef]
28. Bamrungsap, S.; Zhao, Z.; Chen, T.; Wang, L.; Li, C.; Fu, T.; Tan, W. Nanotechnology in therapeutics: A focus on nanoparticles as a drug delivery system. *Nanomedicine* **2012**, *7*, 1253–1271. [CrossRef] [PubMed]
29. Nekkanti, V.; Vabalaboina, V.; Pillai, R. Drug nanoparticles—An overview. In *The Delivery of Nanoparticles*; Hashim, A.A., Ed.; InTech: Rieka, Croatia, 2012; pp. 111–132.
30. Frohlich, E. Cellular targets and mechanisms in the cytotoxic action of non-biodegradable engineered nanoparticles. *Curr. Drug Metab.* **2013**, *14*, 976–988. [CrossRef] [PubMed]
31. Dolez, P.I. *Nanoengineering: Global Approaches to Health and Safety Issues*; Elsevier: Amterdam, The Netherlands, 2015.
32. Busquets, R. *Emerging Nanotechnologies in Food Science*; Elsevier: Amterdam, The Netherlands, 2018.
33. Jampilek, J.; Kralova, K. Nanomaterials applicable in food protection. In *Nanotechnology Applications in Food Industry*; Rai, R.V., Bai, J.A., Eds.; Taylor & Francis Group: Boca Raton, FL, USA, 2018; pp. 75–96.
34. Vaculikova, E.; Grunwaldova, V.; Kral, V.; Dohnal, J.; Jampilek, J. Preparation of candesartan and atorvastatin nanoparticles by solvent evaporation. *Molecules* **2012**, *17*, 13221–13234. [CrossRef]
35. Jampilek, J.; Zaruba, K.; Oravec, M.; Kunes, M.; Babula, P.; Ulbrich, P.; Brezaniova, I.; Triska, J.; Suchy, P. Preparation of silica nanoparticles loaded with nootropics and their in vivo permeation through blood–brain barrier. *Biomed. Res. Int.* **2015**, *2015*, 812673. [CrossRef] [PubMed]
36. Vaculikova, E.; Cernikova, A.; Placha, D.; Pisarcik, M.; Dedkova, K.; Peikertova, P.; Devinsky, F.; Jampilek, J. Cimetidine nanoparticles for permeability enhancement. *J. Nanosci. Nanotechnol.* **2016**, *16*, 7840–7843. [CrossRef]

37. Vaculikova, E.; Cernikova, A.; Placha, D.; Pisarcik, M.; Peikertova, P.; Dedkova, K.; Devinsky, F.; Jampilek, J. Preparation of hydrochlorothiazide nanoparticles for solubility enhancement. *Molecules* **2016**, *21*, 1005. [CrossRef] [PubMed]
38. Pentak, D.; Kozik, V.; Bak, A.; Dybal, P.; Sochanik, A.; Jampilek, J. Methotrexate and cytarabine—Loaded nanocarriers for multidrug cancer therapy. Spectroscopic study. *Molecules* **2016**, *21*, 1689. [CrossRef] [PubMed]
39. Pisarcik, M.; Jampilek, J.; Lukac, M.; Horakova, R.; Devinsky, F.; Bukovsky, M.; Kalina, M.; Tkacz, J.; Opravil, T. Silver nanoparticles stabilised by cationic gemini surfactants with variable spacer length. *Molecules* **2017**, *22*, 1794. [CrossRef]
40. Pisarcik, M.; Lukac, M.; Jampilek, J.; Bilka, F.; Bilkova, A.; Paskova, L.; Devinsky, F.; Horakova, R.; Opravil, T. Silver nanoparticles stabilised with cationic single-chain surfactants. Structure-physical properties-biological activity relationship study. *J. Mol. Liq.* **2018**, *272*, 60–72. [CrossRef]
41. Vaculikova, E.; Pokorna, A.; Placha, D.; Pisarcik, M.; Dedková, K.; Peikertova, P.; Devinsky, F.; Jampilek, J. Improvement of glibenclamide water solubility by nanoparticle preparation. *J. Nanosci. Nanotechnol.* **2019**, *19*, 3031–3034. [CrossRef]
42. Kozik, V.; Bak, A.; Pentak, D.; Hachula, B.; Pytlakowska, K.; Rojkiewicz, M.; Jampilek, J.; Sieron, K.; Jazowiecka-Rakus, J.; Sochanik, A. Derivatives of graphene oxide as potential drug carriers. *J. Nanosci. Nanotechnol.* **2019**, *19*, 2489–2492. [CrossRef]
43. Oehlke, K.; Adamiuk, M.; Behsnilian, D.; Graef, V.; Mayer-Miebach, E.; Walz, E.; Greiner, R. Potential bioavailability enhancement of bioactive compounds using food-grade engineered nanomaterials: A review of the existing evidence. *Food Funct.* **2014**, *5*, 1341–1359. [CrossRef]
44. Akhavan, S.; Assadpour, E.; Katouzian, I.; Jafari, S.M. Lipid nano scale cargos for the protection and delivery of food bioactive ingredients and nutraceuticals. *Trends Food Sci. Technol.* **2018**, *74*, 132–146. [CrossRef]
45. Kumar, D.H.L.; Sarkar, P. Encapsulation of bioactive compounds using nanoemulsions. *Environ. Chem. Lett.* **2018**, *16*, 59–70. [CrossRef]
46. Babazadeh, A.; Ghanbarzadeh, B.; Hamishehkar, H. Formulation of food grade nanostructured lipid carrier (NLC) for potential applications in medicinal-functional foods. *J. Drug Deliv. Sci. Technol.* **2017**, *39*, 50–58. [CrossRef]
47. Simoes, L.D.S.; Madalena, D.A.; Pinheiro, A.C.; Teixeira, J.A.; Vicente, A.A.; Ramos, O.L. Micro- and nano bio-based delivery systems for food applications: In vitro behavior. *Adv. Colloid Interface Sci.* **2017**, *243*, 23–45. [CrossRef] [PubMed]
48. Goncalves, R.F.S.; Martins, J.T.; Duarte, C.M.M.; Vicente, A.A.; Pinheiro, A.C. Advances in nutraceutical delivery systems: From formulation design for bioavailability enhancement to efficacy and safety evaluation. *Trends Food Sci. Technol.* **2018**, *78*, 270–291. [CrossRef]
49. Arora, D.; Jaglan, S. Nanocarriers based delivery of nutraceuticals for cancer prevention and treatment: A review of recent research developments. *Trends Food Sci. Technol.* **2016**, *54*, 114–126. [CrossRef]
50. Katouzian, I.; Jafari, S.M. Nano-encapsulation as a promising approach for targeted delivery and controlled release of vitamins. *Trends Food Sci. Technol.* **2016**, *53*, 34–48. [CrossRef]
51. Chai, J.J.; Jiang, P.; Wang, P.J.; Jiang, Y.M.; Li, D.; Bao, W.E.; Liu, B.X.; Liu, B.; Zhao, L.Y.; Norde, W.; et al. The intelligent delivery systems for bioactive compounds in foods: Physicochemical and physiological conditions, absorption mechanisms, obstacles and responsive strategies. *Trends Food Sci. Technol.* **2018**, *78*, 144–154. [CrossRef]
52. Gleeson, J.P.; Ryan, S.M.; Brayden, D.J. Oral delivery strategies for nutraceuticals: Delivery vehicles and absorption enhancers. *Trends Food Sci. Technol.* **2016**, *53*, 90–101. [CrossRef]
53. Jafari, S.M.; Assaidpoor, E.; Bhandari, B.; He, Y.H. Nano-particle encapsulation of fish oil by spray drying. *Food Res. Int.* **2008**, *41*, 172–183. [CrossRef]
54. Li, Q.; Li, T.; Liu, C.M.; Dai, T.T.; Zhang, R.J.; Zhang, Z.P.; McClements, D.J. Enhancement of carotenoid bioaccessibility from tomatoes using excipient emulsions: Influence of particle size. *Food Biophys.* **2017**, *12*, 172–185. [CrossRef]
55. Bioinicia, Valencia, Spain. Available online: https://bioinicia.com/electrospinning-electrospraying-technology (accessed on 13 February 2019).
56. Lagaron, J.M. *Multifunctional and Nanoreinforced Polymers for Food Packaging*; Woodhead Publishing: Cambridge, UK, 2011.

57. Bhushani, J.A.; Anandharamakrishnan, C. Electrospinning and electrospraying techniques: Potential food based applications. *Trends Food Sci. Technol.* **2014**, *38*, 21–33. [CrossRef]
58. Torres-Giner, S.; Martinez-Abad, A.; Ocio, M.J.; Lagaron, J.M. Stabilization of a nutraceutical omega-3 fatty acid by encapsulation in ultrathin electrosprayed zein prolamine. *J. Food. Sci.* **2010**, *75*, N69–N79. [CrossRef] [PubMed]
59. Perez-Masia, R.; Lopez-Nicolas, R.; Periago, M.J.; Ros, G.; Lagaron, J.M.; Lopez-Rubio, A. Encapsulation of folic acid in food hydrocolloids through nanospray drying and electrospraying for nutraceutical applications. *Food Chem.* **2015**, *168*, 124–133. [CrossRef]
60. Nagarajan, S.; Soussan, L.; Bechelany, M.; Teyssier, C.; Cavailles, V.; Pochat-Bohatier, C.; Miele, P.; Kalkura, N.; Janota, J.M.; Balme, S. Novel biocompatible electrospun gelatin fiber mats with antibiotic drug delivery properties. *J. Mater. Chem. B* **2016**, *4*, 1134–1141. [CrossRef]
61. Khorasani, S.; Danaei, M.; Mozafari, M.R. Nanoliposome technology for the food and nutraceutical industries. *Trends Food Sci. Technol.* **2018**, *79*, 106–115. [CrossRef]
62. Ghanbarzadeh, B.; Babazadeh, A.; Hamishehkar, H. Nano-phytosome as a potential food-grade delivery system. *Food Biosci.* **2016**, *15*, 126–135. [CrossRef]
63. Babazadeh, A.; Ghanbarzadeh, B.; Hamishehkar, H. Phosphatidylcholine-rutin complex as a potential nanocarrier for food applications. *J. Funct. Foods* **2017**, *33*, 134–141. [CrossRef]
64. Bochicchio, S.; Barba, A.A.; Grassi, G.; Lamberti, G. Vitamin delivery: Carriers based on nanoliposomes produced via ultrasonic irradiation. *LWT Food Sci. Technol.* **2016**, *69*, 9–16. [CrossRef]
65. Azzi, J.; Jraij, A.; Auezova, L.; Fourmentin, S.; Greige-Gerges, H. Novel findings for quercetin encapsulation and preservation with cyclodextrins, liposomes, and drug-in-cyclodextrin-in-liposomes. *Food Hydrocoll.* **2018**, *81*, 328–340. [CrossRef]
66. Li, Z.L.; Peng, S.F.; Chen, X.; Zhu, Y.Q.; Zou, L.Q.; Liu, W.; Liu, C.M. Pluronics modified liposomes for curcumin encapsulation: Sustained release, stability and bioaccessibility. *Food Res. Int.* **2018**, *108*, 246–253. [CrossRef]
67. Semenova, M.G.; Antipova, A.S.; Zelikina, D.V.; Martirosova, E.I.; Plashchina, I.G.; Palmina, N.P.; Binyukov, V.I.; Bogdanova, N.G.; Kasparov, V.V.; Shumilina, E.A. Biopolymer nanovehicles for essential polyunsaturated fatty acids: Structure-functionality relationships. *Food Res. Int.* **2016**, *88*, 70–78. [CrossRef] [PubMed]
68. Dey, T.K.; Banerjee, P.; Chatterjee, R.; Dhar, P. Designing of ω-3 PUFA enriched biocompatible nanoemulsion with sesame protein isolate as a natural surfactant: Focus on enhanced shelf-life stability and biocompatibility. *Colloids Surf. A Physicochem. Eng. Asp.* **2018**, *538*, 36–44. [CrossRef]
69. Hategekimana, J.; Chamba, M.V.M.; Shoemaker, C.F.; Majeed, H.; Zhong, F. Vitamin E nanoemulsions by emulsion phase inversion: Effect of environmental stress and long-term storage on stability and degradation in different carrier oil types. *Colloids Surf. A Physicochem. Eng. Asp.* **2015**, *483*, 70–80. [CrossRef]
70. Guttoff, M.; Saberi, A.H.; McClements, D.J. Formation of vitamin D nanoemulsion-based delivery systems by spontaneous emulsification: Factors affecting particle size and stability. *Food Chem.* **2015**, *171*, 117–122. [CrossRef] [PubMed]
71. Salvia-Trujillo, L.; McClements, D.J. Improvement of β-carotene bioaccessibility from dietary supplements using excipient nanoemulsions. *J. Agric. Food Chem.* **2016**, *64*, 4639–4647. [CrossRef] [PubMed]
72. Liu, X.J.; Zhang, R.J.; McClements, D.J.; Li, F.; Liu, H.; Cao, Y.; Xiao, H. Nanoemulsion-based delivery systems for nutraceuticals: Influence of long-chain triglyceride (LCT) type on in vitro digestion and astaxanthin bioaccessibility. *Food Biophys.* **2018**, *13*, 412–421. [CrossRef]
73. Saxena, V.; Hasan, A.; Sharma, S.; Pandey, L.M. Edible oil nanoemulsion: An organic nanoantibiotic as a potential biomolecule delivery vehicle. *Int. J. Polym. Mater.* **2018**, *67*, 410–419. [CrossRef]
74. Parthasarathi, S.; Muthukumar, S.P.; Anandharamakrishnan, C. The influence of droplet size on the stability, in vivo digestion, and oral bioavailability of vitamin E emulsions. *Food Funct.* **2016**, *7*, 2294–2302. [CrossRef]
75. Ozturk, B.; Argin, S.; Ozilgen, M.; McClements, D.J. Formation and stabilization of nanoemulsion-based vitamin E delivery systems using natural surfactants: Quillaja saponin and lecithin. *J. Food Eng.* **2014**, *142*, 57–63. [CrossRef]
76. Peng, S.F.; Li, Z.L.; Zou, L.Q.; Liu, W.; Liu, C.M.; McClements, D.J. Improving curcumin solubility and bioavailability by encapsulation in saponin-coated curcumin nanoparticles prepared using a simple pH-driven loading method. *Food Funct.* **2018**, *9*, 1829–1839. [CrossRef]

77. Zheng, B.J.; Peng, S.F.; Zhang, X.Y.; McClements, D.J. Impact of delivery system type on curcumin bioaccessibility: Comparison of curcumin-loaded nanoemulsions with commercial curcumin supplements. *J. Agric. Food Chem.* **2018**, *66*, 10816–10826. [CrossRef] [PubMed]
78. Braithwaite, M.C.; Choonara, Y.E.; Kumar, P.; Tomar, L.K.; Du Toit, L.C.; Pillay, V. A novel bile salts-lipase polymeric film-infused minitablet system for enhanced oral delivery of cholecalciferol. *Pharm. Dev. Technol.* **2016**, *21*, 832–846. [CrossRef] [PubMed]
79. Liu, F.G.; Ma, C.C.; Zhang, R.J.; Gao, Y.X.; McClements, D.J. Controlling the potential gastrointestinal fate of β-carotene emulsions using interfacial engineering: Impact of coating lipid droplets with polyphenol-protein-carbohydrate conjugate. *Food Chem.* **2017**, *221*, 395–403. [CrossRef] [PubMed]
80. Liu, X.; Bi, J.F.; Xiao, H.; McClements, D.J. Enhancement of nutraceutical bioavailability using excipient nanoemulsions: Role of lipid digestion products on bioaccessibility of carotenoids and phenolics from mangoes. *J. Food Sci.* **2016**, *81*, N754–N761. [CrossRef] [PubMed]
81. Sharifi, F.; Jahangiri, M. Investigation if the stability of vitamin D in emulsion-based delivery systems. *Chem. Ind. Chem. Eng.* **2018**, *24*, 157–167. [CrossRef]
82. Esfanjani, A.F.; Assadpour, E.; Jafari, S.M. Improving the bioavailability of phenolic compounds by loading them within lipid-based nanocarriers. *Trends Food Sci. Technol.* **2018**, *76*, 56–66. [CrossRef]
83. Gonzalez-Reza, R.M.; Quintanar-Guerrero, D.; Del Real-Lopez, A.; Pinon-Segundo, E.; Zambrano-Zaragoza, M.L. Effect of sucrose concentration and pH onto the physical stability of β-carotene nanocapsules. *LWT-Food Sci. Technol.* **2018**, *90*, 354–361. [CrossRef]
84. Dan, N. Compound release from nanostructured lipid carriers (NLCs). *J. Food Eng.* **2016**, *171*, 37–43. [CrossRef]
85. Kong, R.; Xia, Q.; Liu, G.Y. Preparation and characterization of vitamin A palmitate-loaded nanostructured lipid carriers as delivery systems for food products. *Adv. Mater. Res.* **2011**, *236–238*, 1818–1823. [CrossRef]
86. Paucar, O.C.; Tulini, F.L.; Thomazini, M.; Balieiro, J.C.C.; Pallone, E.M.J.A.; Favaro-Trindade, C.S. Production by spray chilling and characterization of solid lipid microparticles loaded with vitamin D_3. *Food Bioprod. Process.* **2016**, *100*, 344–350. [CrossRef]
87. Ramalingam, P.; Yoo, S.W.; Ko, Y.T. Nanodelivery systems based on mucoadhesive polymer coated solid lipid nanoparticles to improve the oral intake of food curcumin. *Food Res. Int.* **2016**, *84*, 113–119. [CrossRef]
88. Nunes, S.; Madureira, A.R.; Campos, D.; Sarmento, B.; Gomes, A.M.; Pintado, M.; Reis, F. Solid lipid nanoparticles as oral delivery systems of phenolic compounds: Overcoming pharmacokinetic limitations for nutraceutical applications. *Crit. Rev. Food Sci. Nutr.* **2017**, *57*, 1863–1873. [CrossRef] [PubMed]
89. Sampathkumar, K.; Loo, S.C.J. Targeted gastrointestinal delivery of nutraceuticals with polysaccharide-based coatings. *Macromol. Biosci.* **2018**, *18*, 1700363. [CrossRef] [PubMed]
90. Hasanvand, E.; Fathi, M.; Bassiri, A. Production and characterization of vitamin D_3 loaded starch nanoparticles: Effect of amylose to amylopectin ratio and sonication parameters. *J. Food Sci. Tech. Mys.* **2018**, *55*, 1314–1324. [CrossRef] [PubMed]
91. Hategekirnana, J.; Masamba, K.G.; Ma, J.G.; Zhong, F. Encapsulation of vitamin E: Effect of physicochemical properties of wall material on retention and stability. *Carbohydr. Polym.* **2015**, *124*, 172–179. [CrossRef] [PubMed]
92. Khan, A.; Wen, Y.B.; Huq, T.; Ni, Y.H. Cellulosic nanomaterials in food and nutraceutical applications: A review. *J. Agric. Food Chem.* **2018**, *66*, 8–19. [CrossRef]
93. Huq, T.; Fraschini, C.; Khan, A.; Riedl, B.; Bouchard, J.; Lacroix, M. Alginate based nanocomposite for microencapsulation of probiotic: Effect of cellulose nanocrystal (CNC) and lecithin. *Carbohydr. Polym.* **2017**, *168*, 61–69. [CrossRef]
94. Guo, C.J.; Yin, J.G.; Chen, D.Q. Co-encapsulation of curcumin and resveratrol into novel nutraceutical hyalurosomes nano-food delivery system based on oligo-hyaluronic acid-curcumin polymer. *Carbohydr. Polym.* **2018**, *181*, 1033–1037. [CrossRef]
95. Xia, S.Q.; Tan, C.; Xue, J.; Lou, X.W.; Zhang, X.M.; Feng, B.A. Chitosan/tripolyphosphate-nanoliposomes core-shell nanocomplexes as vitamin E carriers: Shelf-life and thermal properties. *Int. J. Food Sci. Technol.* **2014**, *49*, 1367–1374. [CrossRef]
96. Ge, J.; Yue, P.X.; Chi, J.P.; Liang, J.; Gao, X.L. Formation and stability of anthocyanins-loaded nanocomplexes prepared with chitosan hydrochloride and carboxymethyl chitosan. *Food Hydrocoll.* **2018**, *74*, 23–31. [CrossRef]

97. Acevedo-Fani, A.; Soliva-Fortuny, R.; Martin-Belloso, O. Photo-protection and controlled release of folic acid using edible alginate/chitosan nanolaminates. *J. Food Eng.* **2018**, *229*, 72–82. [CrossRef]
98. Sun, Q.C.; Zhang, Z.P.; Zhang, R.J.; Gao, R.C.; McClements, D.J. Development of functional or medical foods for oral administration of insulin for diabetes treatment: Gastroprotective edible microgels. *J. Agric. Food Chem.* **2018**, *66*, 4820–4826. [CrossRef] [PubMed]
99. Papagiannopoulos, A.; Vlassi, E. Stimuli-responsive nanoparticles by thermal treatment of bovine serum albumin inside its complexes with chondroitin sulfate. *Food Hydrocoll.* **2019**, *87*, 602–610. [CrossRef]
100. Dai, L.; Wei, Y.; Sun, C.X.; Mao, L.K.; McClements, D.J.; Gao, Y.X. Development of protein-polysaccharide-surfactant ternary complex particles as delivery vehicles for curcumin. *Food Hydrocoll.* **2018**, *85*, 75–85. [CrossRef]
101. Fathi, M.; Donsi, F.; McClements, D.J. Protein-based delivery systems for the nanoencapsulation of food ingredients. *Compr. Rev. Food Sci.* **2018**, *17*, 920–936. [CrossRef]
102. Ramos, O.L.; Pereira, R.N.; Martins, A.; Rodrigues, R.; Fucinos, C.; Teixeira, J.A.; Pastrana, L.; Malcata, F.X.; Vicente, A.A. Design of whey protein nanostructures for incorporation and release of nutraceutical compounds in food. *Crit. Rev. Food Sci. Nutr.* **2017**, *57*, 1377–1393. [CrossRef] [PubMed]
103. Parthasarathi, S.; Anandharamakrishnan, C. Enhancement of oral bioavailability of vitamin E by spray-freeze drying of whey protein microcapsules. *Food Bioprod. Process.* **2016**, *100*, 469–476.
104. Cheng, C.J.; Ferruzzi, M.; Jones, O.G. Fate of lutein-containing zein nanoparticles following simulated gastric and intestinal digestion. *Food Hydrocoll.* **2019**, *87*, 229–236. [CrossRef]
105. Arzeni, C.; Perez, O.E.; LeBlanc, J.G.; Pilosof, A.M.R. Egg albumin-folic acid nanocomplexes: Performance as a functional ingredient and biological activity. *J. Funct. Foods* **2015**, *18*, 379–386. [CrossRef]
106. Zema, P.; Pilosof, A.M.R. On the binding of folic acid to food proteins performing as vitamin micro/nanocarriers. *Food Hydrocoll.* **2018**, *79*, 509–517. [CrossRef]
107. Madalena, D.A.; Ramos, O.L.; Pereira, R.N.; Bourbon, A.I.; Pinheiro, A.C.; Malcata, F.X.; Teixeira, J.A.; Vicente, A.A. In vitro digestion and stability assessment of β-lactoglobulin/riboflavin nanostructures. *Food Hydrocoll.* **2016**, *58*, 89–97. [CrossRef]
108. Ochnio, M.E.; Martinez, J.H.; Allievi, M.C.; Palavecino, M.; Martinez, K.D.; Perez, O.E. Proteins as nano-carriers for bioactive compounds. The case of 7S and 11S soy globulins and folic acid complexation. *Polymers* **2018**, *10*, 149. [CrossRef]
109. Rubio, A.P.D.; Martinez, J.H.; Casillas, D.C.M.; Leskow, F.C.; Piuri, M.; Perez, O.E. *Lactobacillus casei* BL23 produces microvesicles carrying proteins that have been associated with its probiotic effect. *Front. Microbiol.* **2017**, *8*, 1783. [CrossRef] [PubMed]
110. Peng, S.F.; Li, Z.L.; Zou, L.Q.; Liu, W.; Liu, C.M.; McClements, D.J. Enhancement of curcumin bioavailability by encapsulation in sophorolipid-coated nanoparticles: An in vitro and in vivo study. *J. Agric. Food Chem.* **2018**, *66*, 1488–1497. [CrossRef]
111. Liu, G.Y.; Huang, W.J.; Babii, O.; Gong, X.Y.; Tian, Z.G.; Yang, J.Q.; Wang, Y.X.; Jacobs, R.L.; Donna, V.; Lavasanifar, A.; et al. Novel protein-lipid composite nanoparticles with an inner aqueous compartment as delivery systems of hydrophilic nutraceutical compounds. *Nanoscale* **2018**, *10*, 10629–10640. [CrossRef]
112. Lin, Y.; Wang, Y.H.; Yang, X.Q.; Guo, J.; Wang, J.M. Corn protein hydrolysate as a novel nano-vehicle: Enhanced physicochemical stability and in vitro bioaccessibility of vitamin D_3. *LWT Food Sci. Technol.* **2016**, *72*, 510–517. [CrossRef]
113. David, S.; Livney, Y.D. Potato protein based nanovehicles for health promoting hydrophobic bioactives in clear beverages. *Food Hydrocoll.* **2016**, *57*, 229–235. [CrossRef]
114. Cohen, Y.; Levi, M.; Lesmes, U.; Margier, M.; Reboul, E.; Livney, Y.D. Re-assembled casein micelles improve in vitro bioavailability of vitamin D in a Caco-2 cell model. *Food Funct.* **2017**, *8*, 2133–2141. [CrossRef]
115. Ghayour, N.; Hosseini, S.M.H.; Eskandari, M.H.; Esteghlal, S.; Nekoei, A.R.; Gahruie, H.H.; Tatar, M.; Naghibalhossaini, F. Nanoencapsulation of quercetin and curcumin in casein-based delivery systems. *Food Hydrocoll.* **2019**, *87*, 394–403. [CrossRef]
116. Yerramilli, M.; Longmore, N.; Ghosh, S. Stability and bioavailability of curcumin in mixed sodium caseinate and pea protein isolate nanoemulsions. *J. Am. Oil Chem. Soc.* **2018**, *95*, 1013–1026. [CrossRef]
117. Moeller, H.; Martin, D.; Schrader, K.; Hoffmann, W.; Lorenzen, P.C. Spray- or freeze-drying of casein micelles loaded with vitamin D2: Studies on storage stability and in vitro digestibility. *LWT-Food Sci. Technol.* **2018**, *97*, 87–93. [CrossRef]

118. Penalva, R.; Esparza, I.; Agueeros, M.; Gonzalez-Navarro, C.J.; Gonzalez-Ferrero, C.; Irache, J.M. Casein nanoparticles as carriers for the oral delivery of folic acid. *Food Hydrocoll.* **2015**, *44*, 399–406. [CrossRef]
119. Trofimov, A.D.; Ivanova, A.A.; Zyuzin, M.V.; Timin, A.S. Porous inorganic carriers based on silica, calcium carbonate and calcium phosphate for controlled/modulated drug delivery: Fresh outlook and future perspectives. *Pharmaceutics* **2018**, *10*, 167. [CrossRef] [PubMed]
120. Sayed, E.; Haj-Ahmad, R.; Ruparelia, K.; Arshad, M.S.; Chang, M.W.; Ahmad, Z. Porous inorganic drug delivery systems—A review. *AAPS PharmSciTech* **2017**, *18*, 1507–1525. [CrossRef] [PubMed]
121. Mishra, G.; Dash, B.; Pandey, S. Layered double hydroxides: A brief review from fundamentals to application as evolving biomaterials. *Appl. Clay Sci.* **2018**, *153*, 172–186. [CrossRef]
122. Perez-Esteve, E.; Ruiz-Rico, M.; de la Torre, C.; Villaescusa, L.A.; Sancenon, F.; Marcos, M.D.; Amoros, P.; Martinez-Manez, R.; Barat, J.M. Encapsulation of folic acid in different silica porous supports: A comparative study. *Food Chem.* **2016**, *196*, 66–75. [CrossRef] [PubMed]
123. Perez-Esteve, E.; Fuentes, A.; Coll, C.; Acosta, C.; Bernardos, A.; Amoros, P.; Marcos, M.D.; Sancenon, F.; Martinez-Manez, R.; Barat, J.M. Modulation of folic acid bioaccessibility by encapsulation in pH-responsive gated mesoporous silica particles. *Micropor. Mesopor. Mater.* **2015**, *202*, 124–132. [CrossRef]
124. Perez-Esteve, E.; Ruiz-Rico, M.; Fuentes, A.; Marcos, M.D.; Sancenon, F.; Martinez-Manez, R.; Barat, J.M. Enrichment of stirred yogurts with folic acid encapsulated in pH-responsive mesoporous silica particles: Bioaccessibility modulation and physico-chemical characterization. *LWT Food Sci. Technol.* **2016**, *72*, 351–360. [CrossRef]
125. Ruiz-Rico, M.; Perez-Esteve, E.; Lerma-Garcia, M.J.; Marcos, M.D.; Martinez-Manez, R.; Barat, J.M. Protection of folic acid through encapsulation in mesoporous silica particles included in fruit juices. *Food Chem.* **2017**, *218*, 471–478. [CrossRef]
126. Juere, E.; Florek, J.; Bouchoucha, M.; Jambhrunkar, S.; Wong, K.Y.; Popat, A.; Kleitz, F. In vitro dissolution, cellular membrane permeability, and anti-inflammatory response of resveratrol-encapsulated mesoporous silica nanoparticles. *Mol. Pharm.* **2017**, *14*, 4431–4441. [CrossRef]
127. Summerlin, N.; Qu, Z.; Pujara, N.; Sheng, Y.; Jambhrunkar, S.; McGuckin, M.; Popat, A. Colloidal mesoporous silica nanoparticles enhance the biological activity of resveratrol. *Colloids Surf. B Biointerfaces* **2016**, *144*, 1–7. [CrossRef]
128. Singh, S.; Rathi, N.; Angal, A.; Parida, P.; Rautaray, D. Biofortification of food with minerals and vitamins encapsulated in silica. In *Nanoscience in Food and Agriculture 2. Sustainable Agriculture Reviews*; Ranjan, S., Dasgupta, N., Lichtfouse, E., Eds.; Springer: Cham, Germany, 2016; Volume 21, pp. 157–206.
129. Pagano, C.; Tiralti, M.C.; Perioli, L. Nanostructured hybrids for the improvement of folic acid biopharmaceutical properties. *J. Pharm. Pharmacol.* **2016**, *68*, 1384–1395. [CrossRef] [PubMed]
130. Constantinescu-Aruxandei, D.; Frincu, R.M.; Capra, L.; Oancea, F. Selenium analysis and speciation in dietary supplements based on next-generation selenium ingredients. *Nutrients* **2018**, *10*, 1466. [CrossRef] [PubMed]
131. De Villiers, M.M. Antioxidants. In *A Practical Guide to Contemporary Pharmacy Practice*, 3rd ed.; Thompson, J.E., Ed.; Lippincott Williams & Wilkins: Baltiomor, MD, USA, 2009; pp. 216–223.
132. Aditya, N.P.; Espinosa, Y.G.; Norton, I.T. Encapsulation systems for the delivery of hydrophilic nutraceuticals: Food application. *Biotechnol. Adv.* **2017**, *35*, 450–457. [CrossRef]
133. Pisoschi, A.M.; Pop, A.; Cimpeanu, C.; Turcus, V.; Predoi, G.; Iordache, F. Nanoencapsulation techniques for compounds and products with antioxidant and antimicrobial activity—A critical view. *Eur. J. Med. Chem.* **2018**, *157*, 1326–1345. [CrossRef]
134. Hou, M.N.; Li, Q.; Liu, X.X.; Lu, C.; Li, S.; Wang, Z.Z.; Dang, L.P. Substantial enhancement of the antioxidant capacity of an α-linolenic acid loaded microemulsion: Chemical manipulation of the oil-water interface by carbon dots and its potential application. *J. Agric. Food Chem.* **2018**, *66*, 6917–6925. [CrossRef]
135. Kaur, K.; Kaur, J.; Kumar, R.; Mehta, S.K. Formulation and physiochemical study of α-tocopherol based oil in water nanoemulsion stabilized with non toxic, biodegradable surfactant: Sodium stearoyl lactate. *Ultrason. Sonochem.* **2017**, *38*, 570–578. [CrossRef] [PubMed]
136. Tamjidi, F.; Shahedi, M.; Varshosaz, J.; Nasirpour, A. Stability of astaxanthin-loaded nanostructured lipid carriers in beverage systems. *J. Sci. Food Agric.* **2018**, *98*, 511–518. [CrossRef]

137. Liu, F.G.; Ma, D.; Luo, X.; Zhang, Z.Y.; He, L.L.; Gao, Y.X.; McClements, D.J. Fabrication and characterization of protein-phenolic conjugate nanoparticles for co-delivery of curcumin and resveratrol. *Food Hydrocoll.* **2018**, *79*, 450–461. [CrossRef]
138. Tapia-Hernandez, J.A.; Rodriguez-Felix, F.; Juarez-Onofre, J.E.; Ruiz-Cruz, S.; Robles-Garcia, M.A.; Borboa-Flores, J.; Wong-Corral, F.J.; Cinco-Moroyoqui, F.J.; Castro-Enriquez, D.D.; Del-Toro-Sanchez, C.L. Zein-polysaccharide nanoparticles as matrices for antioxidant compounds: A strategy for prevention of chronic degenerative diseases. *Food Res. Int.* **2018**, *111*, 451–471. [CrossRef]
139. Paulo, F.; Santos, L. Inclusion of hydroxytyrosol in ethyl cellulose microparticles: In vitro release studies under digestion conditions. *Food Hydrocoll.* **2018**, *84*, 104–116. [CrossRef]
140. Hu, Y.; Zhang, W.; Ke, Z.; Li, Y.; Zhou, Z. In vitro release and antioxidant activity of Satsuma mandarin (*Citrus reticulata* Blanco cv. unshiu) peel flavonoids encapsulated by pectin nanoparticles. *Int. J. Food Sci. Technol.* **2017**, *52*, 2362–2373. [CrossRef]
141. Huang, X.X.; Huang, X.L.; Gong, Y.S.; Xiao, H.; McClements, D.J.; Hu, K. Enhancement of curcumin water dispersibility and antioxidant activity using core-shell protein-polysaccharide nanoparticles. *Food Res. Int.* **2016**, *87*, 1–9. [CrossRef] [PubMed]
142. Jiao, Z.; Wang, X.D.; Yin, Y.T.; Xia, J.X.; Mei, Y.N. Preparation and evaluation of a chitosan-coated antioxidant liposome containing vitamin C and folic acid. *J. Microencapsul.* **2018**, *35*, 272–280. [CrossRef] [PubMed]
143. Silva, H.D.; Poejo, J.; Pinheiro, A.C.; Donsi, F.; Serra, A.T.; Duarte, C.M.M.; Ferrari, G.; Cerqueira, M.A.; Vicente, A.A. Evaluating the behaviour of curcumin nanoemulsions and multilayer nanoemulsions during dynamic in vitro digestion. *J. Funct. Foods* **2018**, *48*, 605–613. [CrossRef]
144. Kunwar, A.; Priyadarsini, K.I. Free radicals, oxidative stress and importance of antioxidants in human health. *J. Med. Allied. Sci.* **2011**, *1*, 53–60.
145. Pizzino, G.; Irrera, N.; Cucinotta, M.; Pallio, G.; Mannino, F.; Arcoraci, V.; Squadrito, F.; Altavilla, D.; Bitto, A. Oxidative stress: Harms and benefits for human health. *Oxid. Med. Cell. Longev.* **2017**, *2017*, 8416763. [CrossRef]
146. Galati, G.; Sabzevari, O.; Wilson, J.X.; O'Brien, P.J. Prooxidant activity and cellular effects of the phenoxyl radicals of dietary flavonoids and other polyphenolics. *Toxicology* **2002**, *177*, 91–104. [CrossRef]
147. Herbert, V. The antioxidant supplement myth. *Am. J. Clin. Nutr.* **1994**, *60*, 157–168. [CrossRef]
148. Chan, T.S.; Galati, G.; Pannala, A.S.; Rice-Evans, C.; O'Brien, P.J. Simultaneous detection of the antioxidant and pro-oxidant activity of dietary polyphenolics in a peroxidase system. *Free Radic Res.* **2003**, *37*, 787–794. [CrossRef]
149. Rezaei, A.; Fathi, M.; Jafari, S.M. Nanoencapsulation of hydrophobic and low-soluble food bioactive compounds within different nanocarriers. *Food Hydrocoll.* **2019**, *88*, 146–162. [CrossRef]
150. Aadinath, W.; Bhushani, A.; Anandharamakrishnan, C. Synergistic radical scavenging potency of curcumin-in-β-cyclodextrin-in-nanomagnetoliposomes. *Mater. Sci. Eng. C Mater. Biol. Appl.* **2016**, *64*, 293–302. [CrossRef] [PubMed]
151. Yi, J.; Li, Y.; Zhong, F.; Yokoyama, W. The physicochemical stability and in vitro bioaccessibility of β-carotene in oil-in-water sodium caseinate emulsions. *Food Hydrocoll.* **2014**, *35*, 19–27. [CrossRef]
152. Mao, L.K.; Wang, D.; Liu, F.G.; Gao, Y.X. Emulsion design for the delivery of beta-carotene in complex food systems. *Crit. Rev. Food Sci. Nutr.* **2018**, *58*, 770–784. [CrossRef] [PubMed]
153. Brito-Oliveira, T.C.; Molina, C.V.; Netto, F.M.; Pinho, S.C. Encapsulation of β-carotene in lipid microparticles stabilized with hydrolyzed soy protein isolate: Production parameters, α-tocopherol coencapsulation and stability under stress conditions. *J. Food Sci.* **2017**, *82*, 659–669. [CrossRef] [PubMed]
154. Wang, W.Y.; Sun, C.X.; Mao, L.K.; Ma, P.H.; Liu, F.G.; Yang, J.; Gao, Y.X. The biological activities, chemical stability, metabolism and delivery systems of quercetin: A review. *Trends Food Sci. Technol.* **2016**, *56*, 21–38. [CrossRef]
155. Zhang, J.M.; Wang, D.; Wu, Y.H.; Li, W.; Hu, Y.; Zhao, G.; Fu, C.M.; Fu, S.; Zou, L. Lipid-polymer hybrid nanoparticles for oral delivery of tartary buckwheat flavonoids. *J. Agric. Food Chem.* **2018**, *66*, 4923–4932. [CrossRef]
156. Tavakoli, H.; Hosseini, O.; Jafari, S.M.; Katouzian, I. Evaluation of physicochemical and antioxidant properties of yogurt enriched by olive leaf phenolics within nanoliposomes. *J. Agric. Food Chem.* **2018**, *66*, 9231–9240. [CrossRef]

157. Huang, X.L.; Dai, Y.Q.; Cai, J.X.; Zhong, N.J.; Xiao, H.; McClements, D.J.; Hu, K. Resveratrol encapsulation in core-shell biopolymer nanoparticles: Impact on antioxidant and anticancer activities. *Food Hydrocoll.* **2017**, *64*, 157–165. [CrossRef]
158. Liu, Y.X.; Fan, Y.T.; Gao, L.Y.; Zhang, Y.Z.; Yi, J. Enhanced pH and thermal stability, solubility and antioxidant activity of resveratrol by nanocomplexation with α-lactalbumin. *Food Funct.* **2018**, *9*, 4781–4790. [CrossRef]
159. Alarcon-Alarcon, C.; Inostroza-Riquelme, M.; Torres-Gallegos, C.; Araya, C.; Miranda, M.; Sanchez-Caamano, J.C.; Moreno-Villoslada, I.; Oyarzun-Ampuero, F.A. Protection of astaxanthin from photodegradation by its inclusion in hierarchically assembled nano and microstructures with potential as food. *Food Hydrocoll.* **2018**, *83*, 36–44. [CrossRef]
160. Tamjidi, F.; Shahedi, M.; Varshosaz, J.; Nasirpour, A. Stability of astaxanthin-loaded nanostructured lipid carriers as affected by pH, ionic strength, heat treatment, simulated gastric juice and freeze-thawing. *J. Food Sci. Tech. Mysore* **2017**, *54*, 3132–3141. [CrossRef] [PubMed]
161. Khader, M.; Eckl, P.M. Thymoquinone: An emerging natural drug with a wide range of medical applications. *Iran. J. Basic. Med. Sci.* **2014**, *17*, 950–957. [PubMed]
162. El-Far, A.H.; Al Jaouni, S.K.; Li, W.K.; Mousa, S.A. Protective roles of thymoquinone nanoformulations: Potential nanonutraceuticals in human diseases. *Nutrients* **2018**, *10*, 1369. [CrossRef] [PubMed]
163. De Farias, S.S.; Siqueira, S.M.C.; Cunha, A.P.; de Souza, C.A.G.; Fontenelle, R.O.D.; de Araujo, T.G.; de Amorim, A.F.V.; de Menezes, J.E.S.A.; de Morais, S.M.; Ricardo, N.M.P.S. Microencapsulation of riboflavin with galactomannan biopolymer and F127: Physico-chemical characterization, antifungal activity and controlled release. *Ind. Crops Prod.* **2018**, *118*, 271–281. [CrossRef]
164. Cheong, A.M.; Tan, K.W.; Tan, C.P.; Nyam, K.L. Kenaf (*Hibiscus cannabinus* L.) seed oil-in-water pickering nanoemulsions stabilised by mixture of sodium caseinate, Tween 20 and β-cyclodextrin. *Food Hydrocolloids* **2016**, *52*, 934–941. [CrossRef]
165. Pandey, K.R.; Naik, S.R.; Vakil, B.V. Probiotics, prebiotics and synbiotics—A review. *J. Food Sci. Technol.* **2015**, *52*, 7577–7587. [CrossRef] [PubMed]
166. Markowiak, P.; Slizewska, K. Effects of probiotics, prebiotics, and synbiotics on human health. *Nutrients* **2017**, *9*, 1021. [CrossRef] [PubMed]
167. Kerry, R.G.; Patra, J.K.; Gouda, S.; Park, Y.; Shin, H.S.; Das, G. Benefaction of probiotics for human health: A review. *J. Food Drug Anal.* **2018**, *26*, 927–939. [CrossRef]
168. Gbassi, G.K.; Vandamme, T. Probiotic encapsulation technology: From microencapsulation to release into the gut. *Pharmaceutics* **2012**, *4*, 149–163. [CrossRef]
169. Sathyabama, S.; Kumar, M.R.; Devi, P.B.; Vijayabharathi, R.; Priyadharisini, V.B. Co-encapsulation of probiotics with prebiotics on alginate matrix and its effect on viability in simulated gastric environment. *Food Sci. Technol.* **2014**, *57*, 419–425. [CrossRef]
170. Kuo, S.M.; Merhige, P.M.; Hagey, L.R. The effect of dietary prebiotics and probiotics on body weight, large intestine indices, and fecal bile acid profile in wild type and IL10-/- mice. *PLoS ONE* **2013**, *8*, 60270. [CrossRef] [PubMed]
171. Yao, M.F.; Li, B.; Ye, H.W.; Huang, W.H.; Luo, Q.X.; Xiao, H.; McClements, D.J.; Li, L.J. Enhanced viability of probiotics (*Pediococcus pentosaceus* Li05) by encapsulation in microgels doped with inorganic nanoparticles. *Food Hydrocoll.* **2018**, *83*, 246–252. [CrossRef]
172. Atia, A.; Gomaa, A.; Fliss, I.; Beyssac, E.; Garrait, G.; Subirade, M. A prebiotic matrix for encapsulation of probiotics: Physicochemical and microbiological study. *J. Microencapsul.* **2016**, *33*, 89–101. [CrossRef] [PubMed]
173. Peredo, A.G.; Beristain, C.I.; Pascual, L.A.; Azuara, E.; Jimenez, M. The effect of prebiotics on the viability of encapsulated probiotic bacteria. *Food Sci. Technol.* **2016**, *73*, 191–196. [CrossRef]
174. Mishra, S.S.; Behera, P.K.; Kar, B.; Ray, R.C. Advances in probiotics, prebiotics and nutraceuticals. In *Innovations in Technologies for Fermented Food and Beverage Industries*; Panda, S., Shetty, P., Eds.; Springer: Cham, Germany, 2018; pp. 121–141.
175. Salami, A.; Seydi, E.; Pourahmad, J. Use of nutraceuticals for prevention and treatment of cancer. *Iran. J. Pharm. Res.* **2013**, *12*, 219–220. [PubMed]
176. Lefranc, F.; Tabanca, N.; Kiss, R. Assessing the anticancer effects associated with food products and/or nutraceuticals using in vitro and in vivo preclinical development-related pharmacological tests. *Sem. Cancer Biol.* **2017**, *46*, 14–32. [CrossRef] [PubMed]

177. Fritz, H.; Seely, D.; Flower, G.; Skidmore, B.; Fernandes, R.; Vadeboncoeur, S.; Kennedy, D.; Cooley, K.; Wong, R.; Sagar, S.; et al. Soy, red clover, and isoflavones and breast cancer: A systematic review. *PLoS ONE* **2013**, *8*, e81968. [CrossRef]
178. Lotha, R.; Sivasubramanian, A. Flavonoids nutraceuticals in prevention and treatment of cancer: A review. *Asian J. Pharm. Clin. Res.* **2018**, *11*, 42–47. [CrossRef]
179. Chikwere, P. Functional foods and nutraceuticals, wonders in cancer risks—A review. *World Sci. News* **2017**, *64*, 18–33.
180. Wargovich, M.J.; Morris, J.; Brown, V.; Ellis, J.; Logothetis, B.; Weber, R. Nutraceutical use in late-stage cancer. *Cancer Metastasis Rev.* **2010**, *29*, 503–510. [CrossRef] [PubMed]
181. McClements, D.J.; Xiao, H. Designing food structure and composition to enhance nutraceutical bioactivity to support cancer inhibition. *Semin. Cancer Biol.* **2017**, *46*, 215–226. [CrossRef] [PubMed]
182. Liu, L.; Gao, Y.X.; McClements, D.J.; Decker, E.A. Role of continuous phase protein, (-)-epigallocatechin-3-gallate and carrier oil on beta-carotene degradation in oil-in-water emulsions. *Food Chem.* **2016**, *210*, 242–248. [CrossRef]
183. Granja, A.; Frias, I.; Neves, A.R.; Pinheiro, M.; Reis, S. Therapeutic potential of epigallocatechin gallate nanodelivery systems. *Biomed. Res. Int.* **2017**, *2017*, 5813793. [CrossRef]
184. Hu, K.; Huang, X.X.; Gao, Y.Q.; Huang, X.L.; Xiao, H.; McClements, D.J. Core-shell biopolymer nanoparticle delivery systems: Synthesis and characterization of curcumin fortified zein-pectin nanoparticles. *Food Chem.* **2015**, *182*, 275–281. [CrossRef] [PubMed]
185. Quagliariello, V.; Vecchione, R.; Coppola, C.; Di Cicco, C.; De Capua, A.; Piscopo, G.; Paciello, R.; Narciso, V.; Formisano, C.; Taglialatela-Scafati, O.; et al. Cardioprotective effects of nanoemulsions loaded with anti-inflammatory nutraceuticals against doxorubicin-induced cardiotoxicity. *Nutrients* **2018**, *10*, 1304. [CrossRef] [PubMed]
186. Meghani, N.; Patel, P.; Kansara, K.; Ranjan, S.; Dasgupta, N.; Ramalingam, C.; Kumar, A. Formulation of vitamin D encapsulated cinnamon oil nanoemulsion: Its potential anti-cancerous activity in human alveolar carcinoma cells. *Colloids Surf. B Biointerfaces* **2018**, *166*, 349–357. [CrossRef] [PubMed]
187. Alaarg, A.; Jordan, N.Y.; Verhoef, J.J.F.; Metselaar, J.M.; Storm, G.; Kok, R.J. Docosahexaenoic acid liposomes for targeting chronic inflammatory diseases and cancer: An in vitro assessment. *Int. J. Nanomed.* **2016**, *11*, 5027–5040. [CrossRef] [PubMed]
188. Skibinski, C.G.; Das, A.; Chen, K.M.; Liao, J.; Manni, A.; Kester, M.; El-Bayoumy, K. A novel biologically active acid stable liposomal formulation of docosahexaenoic acid in human breast cancer cell lines. *Chem. Biol. Interact.* **2016**, *252*, 1–8. [CrossRef]
189. Gokmen, V.; Mogol, B.A.; Lumaga, R.B.; Fogliano, V.; Kaplun, Z.; Shimoni, E. Development of functional bread containing nanoencapsulated ω-3 fatty acids. *J. Food Eng.* **2011**, *105*, 585–591. [CrossRef]
190. Bhatt, P.C.; Pathak, S.; Kumar, V.; Panda, B.P. Attenuation of neurobehavioral and neurochemical abnormalities in animal model of cognitive deficits of Alzheimer's disease by fermented soybean nanonutraceutical. *Inflammopharmacology* **2018**, *26*, 105–118. [CrossRef]
191. Zempleni, J.; Aguilar-Lozano, A.; Sadri, M.; Sukreet, S.; Manca, S.; Wu, D.; Zhou, F.; Mutai, E. Biological activities of extracellular vesicles and their cargos from bovine and human milk in humans and implications for infants. *J. Nutr.* **2017**, *147*, 3–10. [CrossRef] [PubMed]
192. Rigacci, S.; Stefani, M. Nutraceuticals and amyloid neurodegenerative diseases: A focus on natural phenols. *Expert Rev. Neurother.* **2015**, *1*, 41–52. [CrossRef] [PubMed]
193. Aalinkeel, R.; Kutscher, H.L.; Singh, A.; Cwiklinski, K.; Khechen, N.; Schwartz, S.A.; Prasad, P.N.; Mahajan, S.D. Neuroprotective effects of a biodegradable poly(lactic-co-glycolic acid)-ginsenoside Rg3 nanoformulation: A potential nanotherapy for Alzheimer's disease? *J. Drug Target.* **2018**, *26*, 182–193. [CrossRef] [PubMed]
194. Kumar, S.A.; Brown, L. Alginates in metabolic syndrome. In *Aginates and Theirbiomedical Applications*; Rehm, B.H.A., Moradali, M.F., Eds.; Springer: Singapore, 2018; Volume 11, pp. 223–235.
195. Kar, S.K.; Jansman, A.J.M.; Boeren, S.; Kruijt, L.; Smits, M.A. Protein, peptide, amino acid composition, and potential functional properties of existing and novel dietary protein sources for monogastrics. *J. Anim. Sci.* **2016**, *94*, 30–39. [CrossRef]

196. Pham, T.M.; Ekwaru, J.P.; Mastroeni, S.S.; Mastroeni, M.F.; Loehr, S.A.; Veugelers, P.J. The effect of serum 25-hydroxyvitamin D on elevated homocysteine concentrations in participants of a preventive health program. *PLoS ONE* **2016**, *11*, 0161368. [CrossRef] [PubMed]
197. Xie, C.L.; Lee, S.S.; Choung, S.Y.; Kang, S.S.; Choi, Y.J. Preparation and optimisation of liposome-in-alginate beads containing oyster hydrolysate for sustained release. *Int. J. Food Sci. Technol.* **2016**, *51*, 2209–2216. [CrossRef]
198. Feng, T.; Wang, K.; Liu, F.F.; Ye, R.; Zhu, X.; Zhuang, H.N.; Xue, Z.M. Structural characterization and bioavailability of ternary nanoparticles consisting of amylose, α-linoleic acid and β-lactoglobulin complexed with naringin. *Int. J. Biol. Macromol.* **2017**, *99*, 365–374. [CrossRef]
199. Mahmoud, M.H.; Badr, G.; El Shinnawy, N.A. Camel whey protein improves lymphocyte function and protects against diabetes in the offspring of diabetic mouse dams. *Int. J. Immunopathol. Pharmacol.* **2016**, *29*, 632–646. [CrossRef]
200. Paul, D.; Dey, T.K.; Mukherjee, S.; Ghosh, M.; Dhar, P. Comparative prophylactic effects of alpha-eleostearic acid rich nano and conventional emulsions in induced diabetic rats. *J. Food Sci. Tech. Mysore* **2014**, *51*, 1724–1736. [CrossRef]
201. Tarighat-Esfanjani, A.; Fallahnejad, H.; Omidi, H.; Jafarabadi, M.A.; Abbasi, M.M.; Khorram, S. The effects of natural nano-sized clinoptilolite and metformin on the levels of serum glucose, lipid profile, and minerals in rats with type 2 diabetes mellitus. *Iran. Red Crescent Med. J.* **2018**, *20*, 74365. [CrossRef]
202. Nia, B.H.; Khorram, S.; Rezazadeh, H.; Safaiyan, A.; Tarighat-Esfanjani, A. The effects of natural clinoptilolite and nano-sized clinoptilolite supplementation on glucose levels and oxidative stress in rats with type 1 diabetes. *Can. J. Diabetes.* **2018**, *42*, 31–35.
203. Hossein-Nia, B.; Khorram, S.; Rezazadeh, H.; Safaiyan, A.; Ghiasi, R.; Tarighat-Esfanjani, A. The effects of natural clinoptilolite and nano-sized clinoptilolite supplementation on lipid profile, food intakes and body weight in rats with streptozotocin-induced diabetes. *Adv. Pharm. Bull.* **2018**, *8*, 211–216. [CrossRef] [PubMed]
204. Perumal, V.; Manickam, T.; Bang, K.S.; Velmurugan, P.; Oh, B.T. Antidiabetic potential of bioactive molecules coated chitosan nanoparticles in experimental rats. *Int. J. Biol. Macromol.* **2016**, *92*, 63–69. [CrossRef] [PubMed]
205. Liu, Y.T.; Zeng, S.G.; Liu, Y.X.; Wu, W.J.; Shen, Y.B.; Zhang, L.; Li, C.; Chen, H.; Liu, A.P.; Shen, L. Synthesis and antidiabetic activity of selenium nanoparticles in the presence of polysaccharides from *Catathelasma ventricosum*. *Int. J. Biol. Macromol.* **2018**, *114*, 632–639. [CrossRef]
206. Sechi, M.; Syed, D.N.; Pala, N.; Mariani, A.; Marceddu, S.; Brunetti, A.; Mukhtar, H.; Sanna, V. Nanoencapsulation of dietary flavonoid fisetin: Formulation and in vitro antioxidant and α-glucosidase inhibition activities. *Mater. Sci. Eng. C Mater. Biol. Appl.* **2016**, *68*, 594–602. [CrossRef]
207. Bagherpour, S.; Alizadeh, A.; Ghanbarzadeh, S.; Mohammadi, M.; Hamishehkar, H. Preparation and characterization of Betasitosterol-loaded nanostructured lipid carriers for butter enrichment. *Food Biosci.* **2017**, *20*, 51–55. [CrossRef]
208. Nakada, H.; Sakae, T.; Watanabe, T.; Takahashi, T.; Fujita, K.; Tanimoto, Y.; Teranishi, M.; Kato, T.; Kawai, Y. A new osteoporosis prevention supplements-diet improve bone mineral density in ovariectomized rats on micro-CT. *J. Hard Tissue Biol.* **2014**, *23*, 1–8. [CrossRef]
209. Khashayar, P.; Keshtkar, A.; Ebrahimi, M.; Larijani, B. Nano calcium supplements: Friends or foes? *J. Bone Biol. Osteoporosis* **2015**, *1*, 32–33.
210. Park, H.S.; Jeon, B.J.; Ahn, J.; Kwak, H.S. Effects of nanocalcium supplemented milk on bone calcium metabolism in ovariectomized rats. *Asian-Aust. J. Anim. Sci.* **2007**, *20*, 1266–1271. [CrossRef]
211. Choi, H.S.; Han, J.H.; Chung, S.; Hong, Y.H.; Suh, H.J. Nano-calcium ameliorates ovariectomy-induced bone loss in female rats. *Korean J. Food Sci. Anim. Res.* **2013**, *33*, 515–521. [CrossRef]
212. Huang, S.; Chen, J.C.; Hsu, C.W.; Chang, W.H. Effects of nano calcium carbonate and nano calcium citrate on toxicity in ICR mice and on bone mineral density in an ovariectomized mice model. *Nanotechnology* **2009**, *20*, 375102. [CrossRef] [PubMed]
213. Erfanian, A.; Mirhosseini, H.; Abd Manap, M.Y.; Rasti, B.; Hair-Bejo, M. Influence of nano-size reduction on absorption and bioavailability of calcium from fortified milk powder in rats. *Food Res. Int.* **2014**, *66*, 1–11. [CrossRef]

214. Erfanian, A.; Mirhosseini, H.; Rasti, B.; Hair-Bejo, M.; Bin Mustafa, S.; Abd Manap, M.Y. Absorption and bioavailability of nano-size reduced calcium citrate fortified milk powder in ovariectomized and ovariectomized-osteoporosis rats. *J. Agric. Food Chem.* **2015**, *63*, 5795–57804. [CrossRef]
215. Erfanian, A.; Rasti, B.; Manap, Y. Comparing the calcium bioavailability from two types of nano-sized enriched milk using in-vivo assay. *Food Chem.* **2017**, *214*, 606–613. [CrossRef] [PubMed]
216. Guo, H.H.; Hong, Z.A.; Yi, R.Z. Core-shell collagen peptide chelated calcium/calcium alginate nanoparticles from fish scales for calcium supplementation. *J. Food Sci.* **2015**, *80*, N1595–N1601. [CrossRef] [PubMed]
217. Cai, X.X.; Zhao, L.N.; Wang, S.Y.; Rao, P.F. Fabrication and characterization of the nano-composite of whey protein hydrolysate chelated with calcium. *Food Funct.* **2015**, *6*, 816–823.
218. Noor, Z. Nanohydroxyapatite application to osteoporosis management. *J. Osteoporosis* **2013**, *2013*, 679025. [CrossRef]
219. Zhang, X.; Zhu, L.; Lv, H.; Cao, Y.; Liu, Y.; Xu, Y.; Ye, W.; Wang, J. Repair of rabbit femoral condyle bone defects with injectable nanohydroxyapatite/chitosan composites. *J. Mater. Sci. Mater. Med.* **2012**, *23*, 1941–1949. [CrossRef]
220. Severin, A.V.; Mazina, S.E.; Melikhov, I.V. Physicochemical aspects of the antiseptic action of nanohydroxyapatite. *Biophysics* **2009**, *54*, 701–705. [CrossRef]
221. Chakraborty, A.P. Chicken eggshell as calcium supplement tablet. *Int. J. Sci. Eng. Manag.* **2016**, *1*, 45–49.
222. Ray, S.; Barman, A.K.; Roy, P.K.; Singh, B.K. Chicken eggshell powder as dietary calcium source in chocolate cakes. *Pharma Innov. J.* **2017**, *6*, 1–4.
223. Mijan, M.A.; Lee, Y.K.; Kwak, H.S. Effects of nanopowdered eggshell on postmenopausal osteoporosis: A rat study. *Food Sci. Biotechnol.* **2014**, *23*, 1667–1676. [CrossRef]
224. El-Shibiny, S.; Abd El-Gawad, M.A.M.; Assem, F.M.; El-Sayed, S.M. The use of nano-sized eggshell powder for calcium fortification of cow's and buffalo's milk yogurts. *Acta Sci. Pol. Technol. Aliment.* **2018**, *17*, 37–49. [PubMed]
225. Zanella, D.; Bossi, E.; Gornati, R.; Bastos, C.; Faria, N.; Bernardini, G. Iron oxide nanoparticles can cross plasma membranes. *Sci. Rep.* **2017**, *7*, 11413. [CrossRef] [PubMed]
226. Hosny, K.M.; Banjar, Z.M.; Hariri, A.H.; Hassan, A.H. Solid lipid nanoparticles loaded with iron to overcome barriers for treatment of iron deficiency anemia. *Drug Des. Dev. Ther.* **2015**, *9*, 313–320. [CrossRef] [PubMed]
227. Gornati, R.; Pedretti, E.; Rossi, F.; Cappellini, F.; Zanella, M.; Olivato, I.; Sabbioni, E.; Bernardini, G. Zerovalent Fe, Co and Ni nanoparticle toxicity evaluated on SKOV-3 and U87 cell lines. *J. Appl. Toxicol.* **2016**, *36*, 385–393. [CrossRef]
228. Lonnerdal, B.; Bryant, A.; Liu, X.; Theil, E.C. Iron absorption from soybean ferritin in nonanemic women. *Am. J. Clin. Nutr.* **2006**, *83*, 103–107. [CrossRef]
229. Powell, J.J.; Bruggraber, S.F.A.; Faria, N.; Poots, L.K.; Hondow, N.; Pennycook, T.J.; Latunde-Dada, G.O.; Simpson, R.J.; Brown, A.P.; Pereira, D.I.A. A nano-disperse ferritin-core mimetic that efficiently corrects anemia without luminal iron redox activity. *Nanomedicine* **2014**, *10*, 1529–1538. [CrossRef]
230. Pereira, D.I.A.; Bruggraber, S.F.A.; Faria, N.; Poots, L.K.; Tagmount, M.A.; Aslam, M.F.; Frazer, D.M.; Vulpe, C.D.; Anderson, G.J.; Powell, J.J. Nanoparticulate iron(III) oxo-hydroxide delivers safe iron that is well absorbed and utilised in humans. *Nanomedicine* **2014**, *10*, 1877–1886. [CrossRef]
231. Pereira, D.I.A.; Mohammed, N.I.; Ofordile, O.; Camara, F.; Baldeh, B.; Mendy, T.; Sanyang, C.; Jallow, A.T.; Hossain, I.; Wason, J.; et al. A novel nano-iron supplement to safely combat iron deficiency and anaemia in young children: The IHAT-GUT double-blind, randomised, placebo-controlled trial protocol. *Gates Open Res.* **2018**, *2*, 48. [CrossRef] [PubMed]
232. Hilty, F.M.; Arnold, M.; Hilbe, M.; Teleki, A.; Knijnenburg, J.T.; Ehrensperger, F.; Hurrell, R.F.; Pratsinis, S.E.; Langhans, W.; Zimmermann, M.B. Iron from nanocompounds containing iron and zinc is highly bioavailable in rats without tissue accumulation. *Nat. Nanotechnol.* **2010**, *5*, 374–380. [CrossRef] [PubMed]
233. Srinivasu, B.Y.; Mitra, G.; Muralidharan, M.; Srivastava, D.; Pinto, J.; Thankachan, P.; Suresh, S.; Shet, A.; Rao, S.; Ravikumar, G.; et al. Beneficiary effect of nanosizing ferric pyrophosphate as food fortificant in iron deficiency anemia: Evaluation of bioavailability, toxicity and plasma biomarker. *RSC Adv.* **2015**, *5*, 61678–61687. [CrossRef]
234. Salaheldin, T.A.; Regheb, E.M. In-Vivo nutritional and toxicological evaluation of nano iron fortified biscuits as food supplement for iron deficient anemia. *J. Nanomed. Res.* **2016**, *3*, 00049. [CrossRef]

235. Center for Veterinary Medicine Nanotechnology Programs. Available online: https://www.fda.gov/ScienceResearch/SpecialTopics/Nanotechnology/ucm309682.htm (accessed on 1 December 2018).
236. Animal & Veterinary. Available online: https://www.fda.gov/AnimalVeterinary/UCM2005229 (accessed on 1 December 2018).
237. Robinson, N.G. Nutraceuticals and Dietary Supplements. MSD Veterinary Manual. 2018. Available online: https://www.msdvetmanual.com/management-and-nutrition/complementary-and-alternative-veterinary-medicine/nutraceuticals-and-dietary-supplements (accessed on 1 December 2018).
238. Valpotic, H.; Gracner, D.; Turk, R.; Duricic, D.; Vince, S.; Folnozic, I.; Lojkic, M.; Zaja, I.Z.; Bedrica, L.; Macesic, N.; et al. Zeolite clinoptilolite nanoporous feed additive for animals of veterinary importance: Potentials and limitations. *Period. Biol.* **2017**, *119*, 159–172. [CrossRef]
239. Swain, P.S.; Rao, S.B.N.; Rajendran, D.; Dominic, G.; Selvaraju, S. Nano zinc, an alternative to conventional zinc as animal feed supplement: A review. *Anim. Nutr.* **2016**, *2*, 134–141. [CrossRef] [PubMed]
240. Yan, R.; Zhang, L.; Yang, X.; Wen, C.; Zhou, Y. Bioavailability evaluation of zinc-bearing palygorskite as a zinc source for broiler chickens. *Appl. Clay Sci.* **2016**, *119*, 155–160. [CrossRef]
241. Tsai, Y.H.; Mao, S.Y.; Li, M.Z.; Huang, J.T.; Lien, T.F. Effects of nanosize zinc oxide on zinc retention, eggshell quality, immune response and serum parameters of aged laying hens. *Anim. Feed Sci. Technol.* **2016**, *213*, 99–107. [CrossRef]
242. Chrastinova, L.; Cobanova, K.; Chrenkova, M.; Polacikova, M.; Foemelova, Z.; Laukova, L.; Ondruska, A.; Pogany, S.M.; Strompfova, V.; Mlynekova, Z.; et al. Effect of dietary zinc supplementation on nutrients digestibility and fermentation characteristics of caecal content in physiological experiment with young rabbits. *Slovak J. Anim. Sci.* **2016**, *49*, 23–31.
243. Swain, P.S.; Rajendran, D.; Rao, S.B.; Dominic, G. Preparation and effects of nano mineral particle feeding in livestock: A review. *Vet. World.* **2015**, *8*, 888–891. [CrossRef]
244. Debski, B. Supplementation of pigs diet with zinc and copper as alternative to conventional antimicrobials. *Pol. J. Vet. Sci.* **2016**, *19*, 917–924. [CrossRef]
245. Yin, J.; Li, X.; Li, D.; Yue, T.; Fang, Q.; Ni, J.; Zhou, X.; Wu, G. Dietary supplementation with zinc oxide stimulates ghrelin secretion from stomach of young pigs. *J. Nutr. Biochem.* **2009**, *20*, 783–790. [CrossRef] [PubMed]
246. Li, X.; Yin, J.; Li, D.; Chen, X.; Zang, J.; Zhou, X. Dietary supplementation with zinc oxide increases Igf-I and Igf-I receptor gene expression in the small intestine of weanling piglets. *J. Nutr.* **2006**, *136*, 1786–1791. [CrossRef] [PubMed]
247. Zhou, W.; Kornegay, E.T.; Lindermann, M.D.; Swinkels, J.W.; Welten, M.K.; Wong, E.A. Stimulation of growth by intravenous injection of copper in weanling pigs. *J. Anim. Sci.* **2014**, *72*, 2395–2403. [CrossRef]
248. Jacela, J.Y.; De Rouchey, J.M.; Tokach, M.D.; Goodband, R.D.; Nelssen, J.L.; Renter, D.G.; Dritz, S.S. Feed additives for swine: Fact sheets-high dietary levels of copper and zinc for young pigs, and phytase. *J. Swine Health Prod.* **2010**, *18*, 87–92. [CrossRef]
249. Saha, U.; Fayiga, A.; Hancock, D.; Sonon, L. Selenium in animal nutrition: Deficiencies in soils and forages, requirements, supplementation and toxicity. *Int. J. Appl. Agric. Sci.* **2016**, *2*, 112–125. [CrossRef]
250. Hosnedlova, B.; Kepinska, M.; Skalickova, S.; Fernandez, C.; Ruttkay-Nedecky, B.; Peng, Q.M.; Baron, M.; Melcova, M.; Opatrilova, R.; Zidkova, J.; et al. Nano-selenium and its nanomedicine applications: A critical review. *Int. J. Nanomed.* **2018**, *13*, 2107–2128. [CrossRef] [PubMed]
251. Bai, D.P.; Lin, X.Y.; Huang, Y.F.; Zhang, X.F. Theranostics aspects of various nanoparticles in veterinary medicine. *Int. J. Mol. Sci.* **2018**, *19*, 3299. [CrossRef]
252. Hill, E.K.; Li, J. Current and future prospects for nanotechnology in animal production. *J. Anim. Sci. Biotechnol.* **2017**, *8*, 26. [CrossRef]
253. Nikonov, I.N.; Folmanis, Y.G.; Folmanis, G.E.; Kovalenko, L.V.; Laptev, G.Y.; Egorov, I.A.; Fisinin, V.I.; Tananaev, I.G. Iron nanoparticles as a food additive for poultry. *Dokl. Biol. Sci.* **2011**, *440*, 328–331. [CrossRef]
254. Izquierdo, M.S.; Ghrab, W.; Roo, J.; Hamre, K.; Hernandez-Cruz, C.M.; Bernardini, G.; Terova, G.; Saleh, R. Organic, inorganic and nanoparticles of Se, Zn and Mn in early weaning diets for gilthead seabream (*Sparus aurata*; Linnaeus, 1758). *Aqua. Res.* **2017**, *48*, 2852–2867. [CrossRef]
255. Chris, O.U.; Singh, N.B.; Agarwal, A. Nanoparticles as feed supplement on growth behaviour of cultured catfish (*Clarias gariepinus*) fingerlings. *Appl. Mater. Today* **2018**, *5*, 9076–9081. [CrossRef]

256. Zadmajid, V.; Mohammadi, C. Dietary thyme essential oil (*Thymus vulgaris*) changes serum stress markers, enzyme activity, and hematological parameters in gibel carp (*Carassius auratus gibelio*) exposed to silver nanoparticles. *Iran. J. Fish. Sci.* **2017**, *16*, 1063–1084.
257. Rohani, S.M.; Haghighi, M.; Moghaddam, B.S. Study on nanoparticles of *Aloe vera* extract on growth performance, survival rate and body composition in Siberian sturgeon (*Acipenser baerii*). *Iran. J. Fish. Sci.* **2017**, *16*, 457–468.
258. Alishahi, A.; Mirvaghefi, A.; Tehrani, M.R.; Farahmand, H.; Koshio, S.; Dorkoosh, F.A.; Elsabee, M.Z. Chitosan nanoparticle to carry vitamin C through the gastrointestinal tract and induce the non-specific immunity system of rainbow trout (*Oncorhynchus mykiss*). *Carbohydr. Polym.* **2011**, *86*, 142–146. [CrossRef]
259. Martins, A.C.D.; Flores, J.A.; Porto, C.; Romano, L.A.; Wasielesky, J.W.; Caldas, S.S.; Primel, E.G.; Kulkamp-Guerreiro, I.; Monserrat, J.M. Antioxidant effects of nanoencapsulated lipoic acid in tissues and on the immune condition in haemolymph of Pacific white shrimp *Litopenaeus vannamei* (Boone, 1931). *Aquac. Nutr.* **2018**, *24*, 1255–1262. [CrossRef]
260. El Basuini, M.F.; El-Hais, A.M.; Dawood, M.A.O.; Abou-Zeid, A.E.S.; EL-Damrawy, S.Z.; Khalafalla, M.M.E.S.; Koshio, S.; Ishikawa, M.; Dossou, S. Effects of dietary copper nanoparticles and vitamin C supplementations on growth performance, immune response and stress resistance of red sea bream, *Pagrus major*. *Aquac. Nutr.* **2017**, *23*, 1329–1340. [CrossRef]
261. Wang, H.; Zhu, H.Y.; Wang, X.D.; Li, E.C.; Du, Z.Y.; Qin, J.G.; Chen, L.Q. Comparison of copper bioavailability in copper-methionine, nano-copper oxide and copper sulfate additives in the diet of Russian sturgeon *Acipenser gueldenstaedtii*. *Aquaculture* **2018**, *482*, 146–154. [CrossRef]
262. Kumar, N.; Krishnani, K.K.; Gupta, S.K.; Sharma, R.; Baitha, R.; Singh, D.K.; Singh, N.P. Immuno-protective role of biologically synthesized dietary selenium nanoparticles against multiple stressors in Pangasinodon hypophthalrnus. *Fish. Shellfish Immunol.* **2018**, *78*, 289–298.
263. Anjugam, M.; Vaseeharan, B.; Iswarya, A.; Gobi, N.; Divya, M.; Thangaraj, M.P.; Elumalai, P. Effect of β-1, 3 glucan binding protein based zinc oxide nanoparticles supplemented diet on immune response and disease resistance in *Oreochromis mossambicus* against *Aeromonas hydrophila*. *Fish. Shellfish Immunol.* **2018**, *76*, 247–259. [CrossRef] [PubMed]
264. Shaphar, Z.; Johari, S.A. Effects of dietary organic, inorganic, and nanoparticulate zinc on rainbow trout, *Oncorhynchus mykiss* larvae. *Biol. Trace Elem. Res.* **2018**, in press. [CrossRef]
265. Saffari, S.; Keyvanshokooh, S.; Zakeri, M.; Johari, S.A.; Pasha-Zanoosi, H.; Mozanzadeh, M.T. Effects of dietary organic, inorganic, and nanoparticulate selenium sources on growth, hemato-immunological, and serum biochemical parameters of common carp (*Cyprinus carpio*). *Fish. Physiol. Biochem.* **2018**, *44*, 1087–1097. [CrossRef] [PubMed]
266. Zhou, X.X.; Wang, Y.B.; Gu, Q.; Li, W.F. Effects of different dietary selenium sources (selenium nanoparticle and selenomethionine) on growth performance, muscle composition and glutathione peroxidase enzyme activity of crucian carp (*Carassius auratus gibelio*). *Aquaculture* **2009**, *291*, 78–81. [CrossRef]
267. Khan, K.U.; Zuberi, A.; Nazir, S.; Fernandes, J.B.K.; Jamil, Z.; Sarwar, H. Effects of dietary selenium nanoparticles on physiological and biochemical aspects of juvenile *Tor putitora*. *Turk. J. Zool.* **2016**, *40*, 704–712. [CrossRef]
268. Qin, F.J.; Shi, M.M.; Yuan, H.X.; Yuan, L.X.; Lu, W.H.; Zhang, J.; Tong, J.; Song, X.H. Dietary nano-selenium relieves hypoxia stress and, improves immunity and disease resistance in the Chinese mitten crab (*Eriocheir sinensis*). *Fish. Shellfish Immunol.* **2016**, *54*, 481–488. [CrossRef] [PubMed]
269. Naderi, M.; Keyvanshokooh, S.; Salati, A.P.; Ghaedi, A. Combined or individual effects of dietary vitamin E and selenium nanoparticles on humoral immune status and serum parameters of rainbow trout (*Oncorhynchus mykiss*) under high stocking density. *Aquaculture* **2017**, *474*, 40–47. [CrossRef]
270. Kumar, N.; Krishnani, K.K.; Singh, N.P. Effect of dietary zinc-nanoparticles on growth performance, anti-oxidative and immunological status of fish reared under multiple stressors. *Biol. Trace Elem. Res.* **2018**, *186*, 267–278. [CrossRef]
271. Gangadoo, S.; Stanley, D.; Hughes, R.J.; Moore, R.J.; Chapman, J. Nanoparticles in feed: Progress and prospects in poultry research. *Trends Food Sci. Technol.* **2016**, *58*, 115–126. [CrossRef]
272. Song, Z.G.; Lv, J.D.; Sheikhahmadi, A.; Uerlings, J.; Everaert, N. Attenuating effect of zinc and vitamin E on the intestinal oxidative stress induced by silver nanoparticles in broiler chickens. *Biol. Trace Elem. Res.* **2017**, *180*, 306–313. [CrossRef]

273. Sawosz, F.; Pineda, L.; Hotowy, A.; Jaworski, S.; Prasek, M.; Sawosz, E.; Chwalibog, A. Nano-nutrition of chicken embryos. The effect of silver nanoparticles and ATP on expression of chosen genes involved in myogenesis. *Arch. Anim. Nutr.* **2013**, *67*, 347–355. [CrossRef]
274. Scott, A.; Vadalasetty, K.P.; Lukasiewicz, M.; Jaworski, S.; Wierzbicki, M.; Chwalibog, A.; Sawosz, E. Effect of different levels of copper nanoparticles and copper sulphate on performance, metabolism and blood biochemical profiles in broiler chicken. *J. Anim. Physiol. Anim. Nutr. (Berl.)* **2018**, *102*, E364–E373. [CrossRef] [PubMed]
275. Ognik, K.; Sembratowicz, I.; Cholewinska, E.; Jankowski, J.; Kozlowski, K.; Juskiewicz, J.; Zdunczyk, Z. The effect of administration of copper nanoparticles to chickens in their drinking water on the immune and antioxidant status of the blood. *Anim. Sci. J.* **2018**, *89*, 579–588. [CrossRef] [PubMed]
276. Joshua, P.P.; Valli, C.; Balakrishnan, V. Effect of in ovo supplementation of nano forms of zinc, copper, and selenium on post-hatch performance of broiler chicken. *Vet. World* **2016**, *9*, 287–294. [CrossRef] [PubMed]
277. Abedini, M.; Shariatmadari, F.; Torshizi, M.A.K.; Ahmadi, H. Effects of zinc oxide nanoparticles on the egg quality, immune response, zinc retention, and blood parameters of laying hens in the late phase of production. *J. Anim. Physiol. Anim. Nutr. (Berl.)* **2018**, *102*, 736–745. [CrossRef] [PubMed]
278. Mao, S.Y.; Lien, T.F. Effects of nanosized zinc oxide and -polyglutamic acid on eggshell quality and serum parameters of aged laying hens. *Arch. Anim. Nutr.* **2017**, *71*, 373. [CrossRef] [PubMed]
279. Cai, S.J.; Wu, C.X.; Gong, L.M.; Song, T.; Wu, H.; Zhang, L.Y. Effects of nano-selenium on performance, meat quality, immune function, oxidation resistance, and tissue selenium content in broilers. *Poultry Sci.* **2012**, *91*, 2532–2539. [CrossRef] [PubMed]
280. Boostani, A.; Sadeghi, A.A.; Mousavi, S.N.; Chamani, M.; Kashan, N. The effects of organic, inorganic, and nano-selenium on blood attributes in broiler chickens exposed to oxidative stress. *Acta Sci. Vet.* **2015**, *43*, 1264.
281. Ahmadi, M.; Ahmadian, A.; Seidavi, A.R. Effect of different levels of nano-selenium on performance, blood parameters, immunity and carcass characteristics of broiler chickens. *Poult. Sci. J.* **2018**, *6*, 99–108.
282. Rahmatollah, D.; Farzinpour, A.; Vaziry, A.; Sadeghi, G. Effect of replacing dietary $FeSO_4$ with cysteine-coated Fe_3O_4 nanoparticles on quails. *Ital. J. Anim. Sci.* **2018**, *17*, 121–127. [CrossRef]
283. Lin, Y.C.; Huang, J.T.; Li, M.Z.; Cheng, C.Y.; Lien, T.F. Effects of supplemental nanoparticle trivalent chromium on the nutrient utilization, growth performance and serum traits of broilers. *J. Anim. Physiol. Anim. Nutr. (Berl.)* **2015**, *99*, 59–65. [CrossRef]
284. Xia, T.; Lai, W.Q.; Han, M.M.; Han, M.; Ma, X.; Zhang, L.Y. Dietary ZnO nanoparticles alters intestinal microbiota and inflammation response in weaned piglets. *Oncotarget* **2017**, *8*, 64878–64891. [CrossRef] [PubMed]
285. Li, M.Z.; Huang, J.T.; Tsai, Y.H.; Mao, S.Y.; Fu, C.M.; Lien, T.F. Nanosize of zinc oxide and the effects on zinc digestibility, growth performances, immune response and serum parameters of weanling piglets. *Anim. Sci. J.* **2016**, *87*, 1379–1385. [CrossRef] [PubMed]
286. Kosla, T.; Lasocka, I.; Skibniewska, E.M.; Kolnierzak, M.; Skibniewski, M. Trivalent chromium (Cr III) as a trace element essential for animals and humans. *Med. Weter.* **2018**, *74*, 560–567.
287. Hung, A.T.; Leury, B.J.; Sabin, M.A.; Collins, C.L.; Dunshea, F.R. Dietary nano-chromium tripicolinate increases feed intake and decreases plasma cortisol in finisher gilts during summer. *Trop. Anim. Health Prod.* **2014**, *46*, 1483–1489. [CrossRef] [PubMed]
288. Wang, M.Q.; Wang, C.; Du, Y.J.; Li, H.; Tao, W.J.; Ye, S.S.; He, Y.D.; Chen, S.Y. Effects of chromium-loaded chitosan nanoparticles on growth, carcass characteristics, pork quality, and lipid metabolism in finishing pigs. *Livest. Sci.* **2014**, *161*, 123–129. [CrossRef]
289. Wang, M.Q.; Xu, Z.R.; Zha, L.Y.; Lindemann, M.D. Effects of chromium nanocomposite supplementation on blood metabolites, endocrine parameters and immune traits in finishing pigs. *Anim. Feed Sci. Technol.* **2007**, *139*, 69–80. [CrossRef]
290. Duffy, C.; O'Riordan, D.; O'Sullivan, M.; Jacquier, J.C. In vitro evaluation of chitosan copper chelate gels as a multimicronutrient feed additive for cattle. *J. Sci. Food Agric.* **2018**, *98*, 4177–4183. [CrossRef]
291. Kojouri, G.A.; Jahanabadi, S.; Shakibaie, M.; Ahadi, A.M.; Shahverdi, A.R. Effect of selenium supplementation with sodium selenite and selenium nanoparticles on iron homeostasis and transferrin gene expression in sheep: A preliminary study. *Res. Vet. Sci.* **2012**, *93*, 275–278. [CrossRef]

292. Shi, L.G.; Xun, W.J.; Yue, W.B.; Zhang, C.X.; Ren, Y.S.; Liu, Q.A.; Wang, Q.A.; Shi, L. Effect of elemental nano-selenium on feed digestibility, rumen fermentation, and purine derivatives in sheep. *Anim. Feed Sci. Technol.* **2011**, *163*, 136–142. [CrossRef]
293. El-Sherbiny, M.; Cieslak, A.; Szczechowiak, J.; Kolodziejski, P.; Szulc, P.; Szumacher-Strabel, M. Effect of nanoemulsified oils addition on rumen fermentation and fatty acid proportion in a rumen simulation technique. *J. Anim. Feed Sci.* **2016**, *25*, 116–124. [CrossRef]
294. Refaie, A.M.; Ghazal, M.N.; Easa, F.M.; Barakat, S.A.; Morsy, W.A.; Younan, G.E.; Eisa, W.H. Nano-copper as a new growth promoter in the diet of growing New Zealand white rabbits. *Egypt. J. Rabbit Sci.* **2015**, *25*, 39–57.
295. Hassan, F.A.M.; Mahmoud, R.; El-Araby, I.E. Growth performance, serum biochemical, economic evaluation and IL6 gene expression in growing rabbits fed diets supplemented with zinc nanoparticles. *Zagazig Vet. J.* **2017**, *45*, 238–249. [CrossRef]
296. Ismail, H.T.H.; El-Araby, I.E. Effect of dietary zinc oxide nanoparticles supplementation on biochemical, hematological and genotoxucity parameters in rabbits. *Int. J. Curr. Adv. Res.* **2017**, *6*, 2108–2115.

nanomaterials

MDPI

Article

Use of Electrosprayed Agave Fructans as Nanoencapsulating Hydrocolloids for Bioactives

Jorge A. Ramos-Hernández [1], Juan A. Ragazzo-Sánchez [1,*], Montserrat Calderón-Santoyo [1], Rosa I. Ortiz-Basurto [1], Cristina Prieto [2] and Jose M. Lagaron [2]

1 Laboratorio Integral de Investigación en Alimentos, Tecnológico Nacional de México/Instituto Tecnológico de Tepic, Av. Tecnológico 2595, C.P. Tepic 63175, Nayarit, Mexico; jorgeramos17@outlook.com (J.A.R.-H.); montserratcalder@gmail.com (M.C.-S.); riobasurt@ittepic.edu.mx (R.I.O.-B.)

2 Novel Materials and Nanotechnology Group, IATA-CSIC, Calle Catedrático Agustín Escardino Benlloch 7, 46980 Paterna, Spain; cprieto@iata.csic.es (C.P.); lagaron@iata.csic.es (J.M.L.)

* Correspondence: jragazzo@ittepic.edu.mx or arturoragazzo@hotmail.com; Tel.: +52-311-2119400

Received: 25 September 2018; Accepted: 12 October 2018; Published: 23 October 2018

Abstract: High degree of polymerization Agave fructans (HDPAF) are presented as a novel encapsulating material. Electrospraying coating (EC) was selected as the encapsulation technique and β-carotene as the model bioactive compound. For direct electrospraying, two encapsulation methodologies (solution and emulsion) were proposed to find the formulation which provided a suitable particle morphology and an adequate concentration of β-carotene encapsulated in the particles to provide a protective effect of β-carotene by the nanocapsules. Scanning electron microscopy (SEM) images showed spherical particles with sizes ranging from 440 nm to 880 nm depending on the concentration of HDPAF and processing parameters. FTIR analysis confirmed the interaction and encapsulation of β-carotene with HDPAF. The thermal stability of β-carotene encapsulated in HDPAF was evidenced by thermogravimetric analysis (TGA). The study showed that β-carotene encapsulated in HDPAF by the EC method remained stable for up to 50 h of exposure to ultraviolet (UV) light. Therefore, HDPAF is a viable option to formulate nanocapsules as a new encapsulating material. In addition, EC allowed for increases in the ratio of β-carotene:polymer, as well as its photostability.

Keywords: HDPAF; β-carotene; electrospraying; encapsulation; photoprotection

1. Introduction

Fructans from *Agave tequilana* consist of a complex mixture of fructooligosaccharides (fructose polymer obtained by enzymatic hydrolysis of high polymerization degree Agave fructans (HDPAF) by fructan exohydrolase (FEH) and 1-fructosyl transferase (1-FFT enzymes) containing principally β-(2→1) fructosyl–fructose linkages, but also β-(2→6) and branch moieties) [1,2]. The physico-chemical and functional properties of fructans are linked to the degree of polymerization (DP) as well as the presence of branches. The short-chain fraction, oligofructose, is much more soluble and sweeter than native and long-chain fructans, and can contribute to improve mouthfeel because its properties are closely related to those of other sugars. The high DP (>40 fructose units, $Mw = 3259.95 \pm 181.75$ g/mol) fraction can be used as a fat substitute in low-fat or reduced-fat products (i.e., baking, ice-cream, beverages and yoghurt) since it is less soluble, more viscous and more thermostable than native fructans, which allows for modification of the rheological and sensorial properties of dairy products. In this case, fructans act as a filler or as a breaker of structure in the same way as fat globules do [1,3,4].

To our knowledge, little work has been done on the exploration of the technological applications of fructans. In this way, Furlán et al. [5] evaluated high, medium and low polymerization degree Agave fructans from *Agave tequilana* Weber as lyoprotectant agents on bovine plasma proteins during spray drying and storage. They concluded that the Agave fructans were able to cryoprotect food proteins. Thus, Agave fructans are a valuable alternative as a functional ingredient for food formulation. Ortiz-Basurto et al. [6] studied the characteristics and applications of medium and high polymerization degree Agave fructans from *Agave tequilana* Weber as microencapsulating materials of pitanga or Surinam cherry (*Eugenia uniflora* L.) juice by spray drying. The powders from both fractions were stable and able to protect the bioactive compound during and after the spray-drying process. These good results, together with its characteristics as a biopolymer (classified as biodegradable and Generally Recognized as Safe GRAS [7]), make fructans a really interesting encapsulating material for food, pharma and cosmetic applications.

Up until now, several techniques have been used to encapsulate bioactive components for the food industry, such as extrusion methods [8], fluidized bed coating [9], spray cooling [10] or spray drying [11]. Nowadays, spray drying is the most common and cheapest technology in the food industry to produce microencapsulated additives for food applications [12]. The electrohydrodynamic processing, including both electrospinning and electrospraying techniques, has recently arisen as an alternative technology that can also be used for encapsulation [13,14]. The basic setup for electrospraying consists of four main components: (1) a high-voltage source (1–30 kV), usually operated in direct current mode, though alternating current mode is also possible, (2) a blunt-ended stainless steel needle or capillary, (3) a syringe pump, and (4) a grounded collector in the form of a flat plate. The electrospraying process involves the application of a strong electrostatic field between two electrodes and imposed on a polymer solution. When increasing the electrostatic field up to a critical value, charges on the surface of a pendant drop destabilize the shape of the solution from partially spherical to conical, i.e., the so-called Taylor's cone effect. As the charged jet accelerates toward regions of lower potential, the solvent is evaporated [15]. Besides being a very simple technique, the solvent is evaporated at room temperature; thus, it constitutes an ideal method for protecting sensitive encapsulated ingredients.

The aim of this work was to study the ability of fructans to form capsules by electrospraying and to asses, as an example, the viability of this polysaccharide as encapsulating material. For that purpose, β-carotene was selected as a model substance. The produced particles were characterized in terms of morphology and photoprotective effect.

2. Materials and Methods

2.1. Materials

High polymerization degree Agave fructans (HDPAF) were purchased from Campos Azules Co., (Ciudad de Mexico, Mexico). TEGO SML (sorbitan fatty acid ester) was purchased from Evonik Inc., (Essen, Germany). HPLC grade methanol, absolute ethanol and β-carotene were purchased from Sigma-Aldrich (Steinheim, Germany). Deionized water was used throughout the study.

2.2. Preparation of Formulation

In order to demonstrate the ability of HDPAF to form nanocapsules, different solutions and emulsions were prepared. Solutions contained different concentrations of HDPAF (5%, 10%, 20%, 30%, 40% and 50% w/w), TEGO SML (1%) as a surfactant and a hydroalcoholic solution (water–ethanol, 9:1) as a solvent. They were prepared under magnetic stirring at 350 RPM for 5 min (Agimatic-S model 7000242). Oil in water emulsions (O/W) were formulated at a ratio of 10:90. The continuous phase consisted of different HDPAF concentrations (4%, 9%, 19%, 29%, 39% and 49%) dissolved in the hydroalcoholic solution (water–ethanol, 9:1). The dispersed phase consisted of extra virgin olive oil and was used without further processing. TEGO SML (5% of total emulsion volume) was used as a surfactant to aid the emulsion stability and decrease surface tension. The two phases were first mixed in a high-shear mixer at 16,800 RPM for 2 min (Ultra Turrax T25, IKA, Staufen, Germany) in order to prepare the pre-emulsion. The emulsion process was carried out with an ultrasonic homogenizer model

Sonopuls 2200 (Bandelin Electronic Gmbh & Co., Berlin, Germany) at 20 kHz for 1 min, according to Paximada et al. [16]. The temperature was maintained at 25 ± 1 °C using an ice bath.

The ability to electrospray the solutions and emulsions was evaluated, parameters (flow-rate (30–50 μL/h), voltage (10–20 KV) and the tip-to-collector distance (10–25 cm)) were varied one at a time until the Taylor's cone was visible, and then particle morphology and size were analyzed to select the most adequate solution and emulsion for the photoprotection study.

For solutions containing β-carotene, β-carotene (0.1%) was incorporated in ethanol and then mixed with the solution containing water, TEGO SML and HDPAF. The mixtures were homogenized under continuous stirring at 350 RPM for 30 min. For the emulsions, β-carotene (1%) was previously incorporated in dichloromethane (1 mL) and was gradually added to the olive oil. When the oily phase was saturated with β-carotene, dichloromethane was separated for 24 h by natural evaporation in the extraction chamber, and then the mix was incorporated into the ethanol. The oily phase was added to the solution containing water, TEGO SML and HDPAF, following the same emulsion preparation procedure previously stated.

2.3. *Characterization of Different Solutions and Emulsions*

The apparent viscosity (η) was determined using a rotational viscosimeter Visco Basic Plus L from Fungilab S.A. (San Feliu de Llobregat, Spain) with a Low Viscosity Adapter (LCP). The LCP spindle was placed in the runner bar of the viscometer and 50 mL of sample was placed in a Falcon tube and put in contact with the spindle to obtain the η value. The surface tension was measured using the Wilhemy plate method in an EasyDyne K20 tensiometer (Krüss GmbH, Hamburg, Germany). Twenty-five milliliters of sample was placed in a vessel, then the Wilhemy plate is burned and suspended from the pendulum; the vessel is then placed on the platform to be analyzed. The conductivity was measured using a conductivity meter XS Con6 (Labbox, Barcelona, Spain). The probe was submerged in 10 mL of sample in a Falcon tube until the sensors were covered and stabilized. All measurements were made at 25 °C in triplicate.

2.4. *Preparation of Capsules by Electrospraying*

The electrospraying apparatus, equipped with a variable high-voltage 0–30 kV power supply, was a Fluidnatek® LE-10 from BioInicia S.L. (Valencia, Spain). Solutions and emulsions with and without β-carotene were introduced in a 12 mL plastic syringe and were electrospun under a steady flow rate using a stainless-steel needle of 700 μm diameter. The needle was connected through a Polytetrafluoroethylene (PTFE) tube to the syringe. The syringe was lying on a digitally controlled syringe pump while the needle was horizontal towards the collector. The electrospraying conditions of the solutions and emulsions for obtaining the capsules were optimized and fixed at 0.1 mL/h of flow-rate, 17 kV of voltage and a tip-to-collector distance of 22 cm. The samples were stored in darkness until analysis.

Additionally, a different encapsulation strategy, named electrospraying coating (EC), patented by Lagaron et al. [17] and reported by Librán et al. [18], was used. The coating was a three-step process carried out at room temperature. In the first step, an initial layer of fructans were electrosprayed over the collector. Secondly, 2% of β-carotene with respect to the solution electrosprayed was spread out over the initial electrosprayed material layer. Finally, a top coating layer of fructans was electrosprayed directly over the material to achieve full encapsulation. The capsules were then collected and mechanically mixed and homogenized. The basic setup of a Fluidnatek™ LE10 (Bioinicia S.L., Valencia, Spain) was used to conduct the electrospraying process. The collected nanocapsules were stored in a desiccator at 0% relative humidity (RH) and protected from light for subsequent analysis.

2.5. *Scanning Electron Microscopy (SEM)*

The morphology and size of the encapsulation structures were examined using SEM on a Hitachi microscope (Hitachi S-4100, Tokyo, Japan) after having been sputtered with a gold–palladium mixture under vacuum for 3 min (SC7640, Polaron, Kent, UK). All SEM experiments were carried out with 1–2 mg of sample at 10 kV, obtaining three micrographs per sample. Capsule diameters were measured by means of the Adobe Photoshop CS3 software from the SEM micrographs in their original magnification.

2.6. Fourier Transform Infrared Spectroscopy

Attenuated Total Reflectance Fourier Transform Infrared spectroscopy (ATR-FTIR) (Thermo Scientific Nicolet, iS5 iD5, Waltham, USA) was used to evaluate β-carotene, empty HDPAF nanocapsules and nanocapsuled β-carotene. The samples were placed onto the ATR crystal and all the spectra were recorded from 600 cm^{-1} to 4000 cm^{-1} with a resolution of 8 cm^{-1}.

2.7. Thermogravimetric Analysis (TGA)

Thermogravimetric analyses of free β-carotene and HDPAF nanocapsules without and with β-carotene were done in triplicate using TGA 550 equipment (TA Instruments, New Castle, USA) and TRIOS 4.3.0.38388 was the analysis software used. The analyses were conducted under the following conditions: 3–6 mg of sample, heating from 25 °C to 500 °C, at a heating rate of 5 °C/min under nitrogen flow.

2.8. Ultraviolet (UV) Photostability

With the aim of accelerating the oxidation of β-carotene and simulating the radiation of natural sunlight, an Osram Ultra-Vitalux (300 W) lamp (OSRAM, Múnich, Germany) was used. This blend of radiation is generated by a quartz discharge tube and a tungsten filament [19]. Nanocapsules with β-carotene and free β-carotene were exposed to the UV radiation (13.6 W) at 37 °C. After irradiation at different times (0 h, 6 h, 12 h, 24 h and 48 h), extraction of β-carotene from 2.5 mg of nanocapsules was carried out. The polymeric capsule wall was opened with water (1 mL) under magnetic stirring (200 RPM, 1 min). β-carotene was extracted from the mixture by adding 0.75 mL of chloroform and separated by centrifugation (10,000 RPM, 1 min). An aliquot of the organic phase was taken and the absorbance was measured at 466 nm in a spectrophotometer (Spectrophotometer UV/VIS4000, DINKO instruments, Barcelona, Spain). Chloroform was used as a blank. Oxidation was reported as a function of the relative β-carotene content (% absorbance). Analyses were made in triplicate.

3. Results and Discussion

3.1. Solution Properties

The successful development of encapsulation structures using electrospraying technology strongly depends on the solution properties and, hence, an initial characterization of solution viscosity and viscoelasticity was carried out. From a screening study, it was seen that HDPAF solutions led to very low apparent viscosity values at 5% (see Table 1) due to the low polymer (HDPAF) concentration, which resulted in unstable jetting and no capsules were formed from these solutions. In order to increase the viscosity of the solution, HDPAF concentration was increased. Viscosity values at 10% to 30% HDPAF concentration provided viscosity values previously reported as adequate for electrospraying (1 cP to 10 cP) [17].

Table 1. Physical properties (conductivity, surface tension, and viscosity) of solutions and emulsions at different high degree of polymerization Agave fructans (HDPAF) concentrations.

Concentration (% *w/w*)	Viscosity (cP)	Surface Tension (mN/m)	Conductivity (μS/cm)
		Solutions	
5	1.61 ± 0.05 [a]	24.35 ± 0.05 [a]	69.39 ± 0.03 [a]
10	2.37 ± 0.07 [b]	24.37 ± 0.02 [a]	82.14 ± 0.03 [b]
20	3.42 ± 0.01 [c]	24.85 ± 0.05 [b]	101.20 ± 0.05 [c]
30	6.82 ± 0.02 [d]	23.65 ± 0.05 [c]	93.30 ± 0.06 [d]
40	46.05 ± 0.05 [e]	23.51 ± 0.05 [d]	76.73 ± 0.05 [e]
50	162.22 ± 0.60 [f]	23.46 ± 0.05 [d]	52.86 ± 0.01 [f]
		Emulsions	
5	2.65 ± 0.03 [a]	22.42 ± 0.04 [a]	41.81 ± 0.04 [a]
10	3.40 ± 0.03 [b]	22.91 ± 0.02 [b]	45.82 ± 0.04 [b]
20	8.83 ± 0.08 [c]	24.05 ± 0.03 [c]	52.70 ± 0.02 [c]
30	12.26 ± 0.09 [d]	23.20 ± 0.02 d	50.89 ± 0.02 [d]
40	45.70 ± 0.12 [e]	22.73 ± 0.03 [e]	39.11 ± 0.04 [e]
50	93.54 ± 0.24 [f]	21.13 ± 0.03 [f]	30.26 ± 0.01 [f]

[a–f]: Different superscripts within the same column indicate significant differences among the samples ($p < 0.05$).

Formulations with 40% of HDPAF produced a significant increase in viscosity. This could be attributed to the high amount of HDPAF added (40–50% *w/w*), but also to the high molecular weight of HDPAF, since they consist of a mixture of long polymers and fructooligosaccharides [6], which have been reported to contribute to increased viscosity [18]. In this case, the instability of the Taylor cone resulted in higher voltage values being needed to overcome the surface tension and achieve atomization.

Similar values of surface tension in all formulations were obtained (Table 1). It was also observed that even though high HDPAF concentrations (30–50%) were used, the profile of surface tension values of the aqueous solutions decreased, but the range was still adequate. This behavior allows solutions and emulsions to be processed by electrospraying to obtain capsules. It has been previously reported that solutions with low surface tension favor the electrospraying process [20] and, thus, capsule formation [17], because the intensity of the electrical field must overcome the solution surface tension, expelling an electrified jet from the Taylor's cone formed on the needle tip [21]. Therefore, during drying by electrospraying, the Taylor's cone was held stable for formulations with low surface tension (~20 mN/m), which agrees with the work reported by Jaworek [22], who affirmed that solutions with surface tension above 50 mN/m cannot be electrosprayed, independently of the polymer used. This decrease in surface tension could be attributed to the ethanol addition to solubilize β-carotene in the formulations, because ethanol surface tension is lower than water surface tension [23]. The conductivity values increased when HDPAF increased from 5 to 20% in the solution, but conductivity decreased at concentrations of HDPAF above 30% (Table 1). However, conductivity values were always lower than values reported as adequate to be processed by electrospraying process (<2200.00 μS/cm) [18]. Finally, electrical conductivity should not exceed this value to avoid the destabilization of the electrospraying jet [23]. Emulsion and solution properties showed the same behavior with respect to the physical properties (viscosity, surface tension and conductivity) evaluated (Table 1). This can be attributed to the similar components in formulations. A technological advantage is that depending on the active compound polarity to be encapsulated, it can be selected between emulsions or solutions to incorporate as much compound as possible.

3.2. Capsule Morphology

SEM images demonstrated the ability of HDPAF to form capsules when HDPAF concentrations between 10% and 50% (*w*/*w*) were used as shown in Figure 1. Nanocapsules with spherical morphology and sizes between 650 nm and 760 nm, without cracks, dents or deformations and without being agglomerated (Figure 1) were obtained. The absence of pores or cracks on the capsule surface is important to ensure low oxygen permeability which could lead to the degradation of the encapsulated antioxidant compounds [24]. However, Ortiz-Basurto et al. [6] observed an irregular surface and several indentations on the microparticles of HDPAF encapsulating pitanga juice obtained by spray drying.

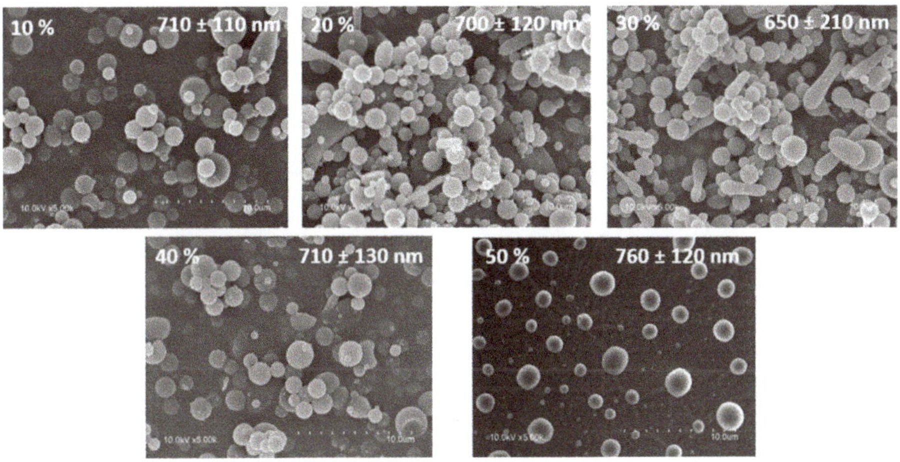

Figure 1. Micrographs obtained by scanning electron microscopy (SEM) of HDPAF nanocapsules at different HDPAF concentrations (% *w*/*w*: 10, 20, 30, 40 and 50) obtained by electrospraying.

3.3. β-Carotene Encapsulation

The encapsulation of the β-carotene supposes a technical challenge due to its high instability to light, its hydrophobicity and its low solubility in common organic solvents [25–27]. The great interest of the food industry in this compound has motivated researchers to try the encapsulation of β-carotene by several methods. Tan and Nakajima [25] suggested the nanodispersion of β-carotene by the solvent evaporation method. Ribeiro et al. [26] proposed the encapsulation in Polylactide(PLA) and Poly lactic-*co*-glycolic acid (PLGA) by the solvent displacement method and Astete et al. [27] proposed the encapsulation in calcium alginate. However, some of these proposals presented the disadvantage of using organic solvents such as acetone [26], hexane [25] or chloroform [27]. Traces of these solvents would make the encapsulates unsuitable for food applications, and, therefore, it is of great interest to find a methodology to obtain nanocapsules based on the use of eco-friendly ingredients with a high encapsulation efficiency, which could reach the status of "generally recognized as safe (GRAS)" granted by the FDA.

Our first proposal was to use a solution to encapsulate the β-carotene. Nevertheless, the low solubility of β-carotene in conventional solvents prevented the obtaining of capsules with β-carotene concentrations over 0.1%. The second attempt was to use an emulsion, but for that option, the use of dichloromethane was required. The residual organic solvent concentration in the capsules was evaluated by headspace-solid-phase microextraction–gas chromatography (HS-SPME-GC) according to Camelo-Méndez et al., [28] with some modifications, and no traces of dichloromethane were detected in samples. Despite this good result, the concentration of the β-carotene in the particles was less than 1%. On the other hand, β-carotene encapsulated by the EC method allowed a higher concentration of β-carotene and consequently, the whole study was focused on this option.

3.4. FTIR Analysis of the Encapsulation Structures

Interactions between β-carotene and HDPAF nanocapsules were evaluated by infrared spectroscopy (FTIR) according to Peinado et al. [29]. The FTIR spectra of β-carotene showed a broad peak at 3411 cm^{-1} that represents the presence of O–H stretching of the hydroxyl group, which is likely due to the interaction of β-carotene with oxygen in the air [30]. The peaks at 2929 cm^{-1} and 2869 cm^{-1} indicate the CH_2 asymmetry and symmetry stretching, respectively (Figure 2a). The presence of carbonyl groups and the stretching symmetry of the C–H bond group was evidenced in peaks at 1717 cm^{-1} and 1366 cm^{-1}, respectively. The sharp peak at 965 cm^{-1} marks the deformation mode of trans-conjugate alkenes as the specific areas of trans=CH (1 in Figure 2a) used for identification of β-carotene [30,31].

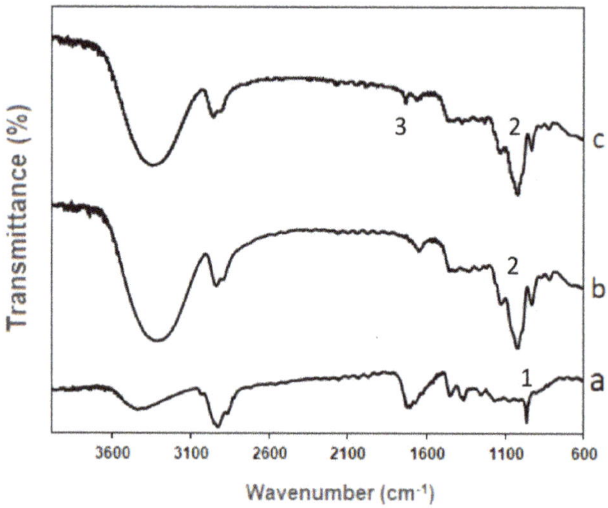

Figure 2. FTIR spectra. β-carotene (**a**), HDPAF nanocapsules (**b**) and HDPAF/β-carotene nanocapsules produced by the electrospraying coating (EC) process(**c**).

The FTIR spectra of HDPAF nanocapsules (Figure 2b) showed the most intensive broad band, with the maximum at 1050 cm^{-1} and two shoulders at 940 cm^{-1} and 1130 cm^{-1}. The bands in the region 900–1153 cm^{-1} have been assigned to C–O and C–C stretching modes (2 in Figure 2b,c). These bands are characteristic of carbohydrates. Moreover, the two overlapped bands at 2930 cm^{-1} and 2870 cm^{-1} are characteristic of carbohydrates too [32]. The band from 2800 cm^{-1} to 3000 cm^{-1} is similar to the inulin spectra reported by Grube et al. [32] and Apolinário et al. [33]; this band is attributed to C–H stretching. The broad stretching peak around 3492 cm^{-1} indicated the presence of hydroxyl groups (–OH) of carbohydrates [33].

The comparison of nanocapsules of HDPAF and HDPAF/β-carotene obtained by the EC process (Figure 2b,c) proved HDPAF as the dominating component. The main differences in the nanocapsules of HDPAF and HDPAF/β-carotene spectra appeared in the 1700–1800 cm^{-1} region, which indicates the C=O interaction of fructose molecules with β-carotene, presenting as a stretching of the peak (3 in Figure 2c). The low intensity of β-carotene suggests that only a slight amount is located on the surface of the HDPAF nanocapsules [29]. Nanocapsules with a low surface intensity observed by FTIR (Figure 2c) suggest a centripetal distribution of β-carotene, where the highest concentration is in the core of the nanocapsule. Such confinement, likely due to the hydrophobicity of β-carotene, is desired, as it would create a barrier against oxygen and protection against thermal decomposition processes [29].

3.5. Thermal Stability of β-Carotene and HDPAF Nanocapsules

The purpose of the thermogravimetric analysis was to evaluate the thermal resistance to degradation of β-carotene encapsulated in HDPAF nanocapsules. Thermograms shown, a termal decomposition of pure β-carotene between 150.58 °C and 354.16 °C (Figure 3a), similar decomposition temperature range (150–450 °C) was reported by Busolo and Lagaron (2015) [34]. HDPAF nanocapsules degraded between 205.48 °C and 257.70 °C (Figure 3b). These differences in stability can be associated to the structure of the molecules, since the HDPAF is a complex mixture of fructooligosaccharides [1,2] and may have functional properties linked to the degree of polymerization.

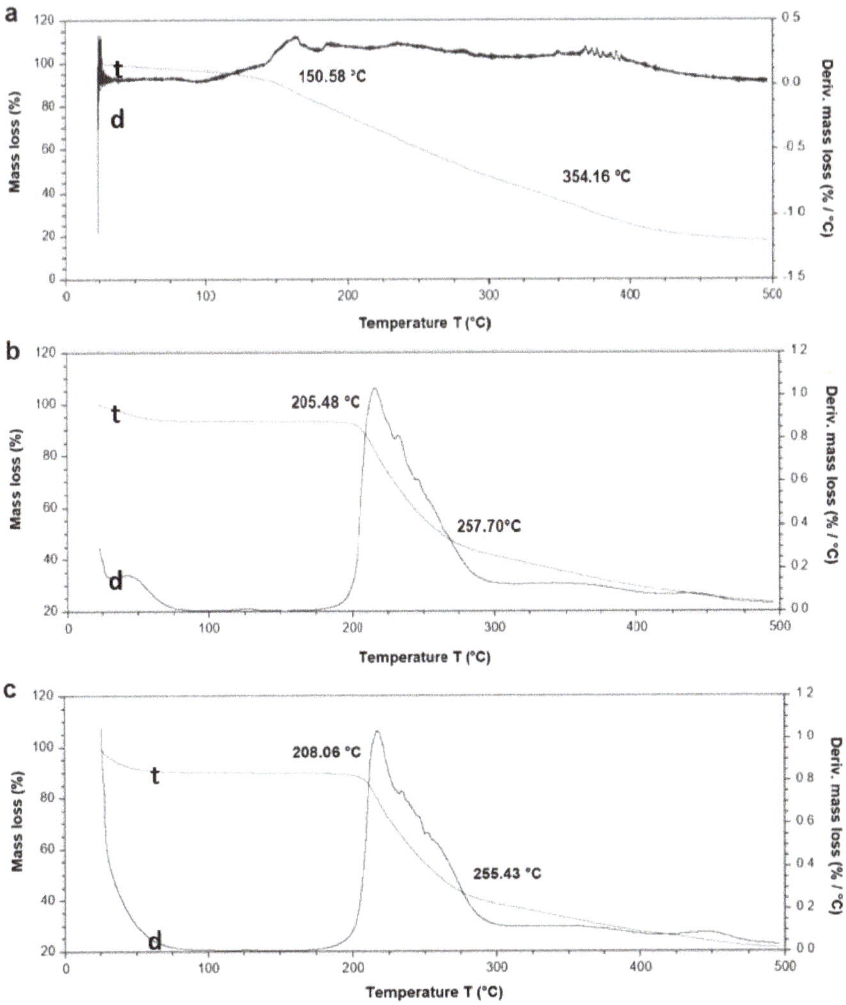

Figure 3. Thermogravimetric profile. β-carotene (**a**), HDPAF nanocapsules (**b**) and HDPAF/β-carotene nanocapsules obtained by EC (**c**). Curve t represents the thermogram and d the thermogram derivate.

β-carotene encapsulated in HDPAF nanocapsules was decomposed between 208.60 °C and 255.43 °C (Figure 3c). This result supports the thermal protective effect of the HDPAF nanocapsule on β-carotene, similar to that reported by Peinado et al. [29] for the encapsulation of β-carotene in electrospun nanofibers of poly(ethylene oxide). The thermal stability of antioxidants as β-carotene depends on whether the molecules are totally encapsulated in the nanocapsules or on the surface [34]. Thermal stability of HDPAF/nanocapsules with and without β-carotene did not show a difference (Figure 3). Therefore, HDPAF exerts a protective role against the thermal degradation of β-carotene.

3.6. Ultraviolet (UV) Photostability of Encapsulated β-Carotene

β-carotene is highly susceptible to photooxidation (oxidation or isomerization) due to the presence of conjugated double bonds in the molecule [19]. The exposure of β-carotene to UV light led to damage in the molecule, producing a decrease in the absorbance (measured at 466 nm) (Figure 4). Degradation of unprotected β-carotene has been also reported by Fernandez et al. [19] and de Freitas Zômpero et al. [21]. However, the β-carotene encapsulated in HDPAF showed a higher stability to UV light even after 48 h of exposure (Figure 4), attributed to the structure of the fructooligosaccharide mixtures.

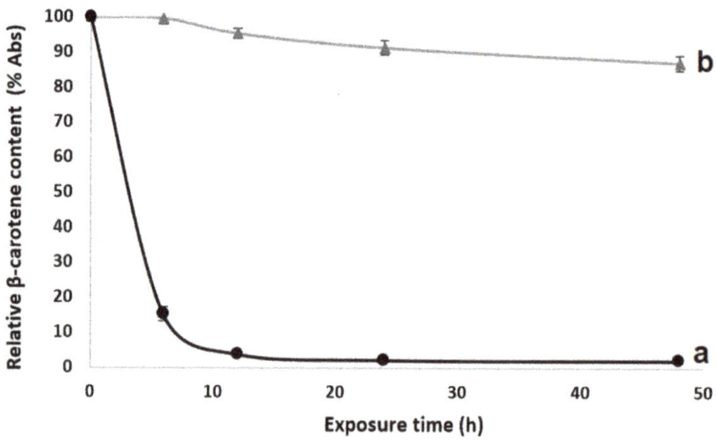

Figure 4. Relative decay in absorbance percentage (% Abs), as a function of exposure time to UV (a) β-carotene and (b) Nanoapsules with HDPAF and β-carotene by EC.

Photoisomerization under UV light exposure is thought to be able to take place in free bioactive compounds, but not very readily in dried particles [19]. López-Rubio and Lagaron [35] produced hydrocolloid films (whey protein concentrate, zein, soy protein and gelatin) containing β-carotene which were able to maintain the β-carotene stability even after 50 h of UV light exposure. De Freitas Zômpero et al. [21] reported that a double encapsulation (nanoliposome + polymeric fiber) by electrospinning was useful to guarantee the β-carotene stability during 6 h of UV light exposure. Therefore, the utilization of the HDPAF as an encapsulating material could be a novel option to be used in nanocapsule manufacture to protect active compounds. In this case, the β-carotene loaded in HDPAF presented with similar behaviors when compared with other polymers/hydrocolloids used before [21,35]. However, this behavior was obtained with a low HDPAF concentration in the particle.

4. Conclusions

In this paper, high degree of polymerization Agave fructans (HDPAF) are presented as a novel encapsulating material. First, their ability to form capsules by electrospraying was tested. The best results, in terms of morphology and capsule size, were obtained when concentrations of 30% and 40% of fructans were used. β-carotene was encapsulated in HDPAF by direct electrospraying and by EC. However, the EC method presented advantages in comparison with emulsion or solution direct electrospraying, since it was possible to obtain particles with higher bioactive:polymer ratios. Moreover, the particles obtained by the EC method showed good photoprotection. Results shown in this work evidence that HDPAF have the capacity to improve the stability of β-carotene. Additionally, HDPAF are appropriate for human consumption, therefore they could be a really interesting encapsulation polymer for the food industry.

Author Contributions: Conceptualization was devised by J.A.R.-S. and J.M.L.; methodology, validation, and formal analysis was carried out by J.A.R.-H., J.A.R.-S., M.C.-S., R.I.O.-B., C.P., and J.M.L.; investigation, resources, data curation, writing—original draft preparation and writing—review and editing was performed by J.A.R.-H., J.A.R.-S., M.C.-S., R.I.O.-B., C.P., and J.M.L.; visualization, supervision, project administration, funding acquisition was carried out by J.A.R.-S. and J.M.L.

Funding: This research was funded by MINECO project grant number AGL2015-63855-C2-1-R.

Acknowledgments: The authors would like to thank the "Plan de Movilidad AGARED 2017" ("AGARED-Red Temática Mexicana Aprovechamiento Integral Sustentable y Biotecnología de los Agaves"; the MINECO project AGL2015-63855-C2-1-R for funding and the CONACYT (Mexico) for the scholarship granted to Jorge Alberto Ramos-Hernández.

Conflicts of Interest: The authors declare no conflict of interest.

References

1. Lopez, M.G.; Mancilla-Margalli, N.A.; Mendoza-Diaz, G. Molecular structures of fructans from agave tequilana weber var. azul. *J. Agric. Food Chem.* **2003**, *51*, 7835–7840. [CrossRef] [PubMed]
2. Ritsema, T.; Smeekens, S.C. Engineering fructan metabolism in plants. *J. Plant. Physiol.* **2003**, *160*, 811–820. [CrossRef] [PubMed]
3. Palatnik, D.R.; Aldrete Herrera, P.; Rinaldoni, A.N.; Ortiz Basurto, R.I.; Campderrós, M.E. Development of reduced-fat cheeses with the addition of agave fructanos. *Dairy Technol.* **2016**, *70*, 212–219. [CrossRef]
4. Guggisberg, D.; Cuthbert-Steven, J.; Piccinali, P.; Bütikofer, U.; Eberhard, P. Rheological, microstructural and sensory characterization of low-fat and whole milk set yoghurt as influenced by inulin addition. *Int. Dairy J.* **2009**, *19*, 107–115. [CrossRef]
5. Furlán, L.T.R.; Herrera, P.A.; Padilla, A.P.; Basurto, R.I.O.; Campderrós, M.E. Assessment of agave fructans as lyoprotectants of bovine plasma proteins concentrated by ultrafiltration. *Food Res. Int.* **2014**, *56*, 146–158. [CrossRef]
6. Ortiz-Basurto, R.I.; Rubio-Ibarra, M.E.; Ragazzo-Sanchez, J.A.; Beristain, C.I.; Jiménez-Fernández, M. Microencapsulation of *Eugenia uniflora* L. juice by spray drying using fructans with different degrees of polymerisation. *Carbohydr. Polym.* **2017**, *175*, 603–609. [CrossRef] [PubMed]
7. FDA. GRAS Notification—Premium Agave Inulin-Resubmission. GRAS Notice (GRN) No. 687. Available online: http://www.fda.gov/Food/IngredientsPackagingLabeling/GRAS/NoticeInventory/default.htm (accessed on 16 December 2016).
8. Li, X.Y.; Chen, X.G.; Sun, Z.W.; Park, H.J.; Cha, D.S. Preparation of alginate/chitosan/carboxymethyl chitosan complex microcapsules and application in *Lactobacillus casei* ATCC 393. *Carbohydr. Polym.* **2011**, *83*, 1479–1485. [CrossRef]
9. Zuidam, N.J.; Shimoni, E. Overview of microencapsulates for use in food products or processes and methods to make them. In *Encapsulation Technologies for Active Food Ingredients and Food Processing*; Zuidam, N.J., Nedovic, V., Eds.; Springer: New York, NY, USA, 2010; Volume 1, pp. 3–29, ISBN 978-1-4419-1007-3.
10. Nedovic, V.; Kalusevic, A.; Manojlovic, V.; Levic, S.; Bugarski, B. An overview of encapsulation technologies for food applications. *Procedia Food Sci.* **2011**, *1*, 1806–1815. [CrossRef]
11. Murugesan, R.; Orsat, V. Spray drying for the production of nutraceutical ingredients–A review. *Food Bioprocess Technol.* **2011**, *5*, 3–14. [CrossRef]
12. Gharsallaoui, A.; Roudaut, G.; Chambin, O.; Voilley, A.; Saurel, R. Applications of spray-drying in microencapsulation of food ingredients: An overview. *Food Res. Int.* **2007**, *40*, 1107–1121. [CrossRef]
13. Torres-Giner, S.; Martinez-Abad, A.; Ocio, M.J.; Lagaron, J.M. Stabilization of a nutraceutical ω-3 fatty acid by encapsulation in ultrathin electrosprayed zein prolamine. *J. Food Sci.* **2010**, *75*, 69–79. [CrossRef] [PubMed]
14. Lopez-Rubio, A.; Lagaron, J.M. Whey protein capsules obtained through electrospraying for the encapsulation of bioactives. *Innov. Food Sci. Emerg. Technol.* **2011**, *13*, 200–206. [CrossRef]

15. Drosou, C.G.; Krokida, M.K.; Biliaderis, C.G. Encapsulation of bioactive compounds through electrospinning/electrospraying and spray drying: A comparative assessment of food-related applications. *Drying Technol.* **2016**, *35*, 139–162. [CrossRef]
16. Paximada, P.; Echegoyen, Y.; Koutinas, A.A.; Mandala, I.G.; Lagaron, J.M. Encapsulation of hydrophilic and lipophilized catechin into nanoparticles through emulsion electrospraying. *Food Hydrocoll.* **2016**, *64*, 123–132. [CrossRef]
17. Lagaron, J.M.; Pérez-Masia, R.; López-Rubio, A. Procedimiento de Protección de Material Biológico y Compuestos Termolábiles Para Posibles Aplicaciones Industriales. Patent Application Number ES P201430034, 23 July 2015.
18. Librán, C.M.; Castro, S.; Lagaron, J.M. Encapsulation by electrospray coating atomization of probiotic strains. *Innov. Food Sci. Emerg. Technol.* **2016**, *39*, 216–222. [CrossRef]
19. Fernández, A.; Torres-Giner, S.; Lagaron, J.M. Novel route to stabilization of bioactive antioxidants by encapsulation in electrospun fibers of zein prolamine. *Food Hydrocoll.* **2009**, *23*, 1427–1432. [CrossRef]
20. Ramakrishna, S.; Fujihara, K.; Teo, W.E.; Yong, T.; Ma, Z.; Ramaseshan, R. Electrospun nanofibers: Solving global issues. *Mater. Today* **2006**, *9*, 40–50. [CrossRef]
21. De Freitas Zômpero, R.H.; López-Rubio, A.; de Pinho, S.C.; Lagaron, J.M.; de la Torre, L.G. Hybrid encapsulation structures based on β-carotene-loaded nanoliposomes within electrospun fibers. *Coll. Surf. B* **2015**, *134*, 475–482. [CrossRef] [PubMed]
22. Jaworek, A. Micro- and nanoparticle production by electrospraying. *Powder Technol.* **2007**, *176*, 18–35. [CrossRef]
23. Gómez-Mascaraque, L.G.; Perez-Masiá, R.; González-Barrio, R.; Periago, M.J.; López-Rubio, A. Potential of microencapsulation through emulsion-electrospraying to improve the bioaccesibility of β-carotene. *Food Hydrocoll.* **2017**, *73*, 1–12. [CrossRef]
24. Fernandes, R.V.; Borges, S.V.; Botrel, D.A. Gum arabic/starch/maltodextrin/inulin as wall materials on the microencapsulation of rosemary essential oil. *Carbohydr. Polym.* **2013**, *101*, 524–532. [CrossRef] [PubMed]
25. Tan, C.P.; Nakajima, M. Beta-carotene nanodispersions: Preparation, characterization and stability evaluation. *Food Chem.* **2005**, *92*, 661–671. [CrossRef]
26. Ribeiro, H.S.; Chu, B.-S.; Ichikawa, S.; Nakajima, M. Preparation of nanodispersions containing β-carotene by solvent displacement method. *Food Hydrocoll.* **2008**, *22*, 12–17. [CrossRef]
27. Astete, C.E.; Sabliov, C.M.; Watanabe, F.; Biris, A. Ca^{2+} cross-linked alginic acid nanoparticles for solubilization of lipophilic natural colorants. *J. Agric. Food Chem.* **2009**, *57*, 7505–7512. [CrossRef] [PubMed]
28. Camelo-Méndez, G.A.; Ragazzo-Sánchez, J.A.; Jiménez-Aparicio, A.R.; Vanegas-Espinoza, P.E.; Paredes-López, O.; Del Villar-Martínez, A.A. Comparative study of anthocyanin and aromatic compounds content of four varieties of Mexican roselle (*Hibiscus sabdariffa* L.) by multivariable analysis. *Plant. Foods Hum. Nutr.* **2013**, *68*, 229–234.
29. Peinado, I.; Mason, M.; Romano, A.; Biasioli, F.; Scampicchio, M. Stability of β-carotene in polyethylene oxide electrospun nanofibers. *Appl. Surf. Sci.* **2016**, *370*, 111–116. [CrossRef]
30. Ammawath, W.; Yaakob, C.M. A rapid method for determination of commercial β-carotene in RBD palm olein by Fourier transform infrared spectroscopy. *Asian. J. Food Agro-Ind.* **2010**, *3*, 443–452.
31. Reksamunandar, R.P.; Edikresnha, D.; Munir, M.M.; Damayanti, S. Encapsulation of β-carotene in poly(vinylpyrrolidone) (PVP) by electrospinning Technique. *Procedia Eng.* **2017**, *170*, 19–23. [CrossRef]
32. Grube, M.; Bekers, M.; Upite, D.; Kaminska, E. Infrared spectra of some fructans. *Spectroscopy* **2002**, *16*, 289–296. [CrossRef]
33. Apolinário, A.C.; de Carvalho, E.M.; de Lima Damasceno, B.P.G.; da Silva, P.C.D.; Converti, A.; Pessoa, A., Jr.; da Silva, J.A. Extraction, isolation and characterization of inulin from Agave sisalana boles. *Ind. Crops Prod.* **2017**, *108*, 355–362. [CrossRef]
34. Busolo, M.A.; Lagaron, J.M. Antioxidant polyethylene films based on a resveratrol containing clay of interest in food packaging applications. *Food Packag. Shelf Life* **2015**, *6*, 30–41. [CrossRef]
35. López-Rubio, A.; Lagaron, J.M. Improved incorporation and stabilisation of β-carotene in hydrocolloids using glycerol. *Food Chem.* **2010**, *125*, 997–1004. [CrossRef]

MDPI
St. Alban-Anlage 66
4052 Basel
Switzerland
Tel. +41 61 683 77 34
Fax +41 61 302 89 18
www.mdpi.com

Nanomaterials Editorial Office
E-mail: nanomaterials@mdpi.com
www.mdpi.com/journal/nanomaterials

www.ingramcontent.com/pod-product-compliance
Lightning Source LLC
Chambersburg PA
CBHW041016270926
51850CB00024B/683